Mediterranean Architectural Heritage
RIPAM10

10th International Meeting on Mediterranean Architectural Heritage (RIPAM10), held from November 2-4, 2023, Er-Rachidia, Morocco

Editors
AZROUR Mohamed[1], BABA Khadija[2], BATAN Abdelkrim[3], EL RHAFFARI Younes[4], FRATINI Fabio[5], GONZALEZ Filipe[6], HADDAD Mustapha[7], IBNOUSSINA Mounsif[8], NOUNAH Abderrahman[9], PITTALUGA Daniela[10], TILIOUA Amine[11]

[1] FST Errachidia, Moulay Ismail University, Morocco
[2] Mohamed V University, EST Salé, Rabat Morocco
[3] FST Errachidia, Morocco
[4] University of Mohamed V, EST Salè, Rabat, Morocco, Morocco
[5] CNR-Firenze, Italy
[6] Faculty of Architecture, University of Lisbon, Portugal
[7] Faculty of Sciences of Meknes, Moulay Ismail University, Morocco
[8] Cadi Ayyad University, Marrakech, Morocco
[9] EST of Salé, Mohamed V University of Rabat, Morocco
[10] DAD- University of Genoa, Italy
[11] Moulay Ismail University, Errachidia-Meknes

Peer review statement

All papers published in this volume of "Materials Research Proceedings" have been peer reviewed. The process of peer review was initiated and overseen by the above proceedings editors. All reviews were conducted by expert referees in accordance to Materials Research Forum LLC high standards.

Published under License by **Materials Research Forum LLC**
Millersville, PA 17551, USA

Published as part of the proceedings series
Materials Research Proceedings
Volume 40 (2024)

ISSN 2474-3941 (Print)
ISSN 2474-395X (Online)

ISBN 978-1-64490-310-0 (Print)
ISBN 978-1-64490-311-7 (eBook)

This book contains information obtained from authentic and highly regarded sources. Reasonable efforts have been made to publish reliable data and information, but the author and publisher cannot assume responsibility for the validity of all materials or the consequences of their use. The authors and publishers have attempted to trace the copyright holders of all material reproduced in this publication and apologize to copyright holders if permission to publish in this form has not been obtained. If any copyright material has not been acknowledged please write and let us know so we may rectify in any future reprint.

Distributed worldwide by

Materials Research Forum LLC
105 Springdale Lane
Millersville, PA 17551
USA
https://www.mrforum.com

Manufactured in the United State of America
10 9 8 7 6 5 4 3 2 1

Table of Contents

Keywords

Editorial

The International Meetings on Mediterranean Architectural Heritage (RIPAM) are significant scientific events that bring together a distinguished community of researchers, scholars, historians, architects, heritage scientists, experts, curators, and professionals from various countries and institutions around the Mediterranean basin, including Morocco, Italy, Portugal, France, Spain, Algeria, and Tunisia.

Since their inauguration in 2005, RIPAM has been held biennially, alternating between the two shores of the Mediterranean. In addition to this year's edition planned in Errachidia, Morocco has had the honor of hosting three editions respectively in Meknes, Marrakech, and Rabat in 2005, 2007, and 2019. Other editions of RIPAM have shone in Lisbon in 2009, M'Sila in 2012, Marseille in 2013, Tunisia in 2015, Genoa in 2017, and also in Lisbon in 2022. Each edition of RIPAM embodies a celebration of the Mediterranean architectural heritage, a forum for sharing knowledge, and an invaluable platform for international collaboration.

The Mediterranean region is distinguished by an architectural heritage of great richness and diversity, reflecting the ingenuity and creativity of artisans, as well as the cultural depth that characterizes this part of the globe. Recognizing that the building and construction sector is one of the largest emitters of greenhouse gases, accounting for about 39% of carbon emissions and 33% of waste production, stakeholders in this sector have been prompted to replace traditional materials with environmentally friendly materials.

Currently, concerns about energy and environmental issues in the construction industry have led sector stakeholders to prioritize the use of eco-materials instead of conventional materials. This trend has sparked renewed interest in construction using local materials in recent years (including earth, stone, wood, among others), due to their sustainability and highly favorable environmental footprint.

Similar to the oasis areas, the Drâa-Tafilalet region is home to an architectural heritage in earth of remarkable richness and diversity, encompassing structures such as ksars, kasbahs, mosques, fortifications, and granaries. The restoration of this heritage, beyond its immediate impact on the local community, holds substantial significance touching regional economy, cultural legacy, and the tourism sector. Rehabilitation projects frequently require the involvement of specialized local labor for construction, restoration, and rehabilitation work. Once restored, these structures become tourist attractions, generating economic benefits for local and regional businesses in the tourism, industrial, and agricultural sectors. They thus contribute to the preservation and enhancement of the architectural heritage in earth of invaluable richness specific to the region.

Thus, the city of Errachidia, capital of the region, becomes a conducive setting to host the tenth edition of the International Meeting on Mediterranean Architectural Heritage (RIPAM10) from November 2 to 4, 2023.

We are deeply thankful to the Minister of Culture, Youth, and Communication, as well as the President of Moulay Ismail University of Meknes. Their steadfast backing has played a pivotal role in the triumph of the RIPAM10 congress, nurturing global cooperation, streamlining the spread of research, and encouraging a vibrant exchange of ideas.

Committees

Editorial Committee
AZROUR Mohamed, FST Errachidia, Moulay Ismail University, Morocco
BABA Khadija, Mohamed V University, EST Salé, Rabat Morocco
BATAN Abdelkrim, FST Errachidia, Morocco
EL RHAFFARI Younes, University of Mohamed V, EST Salè, Rabat, Morocco, Morocco
FRATINI Fabio, CNR-Firenze, Italy
GONZALEZ Filipe, Faculty of Architecture, University of Lisbon, Portugal
HADDAD Mustapha, Faculty of Sciences of Meknes, Moulay Ismail University, Morocco
IBNOUSSINA Mounsif, Cadi Ayyad University, Marrakech, Morocco
NOUNAH Abderrahman, EST of Salé, Mohamed V University of Rabat, Morocco
PITTALUGA Daniela, DAD- University of Genoa, Italy
TILIOUA Amine, Moulay Ismail University, Errachidia-Meknes

Honorary Committee
Ministry of Cultural Affairs and Youth and Communication
Wali of the Drâa Tafilalte Region
President of the Drâa Tafilalte Region
President of the Moulay Ismail University of Meknes
President of the Provincial Council of Errachidia
General Director of ANDOZOA
Director of Higher School of Technology Sale
President of the Errachidia municipal council
Director of CRI Draa-Tafilalet
Director of the Regional Agricultural Development Office of Tafilalet and Er-Rachidia
President of the Urban Community of Errachidia
President of the Draa Tafilalte artisanal chamber
Dean of the Faculty of Sciences and Technology of Errachidia
Director of the Errachidia Urban Agency
Regional Director of the Ministry of Cultural Affairs and Youth
Director of CERKAS Ouarzazate

Permanent Committee
ALEGRIA José Alberto, Architect, Lisbon, Portugal
AZROUR Mohamed, FST Errachidia, Moulay Ismaïl University of Meknès, Morocco
BEJJIT Lahcen, EST, Moulay Ismaïl University of Meknès, Morocco
BELHARETH Taouik, ENAU- University of Carthage, Tunisia
BROMBLET Philippe, CICRP- Marseille, France
BUGINI Roberto, CNR- Gino Bozza Milano, Italy
EL RHAFFARI Younes, EST Salé, Mohamed V University of Rabat, Morocco
FRATINI Fabio, CNR-Firenze, Italy
GONZALEZ Filipe, Faculty of Architecture, University of Lisbonne, Portugal
HADDAD Mustapha, FS, Moulay Ismaïl University of Meknès, Morocco
IBNOUSSINA Mounsif, Cadi Ayyad University of Marrakech, Morocco
KAMEL Said, FS, Moulay Ismaïl University, Morocco
KHALFALLAH Boudjemaa, IGTU- University of M'sila, Algeria
MAI Roland, CICRP-Marseille, France
MILI Mohamed, IGTU- University of M'sila, Algeria

NOCAIRI Mohamed, Cadi Ayyad University of Marrakech, Morocco
PITTALUGA Daniela, DAD- University of Genoa, Italy
SEBAI Abid, ENAU- University of Carthage, Tunisia

Coordination Committee
AZROUR Mohamed, FST Errachidia, Moulay Ismaïl University of Meknès, Morocco
BABA Khadija, EST Salé, Mohamed V University of Rabat, Morocco
BATAN Abdelkrim, FST Errachidia, Moulay Ismaïl University of Meknès, Morocco
BOUSSALH Mohamed, CERKAS Ouarzazate, Morocco
EL RHAFFARI Younes, EST Salé, Mohamed V University of Rabat, Morocco
HADDAD Mustapha, Faculty of Sciences Moulay Ismaïl University of Meknès, Morocco
IBNOUSSINA Mounsif, Cadi Ayyad University of Marrakech, Morocco
TILIOUA Amine, FST Errachidia, Moulay Ismaïl University of Meknès, Morocco

Scientific Committee
AIT HOU Ahmed, Moulay Ismail University, FST., Errachidia, Morocco
ALEGRIA José Alberto, Darquita, Albufeira, Portugal
AMEUR Elamrani, Moulay Ismail University, EST- Meknes, Morocco
ANDREAS Krewet, Terre Construction, building and sustainable energies, France
AZROUR Mohamed, Moulay Ismail University, FST., Errachidia, Morocco
BABA Khadija, Mohamed V University, EST Salé, Rabat, Morocco
BATAN Abdelkrim, Moulay Ismail University, FST., Errachidia, Morocco
BEJJIT Lahcen, Moulay Ismail University, EST- Meknes, Morocco
BELHARETH Taouik, ENAU - University of Carthage, Tunisia
BEN N'CER Abdelouahed, INSAP, Rabat, Morocco
BENYAICH Fouad, Moulay Ismail University, FS-Meknes, Morocco
BENZZAROUK Amar, University of Picardie Jules Verne Amiens, France
BIH Lahcen, Moulay Ismail University, ENSAM, Meknes, Morocco
BOUSALAH Mohamed, CERCAS, Ouarzazate, Morocco
BROMBLET Philippe, CICRP- Marseille, France
BUGINI Roberto, CNR- Gino Bozza Milan, Italy
CARAZAS AEDO W. Architect Consultant-Habiat, Peru
CHAOUKI SADIK, LCPMC, Hassan II University Casablanca Morocco
CHEBAB Driss, Moulay Ismail University, FST, Errachidia, Morocco
CHERRADI Fayçal, MCJC, Rabat, Morocco
EL AMRAOUI Mohamed, FS-Meknesl, Morocco
EL BAKKALI Abdelmjid, Moulay Ismail University, FST., Errachidia Morocco
El GANAOUI Mohamed, University of Lorraine, Nancy, France
El HARROUNI Khalid, National School of Architecture, Rabat, Morocco
EL RHAFFARI Younès, Mohamed V University, EST Salé, Rabat Morocco
ELBOUARI Abdslam, FS Ben M'Sik, Hassan II University, Casablanca Morocco
ELMINOR Hassan, Ibnu Zohr University, ENSA, Agadir, Morocco
FRATINI Fabio, University of Florence, Dept. Architecture DIDA, Italy
GAROUM Mohammed, Mohamed V University, EST Salé, Rabat, Morocco
GHARIBI El Khadir, Mohamed Premier University, FS Oujda, Morocco
GONZALEZ Filipe, FA, Lisbon, Portugal
HABOUBI Khadija ENSA Al-Hoceima Abdelmalek Essaâdi University Morocco
HADDAD Mustapha, Moulay Ismail University, FS-Meknes, Morocco
HANNACH Hassan, FS Ben M'Sik, Hassan II University, Casablanca Morocco

IBNOUSSINA Mounsif, Cadi Ayad University, FSSM, Marrakech, Morocco
JEMJAMI Saloua, FST, Settat, Hassan I University, Settat, Morocco
KASIMI Rachida, Cadi Ayad FSSM University, Marrakech, Morocco
KHALFALLAH Boudjemaa, IGTU- University of M'sila, Algeria
KHARMICH Hassane, National School of Architecture, Rabat, Morocco
LAAROUSSI Najma, EST of Salé, Mohamed V University, Rabat, Morocco
LAHMAR Abdelilah, University of Picardie Jules Verne Amiens, France
LIMAM Ali, INSA, Lyon, France
MAI Roland, CICRP-Marseille, France
MANOUN Bouchaib, FST, Settat, Hassan I University, Settat, Morocco
MANSOUR Majid, National School of Architecture, Rabat, Morocco
MATTONE Manuela, CNRI, Turin, Italy
MILI Mohamed, IGTU - M'sila University, Algeria
MOCERINO Consiglia, Faculty of Architecture, La Sapienza University of Rome, Italy
MORISET Sébastien. CRATerre, Grenoble, France
MOUKMIR Oussama, Architect, Marrakech Morocco
MOUSSAOUI Fouad, FST Errachidia, Moulay Ismail University, Morocco
NOCAIRI Mohamed, FSSM, Marrakech, Morocco
NOUNAH Abderrahman, EST of Salé, Mohamed V University, Rabat, Morocco
ODGHIRI ABDELLAH, Architect, Germany
OUADIF Latifa, Mohammadia School of Engineering, Mohammed V, University of Rabat,
Morocco
OUBALLOUK Elmoustapha., Urban planner, Urban Agency of Errachidia
PITTALUGA Daniella, FA, Genoa, Italy
RABACH Lahcen, CRI, Draa Tafilalet, Errachidia, Morocco
ROVERO Luisa, FA, Florence, Italy
SADKI Driss, Moulay Ismail University, Meknes, Morocco
SAVERIO Mecca, DIDA, Florence, Italy
SEBAI Abid, ENAU - University of Carthage, Tunisia
TAMRAOUI Youssef - Mohammed VI Polytechnic University of Bengerir, Morocco
TILIOUA Amine, Moulay Ismail University, FST., Errachidia, Morocco
TONIETTI Ugo, Faculty of Architecture, Florence, Italy
TURCHANINA Oksana, FA, Lisbon, Portugal
VARUM Humberto, College of Engineering, University of Porto, Portugal
ZNINI Mohamed, Moulay Ismail University, FST., Errachidia, Morocco
ZUCCHIATTI Alessandro, CIC, Genoa, Italy

National organizing committee
AZROUR Mohamed, FST Errachidia, Moulay Ismaïl University of Meknès, Morocco
BABA Khadija, EST Salé, Mohamed V University of Rabat, Morocco
BATAN Abdelkrim, FST Errachidia, Moulay Ismaïl University of Meknès, Morocco
BEJJIT Lahcen, EST, Moulay Ismaïl University of Meknès, Morocco
BIH Lahcen, ENSAM, Moulay Ismaïl University of Meknès, Morocco
EL AMRAOUI Mohamed, FS, Moulay Ismaïl University of Meknès, Morocco
EL RHAFFARI Younes, EST Salé, Mohamed V University of Rabat, Morocco
ELAMRANI Aumeur, EST, Moulay Ismaïl University of Meknès, Morocco
HADDAD Mustapha, FS, Moulay Ismaïl University of Meknès, Morocco
IBNOUSSINA Mounsif, Cadi Ayyad University of Marrakech, Morocco
KASIMI Rachida, Cadi Ayyad University of Marrakech, Morocco

MANOUN Bouchaib, FST of Settat, Hassan I University of Settat, Morocco
NOCAIRI Mohamed, Cadi Ayyad University of Marrakech, Morocco
NOUNAH Abderrahman, EST Salé, Mohamed V University of Rabat, Morocco
TILIOUA Amine, FST Errachidia, Moulay Ismaïl University of Meknès, Morocco

Local organizing committee
ACHENANI Youssef, FST Errachidia, Moulay Ismaïl University of Meknès, Morocco
AIT HOU Ahmed, FST Errachidia, Moulay Ismaïl University of Meknès, Morocco
AIT SIDI MOU Abdelaziz, FST Errachidia, Moulay Ismaïl University of Meknès, Morocco
AZDOUZ M'Barek, FST Errachidia, Moulay Ismaïl University of Meknès, Morocco
AZROUR Mohamed, FST Errachidia, Moulay Ismaïl University of Meknès, Morocco
BATAN Abdelkrim, FST Errachidia, Moulay Ismaïl University of Meknès, Morocco
BEN BAAZIZ Meryem, FST Errachidia, Moulay Ismaïl University of Meknès, Morocco
BOUHAZMA Sarra, FST Errachidia, Moulay Ismaïl University of Meknès, Morocco
CHEBABE Driss, FST Errachidia, Moulay Ismaïl University of Meknès, Morocco
ECHIHI Siham, FST Errachidia, Moulay Ismaïl University of Meknès, Morocco
EL BAKKALI Abdelmajid, FST Errachidia, Moulay Ismaïl University of Meknès, Morocco
ELGHOZLANI Mohammed, FST Errachidia, Moulay Ismaïl University of Meknès, Morocco
KARIMI Said, FP Errachidia, Moulay Ismaïl University of Meknès, Morocco
LAGHZIL Mohamed, FP Errachidia, Moulay Ismaïl University of Meknès, Morocco
MABROUK El Houssine, FST Errachidia, Moulay Ismaïl University of Meknès, Morocco
MOUSSAOUI Fouad, FST Errachidia, Moulay Ismaïl University of Meknès, Morocco
MRANI Alaoui Mohamed, FP Errachidia, Moulay Ismaïl University of Meknès, Morocco
NOU Mohamed, FST Errachidia, Moulay Ismaïl University of Meknès, Morocco
OUALLAL Hassan, FST Errachidia, Moulay Ismaïl University of Meknès, Morocco
OUBAIR Ahmad, FST Errachidia, Moulay Ismaïl University of Meknès, Morocco
OUBALLOUK El Mustapha, Urban planner, Errachidia urban agency, Morocco
RABACH Lahcen, CRI, Drâa-Tafilalet, Errachidia, Morocco
SOUINIDA Laidi, FST Errachidia, Moulay Ismaïl University of Meknès, Morocco
TILIOUA Amine, FST Errachidia, Moulay Ismaïl University of Meknès, Morocco
ZNINI Mohamed, FST Errachidia, Moulay Ismaïl University of Meknès, Morocco

Junior committee
CHRACHMY Mohammed, FST Errachidia, Moulay Ismaïl University of Meknès, Morocco
DOUGHMI Khaoula, LGCE/ ESTS/EMI, Mohammed V University of Rabat, Morocco
EL MESKI Mohamed, FST Errachidia, Moulay Ismaïl University of Meknès, Morocco
KHLIFATI Oumaima, LGCE/ ESTS/EMI, Mohammed V University of Rabat, Morocco
LECHHEB Mahdi, FST Errachidia, Moulay Ismaïl University of Meknès, Morocco
MASROUR Ilham, LGCE/ ESTS/EMI, Mohammed V University of Rabat, Morocco
YOUSSEFI Youssef, FST Errachidia, Moulay Ismaïl University of Meknès, Morocco
ZGUENI Hicham, FST Errachidia, Moulay Ismaïl University of Meknès, Morocco
GHIBATE Rajae, FST Errachidia, Moulay Ismaïl University of Meknès, Morocco

Mediterranean Architectural Heritage - RIPAM10   Materials Research Forum LLC
Materials Research Proceedings 40 (2024) 1-11   https://doi.org/10.21741/9781644903117-1

Tire-Based Anti-Seismic Fibers to Increase the Ductility of Traditional Hydraulic Lime Concrete

Abderrahim BELABID[1,a*], Hajar AKHZOUZ[2,b], Hanane ELMINOR[3,c], Hassan ELMINOR[4,d]

[1] Department of Mechanics, Materials and Civil Engineering, National School of Applied Sciences, Ibn Zohr University, Agadir 80000, Morocco

[2] Departement of Civil Engineering, Euromed University of Fez, Fez 30000, Morocco

[3] Universiaspolis University of Agadir, Agadir 80000, Morocco

[4] Department of Mechanics, Materials and Civil Engineering, National School of Applied Sciences, Ibn Zohr University, Agadir 80000, Morocco

[a*]Abderrahim.belabid@edu.uiz.ac.ma, [b]h.akhzouz@ueuromed.org
[c]hanane.elminor@e-polytechnique.ma, [d]h.elminor@uiz.ac.ma

Keywords: Earthquake-Resistant Fibers, Traditional Lime Concrete, Tire Recycling, Rubber, Ductility

Abstract: Fibers have been widely used in construction since antiquity to reinforce raw earth or lime-based mortar. They prevent the propagation of micro-cracks and improve cohesion and shear strength. High-elasticity and plasticity fibers also enhance the material's capacity to absorb energy. The aim of this paper is to investigate the possibility of improving the seismic behavior of traditional buildings through the incorporation of fibers extracted from waste tires known for their high ductility. The objective is to recycle this non-biodegradable waste and utilize its mechanical characteristics to enhance the seismic performance of traditional buildings. Rubber fibers were incorporated at a rate of 1.5% and 3% on a comparative traditional hydraulic lime concrete. Ductility parameters are estimated from the analysis of stress-strain diagrams obtained by applying uniaxial compression tests in accordance with French standards NF P94-420, and NF P94-425. The results show a significant improvement in the traditional hydraulic lime concrete ductility after the addition of fibers made from tire waste. This method will enable the recycling of tire waste, environmental protection, and enhanced seismic performance of traditional structures. The aspects to be addressed for the development of research fields on earthquake-resistant fiber technology were also formulated.

Nomenclature

Rh	Relative humidity
T	Temperature
ε_e	Elastic limit
ε_f	Fracture strain / maximum strain at rupture
ε_u	Strain at maximum stress
ρ	Hardened density
σ_f	Stress rupture strength
σ_u	Ultimate strength
μe	Micro-Deformation (10^{-6} $\Delta L/L$)

Abbreviations

CRC	Crumbled Rubber Concrete
ERFT	Earthquake-Resistant Fiber Technology

NF	French Standard
ISO	International Organization for Standardization
NSAS	National School of Applied Sciences
EN	European Standard
USC	Stress-Strain Curve
UCS	Unconfined Compressive Stress

1. Introduction

One billion old tires are generated each year, only 5% of which are recycled in the sector of civil engineering, with the remainder being disposed of in landfills [1]. The energy dissipation, ductility, and damping rate of CRC concretes have been clearly improved in previous investigations on the addition of rubber crumb, while the compressive strength has decreased [2], [3]. These properties are frequently advised for structures that are vulnerable to seismic shocks, particularly for traditional structures made of geo- and bio-sourced materials, for which it is extremely challenging to increase stiffness and strength or to create conventional seismic-resistant solutions by incorporating steels.

Before it can be implemented, recycling tire waste in the traditional building industry requires rigorous technical and economic research. If we exclude efforts to improve seismic behavior using glass and carbon fiber reinforcements [4]or the technique of confining walls with wire mesh or polypropylene mesh strips [5], there aren't many viable and effective alternatives for the development of seismic solutions for traditional buildings. Faced with this reality, it would be very useful to take stock of the idea of developing earthquake-resistant fibers based on the concept of hybrid construction technology [6]capable of improving the anti-seismic behavior of traditional constructions by recycling waste materials, in particular tires.

It is proving very difficult to give traditional materials (stone or raw earth) sufficient rigidity and strength to limit deformation without reinforcement, even with stabilization by binders such as cement [6]. Studies on the stabilization of raw earth with binders have shown that there is only a small increase in strength compared to the percentage of binders added [7], so it is important to make sure that the structure has the ductility needed to absorb seismic energy through inelastic deformation and resist without collapsing. The two advantages of this method are the ecological benefit of recycling plastic and tire trash and the technological advantage of providing conventional building materials the ductility to absorb seismic vibrations.

Historical evidence clearly demonstrates how traditional structures built with masonry lime-based mortar resist earthquakes well, such as Hagia Sophia. This is because lime mortar's presence gives masonry its ductility, which enables the structure to absorb seismic energy[8]. In a similar vein, more research on the mobilization of ductility and form is required to enhance the anti-seismic performance of traditional structures.

Studies on the use of tires in concrete have concentrated on using crumb rubber-based aggregates in place of some common aggregates. In contrast, in our study we used fibrous shaped rubber incorporated into hydraulic lime concretes from Marrakech, with the aim of assessing their impact on improving the ductility of lime concrete. Rubber fibers were incorporated at a rate of 1.5% and 3% on a comparative lime concrete. The results of stress-strain diagram analysis obtained by applying uni-axial compression tests for the various specimens prepared show an increase in ductility depending on the ratio of added fibers. The term "earthquake-resistant fiber technology" will be used going forward to describe this technology.

2. Materials and experimental methods

2.1 Materials selection and characteristics

We used Marrakech hydraulic lime for this investigation, which is produced through the traditional firing of marly limestone from the Marrakech region. This process generates a high percentage of

Materials Research Forum LLC
https://doi.org/10.21741/9781644903117-1

unfired marly limestone, which is combined with the silicoclastic fraction to form the aggregates of the mixture [9]. The average density of the used lime is 2,46. Particle size analysis was carried out in accordance with French standard NF P94-056 [10]for the fraction of soil greater than 80 μm, followed by a sedimentation test for fines in accordance with French Standard NF P94-057 [11] (figure 1):

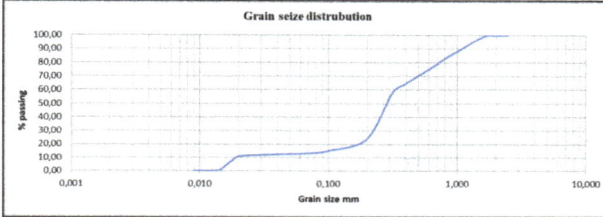

Figure 1 : Grain size distribution of the used Marrakech hydraulic lime

It is common knowledge that 24% of grains have a weighted diameter of less than 200 μm, this portion makes up the binding phase of the mixture [12]. As a result, the combination has a weight ratio of one component binder to three parts aggregate. This is the proportion historically used to formulate the hydraulic lime-based mortar used in old lime concrete constructions [13]. For our study, we will use this mix without any correction for the incorporation of other aggregates, as the main objective is to evaluate the impact of the incorporation of rubber-based fibers on the ductility and not on the strength or compactness of the mix.

Employed tires are sliced into fibrous shape in order to obtain the rubber fibers that were employed in this investigation (Figure 2). The extraction of fibers from used tires is motivated by the desire to recycle this type of waste in the construction industry, with all its attendant ecological benefits as well as for their high mechanical ductility, which can improve the seismic behavior of materials through improved damping and energy dissipation rates. The mechanical properties of the fibers (figure 2) were determined by stress-strain diagram analysis carried out on tire components in accordance with ISO 527, which describes methods for testing the tensile strength of plastics and other resinous materials.

Figure 2 : Fibers extracted from tire waste

The physical and mechanical characteristics of tire-based fibers are listed in Table 1.

Table 1 : Physical and mechanical characteristics of tire-based fibers

Parameters	Values
Density	1.342
Diameter (mm)	3 à 6
Length (cm)	3 à 5
Young's modulus (GPa)	8.561
Rupture stress σ_f (MPa)	2.06
Rupture strain ε_f (%)	69.60

Mediterranean Architectural Heritage - RIPAM10
Materials Research Proceedings 40 (2024) 1-11

Materials Research Forum LLC
https://doi.org/10.21741/9781644903117-1

A ratio of 0.3% to 0.45% is used by Imanzadeh [14]to assess the ductility of earth concrete forced by flax fibers. Studies carried out on the improvement of mechanical and technical performance through the incorporation of different types of fibers such as polypropylene, glass and Basalt fibers have adopted a ratio of between 0.20 and 2% by volume [15]. Weidong [16]used a ratio of 1.2 and 3% by weight to study the properties of asphalt mixtures modified with recycled tire rubber. For our study, we adopted a ratio of 1.5% and 3% by volume on a control mix of lime concrete from Marrakech. Following the example of studies carried out on hydraulic lime mortar, we have adopted a Water/Lime ratio equal to 0.60[15].

2.2 Test methods

2.2.1 Specimen preparation

After determining the physical properties and proportions of the materials used to formulate lime concrete, samples of fiber-reinforced lime concrete were prepared for characterization in the hardened state. The NSAS laboratory served as the location for the mixture preparation. To obtain a homogeneous mixture with evenly distributed fibers, we used a laboratory blender with a capacity of 5l. First, we mixed the Marrakech hydraulic lime with the earthquake-resistant fibers to homogenize the mixture, then we added water successively according to the predefined dosage. Samples were taken in accordance with standard EN 12350-1 [17]. Cylindrical specimens were filled in three layers, hand-tightened with a pitting bar. Each layer of concrete was subjected to 25 blows to eliminate air bubbles. Demolding was carried out after 24 h. The specimens were then numbered, dated and stored for 90 days under controlled laboratory conditions (T=22± 3 and Rh= 65% ± 5) until the day of testing, the optimum conditions for ideal carbonation [18].

2.2.2 Composite characterization tests in the hardened state

Density measurement in the hardened state was carried out in accordance with EN 12390-7 [19], It is obtained by dividing the mass of the specimens by its volume. Ductility parameters are estimated from the analysis of stress-strain diagrams obtained by applying uniaxial compression tests in accordance with French standards NF P94-420, and NF P94-425, [20], [21]on previously prepared samples. The experimental set-up includes a press with a maximum loading capacity of 110 kN and a displacement transducer with an accuracy of ±0.05 mm. The displacement speed chosen for the experiment is 0.1 mm/min[14]. The machine applies an increasing compression force centered on the specimen until it breaks. The machine is connected to a computer, which processes the results obtained in the form of reports, curves and tables (Figure 3).

Figure 3 : Device for determining the stress-strain diagram of samples with extensometers

The concrete press allows measurement of longitudinal displacements using longitudinal strain transducers (extensometers), as shown in figure 4. The stress-strain curve is then plotted to study of the mechanical behavior of the lime concrete in question, especially in the plastic zone considered in this study.

Mediterranean Architectural Heritage - RIPAM10

Materials Research Proceedings 40 (2024) 1-11

Materials Research Forum LLC

https://doi.org/10.21741/9781644903117-1

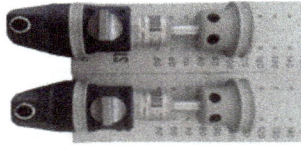

Figure 4 : Longitudinal displacement measuring device (extensometers)

To characterize the ductility of lime concrete, from the stress-strain diagram we will determine strain at maximum stress ε_u , maximum strain at rupture ε_f . A qualitative approach is adopted to conclude on the evolution of the strain-hardening zone; the choice of this approach is justified by the difficulty of accurately determining the elastic limit strain ε_e by the obtained results for this study.

3. Results and discussion

The stress-strain diagrams of the four specimens of traditional lime concrete with 0% fiber incorporated were plotted in Figure 5, and the results were grouped together in Table 2. The stresses at rupture obtained were low, varying between 0.68 and 0.93 MPa, with an average value of 0.795 MPa. This is essentially due to the sandy nature of the aggregates, but also to the traditional procedure followed in the production of traditional hydraulic lime, which results in ineffective quality control of the final product. As a result, traditional lime concrete has to be reinforced with cement when used as mortar or concrete, a course of action that the ecological community does not advise. Hence the interest in using conventional hydraulic limes, not yet available on the Moroccan market, with a rigorous mix design. It should be noted that methods for formulating hydraulic lime concrete are not widely developed in the literature, compared with portland cement-based concrete[9]. Maximum strains obtained ranged from 681 to 1798.8 10^{-6}, with an average absolute value of 1257.35 10^{-6}. These relatively low values correspond well to the low values of stresses at fracture. It should be noted that the device used did not record deformations corresponding to stresses of less than 0.5 MPa, so we cannot draw any dismaying conclusions about the elastic characteristics of the hydraulic lime concretes used.

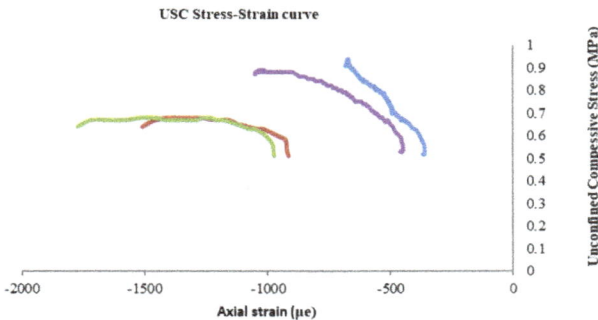

Figure 5 : Summarizing stress-strain diagrams for specimens with 0% rubber fibers

Table 2 : mechanical characteristics of the specimens with 0 % volume content of rubber fibers

	P (g/cm³)	Ultimate strenght σ_u (Mpa)	Fracture stress σ_f (MPa)	Strain at maximum stress ε_u (µe)	Ultimate strain ε_f (µe)
Spicemen 1	1.217	0.93	0.88	-671	-681
Spicemen 2	1.143	0.89	0.82	-1017	-1033
Spicemen 3	1.110	0.68	0.64	-1129.60	-1798.80
Spicemen 4	1.108	0.68	0.60	-1278.80	-1516.60
Average values	1.144	0.795	0.735	-1024.100	-1257.350

The stress-strain diagrams of the four specimens of traditional lime concrete with 1.5% fiber incorporated were plotted in Figure 6, and the results were grouped together in Table 3. The stresses at rupture obtained were low, ranging from 0.77 to 0.92 MPa, with an average value of 0.812 MPa. It can be seen that the incorporation of fibers had no significant impact on the evolution of stress at fracture, since the difference between the mean values for the two cases is of the order of 0.017 MPa. In contrast, there was a significant change in the maximum deformations obtained, with the average value between the two cases rising from 1257.35 to 2041.78 10^{-6}, confirming the basic hypothesis concerning the effect of tire-based fibers on the evolution of the ductility of the composite material as a function of the percentage of fibers added. The same observations were noted for the evolution of strain at maximum stress ε_u, which rose from 1024.1 10^{-6} in absolute value to 1635.88 10^{-6}, with a clear evolution of the necking zone in particular and the plastic zone in general of fiber-reinforced lime concrete curve.

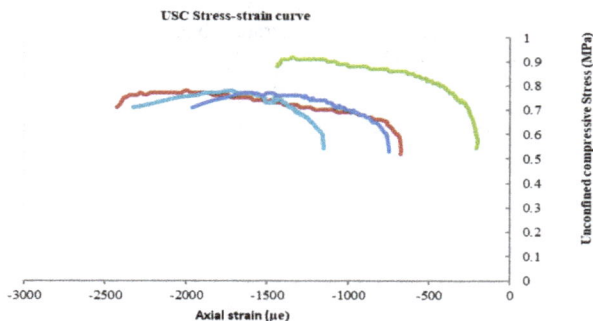

Figure 6 : Summarizing stress-strain diagrams for specimens with 1.5% rubber fibers

Table 3 : mechanical characteristics of the specimens with 1.5 % volume content of rubber fibers

	ρ (g/cm³)	Ultimate strenght σ_u (Mpa)	Fracture stress σ_f (MPa)	Strain at maximum stress ε_u(µe)	Ultimate strain ε_f (µe)
Spicemen 5	1.166	0.77	0.71	-1469	-1965
Spicemen 6	1.109	0.78	0.71	-1724.9	-2329.6
Spicemen 7	1.127	0.92	0.88	-1346.1	-1441.8
Spicemen 8	1.139	0.78	0.71	-2003.5	-2430.7
Average values	1.135	0.812	0.752	-1635.880	-2041.780

The stress-strain diagrams of the four specimens of traditional lime concrete with 3% fiber incorporated were plotted in Figure 7, and the results were summarized in Table 4. Obtained ultimate stresses were low, varying between 0.66 and 1.13 MPa with an average value of 0.82 MPa. It can be seen that the incorporation of fibers had no significant impact on the evolution of stress at fracture, since the difference between the mean values for the three cases did not exceed 0.025MPa as the maximum value. The significant difference between the maximum compressive strength of the different specimens at 3% fiber incorporation, which rose from 0.66Mpa for specimens 11 and 12 to 1.13MPa for specimen 9, can be interpreted by the effect of fiber length on their homogeneous distribution in the mix, so a low strength is a sign of poor fiber distribution in the mix that has impacted strength. In order to avoid this phenomenon, we recommend using fibers with an average length of 2 cm. With the mean value rising from 2041.78 10^{-6} for specimens at 1.5% to 4784.9 10^{-6} in absolute value for specimens at 3%, there was a clear increase in the ultimate strain values obtained as a function of the percentage of fibers incorporated, further validating the basic hypothesis regarding the impact of tire-based fibers on material ductility. Similar observations were noted for the evolution of strain at maximum stress ε_u, which rose from 1635.88 10^{-6} in absolute value to 2937.475 10^{-6}, with a clear evolution of the necking zone in particular and the plastic zone in general of the fiber-reinforced lime concrete strain-stress curve. It is clear that the values of characteristic deformations and ductility will continue to increase as the mechanical characteristics of concretes, the fibers used and their incorporation rate increase [22].

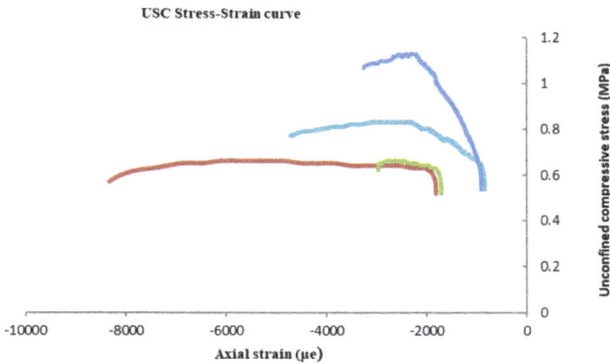

Figure 7 : Summarizing stress-strain diagrams for specimens with 3% rubber fibers

Table 4 : mechanical characteristics of the specimens with 3 % volume content of rubber fibers

	ρ (g/cm3)	Ultimate strenght σ_u (Mpa)	Fracture stress σ_f (MPa)	Strain at maximum stress $\varepsilon_u(\mu e)$	Ultimate strain ε_f (µe)
Spicemen 9	1.133	1.13	1.07	-2196.8	-3252.4
Spicemen 10	1.179	0.83	0.77	-2288.4	-4712.9
Spicemen 11	1.149	0.66	0.57	-2444	-2665
Spicemen 12	1.109	0.66	0.53	-4820.7	-8509.3
Average values	1.142	0.820	0.735	-2937.475	-4784.900

Table 5 provides an overview of the outcomes discovered. As can be seen, the density of the fiberized lime concrete was not significantly affected by the fibers' incorporation. This is essentially due to the very close values of the density of the hardened lime concrete, which is of the order of 1.144, and that of the rubber fibers used, which is of the order of 1.342. The same is true of the tensile stresses, with a very modest improvement that can be explained by the fact that the tensile strength of fibers (2.06 MPa) is higher than that of traditional lime concrete (0.735 MPa). On the positive side, the incorporation of tire-based fibers has been shown to improve the ductility of the composite material. Indeed, the average value of $\Delta_{\varepsilon_u\varepsilon_f}$ characterizing the necking zone increased from 233.255 10^{-6} for lime concrete with 0% of fibers incorporated to 1847.425 10^{-6} for 3% of fibers incorporated. A qualitative analysis of the resulting curves leads to the same conclusion regarding elastic limit and strain-hardening zone, confirming the hypothesis that the addition of tire-based fibers improves ductility. The same hypothesis has been supported by earlier research on the recycling of tire crumbs into cement with a significant reduction in compression resistance [23]. This is due to the significant difference between the mechanical characteristics of cement-based concrete and rubber.

Table 5 : Summary table of obtained results with different rates of rubber fibers incorporation by volume

	ρ (g/cm3)	Ultimate strenght σ_u (Mpa)	Fracture stress σ_f (MPa)	Strain at maximum stress $\varepsilon_u(\mu e)$	Ultimate strain ε_f (μe)	Necking Zone $\Delta_{\varepsilon_u\varepsilon_f}$
0% rubber fibers	1.144	0.795	0.735	-1024.1	-1257.35	233.250
1.5%rubber fibers	1.135	0.8125	0.7525	-1635.88	-2041.78	405.900
3% rubber fibers	1.142	0.820	0.735	-2937.475	-4784.900	1847.425

While biobased fibers have the same advantage of increasing the ductility of traditional lime- or earth-based concretes [24], rubber-based fibers offer a number of additional advantages: they are less affected by moisture, which is an intrinsic characteristic of hygroscopic materials such as hydraulic lime concrete, even though in-depth studies are needed on the effects of lixiviation on the construction and its environment. Similarly, tire recycling reduces the environmental impact of waste and creates economic value and employment [25]. One of the techniques used for seismic reinforcement of traditional buildings is reinforcement with boundary wooden elements [26]. Even this technique can prevent the total collapse of the building and delay the onset of collapse, it just ensures limited structural continuity, not forgetting the difficulty of installation and the extra cost involved in incorporating wood. For the technique of confining walls with wire mesh or polypropylene netting strip [5], it just strengthens the wall as a whole without any effective improvement in the mechanical characteristics of strength, rigidity and ductility of the materials. With earthquake-resistant fiber technology, we can improve the mechanical characteristics of traditional concretes, especially ductility and tensile strength, and consequently improve the structure locally and globally. In addition, this technique proves more suitable for improving the anti-seismic behavior of constructions based on monolithic structure technology, a construction technique recently patented in Morocco which is based on the casting of a single structure with a well-defined shape regenerated from an arched portal frame [27].

Conclusions and recommendations

The present study analyzes the effects of fiber incorporation on improving the ductility of traditional hydraulic lime concrete. Following the example of studies carried out in the literature on the effects of fiber incorporation on ductility improvement, it has also been demonstrated in our study that the incorporation of rubber-based fibers considerably improves this property of the composite. In this study, the incorporation of tire-based fibers (at a rate of 3%) increases significantly the material's ductility. Seismic fiber technology has a number of advantages: it recycles tire waste, protects the environment and improves the seismic behavior of traditional buildings. If this technology is to be successfully implemented, a number of aspects need to be addressed in future studies on the subject, including:

- Verify the outcomes of the full-scale vibration table test.
- Expand research on tire-based fibers in order to comprehend the impact of forms and dimensions on the development of the mechanical properties of lime concretes made from tire fibers.
- Extend studies on earthquake-resistant fibers to other forms of traditional materials such as rammed earth, widely used in Morocco.
- Develop multi-scale numerical models capable of accurately simulating the behavior of earthquake-resistant fiber-based concretes.
- Produce standards relating specifically to seismic-resistant fiber technology in traditional construction.

Consent For Publication

Not applicable

Funding

None

Conflict of Interest

The authors declare no conflict of interest, financial or otherwise.

Acknowledgements

The authors would like to thank the universities for allowing to conduct this research.

References

[1] C. Halsband, L. Sørensen, A. M. Booth, and D. Herzke, "Car Tire Crumb Rubber: Does Leaching Produce a Toxic Chemical Cocktail in Coastal Marine Systems?," *Front Environ Sci*, vol. 8, Jul. 2020. https://doi.org/10.3389/fenvs.2020.00125

[2] O. Youssf, M. A. ElGawady, and J. E. Mills, "Experimental Investigation of Crumb Rubber Concrete Columns under Seismic Loading," *Structures*, vol. 3, pp. 13–27, Aug. 2015. https://doi.org/10.1016/j.istruc.2015.02.005

[3] P. Kara De Maeijer, B. Craeye, J. Blom, and L. Bervoets, "Crumb Rubber in Concrete—The Barriers for Application in the Construction Industry," *Infrastructures (Basel)*, vol. 6, no. 8, p. 116, Aug. 2021. https://doi.org/10.3390/infrastructures6080116

[4] M. Corradi, A. I. Osofero, A. Borri, and G. Castori, "Strengthening of Historic Masonry Structures with Composite Materials," 2015, pp. 257–292. https://doi.org/10.4018/978-1-4666-8286-3.ch008

[5] A. Chourasia, S. Singhal, and J. Parashar, "Experimental investigation of seismic strengthening technique for confined masonry buildings," *Journal of Building Engineering*, vol. 25, p. 100834, Sep. 2019. https://doi.org/10.1016/j.jobe.2019.100834

[6] A. Belabid, H. Elminor, and H. Akhzouz, "The Concept of Hybrid Construction Technology : State of the Art and Future Prospects," *Future Cities and Environment*, vol. 8, no. 1, Dec. 2022. https://doi.org/10.5334/fce.159

[7] K. Ouedraogo, "Stabilisation de matériaux de construction durables et écologiques à base de terre crue par des liants organiques et/ou minéraux à faibles impacts environnementaux (Doctoral dissertation," Université Paul Sabatier-Toulouse III, 2019. Accessed: Jun. 21, 2023. [Online]. Available: http://thesesups.ups-tlse.fr/4537/1/2019TOU30199.pdf

[8] A. Moropoulou, A. Bakolas, and E. Aggelakopoulou, "Evaluation of pozzolanic activity of natural and artificial pozzolans by thermal analysis," *Thermochim Acta*, vol. 420, no. 1–2, pp. 135–140, Oct. 2004. https://doi.org/10.1016/j.tca.2003.11.059

[9] A. Belabid, H. Elminor, and H. Akhzouz, "Study of paramaters influencing the setting of hydraulic lime: an overview," *International Review of Civil Engineering*, vol. 14, no. 3, Sep. 2023.

[10] Standard AFNOR, "NF P94-056, reconnaissance et essais – Analyse granulométrique - Méthode par tamisage à sec après lavage," 1996.

[11] Standard AFNOR, "NF P94-057, Sols : reconnaissance et essais - Analyse granulométrique des sols - Méthode par sédimentation," 1992.

[12] A. El Amrani et al, "From the stone to the lime for Tadelakt: Marrakesh traditional plaster," *Journal of Materials and Environmental Science*, vol. 9, no. 3, pp. 354–362, 2018.

[13] O. Cazalla, C. Rodriguez-Navarro, E. Sebastian, G. Cultrone, and M. J. Torre, "Aging of Lime Putty: Effects on Traditional Lime Mortar Carbonation," *Journal of the American Ceramic Society*, vol. 83, no. 5, pp. 1070–1076, Dec. 2004. https://doi.org/10.1111/j.1151-2916.2000.tb01332.x

[14] S. Imanzadeh, A. Hibouche, A. Jarno, and S. Taibi, "Formulating and optimizing the compressive strength of a raw earth concrete by mixture design," *Constr Build Mater*, vol. 163, pp. 149–159, Feb. 2018. https://doi.org/10.1016/j.conbuildmat.2017.12.088

[15] M. M. Barbero-Barrera, N. Flores Medina, and C. Guardia-Martín, "Influence of the addition of waste graphite powder on the physical and microstructural performance of hydraulic lime pastes," *Constr Build Mater*, vol. 149, pp. 599–611, Sep. 2017. https://doi.org/10.1016/j.conbuildmat.2017.05.156

[16] W. Cao, "Study on properties of recycled tire rubber modified asphalt mixtures using dry process," *Constr Build Mater*, vol. 21, no. 5, pp. 1011–1015, May 2007. https://doi.org/10.1016/j.conbuildmat.2006.02.004

[17] EN 12350-1, "Testing fresh concrete — Part 1 : Sampling," 1999.

[18] M. Chabannes, E. Garcia-Diaz, L. Clerc, J.-C. Bénézet, and F. Becquart, *Lime Hemp and Rice Husk-Based Concretes for Building Envelopes*. Cham: Springer International Publishing, 2018. https://doi.org/10.1007/978-3-319-67660-9

[19] EN 12390-7, "Testing hardened concrete — Part 7: Density of hardened concrete," 2012.

[20] Standard AFNOR, "NF P94-420, Détermination de la résistance à la compression uniaxiale," 2000.

[21] Standard AFNOR, "NF P94-425, Méthodes d'essai pour roches - Détermination du module d'Young et du coefficient de Poisson," 2002.

[22] A. Laborel-Préneron, J. E. Aubert, C. Magniont, C. Tribout, and A. Bertron, "Plant aggregates and fibers in earth construction materials: A review," *Constr Build Mater*, vol. 111, pp. 719–734, May 2016. https://doi.org/10.1016/j.conbuildmat.2016.02.119

Materials Research Forum LLC

https://doi.org/10.21741/9781644903117-1

[23] A. Mohajerani *et al.*, "Recycling waste rubber tyres in construction materials and associated environmental considerations: A review," *Resour Conserv Recycl*, vol. 155, p. 104679, Apr. 2020. https://doi.org/10.1016/j.resconrec.2020.104679

[24] H. M. Hamada, J. Shi, M. S. Al Jawahery, A. Majdi, S. T. Yousif, and G. Kaplan, "Application of natural fibres in cement concrete: A critical review," *Mater Today Commun*, vol. 35, p. 105833, Jun. 2023. https://doi.org/10.1016/j.mtcomm.2023.105833

[25] W. Price and E. D. Smith, "Waste tire recycling: environmental benefits and commercial challenges," *International Journal of Environmental Technology and Management*, vol. 6, no. 3–4, pp. 362–374, 2006.

[26] L. E. Yamin, C. A. Phillips, J. C. Reyes, and D. M. Ruiz, "Seismic behavior and rehabilitation alternatives for adobe and rammed earth buildings," in *In 13th world conference on earthquake engineering*, BC Canada: Vancouver, Aug. 2004, pp. 1–6.

[27] A. Belabid, H. Akhzouz, H. Elminor, and H. Elminor, "Monolithic Structure Technology: A New Construction Process to Enhance Traditional Construction.," *International Journal of Sustainable Construction Engineering and Technology*, vol. 14, no. 1, pp. 42–47, Feb. 2023, Accessed: Jun. 22, 2023. [Online]. Available: https://publisher.uthm.edu.my/ojs/index.php/IJSCET/article/view/12180

Mediterranean Architectural Heritage - RIPAM10
Materials Research Proceedings 40 (2024) 12-18

Materials Research Forum LLC
https://doi.org/10.21741/9781644903117-2

Recycled Glass-Fiber Reinforced Cement (RGFRC) Waste as a Substitute in Concrete Production

Amine NAIM[1], Zineb MOUJOUD[1], Ikrame HATTAB[2], Mohamed ELGHOZLANI[3], Omar TANANE[1], Abdeslam EL BOUARI[1], Reda ELKACMI[2,*]

[1]Laboratory of Physical Chemistry, Materials and Catalysis (LCPMC), Faculty of Sciences Ben M'Sik, Hassan II University of Casablanca, Morocco

[2]Environmental and Agro-Industrial Process Team, Department of Chemistry and Environment, Faculty of Sciences and Technology, University Sultan Moulay Slimane, Beni Mellal, Morocco

[3]Laboratory of Materials Engineering for the Environment & Natural Resources, Team: Natural Substances & Molecular Synthesis and Modeling, Department of Chemistry, Faculty of Sciences and Technics, BP 509 Boutalamine, Errachidia, Moulay Ismail University of Meknes, Morocco

*r.elkacmi@usms.ma

Keywords: Recycled Glass Fiber Reinforced Cement Waste, Concrete, Compressive Strength Behavior, Mechanical Properties

Abstract. Nowadays, a large amount of glass fiber reinforced cement (GFRC) waste from construction industries and demolition activities presents a significant source of major environmental and economic problems. In order to protect the environment, many studies have been conducted to recycle and reuse these wastes in concrete production. The present work also aims to reach this objective and to show technically the possibility of recycling glass fiber reinforced cement waste (RGFRC) as a partial substitution in concrete production. Three concrete mix variations were formulated: one comprising solely natural aggregate (NC) serving as the control, and two others incorporating a blend of natural and recycled glass fiber reinforced cement (RGFRC) with 20% and 40% replacement of recycled aggregate, respectively. The test of compressive strength behavior was performed on the mixes. The results showed that concrete containing 20% RGFRC has the best mechanical properties compared with the control concrete and that using more RGFRC would have a harmful impact on the mechanical characteristics of concrete.

Introduction

The consequent increase in the construction sector is reflected in an ever-increasing need for raw materials [1,2]. The natural deposits of potentially exploitable aggregates are becoming scarcer. Many studies have been published on the development of new materials using waste from demolition and construction sites [3–5].

In recent times, there has been a noticeable trend in using recycled aggregates from industrial or demolition waste in the concrete production [6–10]. This approach not only facilitates waste recycling in modern society but also enables the formulation of more cost-effective concretes with enhanced properties [11].

Many previous studies have reported results on testing the properties of Recycled Glass Fiber Reinforced Concrete (RGFRC) and on examining the effects of its partial or total incorporation in the production of concrete. *In a study carried out by Olorunsogo and Padayachee*, [12], emphasis was placed on the sustainability of concrete produced with different proportions of recycled concrete aggregates. Results indicated a decrease in durability as the quantities of recycled aggregates increased, while durability improved with aging. This phenomenon has been

Materials Research Forum LLC
https://doi.org/10.21741/9781644903117-2

attributed to cracks and fissures the RA during processing, making the aggregate more susceptible to permeation, diffusion and absorption.

Tabsh and Abdelfatah [13] studied the strength of concrete from coarse aggregates of recycled concrete, taking into account both the source of the recycled concrete and the targeted strength of the concrete. Their results revealed that the strength of concrete could be 10 to 25% lower than conventional concrete using coarse natural aggregates, whatever its origin. In a study conducted by *Mališev et al.* [14], the properties of fresh and hardened concrete were examined with three amount of recycled coarse aggregates (0%, 50%, and 100%). The results indicated that concrete incorporating recycled aggregate performed satisfactorily, showing no significant deviation from the performance of natural concrete, irrespective of the replacement ratio. The researchers explained that achieving concrete with good mechanical properties containing recycled concrete aggregate requires the use of large aggregates of high recycled concrete and compliance with specific guidelines for the design and production of this new type of concrete.

Thomas et al. [15] conducted another study to assess the concrete strength and durability properties of crushed concrete aggregates. The test results revealed that replacing natural aggregates with up to 25% CCA had no negative impact on the strength properties of concrete. However, the durability investigation indicated that an increase in CCA percentage leads to a reduction in CCA concrete durability.

This is related to a higher water absorption of the CCA and to the presence of porous mortar on the surface, both influencing the durability of the concrete. The study demonstrated that augmenting the cement content in the mix increases CCA concrete durability by enhancing either the density and the quality of the hydrated cement paste.

Major part of the literature review indicates that employing 50% or more Recycled Concrete Aggregate (RCA) would have a negative impact on the mechanical properties of concrete [16,17].

In Morocco, the acceleration of urbanization and rapid population growth have caused a generation of construction and demolition waste (CDW). Total of CDW production is approximately 30 million tons per year including concrete waste. *The exhaustion of natural resources and the largest demand for natural building materials face Morocco to develop the reuse of RCA as alternative source of concrete production according to the standards regulations* [18]. For this purpose, the aim of this study is to contribute to the valorization of concrete waste in the concrete production; in order to reduce waste, to well manage natural resources, and finally to prevent the environmental pollution. The experimental study consists to examine the influence of the partial substitution (20% - 40%) by concrete waste on the mechanical behavior of the studied concrete.

Material and Methods

Preparation of materials

The used materials in this study are: natural concrete aggregates composed of two fractions (NA 8/16 and NA 14/20), natural fine sand (NS) with fraction 0/5, Portland cement CPJ 45, Water (Drinking water of the network, free of impurities) and the recycled glass fiber reinforced cement (RGFRC). The RGFRC is a concrete waste, collected from a building demolition site, crushed in the laboratory and ground to produce gravel of granular class 0/20.

20% and 40% of this RGFRC waste was utilized to substitute the natural aggregate in concrete production.

In this study, at first, we were tested the properties of the four kinds aggregates prior to mix proportion design. The preliminary tests, namely the particle size analysis test, the Methylene blue test, the sand equivalent, the surface cleanliness, the Los Angeles test, the coefficient of flattening, the water absorption coefficient, real density and the bulk density of the aggregates, were carried out according to international and national standards. The results of the different tests are presented in Table 1.

Materials Research Forum LLC
https://doi.org/10.21741/9781644903117-2

Table 1: Aggregates properties.

Name of Test	Reference Standard	Aggregate			
		NS 0/5	NA 8/16	NA 14/20	RCA 0/20
d/D (mm)	EN 933-1 [19]	0/5	8/16	14/20	0/20
Sand equivalent test (%)	EN 933-08 [20]	70	***	***	71
Surface cleanliness (%)	NM 10.1.169 [21]	***	0,5	0,3	3,0
Methylene blue test (g/Kg)	EN 933-09 [22]	0,7	***	***	0,9
Water Absorption (%)	EN 1097-6 [23]	0,19	1,37	1,05	0,42
Bulk density mg/m3	NM 10.1.707 [24]	1,76	1,40	1,40	1,42
Real density (t/m3)	EN 1097-6 [23]	2,74	2,71	2,71	2,69
Coefficient of flattening (%)	NM 10.1.155 [25]	***	5	6	20
LOS ANGLES(%)	EN 1097-2 [26]	***	20	19	23

Mix proportions

The teste on the Aggregates were followed by incorporating RGFRC waste in the concrete mixture as a fraction substitution of natural aggregates. The proportions of the concrete mixture were calculated using the most widely used method in the industry: **DREUX GORISSE's method** [27].

Three mix compositions were designed. The first mix M0 (reference concrete) was made only of natural aggregates, the second mix M20 was made with 20% substitution of natural aggregates by recycled glass fiber reinforced cement and the third mixture M40 is with 40% RGFRC replacement. Mix design is shown in Table 2.

Table 2: Mix proportion of concretes (kg/m3).

	Composition of natural aggregate (kg/m)			Composition of recycled aggregate (kg/m)	Water (L)	Cement (kg)
	0/5	8/16	14/20			
M0	876	464	520	0	9,450	17,500
M20	700,8	371,2	416,0	372	9,450	17,500
M40	525,6	278,4	312,0	744	9,450	17,500

For each mix design, three samples were prepared to increase the reliability of the tests.

Testing

At age of 7 and 28 days, the densities of all concrete samples were calculated according to *EN 12390-7 standard* [28].

Compressive strength tests were carried out conforming to *EN 12390-3 standard* [29]. The cylindrical specimens tested are loaded under compression until failure and fracture according to *EN 12390-4 standard* [30]. The strength properties were measured after 7 and 28 days of curing.

Mediterranean Architectural Heritage - RIPAM10

Materials Research Proceedings 40 (2024) 12-18

Materials Research Forum LLC

https://doi.org/10.21741/9781644903117-2

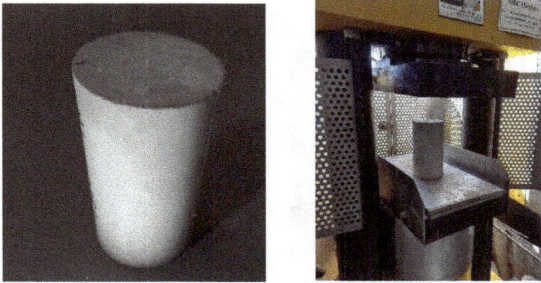

Fig. 1. *Concrete specimens prepared for compressive strength test.*

Results and discussion

Relative density

The relative density for the tow experimental phases of curing with 7 and 28 days of age respectively is presented in Figure 2.

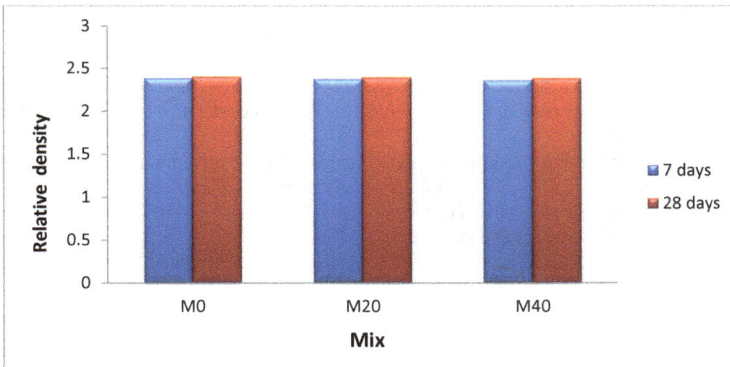

Figure 2: Average density of concretes.

According to the graph, the control concrete M0 (reference formulation based on natural aggregates) has a density of 2.38 in 7 days of curing and 2.40 in 28 days. The average density of the concrete containing 20% RGFRC M20 was about 2.37 in 7 days of curing and 2.39 in 28 days. The concrete based on 40% waste M40 had an average density of 2.36 in 7 days of curing and 2.38 in 28 days. These results show that all the tested mixes present exhibit similar behavior to the density of the reference formulation.

The evolution of the samples density at the age of 28-day is low and similar to the initial values. *This is mainly due to the high quality of recycled aggregate* [31].

Compressive strength Tests

The results of the compressive strength tests on concrete with different substitution degrees and ages, are shown on Figure 3

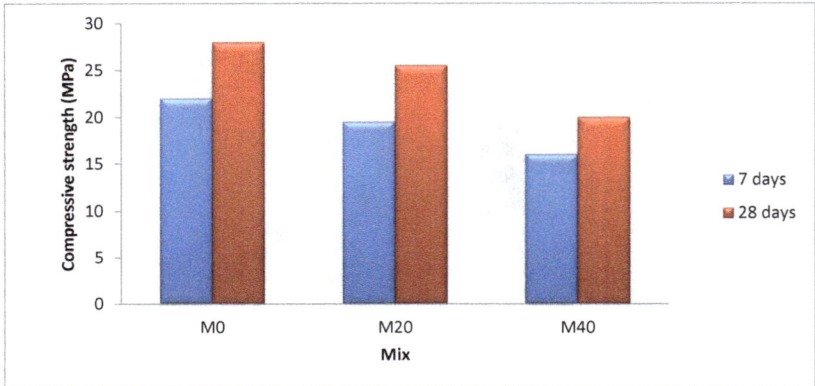

Figure 3: Compressive strength tests of Mix (MPa).

At the age of 7 and 28 days, the value of the compressive strength of the concrete containing RGFRC is less than that of the natural concrete. This might due to the presence of the old mortar attached to the recycled aggregates. *This mortar causes increased porosity of the concrete which diminish the mechanical performances of the new concrete* [19].

Exceeding 28 days of curing, the resistance is more influenced by the incorporation of recycled aggregates than the mixes tested after 7 days. *This observation is explained by the long-term self-cementing effects of the old cement mortar and the interaction of the new cement paste and the old cement mortar* [12,20].

At 7 and 28 days of curing, the compressive strength of concrete containing 20% RGFRC was greater than that containing 40% of RGFRC. *It clearly show that the compressive strength reduces with increasing the amount of recycled aggregates* [12,15]. The use of RGFRC exceeds than 20% has a significant loss of compressive strength.

Based on *the Moroccan Standard NM 10.1.008* [32], the result obtained from the M20 (20% of RGFRC) is acceptable for a concrete class B25.

Conclusion

The experimental work was conducted to study the use of recycled glass fiber reinforced cement as partial substitution of natural aggregates in the concrete production . Properties as grind size distribution, bulk density, water absorption and surface cleanliness were evaluated. These tests were followed by design different mixtures with RGFRC substitution. The most important conclusions can be stated as follows.

- The densities of concrete mixes made with recycled aggregates are the same as the density of the control concrete at a given curing age. This is probably due to the high quality of the recycled aggregates.
- The compressive strength decreases as the percentage of recycled aggregates in the mix increases and, as expected, improves with age due to long-term.
- The substitution of natural aggregates beyond 20% of recycled aggregates decreases the compressive strength.
- The concrete with 20% replacement of natural aggregates is acceptable for a concrete class B25 in the sense of *Moroccan standard NM 10.1.008*.

Compliance with Ethical Standard Conflict of Interest

The authors declare that they have no conflict of interest.

Materials Research Forum LLC
https://doi.org/10.21741/9781644903117-2

References

[1] X. Cui, Examining urban metabolism: A material flow perspective on cities and their sustainability, Journal of Cleaner Production. (2019) 15. https://doi.org/10.1016/j.jclepro.2019.01.021

[2] Y. Kalmykova, Resource consumption drivers and pathways to reduction: economy, policy and lifestyle impact on material flows at the national and urban scale, Journal of Cleaner Production.(2016) 11. https://doi.org/10.1016/j.jclepro.2015.02.027

[3] M. Yeheyis, K. Hewage, M.S. Alam, C. Eskicioglu, R. Sadiq, An overview of construction and demolition waste management in Canada: a lifecycle analysis approach to sustainability, (n.d.)11.

[4] M.S. Aslam, Review of construction and demolition waste management in China and USA,Journal of Environmental Management. (2020) 13. https://doi.org/10.1016/j.jenvman.2020.110445

[5] U.A. Umar, A case study on the effective implementation of the reuse and recycling of construction & demolition waste management practices in Malaysia, (n.d.) 9.

[6] A.T. Gebremariam, Comprehensive study on the most sustainable concrete design made of recycled concrete, glass and mineral wool from C&D wastes, (n.d.) 12.

[7] K.P. Verian, Properties of recycled concrete aggregate and their influence in new concrete production, (2018) 20. https://doi.org/10.1016/j.resconrec.2018.02.005

[8] J. Hu, Z. Wang, Y. Kim, Feasibility study of using fine recycled concrete aggregate in producing self-consolidation concrete, (n.d.) 16.

[9] A. Abbas, Environmental Benefits of Green Concrete, (n.d.) 9.

[10] A. Kirby, R. Gaimster, RECYCLED WASTE CONCRETE European and North American Practice and its Applicability to New Zealand, (n.d.) 9.

[11] S.M. Mohammed, Recycled aggregates in concrete production: engineering properties and environmental impact, MATEC Web of Conferences. (2017) 8.

[12] F.T. Olorunsogo, N. Padayachee, Performance of recycled aggregate concrete monitored by durability indexes, Cement and Concrete Research. 32 (2002) 179–185. https://doi.org/10.1016/S0008-8846(01)00653-6

[13] S.W. Tabsh, A.S. Abdelfatah, Influence of recycled concrete aggregates on strength properties of concrete, Construction and Building Materials. 23 (2009) 1163–1167. https://doi.org/10.1016/j.conbuildmat.2008.06.007

[14] M. Malešev, V. Radonjanin, S. Marinković, Recycled Concrete as Aggregate for Structural Concrete Production, (2010) 23. https://doi.org/10.3390/su2051204

[15] J. Thomas, STRENGTH AND DURABILITY OF CONCRETE CONTAINING CRUSHED CONCRETE AGGREGATES, (n.d.) 64.

[16] C.S. Poon, Z.H. Shui, L. Lam, H. Fok, S.C. Kou, Influence of moisture states of natural and recycled aggregates on the slump and compressive strength of concrete, Cement and Concrete Research.(2004) 6. https://doi.org/10.1016/S0008-8846(03)00186-8

[17] D. Talamona, K.H. Tan, Properties of recycled aggregate concrete for sustainable urban built environment, (n.d.) 11.

[18] Loi N_28-00 relative à la gestion des déchets et à leur élimination promulguée, 2006. http://aut.gov.ma/pdf/Loi_n 28-00_relative_a_la_gestion

[19] AFNOR, Tests for geometrical properties of aggregates- Part 1: Determination of granularity - sieve analysis, NF EN 933-1, (2018)

[20] AFNOR, Tests for geometrical properties of aggregates - Part 8: Assessment of fines – Sand equivalent test, NF EN 933-08, (2015)

[21] IMANOR, "Granulats -Détermination de la Propreté Puperficielle, Morrocan Standards", NM 10.1.169, (1995)

[22] AFNOR, Tests for geometrical properties of aggregates - Part 8: Assessment of fines – Methylene Blue test, NF EN 933-09, (2013)

[23] IMANOR, Tests for mechanical and physical properties of aggregates -Part 6: Determination of particle density and water absorption, NM EN 1097-6, (2018)

[24] IMANOR, Tests for mechanical and physical properties of aggregates: Methods for the determination of bulk density, NM 10.1.707, (2008)

[25] IMANOR, "Granulats - Mesure du coefficient d'aplatissement, Morrocan Standards », NM 10.1.155, (1995).

[26] IMANOR, Tests for mechanical and physical properties of aggregates - Part 2: Methods for the determination of resistance to fragmentation, NM EN 1097-2, (2018)

[27] Dreux G., Gorisse F. and Simonnet J. "Composition des Bétons : Méthode Dreux-Gorisse, Annales de l'Institut Technique du Batiment et des Travaux Publics, France" (1983). https://doi.org/10.51257/a-v1-c2220

[28] AFNOR, Testing hardened concrete - Part 7: Density of hardened concrete, NF EN 12390-7, (2019)

[29] AFNOR, Testing hardened concrete - Part 3: Compressive strength of test specimens, NF EN 12390-3 (2019)

[30] AFNOR, Testing hardened concrete - Part 4: Compressive strength - Specification for testing machines, NF EN 12390-4 (2019)

[31] S.-C. Kou, C.-S. Poon, M. Etxeberria, Influence of recycled aggregates on long term mechanical properties and pore size distribution of concrete, Cement and Concrete Composites. 33 (2011) 286–291. https://doi.org/10.1016/j.cemconcomp.2010.10.003

[32] IMANOR, Concrete: Specification, performance, production and conformity, NM 10.1.008 (2009).

Mediterranean Architectural Heritage - RIPAM10
Materials Research Proceedings 40 (2024) 19-32

Materials Research Forum LLC
https://doi.org/10.21741/9781644903117-3

Assessing the Feasibility, Usability, and Durability of Recycled Construction and Demolition Waste in Road Construction in Morocco

Amine NAIM[1], Ikrame HATTAB[2], Rajaa ZAHNOUNE[2], Mohamed ELGHOZLANI[3], Omar TANANE[1], Abdeslam EL BOUARI[1], Reda ELKACMI[2,*]

1Laboratory of Physical Chemistry, Materials and Catalysis (LCPMC), Faculty of Sciences Ben M'Sik, Hassan II University of Casablanca, Morocco

[2]Environmental and Agro-Industrial Process Team, Department of Chemistry and Environment, Faculty of Sciences and Technology, University Sultan Moulay Slimane, Beni Mellal, Morocco

[3]Laboratory of Materials Engineering for the Environment & Natural Resources, Team: Natural Substances & Molecular Synthesis and Modeling, Department of Chemistry, Faculty of Sciences and Technics, BP 509 Boutalamine, Errachidia, Moulay Ismail University of Meknes, Morocco

*r.elkacmi@usms.ma

Keywords: Construction Waste, Demolition Waste, Recycled Aggregate, Unbound Aggregates, Environmental Performance

Abstract. Growing concerns about environmental sustainability and increasingly restrictive waste management regulations have led to the use of construction and demolition wastes (CDWs) as recycled aggregates for civil engineering projects, such as construction and infrastructure development. In this context, this paper presents an experimental laboratory analysis of the technical and environmental properties of recycled aggregates obtained from selected CDWs, conforming to European standards. The technical evaluation encompassed composition tests, particle size distribution (granulometry), density, water absorption, shape, the Los Angeles test, and the Micro-Deval test. The environmental assessment focused on the presence of potential contaminants, such as PAHs (polycyclic aromatic hydrocarbons), PCBs (polychlorinated biphenyls), and BTEX (benzene, toluene, ethylbenzene, and xylene). The laboratory test results are discussed and compared with current requirements. The paper concludes with key findings and recommendations derived from this investigation.

1. Introduction

Construction and demolition wastes (CDWs) are one of the massive wastes produced in large quantities worldwide [1,2]. Each year, CDWs generation exceeds 3 billion tons worldwide, and the current situation indicates a trend for future increases [2,3]. For example, in Morocco, approximately 14 million tons of construction and demolition wastes are made in 2015, which represents about 50% of total waste and it is estimated to be around 16 million tons in 2030 [4]. These wastes typically come from demolition, modification, extension, construction, reconstruction and maintenance of buildings and other infrastructure as well as the building rubble result of some catastrophic activity [5–7]. Therefore, the majority of CDWs are landfilled without any treatment or control, which increases negative impacts on public health and environment [8–10].

Current environmental policies intend to recycle and use CDWs as a substitution of natural aggregates in different fields. In general, the largest demand for natural aggregates, especially in civil engineering projects, leads to the high CO_2 emissions during mining and depletion of virgin raw aggregates [1,11,12]. Indeed, the use of CDWs aggregates can contribute to a sustainable world and create a trade potential for recycled aggregates [1,5,13–15]. CDWs are mainly

Mediterranean Architectural Heritage - RIPAM10   Materials Research Forum LLC
Materials Research Proceedings 40 (2024) 19-32   https://doi.org/10.21741/9781644903117-3

composed of heterogeneous and persistent material such as concrete, bricks, mortar and other materials (wood, plastics, glass and metals) [5]. The CDWs aggregates' composition can vary considerably from country to country, depending on different factors such as the source of the waste (roads, built structures, etc.), the construction materials (concrete, masonry bricks, wood, etc.), and the demolition technique used [5,16,17]. Nonetheless, the overall waste volume contains between 40 and 85% of the inert fraction [5,18,19], making it suitable to reuse in concrete blocks [16,20,21], cement [22–25], beach regeneration [2] and road construction project [26–30].

A huge amount of experimental studies have shown that the most suitable management application for recycled CDWs aggregates is in embankments, shoulders, road subgrades, foundation layers and unbound bases for road construction, where unbound aggregates are used [5,31–33]. This is supported by the fact that it is easy to obtain quick results compared to other application that require much more intensive research [17]. However, the suitability of recycled CDWs aggregates depends on their physico-chemical and mechanical properties [2,29,34–37]. Therefore, different countries, such as European countries, have developed and implemented standardized limits for controlling the production and use of recycled CDWs aggregates in road construction projects, taking into account heterogeneity, density, water absorption, shape, toughness and hardness of the aggregates [5,38,39].

On the other hand, the environmental performance of CDWs is critical in evaluating their possible reuse. CDWs may include slight proportion of waste materials including heavy metals and organic chemicals, which may impact negatively the environment [40]. When recycled CDWs aggregates are used in roads construction, rain or seepage water can leach these harmful elements, posing a particular risk to the environment and human society [41,42]. Many researchers have found organic contaminants like polyaromatic hydrocarbons (PAH) and polychlorinated biphenyls (PCBs) in CDWs [40,43]. The impact of these pollutants on health, ecology, and the environment is of particular concern because of their specific toxicity [44–46]. Hence, scientists include a risk assessment based on both their residue levels and toxicity in surface and groundwater resources with a comparison to regulatory standards or risk thresholds. Generally, recent studies have revealed that the use of recycled CDWs aggregates for road construction project don't present any greater risk of leaching than natural aggregates, with the exception off sulfates, that must be correctly analyzed [17,47].

This study involved the laboratory characterization of the recycled CDWs aggregates obtained from a demolition site in the region of Casablanca city (Morocco). To test the quality of the aggregates, an environmental characterization of persistent potentially toxic compounds (namely PAHs, PCBs and HMs) in the aggregates was undertaken. Additionally, their physico-chemical and mechanical properties were determined according to the European standard EN 13242 [48]. The test results are then presented and discussed, in light of the use of recycled CDWs aggregates.

2. Materials and methods

2.1 Materials

The construction and demolition wastes (CDWs) used in this study was collected from a demolition site in the region of Casablanca city (Morocco) as shown in **Fig. 1**. It was a technical center designed in a ground and one floor , built in 1992 and demolished in 2018.

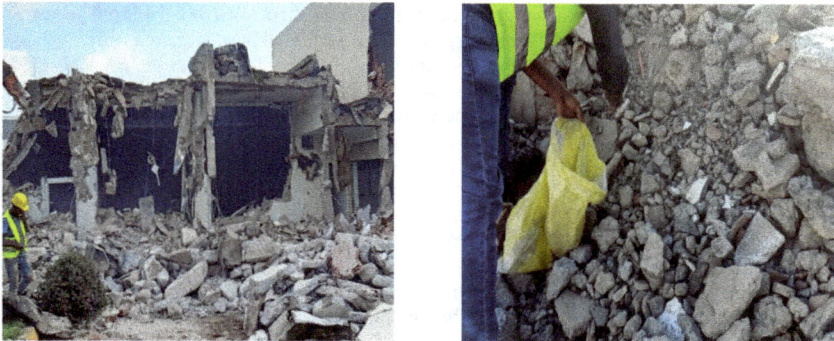

Fig. 1. *The collection of CDWs from a demolition site in Casablanca.*

Several types of materials were found on the demolition site of the building:
- Metallic materials (iron,copper,aluminum,metal alloys:staineless steel);
- Organic materials (wood,paints,coatings);
- Mineral or inorganic materials (rock,glass,brick and masonry);
- Composites materials (reinforced concrete,cement).

Ten (10) samples with a minimum quantity of 80 kg were taken from different points on the site. The CDWs were previously selected and the pollutants (e.g. plastics, glass, metals and wood) were removed. Then, the collected samples were homogenised and reduced in the laboratory using a quartering method described in EN 932-2 standard [49]. The samples were subsequently stored in containers protected from contact with other materials.

2.2 Methods

2.2.1 Environmental performance

Construction and demolition wastes (CDWs) can contain small amounts of hazardous waste that have the potential to adversely affect human health and the environment. Therefore, to ensure that the recycled CDWs aggregates can be used without concern for contamination of the surrounding environment, the samples were tested for their environmental performance. Potential contaminants such as PAHs (polycyclic aromatic hydrocarbons), PCBs (polychlorinated biphenyls), BTEX (Benzene-Toluene-Ethylbenzene-Xylene) and Mineral Oils were determined in compliance with the standards *EN 15934* [50] and *16703* [51]. The tests were conducted by an independent certified commercial laboratory. Each test was performed with three random samples and average values were presented for all tests.

2.2.2 Physical- chemical and mechanical properties

So as to determine the physical-chemical and mechanical characteristics of aggregates; composition test, granulometry, bulk density (ρb), real density (ρr), water absorption (WA), the water-soluble sulphate (SO4%) content and Flattening coefficient were measured according to the conditions and procedures adopted by the *European standard EN 13242* [48]. Mixtures of aggregates recycled from construction and demolition waste are heterogeneous in terms of composition. Therefore, to distinguish the constituent materials and to evaluate their relative tenors , recycled CDWs aggregates were examined according to *the European standard EN 933-11* [52]. The particle size distribution and the fines content were carried out by dry sieving in accordance with the procedure from *EN 933-1* [53].Water absorption (WA) and real density (ρr) were determined by the pycnometer method according to *standard EN 1097-6* [54]. Additionally, the bulk density (ρb) of aggregates was studied according to *standard EN 1097-3* [55]. The water-soluble sulfate content of recycled CDWs aggregates was determined in accordance with *EN 1744-1 standard* [56]. The Flattening coefficient was determined in accordance with the *EN 933-3* [57] standard to identify the shape of the aggregates obtained.

To evaluate the mechanical properties of the CDWs gravels, the Los Angeles and the Micro-Deval tests were performed. The Los Angeles test (LA) was carried out according to *EN 1097-2* [58] to determine the fragmentation resistance of gravels. The standard subjects a coarse aggregate sample to abrasion and crushing in a rotating steel drum containing a specified number of steel spheres. Once the rotation cycle (500 tours) was completed, the crushed aggregates were then sieved through a 1.6 mm sieve. The mass passed through the sieve was expressed as a percentage of the total aggregate mass.

The Micro-Deval test was determined according to the EN 1097-1 standard [59] to check the wear resistance of coarse aggregates and therefore their quality. The test consists to roll the sample of aggregate mixed with water and spherical steel balls in a sealed rotating drum. After a full rotation cycle, the aggregates were sieved through the 1.6 mm sieve and then dried at 105∘C until a constant mass was achieved. The loss of mass (%) was calculated as the Micro-Deval coefficient.

3. Results and Discussion

3.1 Environmental performance

The content of PAHs, PCBs, BTEX and mineral oils in the CDW samples was given in Table 1. As stated in the results attained, the content of total PCBs from CDWs was not detected. Similarly, the tenor of total PAHs from CDWs was not detected, excluding of Naphthalene that contain 0.8 mg/kg, also below the legal limit, which can be considered negligible compared to the specified maximum content (40 mg/kg). Furthermore, the concentration of mineral oils in the leachate was 850 mg/kg, and this concentration was always below the limit value (1000 mg/kg). Consequently, the release of organic compounds to the leachates was not detected.

To sum up, the tenors of inorganic and organic elements in the leachates were much lower than the specifies maximum contents. Thus, it may be said that construction and demolition wastes are inert for the use in road construction project.

Table 1 Purport of PAHs, PCBs, BTEX and mineral oils (mg/kg) from CDWs samples.

Test	Average Content in (mg/kg)	Specified Maximum Content in (mg/kg)	Reference Standard
PolyChlorinated Biphenyls (PCBs)	Not detected	0.5	EN 15934 [50]
Mineral Oils	850	1000	EN 16703 [51]
Volatile Organic Compounds			
Benzene	Not detected	1	
Ethylbenzene	Not detected		
Toluene	Not detected	1.25	EN 15934 [50]
Xylene	Not detected		
Polycyclic Aromatic Hydrocarbons (PAHs)			
Naphthalene	0.8		
Anthracene	Not detected		
Phenanthrene	Not detected		
Fluoride	Not detected	40	EN 15934 [50]
Benzanthracene	Not detected		
Chrysene	Not detected		

Benzofluoranthene	Not detected		
Benzopyrene	Not detected		
Benzoperylene	Not detected		
Endopyrene	Not detected		

3.2 Composition of recycled CDWs aggregates

The CDWs composition varies considerably amongst countries, depending on construction technologies and components (e.g., concrete, brick, ceramic materials, etc.) and the source of the waste (e.g., built structures, bridges, roads, etc.) [15,16]. The proportions of the different constituents present in the collected CDWs were listed in Table 2. The classification of the different constituents follows the specification in EN 13242 [48]. It appears that, about 50% of the CDWs aggregates originate from natural aggregate and 17% of concrete and mortar. The CDWs aggregates were classified as Rc_{17}, Rg_{2-} and $Rcug_{50}$ ($Rcug = Rc + Rb + Rg$). It was also shown that the CDWs samples contain about 32% by weight of other materials (e.g. steel, wood, plastics, etc.) and traces of masonry clay and bituminous materials; and did not contain any traces of glass or float materials. Based on these parameters, the CDWs aggregate used was classified as Rb_{10-}, Ra_{1-}, Rg_{2-}, and FL_{5-}.

Table 2 Classification of recycled CDWs constituents.

Constituents	Description	Contents in %	Category EN 13242 [48]
Rc	Concrete, concrete products, mortar, concrete masonry units	17	Rc_{17}
Rcug	Rcug = Rc + Ru + Rg with Ru : Unbound aggregate, natural stone, hydraulically bound aggregate	67	$Rcug_{50}$
Rb	Clay and cesium silicate masonry units (e.g. bricks and tiles), aerated non-floating concrete	0.44	Rb_{10-}
Ra	Bituminous materials	0.75	Ra_{1-}
Rg	Glass	0	Rg_{2-}
X	Other materials (e.g. steel, plastics, wood, etc.)	32	-
FL	Floating material (cm^3 /kg)	0	FL_{5-}

The physico-mechanical and chemical properties were determined using the standardised methods established in EN 13242 [48]. Table 3 shows the properties of the collected CDWs aggregates.

Table 3 Results of the physical and mechanical tests of recycled CDWs aggregates.

Proprety	Results	Class EN 13242 [48]	Category EN 13242 [48]
Granular Class (mm)	0/40	0/40	-
Fines content < 0.063 mm (%)	9	f9	-
Flattening Coefficient (%)	19	FL20	-
Bulk Density (Kg/m3)	1700	-	-
Real Density (Kg/m3)	2500	-	-
Water Absorption Coefficient (%)	5.2	-	-
Sulphate Content (%)	0.04	-	SS0.2
Los Angeles (%)	40	LA40	E
Micro Deval (%)	37	MDE40	E

The CDWs were crushed and sieved in the laboratory according to EN 933-1 in order to produce a dimensional distribution of the aggregates with a maximum particle size of 40 mm. the grading curve in Fig. 2. shows the results of the sieve analysis performed on the CDWs samples. Recycled aggregates show continual curves, stipulating greater possibilities of synergy between particles and the chance of a greater degree of compaction [17,60]. It was observed that CDWs aggregates have approximately the similar size distribution than natural aggregates [61]. The fine fraction (< 0.063 mm) represents lower than 9%, which classified it as f9 according to EN 13242 [48].

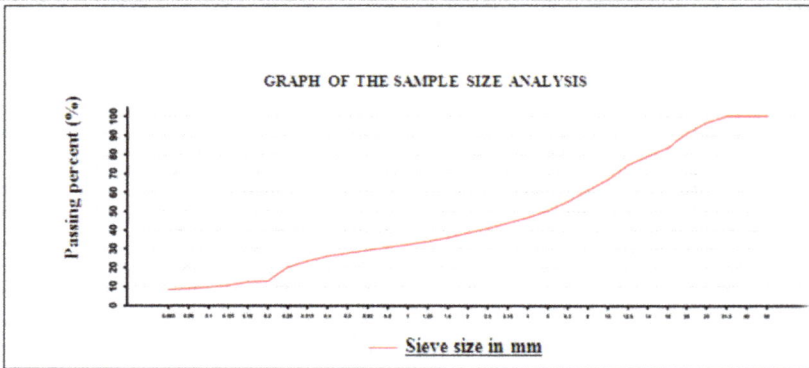

Fig. 2. Particle size distribution curves of recycled CDWs aggregates

The shape of the aggregates was determined by the flattening coefficient, defined in EN 933-3 [57]. The flattening coefficient of the CDWs aggregates was approximately 19%, which classified it as Fl20 according to EN 13242 [48]. Usually, an aggregate favorable shape (spherical, cubic or nodular) has a flattening coefficient generally between 5% and 20% [62,63]. Particles that are too elongated or too flat tend to shatter in their small dimension during compaction [17]. The angular morphology of the ground waste particles can be observed in the SEM images in Fig. 3. A good angularity of the aggregates increases the inter-granular friction that gives the track a very high degree of elasticity and strength [62,63].

Mediterranean Architectural Heritage - RIPAM10

Materials Research Proceedings 40 (2024) 19-32

Materials Research Forum LLC

https://doi.org/10.21741/9781644903117-3

Fig. 3. Scanning electron microscopy (SEM) images of recycled CDWs aggregates.

The bulk (ρb) and real (ρr) densities of the recycled CDWs aggregates were 1700 Kg/m3 and 2500 Kg/m3, respectively. Overall, CDWs aggregates were the less dense than those of natural aggregates (NA) [35,64]. This reduction in CDWs density is a consequence of constitution of CDWs, in particular the crushed natural aggregates and the mortar attached to the surface of the particles [17]. However, it was reported that the size of the recycled CDWs aggregates affect the density. In fact, the smaller the size of the recycled CDWs aggregates is, the higher the amount of attached mortar on the recycled aggregates surface; which means that some of the small particles of recycled CDWs aggregates were entirely mortar [64].

Moreover, the water absorption (WA) of the recycled CDWs aggregates was 7.2%. This result confirms what was observed in the literature review [34,36,65]. Generally, recycled CDWs almost have higher water absorption values than natural aggregates that have water absorption (WA) values between 0.5% and 1.5% [35,66,67]. However, these higher WA values of the recycled CDWs aggregates could be attributed to their lower density and therefore higher porosity compared to natural aggregates [64]. In addition, the high amount of porous materials, such as bonded mortars, included in the recycled aggregates constituents also contributes to the increase in water absorption (WA) values. Furthermore, by comparing the WA results with the WA values recommended in EN 13242 standard [48], it can be said that recycled aggregates belonging to category $WA_{cm}7.2$

In addition to the usual characteristics (e.g. granulometry, density, water absorption, shape ...), the water-soluble sulphate (SO4%) content should necessary be measured. Indeed, in the presence of water, there will may be a reaction between sulphates and hydrated cement, causing ettringite [17]. This mineral have a volume much larger than the initial material, leading to disorders on the structure (risk of expansions) and geometric vulnerability of the section [34]. So, it is therefore necessary to measure the quantities present in the aggregates resulting from the deconstruction and demolition waste (e.g. gypsum, plaster). Furthermore, the EN 13242 standard [48] specifies that the water-soluble sulphate content must remain $\leq 0.7\%$ in mass for aggregates. Table 3 confirms that the specimens hold small amounts of water-soluble sulphate with value of 0.04% far below the limit of max SS0.7 or even the lowest limit SS0.2. Pre-screening to eliminate the gypsum fragments from the CDWs recycling operation decreases the sulphate percentage and increases the aggregates rank.

The mechanical performance of the recycled CDWs aggregates was investigated according to the Los Angeles (LA) test and the Micro-Deval (MDE) test. From the experimental results reported in the Table 3, the values of LA and MDE coefficients were 40% and 37%, respectively. However, these values are acceptable and are in accordance with the requirements of the EN 13242 standard [48]. It was reported that the recycled aggregates has a significantly higher LA and MDE coefficients than the natural ones [66,67]. This can be probably due to the presence of old cement paste surrounding natural aggregates and ceramic materials, etc. in the recycled CDWs aggregates which is less resistant to fragmentation wear resistance [68–70]. Furthermore, the wear resistance of gravels is in a lesser degree governed by the particle shape, whereas the production of fines is more important for recycled aggregates [71]. By analyzing the coding system proposed by the EN 13242 standard [48] and on the basis of the LA and MDE coefficients obtained, it was possible to estimate that the recycled aggregates belonging to category LA40 and MDE40.

Conclusion

The purpose of this paper, as mentioned earlier, was to investigate and promote the utilization of construction and demolition wastes (CDWs) generated from a demolition site in road construction projects. Based on the results obtained, the following conclusions can be drawn:

- The collected CDWs are both cost-effective and environmentally friendly, especially in terms of safeguarding soil and surface or groundwater from potential contamination. Therefore, these recycled CDW aggregates can be categorized as inert materials and can be reused in sustainable road construction projects.
- The water absorption (WA) of the recycled CDW aggregates is higher (7.2%), and their densities (ρb and ρr) are lower compared to natural aggregates due to the porous cement paste adhered to the aggregates.
- The mechanical performance of recycled CDW aggregates was evaluated using the Los Angeles (LA) test and the Micro-Deval (MDE) test. The results indicate that the LA and MDE coefficient values were 40% and 37%, respectively.
- According to the specifications outlined in the EN 13242 standard, the recycled CDW aggregates meet the requirements and are suitable for specific applications, particularly in civil engineering and road construction.

- Based on the physical and mechanical results, the recycled CDW aggregates are suitable for use as unbound granular material (referred to as "GNT") of type 5, provided that the crushing and screening processes are properly managed to obtain aggregates of class 0/31.5.
- According to the Moroccan catalog of standard structures for new pavements, these recycled aggregates can be employed as shoulder materials of type MS1.

In summary, the findings presented in this study support the use of recycled CDW aggregates as a viable substitute for natural aggregates in road construction and various civil projects, contributing to the economic and environmental sustainability of countries

Declaration of Competing Interest

The authors declare that they have no known competing financial interests or personal relationships that could have appeared to influence the work reported in this paper.

Acknowledgments

This work would not exist without the help of a great number of great people. The authors would like to thank all those who contributed to this study.

References

[1] R. Waskow, V. Gonçalves Maciel, R. Tubino, A. Passuello, Environmental performance of construction and demolition waste management strategies for valorization of recycled coarse aggregate, Journal of Environmental Management. 295 (2021) 113094. https://doi.org/10.1016/j.jenvman.2021.113094

[2] B. Arhoun, C. Jiménez, F.X. Niell, J.M. Rodriguez-Maroto, Investigating the physical and chemical characteristics of construction and demolition wastes as filler to regenerate beaches, Resources, Conservation and Recycling. 179 (2022) 106044. https://doi.org/10.1016/j.resconrec.2021.106044

[3] A. Busari, E. Adeyanju, T. Loto, D. Ademola, Recycled Aggregate in Pavement Construction: Review of Literatures, J. Phys.: Conf. Ser. 1378 (2019) 022026. https://doi.org/10.1088/1742-6596/1378/2/022026

[4] Deutsche Gesellschaft für Internationale Zusammenarbeit (GIZ) "Report on the Solid Waste Management in MOROCCO", 2014, pp. 36-37.). https://www.climate-chance.org

[5] G.S. dos Reis, M. Quattrone, W.M. Ambrós, B. Grigore Cazacliu, C. Hoffmann Sampaio, Current Applications of Recycled Aggregates from Construction and Demolition: A Review, Materials. 14 (2021) 1700. https://doi.org/10.3390/ma14071700

[6] M.S. Aslam, B. Huang, L. Cui, Review of construction and demolition waste management in China and USA, Journal of Environmental Management. 264 (2020) 110445. https://doi.org/10.1016/j.jenvman.2020.110445

[7] C.S. Vieira, P.M. Pereira, Use of recycled construction and demolition materials in geotechnical applications: A review, Resources, Conservation and Recycling. 103 (2015) 192–204. https://doi.org/10.1016/j.resconrec.2015.07.023

[8] P. Sormunen, T. Kärki, Recycled construction and demolition waste as a possible source of materials for composite manufacturing, Journal of Building Engineering. 24 (2019) 100742. https://doi.org/10.1016/j.jobe.2019.100742

[9] N.C. Luu, L.H. Nguyen, T.V.N. Tran, Y. Isobe, M. Kawasaki, K. Kawamoto, CONSTRUCTION AND DEMOLITION WASTE ILLEGAL DUMPING: ENVIRONMENTAL, SOCIAL AND ECONOMIC IMPACTS ASSESSMENT FOR A

GROWING CITY, JGS Special Publication. 9 (2021) 148–155.
https://doi.org/10.3208/jgssp.v09.cpeg133

[10] T.K.L. Nguyen, H.H. Ngo, W. Guo, T.L.H. Nguyen, S.W. Chang, D.D. Nguyen, S. Varjani, Z. Lei, L. Deng, Environmental impacts and greenhouse gas emissions assessment for energy recovery and material recycle of the wastewater treatment plant, Science of The Total Environment. 784 (2021) 147135. https://doi.org/10.1016/j.scitotenv.2021.147135

[11] U. Hasan, A. Whyte, H. Al Jassmi, Life cycle assessment of roadworks in United Arab Emirates: Recycled construction waste, reclaimed asphalt pavement, warm-mix asphalt and blast furnace slag use against traditional approach, Journal of Cleaner Production. 257 (2020) 120531. https://doi.org/10.1016/j.jclepro.2020.120531

[12] M.G. Sohail, W. Alnahhal, A. Taha, K. Abdelaal, Sustainable alternative aggregates: Characterization and influence on mechanical behavior of basalt fiber reinforced concrete, Construction and Building Materials. 255 (2020) 119365. https://doi.org/10.1016/j.conbuildmat.2020.119365

[13] L.W. Zhang, A.O. Sojobi, V.K.R. Kodur, K.M. Liew, Effective utilization and recycling of mixed recycled aggregates for a greener environment, Journal of Cleaner Production. 236 (2019) 117600. https://doi.org/10.1016/j.jclepro.2019.07.075

[14] U. Mercado Burciaga, P.V. Sáez, F. Javier Hernández Ayón, Strategies to Reduce CO2 Emissions in Housing Building by Means of CDW, Emerg Sci J. 3 (2019) 274–284. https://doi.org/10.28991/esj-2019-01190

[15] S. Pourkhorshidi, C. Sangiorgi, D. Torreggiani, P. Tassinari, Using Recycled Aggregates from Construction and Demolition Waste in Unbound Layers of Pavements, Sustainability. 12 (2020) 9386. https://doi.org/10.3390/su12229386

[16] V.W.Y. Tam, M. Soomro, A.C.J. Evangelista, A review of recycled aggregate in concrete applications (2000–2017), Construction and Building Materials. 172 (2018) 272–292. https://doi.org/10.1016/j.conbuildmat.2018.03.240

[17] J.R. Jiménez, Recycled aggregates (RAs) for roads, in: Handbook of Recycled Concrete and Demolition Waste, Elsevier, 2013: pp. 351–377. https://doi.org/10.1533/9780857096906.3.351

[18] D.F. Caicedo, G.S. dos Reis, E.C. Lima, I.A.S. De Brum, P.S. Thue, B.G. Cazacliu, D.R. Lima, A.H. dos Santos, G.L. Dotto, Efficient adsorbent based on construction and demolition wastes functionalized with 3-aminopropyltriethoxysilane (APTES) for the removal ciprofloxacin from hospital synthetic effluents, Journal of Environmental Chemical Engineering. 8 (2020) 103875. https://doi.org/10.1016/j.jece.2020.103875

[19] C. Zhang, M. Hu, X. Yang, B. Miranda-Xicotencatl, B. Sprecher, F. Di Maio, X. Zhong, A. Tukker, Upgrading construction and demolition waste management from downcycling to recycling in the Netherlands, Journal of Cleaner Production. 266 (2020) 121718. https://doi.org/10.1016/j.jclepro.2020.121718

[20] M. Panizza, M. Natali, E. Garbin, S. Tamburini, M. Secco, Assessment of geopolymers with Construction and Demolition Waste (CDW) aggregates as a building material, Construction and Building Materials. 181 (2018) 119–133. https://doi.org/10.1016/j.conbuildmat.2018.06.018

[21] G. S, R. N, S. G, Effective Utilisation of Construction and Demolition Waste (Cdw) As Recycled Aggregate in Concrete Construction – A Critical Review, Int. Res. J. Multidiscip. Technovation. (2019) 465–469. https://doi.org/10.34256/irjmtcon65

[22] C. Piña Ramírez, M. del Río Merino, C. Viñas Arrebola, A. Vidales Barriguete, M. Kosior-Kazberuk, Analysis of the mechanical behaviour of the cement mortars with additives of mineral wool fibres from recycling of CDW, Waste Management & Research: The Journal for a Sustainable Circular Economy. 210 (2019) 56–62.
https://doi.org/10.1016/j.conbuildmat.2019.03.062

[23] J. Moreno-Juez, I.J. Vegas, M. Frías Rojas, R. Vigil de la Villa, E. Guede-Vázquez, Laboratory-scale study and semi-industrial validation of viability of inorganic CDW fine fractions as SCMs in blended cements, Construction and Building Materials. 271 (2021) 121823.
https://doi.org/10.1016/j.conbuildmat.2020.121823

[24] M.L.P. Antunes, A.B. de Sá, P.S. Oliveira, E.C. Rangel, Utilization of gypsum from construction and demolition waste in Portland cement mortar, Cerâmica. 65 (2019) 1–6.
https://doi.org/10.1590/0366-6913201965S12588

[25] R.A. Robayo-Salazar, W. Valencia-Saavedra, R. Mejía de Gutiérrez, Construction and Demolition Waste (CDW) Recycling—As Both Binder and Aggregates—In Alkali-Activated Materials: A Novel Re-Use Concept, Sustainability. 12 (2020) 5775.
https://doi.org/10.3390/su12145775

[26] A.C. Freire, J.M.C. Neves, R. Pestana, Analysis of the Properties of Recycled Aggregates for Unbound Granular Asphalt Pavement Layers, (n.d.) 10

[27] J. Li, M. Saberian, B.T. Nguyen, Effect of crumb rubber on the mechanical properties of crushed recycled pavement materials, Journal of Environmental Management. 218 (2018) 291–299. https://doi.org/10.1016/j.jenvman.2018.04.062

[28] G. Tavakoli Mehrjardi, A. Azizi, A. Haji-Azizi, G. Asdollafardi, Evaluating and improving the construction and demolition waste technical properties to use in road construction, Transportation Geotechnics. 23 (2020) 100349. https://doi.org/10.1016/j.trgeo.2020.100349

[29] F. Varela, E. Cerro-Prada, F. Escolano, Preparation, Characterization and Modeling of Unbound Granular Materials for Road Foundations, Applied Sciences. 8 (2018) 1548.
https://doi.org/10.3390/app8091548

[30] A. M. Arisha, A.R. Gabr, S.M. El-Badawy, S.A. Shwally, Performance Evaluation of Construction and Demolition Waste Materials for Pavement Construction in Egypt, J. Mater. Civ. Eng. 30 (2018) 04017270. https://doi.org/10.1061/(ASCE)MT.1943-5533.0002127

[31] EFFECTS OF PARTICLE SIZE AND TYPE OF AGGREGATE ON MECHANICAL PROPERTIES AND ENVIRONMENTAL SAFETY OF UNBOUND ROAD BASE AND SUBBASE MATERIALS: A LITERATURE REVIEW | GEOMATE Journal, (2021).
https://geomatejournal.com/geomate/article/view/200 (accessed February 28, 2022).

[32] J. Tavira, J.R. Jiménez, J. Ayuso, M.J. Sierra, E.F. Ledesma, Functional and structural parameters of a paved road section constructed with mixed recycled aggregates from non-selected construction and demolition waste with excavation soil, Construction and Building Materials. 164 (2018) 57–69. https://doi.org/10.1016/j.conbuildmat.2017.12.195

[33] D. Ciampa, R. Cioffi, F. Colangelo, M. Diomedi, I. Farina, S. Olita, Use of Unbound Materials for Sustainable Road Infrastructures, Applied Sciences. 10 (2020) 3465.
https://doi.org/10.3390/app10103465

[34] F. Agrela, M. Sánchez de Juan, J. Ayuso, V.L. Geraldes, J.R. Jiménez, Limiting properties in the characterisation of mixed recycled aggregates for use in the manufacture of concrete,

Construction and Building Materials. 25 (2011) 3950–3955.
https://doi.org/10.1016/j.conbuildmat.2011.04.027

[35] F. Rodrigues, M.T. Carvalho, L. Evangelista, J. de Brito, Physical–chemical and mineralogical characterization of fine aggregates from construction and demolition waste recycling plants, Journal of Cleaner Production. 52 (2013) 438–445.
https://doi.org/10.1016/j.jclepro.2013.02.023

[36] R.V. Silva, J. de Brito, R.K. Dhir, Properties and composition of recycled aggregates from construction and demolition waste suitable for concrete production, Construction and Building Materials. 65 (2014) 201–217. https://doi.org/10.1016/j.conbuildmat.2014.04.117

[37] N. Cristelo, C.S. Vieira, M. de Lurdes Lopes, Geotechnical and Geoenvironmental Assessment of Recycled Construction and Demolition Waste for Road Embankments, Procedia Engineering. 143 (2016) 51–58. https://doi.org/10.1016/j.proeng.2016.06.007

[38] C. Zhang, M. Hu, F. Di Maio, B. Sprecher, X. Yang, A. Tukker, An overview of the waste hierarchy framework for analyzing the circularity in construction and demolition waste management in Europe, Science of The Total Environment. 803 (2022) 149892.
https://doi.org/10.1016/j.scitotenv.2021.149892.

[39] G. Rodríguez, I. Sáez del Bosque, E. Asensio, M. Sánchez de Rojas, C. Medina, Construction and demolition waste applications and maximum daily output in Spanish recycling plants, Waste Manag Res. 38 (2020) 423–432. https://doi.org/10.1177/0734242X20904437

[40] A.S. Molla, P. Tang, W. Sher, D.N. Bekele, Chemicals of concern in construction and demolition waste fine residues: A systematic literature review, Journal of Environmental Management. 299 (2021) 113654. https://doi.org/10.1016/j.jenvman.2021.113654

[41] A. Podlasek, A. Jakimiuk, M.D. Vaverková, E. Koda, Monitoring and Assessment of Groundwater Quality at Landfill Sites: Selected Case Studies of Poland and the Czech Republic, Sustainability. 13 (2021) 7769. https://doi.org/10.3390/su13147769

[42] X. Li, S. Ning, P. Zhang, W. Yang, Environmental Pollution and Health Risks of Heavy Metals in the Soil Around a Construction Waste Landfill, IJDNE. 15 (2020) 393–399.
https://doi.org/10.18280/ijdne.150312

[43] S. Bosoc, G. Suciu, A. Scheianu, I. Petre, Real-time sorting system for the Construction and Demolition Waste materials, in: 2021 13th International Conference on Electronics, Computers and Artificial Intelligence (ECAI), 2021: pp. 1–6.
https://doi.org/10.1109/ECAI52376.2021.9515117.

[44] C.J. Spreadbury, K.A. Clavier, A.M. Lin, T.G. Townsend, A critical analysis of leaching and environmental risk assessment for reclaimed asphalt pavement management, Science of The Total Environment. 775 (2021) 145741. https://doi.org/10.1016/j.scitotenv.2021.145741

[45] M.A. Imteaz, A. Arulrajah, S. Horpibulsuk, A. Ahsan, Environmental Suitability and Carbon Footprint Savings of Recycled Tyre Crumbs for Road Applications, Int J Environ Res. 12 (2018) 693–702. https://doi.org/10.1007/s41742-018-0126-7

[46] Y.F. Song, B.-M. Wilke, X.Y. Song, P. Gong, Q.X. Zhou, G.F. Yang, Polycyclic aromatic hydrocarbons (PAHs), polychlorinated biphenyls (PCBs) and heavy metals (HMs) as well as their genotoxicity in soil after long-term wastewater irrigation, Chemosphere. 65 (2006) 1859–1868. https://doi.org/10.1016/j.chemosphere.2006.03.076

Mediterranean Architectural Heritage - RIPAM10
Materials Research Proceedings 40 (2024) 19-32

Materials Research Forum LLC
https://doi.org/10.21741/9781644903117-3

[47] C. Alexandridou, G.N. Angelopoulos, F.A. Coutelieris, Physical, Chemical and Mineralogical Characterization of Construction and Demolition Waste Produced in Greece, 8 (2014) 6.

[48] EN 13242 (2002) Aggregates for unbound and hydraulically bound materials for use in civil engineering work and road construction, European Committee for Standardization (CEN).

[49] EN 932-2 (2012) Tests for geometrical properties of aggregates- Part 2: Methods for reducing laboratory samples, European Committee for Standardization (CEN).

[50] EN 15934 (2012) Sludge, treated biowaste, soil and waste - Calculation of dry matter fraction after determination of dry residue or water content, European Committee for Standardization (CEN).

[51] EN ISO 16703 (2011) Soil quality - Determination of content of hydrocarbon in the range C10 to C40 by gas chromatography, European Committee for Standardization (CEN).

[52] EN 933-11 (2012) Tests for geometrical properties of aggregates- Part 11: Classification test for the constituents of coarse recycled aggregate, European Committee for Standardization (CEN).

[53] EN 933-1 (2012) Tests for geometrical properties of aggregates- Part 1: Determination of granularity - sieve analysis, European Committee for Standardization (CEN).

[54] EN 1097-6 (2013) Tests for mechanical and physical properties of aggregates -Part 6: Determination of particle density and water absorption, European Committee for Standardization (CEN).

[55] EN 1097-3 (1998) Tests for mechanical and physical properties of aggregates- Part 3: Determination of loose bulk density and voids, European Committee for Standardization (CEN).

[56] EN 1744-1 (2009) Tests for chemical properties of aggregates. Part 1: Chemical analysis, European Committee for Standardization (CEN).

[57] EN 933-3 (2012) Tests for geometrical properties of aggregates- Part 3: Determination of particle shape-Flattening coefficient, European Committee for Standardization (CEN).

[58] EN 1097-2 (2010) Tests for mechanical and physical properties of aggregates- Part 2: Methods for the determination of resistance to fragmentation, European Committee for Standardization (CEN).

[59] EN 1097-1 (2011) Tests for mechanical and physical properties of aggregates- Part 1: Determination of the resistance to wear (Micro-Deval), European Committee for Standardization (CEN).

[60] J.R. Jiménez, J. Ayuso, F. Agrela, M. López, A.P. Galvín, Utilisation of unbound recycled aggregates from selected CDW in unpaved rural roads, Resources, Conservation and Recycling. 58 (2012) 88–97. https://doi.org/10.1016/j.resconrec.2011.10.012

[61] L. Butler, J. West, S. Tighe, Effect of Recycled Concrete Aggregate Properties on Mixture Proportions of Structural Concrete, Transportation Research Record Journal of the Transportation Research Board. 2290 (2012) 105–114. https://doi.org/10.3141/2290-14

[62] A. Diedhiou, L. Sow, N.M. Diop, Experimental Characterization of Shape of an Aggregate by a Numerical Value—Application to Senegalese Basaltic Aggregates for Rail Transport, OJCE. 10 (2020) 131–142. https://doi.org/10.4236/ojce.2020.102012

Mediterranean Architectural Heritage - RIPAM10
Materials Research Proceedings 40 (2024) 19-32

Materials Research Forum LLC
https://doi.org/10.21741/9781644903117-3

[63] M.L. Chérif Aidara, M. Ba, A. Carter, Impact of Intrinsic Properties of Aggregate and Volumetric Properties of Hot Mixture Asphalt (HMA) in the Influence of the Resistance to Rutting, OJCE. 10 (2020) 187–194. https://doi.org/10.4236/ojce.2020.103016

[64] K.H. Younis, K. Pilakoutas, Strength prediction model and methods for improving recycled aggregate concrete, Construction and Building Materials. 49 (2013) 688–701. https://doi.org/10.1016/j.conbuildmat.2013.09.003

[65] P. Nuaklong, A. Wongsa, V. Sata, K. Boonserm, J. Sanjayan, P. Chindaprasirt, Properties of high-calcium and low-calcium fly ash combination geopolymer mortar containing recycled aggregate, Heliyon. 5 (2019) e02513. https://doi.org/10.1016/j.heliyon.2019.e02513

[66] M. Contreras-Llanes, M. Romero, M.J. Gázquez, J.P. Bolívar, Recycled Aggregates from Construction and Demolition Waste in the Manufacture of Urban Pavements, Materials. 14 (2021) 6605. https://doi.org/10.3390/ma14216605

[67] S. Omary, E. Ghorbel, G. Wardeh, Relationships between recycled concrete aggregates characteristics and recycled aggregates concretes properties, Construction and Building Materials. 108 (2016) 163–174. https://doi.org/10.1016/j.conbuildmat.2016.01.042

[68] R. Kumar, Influence of recycled coarse aggregate derived from construction and demolition waste (CDW) on abrasion resistance of pavement concrete, Construction and Building Materials. 142 (2017) 248–255. https://doi.org/10.1016/j.conbuildmat.2017.03.077

[69] M. Bravo, J. de Brito, J. Pontes, L. Evangelista, Mechanical performance of concrete made with aggregates from construction and demolition waste recycling plants, Journal of Cleaner Production. 99 (2015) 59–74. https://doi.org/10.1016/j.jclepro.2015.03.012

[70] S. Omary, E. Ghorbel, G. Wardeh, Relationships between recycled concrete aggregates characteristics and recycled aggregates concretes properties, Construction and Building Materials. 108 (2016) 163–174. https://doi.org/10.1016/j.conbuildmat.2016.01.042

[71] L. Courard, M. Rondeux, Z. Zhao, F. Michel, Use of Recycled Fine Aggregates from C&DW for Unbound Road Sub-Base, Materials. 13 (2020) 2994. https://doi.org/10.3390/ma13132994

Mediterranean Architectural Heritage - RIPAM10
Materials Research Proceedings 40 (2024) 33-40

Materials Research Forum LLC
https://doi.org/10.21741/9781644903117-4

Adaptation of Traditional Construction Methods for a Sustainable Transition of the Dwelling (Case of Riads in Fez and Hanoks in Seoul)

Rime EL HARROUNI[1,3,a*], Iman BENKIRANE[2,b], Vincent BECUE[3,c]

[1]Euro-Mediterranean School of Architecture, Design and Urbanism, Euro-Mediterranean University of Fez Institution, Morocco

[2]Ecole Nationale d'Architecture de Rabat, BP 6372, Rabat Instituts, Rabat, Morocco

[3]Faculty of Architecture and Urbanism, University of Mons, Belgium

[a]rimaelharrouni@gmail.com, [b]i.benkirane@enarabat.ac.ma, [c]vincent.becue@umons.ac.be

Keywords: Riad, Fez, Hanok, Seoul, Socio-Cultural Transition, Patio, Space Quality, Sustainable Transformation

Abstract. Significant initiatives have been carried out by certain Moroccan organizations regarding safeguarding the Riads and Dars in Fez. Nevertheless, prior studies on the rehabilitation and restoration of traditional houses have only addressed the structural elements and thermal properties of the traditional environment, ignoring the spatial arrangement of the home and how it might be modified to better suit the needs and practices of modern residents. In reality, the production of newly effective technology involving structures and construction taking into account qualitative performances of the traditional dwelling has yet to be satisfactory specifically, as it is sparking a social discourse over the legitimacy of the traditional built design. This study aims to investigate the dissolution of the traditional dwelling in the medina of Fez to the degradation of the construction materials, and the safeguarding and rehabilitation Process of said dwelling. The main problem would be to answer: How to find the balance between the old and the new to provide a better quality of space? The defined method will take the form of a systemic comparative approach with the Korean model of dwelling called Hanok, to identify the similarities in terms of traditional methods of construction and the developed techniques used to transform said dwelling. For the sustainable development of these traditional habitat models, we need to establish specialized tools and a response plan for future Riads technology by comprehending consumer needs through: ongoing research on green technology to assess the usefulness of activities-oriented design in traditional homes and the improvement of natural materials in the rehabilitation process. Overall, this research aims to develop a specific scientific approach to transformation and adaptive reuse for sustainable habitability based on classifications of behavioral factors, technical factors, and contextual factors.

Introduction

The act of demolishing and rebuilding is carried out for the sake of efficiency, safety, and modernization. We might feel the need to reconstruct a building after a natural disaster or because of its insalubrity and contextual localization in an endangered area or on lands deemed obsolete, and inappropriate for modern life and the automobile. This demolition-reconstruction approach is still used today by the National Agency for Urban Renewal (ANRU), the Agency for the Development and Rehabilitation of the Medina of Fez (ADER), the Seoul Urban Solutions Agency (SUSA), and other national governmental institutions to significantly modify neighborhoods with a high concentration of low-income households and to upgrade and revitalize them.

If the reasons for destroying are not lacking, simultaneously the act of transforming invites us not to demolish but to think of sustainable ways to reuse and transform the traditional dwelling to

Mediterranean Architectural Heritage - RIPAM10   Materials Research Forum LLC
Materials Research Proceedings 40 (2024) 33-40   https://doi.org/10.21741/9781644903117-4

accommodate new activities and answers to consumer's current needs. Housing demands originate from a combination of biological necessities/opportunities/restraints and complicated contextual contexts in which the meaning and function of housing, and thus the notion of what quality living is and what the essential conditions for quality living are, might change. This is due not only to the fact that quality standards can differ (for example, in terms of the degree of insulation), but also to the fact that housing has various uses and meanings, and thus quality can be judged using quite different parameters [1].

In this research, we will be focusing on the Hanoks in Seoul and the Riads in Fez as two examples of institutions from Korea and Morocco that have taken crucial initiatives to preserve ancient dwellings. These days, the new lifestyles and altered household structures have caused the traditional dwelling to lose its identity and significance, which has resulted in the abandonment of Riads and Hanoks in favor of apartments and the degradation of those dwellings due to a lack of a comprehensive plan for appropriate rehabilitation and revitalization. Therefore, before advancing in an optic of transformation and reuse we have to tackle the traditional dwelling in terms of prospection of the restoration materials and identifying the old ingenious ways of constructions and new technologies that can be used to develop these models of habitat.

The method used is a comparative systemic approach based on the objective mode: The reflection here is turned towards the composition of forms, construction materials, orders, the topology of structures, spatial juxtaposition…

We will be focusing on the quality of space and comfort (thermal), sensorial perception…

This will help us identify the static physical and sensorial parameters of each model in their traditional state such as natural materials and sustainable traditional techniques that reflect an ingenious way of life and that we can upgrade with new technologies to solve thermal and structural problems of today.

Joined Parameters between the traditional Riads and Hanoks:

Physical Parameters:
-In Fez, the craftsmen used two types of earth, one yellowish, taken from near the neighborhood, in the place called 'Aïn Nuqbi, brought on the back of a donkey in baskets (Gwari), in dwarf palm, the other in compact gray blocks, is bluish and comes from the neighboring hills of the city in the direction of Sïdï Harazem. Once these large blocks of clay (Ndra) are transported, and crushed into small blocks (tuba), the mixing is done following a natural process[1] and after being impregnated with water, the artisans can move on to kneading and mixing in the workshop [2].

After the molding and smoothing operations, the Mzehrya tiles are shaped on the surface and polished and reworked to receive their final shape and are ready to be baked for the first firing (Zellige).

The rectangular pieces of Bejmât are 14,5cm long by 5 cm wide and 2,5 cm thick, and used mainly for paving patios following a process similar to the manufacture of Zellige. The difference here lies in the mold that affects the rectangular truncated pyramid shape of the Bejmât.

To summarize, construction ceramics abound in traditional Dars and Riads from the Merinid period, and they continue to play an important role in Fez today.

---

[1] The crushed blocks of clay are discharged into a circular pit (zaba) dug into the ground in the shape of a cup with the edges reinforced with large round pebbles as well as the bottom which is paved with pebbles. Water is then brought in by means of metal pipes. On the wall opposite the water inlet and at ground level, a small channel leads the liquid to this special pit of 2.20 m in diameter and 0.60 m in depth. While the water fills the basin, the mixture will stay like this for one to two days, the time necessary for the pieces to disintegrate by themselves and be ready for kneading.

The walls are either in Pisé, less than a meter wide, or in earthen-baked bricks, and 40cm thick. The thickness and height of these walls provide good thermal inertia, which is important given the temperature and weather properties of the region.

To reinforce these walls, Wooden chaining could be used the stringers are sometimes coated with plaster to prevent crumbling due to humidity. Fig. 1 shows a section of Dar Lezrak with all the decorative and structural elements found in a typical Riad in Fez.

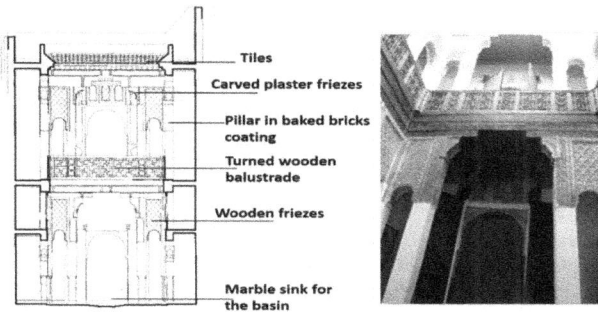

Figure 1: Section of Riad Lezrak in Fez

In a similar way, the transformation of the natural soil using traditional techniques also becomes part of the Hanoks construction process and depends on the condition of the soil. For sandy ground, the existing soil is compacted with water, and in the case of clay soil, steel ash is sprinkled and then compacted. If the distribution of soil particles or the proper input ratio of steel ash is quantified and standardized using the plate-breaking technique, which is a Hanok foundation method, it becomes possible to construct a foundation that is both nature-friendly and as strong as a concrete foundation [3]

Soil is an ideal building material for Hanok construction because it has excellent humidity control and ventilation, and is easily available.

Clay kiln-baked tiles were commonly utilized as roofing materials. For walls, Jeondol (전돌은) was manufactured by baking easily accessible soil at a high temperature, making it robust and resistant to fire and cold, thus compensating for a variety of drawbacks and inadequacies of the primary building materials, wood, and stone. That way we could find two types of walls in Hanok, simbyeok, and stone brick walls (Fig. 2). Other than aggregates, plastering the walls involves raising lattices and applying fodder, chopped straw, combined with mud to prevent the mud wall from splitting and makes it more adhesive [4].

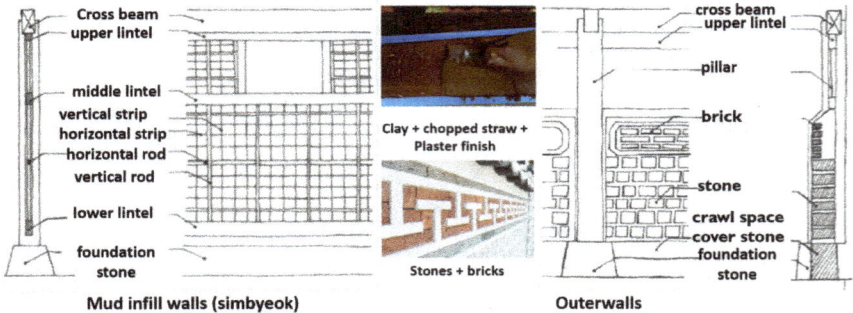

Figure 2: Types of walls in Hanoks

As for woodwork, in Hanoks, Cornerstones and flags receive almost all of the internal load, while walls play the role of partitions receiving only a slight load. Roof tiles are a common structural method. Three layers of flat roof tiles are laid, and round roof tiles are placed on top followed by the one placed for the curved ridge. Two types of purlins are used: Square purlins were used in lower-class homes and round ones were used in upper-class homes. As for round columns, they were restricted from being used in common residences and were typically used in royal palaces [5]. So here, we can see that there is a cultural and social significance even to the materials and structural elements used.

Both models of dwellings also present an architectural design that has remained mostly unchanged in terms of core shapes for several centuries and that responds to ornamentation reflecting nature, using mostly geometric and floral repertory.

Overall, wood is responsible for structural resistance and holding the house's shape but is also used in both models as a decorative element to emphasize the cultural aspect of each historical context. Soil is reported to be used for roofs, walls, and floors to suit the needs of the living environment. The roof soil is in charge of insulation and heat storage, the wall soil controls humidity and deodorizes to create a pleasant and healthy living place, and the floor soil is mixed with spheres to act as a heating function via thermal radiation. Stones and bricks are used as a means to prevent fires from spreading from one residence to another [6].

Table 1 concludes all the natural materials used in both dwellings.

Table 1: Material and their area of use in both Riads and Hanoks

Division	Riads		Hanoks	
	Materials	Area of Use	Materials	Area of Use
Soil	Earthen baked bricks, Lime, glazed bricks	Walls, Patio Covering, Stairs, curved tiles, earthenware tiles (zellige)	Earthen baked bricks, Lime, Pisé	Roof , Wall
Stone	Marble, limestone (fessaia), Granite, Brick shaped Stones	Floor Covering, Door Frame, Bassin's Edges, Wall, Roof	Granite, Roofing Tile (Gi-wa), Brick shaped Stones	Cornerstone,foundation Stylobate, Wall,Door, roof
Wood	Cedar wood	Framework,Eaves, Doors,Windows, balcony railings, Walls, alcove arcades (Bhou),Framed ceiling	Pine	Eaves, Doors, Gong-po, Pillar, windows, wall, Decorations, Framed Ceiling
Grass	-	-	Rice Straw	Roof, Wall
Paper	-	-	Han-Ji	Windows
Metal	wrought iron	Door's Grids, Windows, Decorations	Bronze	Wall, Decorations, Windows

In both traditional dwelling models, the wooden structure defines the galleries and a hierarchical organization of spaces that express sequence, rhythm, and openness traits.

Mediterranean Architectural Heritage - RIPAM10   Materials Research Forum LLC
Materials Research Proceedings 40 (2024) 33-40   https://doi.org/10.21741/9781644903117-4

For Hanoks there is a specific unit for length and width called Kan of 간(間) that considers human behavior and sensibility as a measure. The size of the 간(間), which is the basic concept for all spatial arrangements in a Traditional Hanok, is 3x5=15 cheok (≒4.5m), and there are rules of 3x6, 3x7, 3x8, depending on the status (Fig. 3).

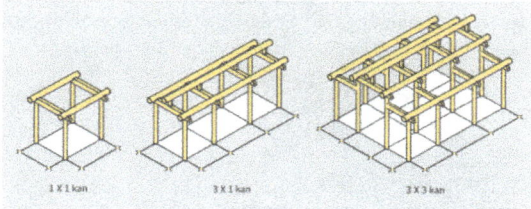

Figure 3: Concept of Kan 간(間)

Here, the number 3 represents the number of heaven, earth, and human universes, and 5 is the average human height. The cosmic number 3 and the human number 5 meet and become the basis for building a house set as the middle universe. The basic module of the 간(間) is functional and comfortable for human life and by juxtaposition to other modules it becomes a room and a house. Unlike the Western spatial perception system, which perceives objects in space and the remaining part is perceived as space, this concept of emptiness as a space is a traditional concept of dimension that perceives space centered on humans.

In the Buddhist culture and following the idea of Lao-tzu (157–141 BC), space is not defined only by walls that remain in a simple state of matter, nor is it defined by emptying that lacks axiomatic benefits. By studying emptiness rather than filling the space, it can be made into a more experiential and diverse space [7].

We find this rhythmic succession in the arcades composing the galleries of the riads, each time defining a space around the patio (iwan, loggia…) (Fig. 4). Most of the time the arcade constitutes an odd number but we can't take this observation as a generality to see if there is deeper cultural meaning behind it. But the resemblance with the Hanoks would reside in the balance between the void and the matter, and the modularity of the spaces generated by the structure.

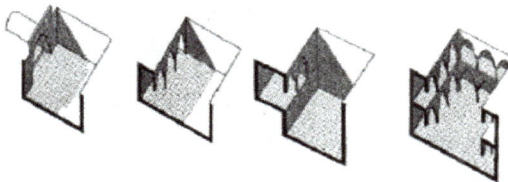

Figure 4: Type of Spatial organization around the Patio

The difference would be in the height and the number of floors constituting each dwelling. The traditional Hanoks mostly have one floor because there was no solution in the past to replicate the andol (natural floor heating) system in a second story.

Whereas Moroccan Riads could go up to three stories, each one usually accommodating a generation of family.

Mediterranean Architectural Heritage - RIPAM10   Materials Research Forum LLC
Materials Research Proceedings 40 (2024) 33-40   https://doi.org/10.21741/9781644903117-4

Sensorial Parameters:

For sensorial parameters, we will focus here on comparing the techniques in each model of dwelling to provide thermal quality. In Moroccan Riads and especially during winter, walls and roof ceilings should be coated with an additional thermal insulation layer of 50-30 mm using for example wood wool or mineral wool [8]. Air penetration is another crucial component of comfort, particularly day ventilation, which helps to enhance the interior temperature in the winter. The Patio transforms a one or two-facade building into a four-solar orientation architecture. It provides natural light, creates an open private space, and helps control the building's thermal behavior (Fig.5)

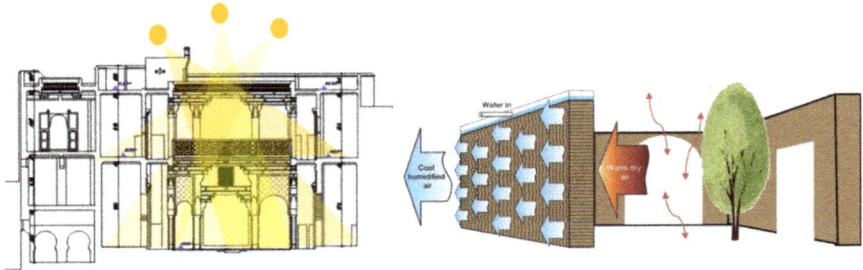

Figure 5: Thermal properties of the Patio inside a Riad

Like in the Riads, the patio helps for cooling the space inside the Hanoks, but the additional feature here would be the Ondol, This Korean floor heating system transfer the heat from the fire of the furnace through the flue channel and warms the stone prop above, so that the heat reaches the floor of the room (Fig. 6).

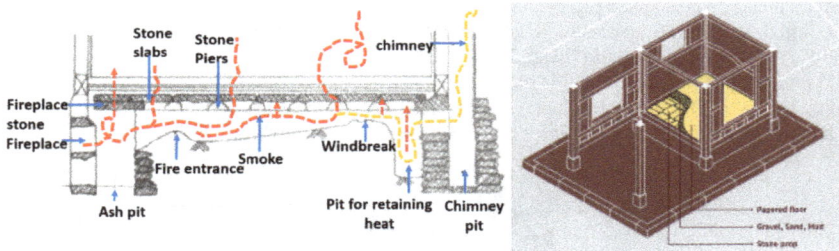

Figure 6: Traditional Korean Heating System (Ondol)

Current Rehabilitation and Revitalization Techniques for Riads and Hanoks:

Morocco has developed over the last decades a great experience in the field of the rehabilitation and Reconversion of traditional architecture. Its experiences were, in most cases, initiated by the Ministry of Culture within the framework of projects relating to World Heritage sites, and monuments or buildings classified at the national level. However, it is generally more a heritage approach of monumental restoration than a real revitalization [9]. That mostly involves restoring the waterproofing of the roof terraces, changing the coating of the walls, replacing the floors and certain load-bearing logs for the structure, and expanding some rooms to accommodate new activities.

Among the architectural systems conceived to adapt to new ways of living, we can find some new technologies added to the original traditional framework of the Hanok, to improve air

Mediterranean Architectural Heritage - RIPAM10   Materials Research Forum LLC
Materials Research Proceedings 40 (2024) 33-40   https://doi.org/10.21741/9781644903117-4

ventilation, and entry of light while providing thermal insulation thanks to smart windows, roof skylights, trusses, roof lightweight body [10]. Reinforcement of the wooden structure, as well as creation of built-in wardrobes and variable wall alcoves as well as built-in storage in Maru (wooden floor) to provide flexible use of space and storage as the family grows.

Table 2 shows the new techniques developed in the Korean Hanoks, which can serve as models to implement in the Moroccan Riads.

Table 2: Necessary functions and new technologies developed in Hanoks.

Discussion of the Results and Conclusion:
While taking into account the different modes of comparative evaluation and the specific characteristics of each traditional dwelling such as physical, behavioral, and sensorial parameters, we find the common settings for Riads and Hanoks :

Riads are located in the historic medina of Fez, which is a UNESCO World Heritage Site.
- Typically found in narrow, winding alleys and streets that are inaccessible to vehicles.
- Often located close to important landmarks, such as mosques, madrasas, and markets.
- Designed to be inward-facing, with a central courtyard and rooms surrounding it.
- Often feature geometric and floral decorative elements such as mosaics, intricate tile work, and carved wood.
- Associated with Islamic cultural tradition.

Hanoks are Located in traditional Korean neighborhoods in Seoul, such as Bukchon Hanok Village and Namsangol Hanok Village.
- Often located on hillside locations with views of the surrounding landscape.
- Typically designed with an inward-facing courtyard or garden.
- Built using natural materials such as wood, stone, and soil, and often feature a tiled roof.
- Often feature geometric and floral decorative elements such as roof tile, woodwork, and furniture.
- Associated with Confucian and Buddhist traditions.

Although there are some variations in construction methods and structural design, Hanoks in Seoul and Riads in Fez Medina both aim to create a quiet, private area in the middle of a busy metropolitan setting, and they frequently include natural and mineral features like gardens and

Mediterranean Architectural Heritage - RIPAM10
Materials Research Proceedings 40 (2024) 33-40

Materials Research Forum LLC
https://doi.org/10.21741/9781644903117-4

fountains/streams. Additionally, they both place a high priority on maintaining traditional architectural designs and cultural traditions.

Both housing styles have multipurpose rooms that allow the space to adapt to changing demands and activities as well as the inhabitant's future needs.

Both have an inward-facing layout with open spaces like the patio and Daecheong that offer cover during the hot summer months to keep the house cool while using natural heating sources during winter.

The two habitat models honor the idea of intimacy and gender segregation by restricting the Riad's external apertures and offering a wing specifically for women (Anchae/안채), which is located far from the Hanok's main entrance.

The following measures were taken to enhance the traditional homes' sensory quality:
-Lengthening the structure along the axis that runs parallel to the north and south façades.
-North-south orientation, or the long axis from east to west, is advised for buildings.
-Moderating the openings' size to a medium range, 25–40% of the total wall surface.
-Upgrading the insulation in the walls and roof and suggesting lightweight or low-temperature alternatives for these building components.

To improve the physical aspect of the traditional dwellings observed:
-Improved construction techniques, including seismic reinforcement and bracing elements
-Develop a modular structure for flexible spatial composition and activities
-Innovative use of architectural elements of the dwelling for storage

References

[1] F. Bellanger. Habitat(s) : questions et hypothèses sur l'évolution de l'habitat. L'aube, 1999.

[2] J. Revault, L. Golvin, A. Amahan, JP. Icheter, M.CH. Fromon. Demeures et palais de Fès. Edition du centre national de la recherche scientifique de Paris, 1985.

[3] National Hanok Center. Building a Hanok. Architecture and Urban Research Institute (AURI) 194, Jeoljae-ro, Sejong-si, Korea, 2017.

[4] J. Jihae Shon. Hanok Interventions, School of Architecture University of Hawai'i at Mānoa, 2011.

[5] JS. Choi. Understanding Koreans and Their Culture. Seoul: Her One Media, 2007.

[6] KM. Lee, SJ. Lee. Hanok Building Standards Guide. Architecture and Urban Research Institute National Hanok Center, 194, Jeoljae-ro, Sejong-si, Korea, 2015.

[7] T. Lao. Tao Te Ching, Broché, 2016

[8] K. El Harrouni, M. Ben Aicha, R. El Harrouni. Parametric Modelling and Traditional Architecture: Improving the thermal comfort of the traditional courtyard house in Morocco. MATEC Web of Conferences 149 (2018): 02051. https://doi.org/10.1051/matecconf/201814902051

[9] X. Casanovas, A. Marou, Q. Wilbaux, F. Cherradi. Réhabilitation et action sociale à Marrakech, Maroc L'amélioration du cadre de vie traditionnel. RéhabiMed. Centre Méditerranéen de l'environnement Marrakech, 2008.

[10] JY. Park, MH. Jun, EK. Cho. The Modern Architectural System Conception for each Part of the New-Hanok Type Public Building. KIEAE Journal, Vol. 18, No. 6, pp.103-110, 2018. https://doi.org/10.12813/kieae.2018.18.6.103

Mediterranean Architectural Heritage - RIPAM10
Materials Research Proceedings 40 (2024) 41-54

Materials Research Forum LLC
https://doi.org/10.21741/9781644903117-5

Experimental Study of the Reinforcement of Unstabilized and Stabilized Local Clay Materials with Date Palm Fibers

Youssef KHRISSI [1,a*], Amine TILIOUA [1,b*], Houda LIFI [2,3,c]

[1]Research Team in Thermal and Applied Thermodynamics (2.T.A.), Mechanics, Energy Efficiency and Renewable Energies Laboratory (L.M.3.E.R.). Department of Physics, Faculty of Sciences and Techniques Errachidia. Moulay Ismael University of Meknès, B.P. 509, Boutalamine, Errachidia, Morocco

[2] Laboratory of Materials, Processes, Environment and Quality, National School of Applied Sciences, Cadi Ayyad University, Safi, Morocco

[3] Moulay Ismail University, Faculty of Science, Laboratory Spectrometry of Materials and Archaeomaterials (LASMAR), Meknes, Morocco

[a] youssefkhrissi@gmail.com, [b] a.tilioua@umi.ac.ma, [c] lifihouda@gmail.com

Keywords: Clay, Spathe, Thermal Characterization, Mechanical Performance, Water Absorption, Density

Abstract. The aim of this study is to experimentally test the stabilization of unexploited clay from the Errachidia region (south-east Morocco) with date palm spathes, with a view to its potential use in construction. The main objective of the present work is to evaluate the thermophysical and mechanical behavior of fiber-stabilized clay blocks. Several samples of spathe-reinforced clay at six different grades (0%, 1%, 2%, 3%, 4%, and 5%) were prepared and tested. Thermal characterization was carried out using the PHYWE House thermal insulation method to determine thermal conductivity and resistance. Mechanical performance was measured in terms of compressive and flexural strength. In addition, the chemical identification of Errachidia clay was studied using the X-ray fluorescence method. The results of the clay identification showed that Errachidia clay meets the minimum requirements for the manufacture of compressed earth bricks and adobe. The results of the thermophysical tests showed that the addition of date palm spathes had a positive influence on the lightness and thermophysical properties of the clay samples stabilized by the spathes. In terms of mechanical test results, the flexural and compressive strengths of clay blocks stabilized with date palm fibers continue to increase up to a fiber content of 3%. After this content, mechanical performance decreases with the addition of spathes and no improvement is detected. Consequently, a fiber content of 3% represents the optimum content for stabilizing Errachidia clay. At this content, stabilized clay blocks show optimal mechanical performance and improved thermal properties compared to reference samples. However, increasing the percentage of fiber mass leads to an increase in water absorption and a decrease in density. Clay compounds reinforced with date palm spathe can be considered as environmentally friendly building materials.

1. Introduction

Owing to the swift urbanization and population expansion, the substantial energy requirements in the foreseeable future pose a looming crisis globally. This escalating demand for energy is driven not only by large-scale development projects but also by routine daily activities. The construction industry, responsible for approximately a quarter of Morocco's yearly energy consumption, is a major contributor to 30% of energy-related $CO2$ emissions and roughly a third of black carbon emissions [1]. Hence, there is an urgent need for substantial advancements in the realm of sustainable construction, with the objective of achieving a 30% reduction in the energy intensity

Mediterranean Architectural Heritage - RIPAM10   Materials Research Forum LLC
Materials Research Proceedings 40 (2024) 41-54   https://doi.org/10.21741/9781644903117-5

of buildings by 2030 in comparison to the 2015 baseline [2]. One of the strategies employed by Moroccan authorities to curtail energy consumption in building air-conditioning systems involves enhancing the thermal performance of the building envelope. This is crucial because conventional building materials, primarily constructed using standard mortar, often lack the requisite thermal insulation properties. This holds particularly true for regions in the Mediterranean, with Morocco being no exception, where conventional building materials are known for their high environmental impact. They contribute significantly to pollution due to their CO_2 emissions during manufacturing and create disposal-related ecological challenges. The adoption of locally sourced materials in the construction sector has emerged as a vital solution to address the economic challenges faced by developing nations [3,4]. Given that the building industry is a substantial consumer of both materials and energy resources, as well as a major contributor to pollution and waste generation, the pursuit of sustainable construction has increasingly shifted focus toward the judicious use of industrial and agro-industrial materials [5]. These materials offer numerous benefits, including ready availability, recyclability, cost-effectiveness, non-toxicity, resistance to wear and tear, biodegradability, and favorable thermo-mechanical performance. Consequently, leveraging waste and renewable resources as alternative construction materials has gained traction as an effective means to address environmental concerns in most developing countries, with researchers now emphasizing environmental preservation as a prerequisite before implementing new technologies. Soil serves as a robust, eco-friendly, and highly thermally retentive building material. In the construction of traditional dwellings and ksars in the cold winters and hot summers of southern Morocco, soil is extensively employed. Enhancing the thermal and mechanical properties of clay can be achieved by integrating natural additives, leading to a more effective insulating composite material. Ongoing research has been dedicated to investigating the impact of stabilization on the mechanical and thermal characteristics, as well as the longevity, of clay-based materials. Therefore, the primary goal of reinforcing soil masses is to enhance structural stability, increase load-bearing capacity, and diminish settlement and lateral deformation [6]. Furthermore, reinforcement encompasses the integration of specific materials possessing desired attributes into other materials that may lack these characteristics. Consequently, soil reinforcement can be described as a method directed at enhancing the technical properties of soil with the intention of refining parameters such as shear strength, compressibility, density, and thermal conductivity [7]. In a study conducted by Ben Mansour and al [8], it was deduced that optimizing the bulk density of compressed earth blocks can effectively achieve the dual purpose of reducing thermal conductivity while providing sufficient compressive strength. Recently, researchers have increasingly directed their attention toward fiber-soil composites as a subject of heightened interest. The Drâa-Tafilalet region in the southeast of Morocco boasts a vast expanse dedicated to date palm cultivation. Among the provinces in this region, Ouarzazate, situated in the Drâa Valley, and Errachidia, covering Tafilalet and Ziz Valley, stand out as the principal hubs for date palm cultivation, collectively constituting the largest regions for date palm cultivation. Ouarzazate province leads the way with 40% of the total date palm trees, followed by Errachidia with 28.24%. In light of this, our study centers on the development and characterization of novel eco-friendly building materials. These materials are derived from soil and reinforced with waste from date palms, which are abundantly available in the Drâa-Tafilalet region. The intention is to employ these materials in housing construction. Several researchers have explored the utilization of palm by-products in building materials due to their exceptional thermal and mechanical properties. Palm fibers exhibit filamentous textures and possess unique attributes, including affordability, local abundance, durability, and lightweight properties [9]. Fibers obtained from deteriorated palm material exhibit brittleness, possess low tensile strength, a low modulus of elasticity, and a high capacity for water absorption [10]. Research conducted by Salehan and Yaacob [11] demonstrated that water absorption levels slightly rise as the palm fiber content increases. Synthetic fibers are a

Mediterranean Architectural Heritage - RIPAM10   Materials Research Forum LLC
Materials Research Proceedings 40 (2024) 41-54   https://doi.org/10.21741/9781644903117-5

commonly employed resource in the field of soil reinforcement, facilitating the augmentation of both compressive and shear strength [12,13]. In a study by Namango, a substantial boost in strength was observed with escalating proportions of sisal fibers, cassava powder, cement, and cement fiber, albeit within specific limits. The findings indicate that exceeding these thresholds for sisal fiber content has an adverse effect on the strength attributes of compressed earth blocks [14]. Minke highlighted the potential of incorporating fibers such as human or animal hair, coir, sisal, agave, bamboo, and straw to mitigate shrinkage by reducing clay content and allowing for some water absorption through the fiber's pores [15]. Similarly, Villamizar and al [16] observed that the utilization of cassava peels notably enhanced the dry strength of mixtures, a property beneficial in addressing challenges associated with handling block-earth construction waste. Nevertheless, Rigassi argued, without presenting specific research data, that while fibers are frequently employed to reinforce adobe, they may not be compatible with the compaction pressures associated with compressed earth blocks (BTC) due to their potential elasticity-inducing properties [17]. It is noteworthy, however, that in the pursuit of environmental conservation, investigations have been conducted on compressed earth blocks infused with recycled synthetic fibers, as exemplified by the work of Eko and al [18]. In their work, Ouakarrouch and al [19] conducted both experimental and numerical analyses to evaluate the thermal comfort of "Ksar Lamaadid" in the Erfoud region, which is constructed using an innovative biocomposite material comprising clay and sisal fibers. Their research demonstrated that incorporating 4% sisal fibers can lead to an approximate 11.2% reduction in the material's thermal conductivity. This reduction has the potential to enhance thermal comfort and reduce greenhouse gas emissions by roughly 62442 kg CO_2 per year. Liuzzi et al. [20] investigated the impact of adding plant fibers on the thermophysical characteristics of clay mixtures, while Calatan and al [21] found that the utilization of hemp fibers positively influenced both thermal and mechanical properties. The study conducted by Ismail and Yaacob [22] revealed that incorporating 3% palm fiber resulted in an improvement in the compressive strength of composite bricks. Additionally, the authors noted a slight uptick in water absorption as the fiber content increased. The incorporation of date palm as a natural additive in construction materials is not a novel concept, as multiple studies have already demonstrated the manifold benefits of utilizing natural fibers in brick production. Taallah and Guettala, in their extensive research [23,24], delved into various compositional factors, including the addition of date palm, and their influence on reducing the thermal conductivity of bricks. In addition to floor bricks, a separate investigation concluded that the inclusion of palm fibers in mortar specimens generally enhances post-crack performance and ductility when compared to specimens lacking palm fibers, thus extending the composite material's resistance to failure [25]. More recently, a study introduced the combination of palm fibers and lime in the development of an eco-friendly insulation material, revealing its advantages in terms of thermo-acoustic properties [26]. Furthermore, Oskouei and al reported a noteworthy enhancement in compressive strength, measuring at 15.6%, resulting from the inclusion of palm fibers within the composition of gravel, sand, and clay [27]. This compositional adjustment also led to a significant 62.5% reduction in final deformation and an extended duration of resistance to water exposure, surpassing the properties of materials consisting solely of gravel, sand, and clay by 23%. In Biskra, Algeria, D. Khoudja et al. conducted a thermomechanical study on raw earth bricks stabilized with lime and fortified with aggregates derived from a blend of waste components from date palm materials. Their research exhibited improved thermal insulation, with a thermal conductivity rating of 0.342 W/(m·K) for bricks containing 10% aggregates, while still meeting the minimum performance standards required for earth construction [28]. Our experimental research centers on the creation of a novel composite material using a blend of date palm fibers and clay. This composite serves as the fundamental component for adobe and brick construction in the southeastern region of Morocco, with the aim of enhancing their thermal and mechanical attributes, as well as their

Mediterranean Architectural Heritage - RIPAM10   Materials Research Forum LLC
Materials Research Proceedings 40 (2024) 41-54   https://doi.org/10.21741/9781644903117-5

hygroscopic behavior. The primary objective of this study is to investigate the impact of integrating date palm waste into the composition of mud bricks concerning their thermal, physical, and mechanical properties. The ultimate goal is to develop bricks with robust mechanical characteristics and improved thermal insulation properties, allowing for the construction of housing with highly insulating walls.

2. Experimental methods

2.1 Materials

2.1.1 Date palm spathe

The date palm spathes used in this study were sourced from the Draa Tafilalt oasis in Morocco. These date palm fibers had been exposed to the natural environment, leading to contamination by significant amounts of sand and dust. The fibers underwent a rigorous cleaning process, which involved washing them with fresh water and manually separating them into fiber bundles. Prior to their use, a high-pressure water wash was performed to eliminate any pollutants. Subsequently, the fibers were left to air-dry for a duration of three days. After the drying process, the fibers were cut to the desired length. For this study, we utilized individual fiber samples, each measuring 5 mm in length, as described below and depicted in Figure 1.

Fig.1 Element of the date palm used in the study.

2.1.2 Sol

The earth used (figure 2) comes from the Errachidia region in south-east Morocco, and is a very high-quality material. It is often used in construction projects because of its properties, which are well suited to certain types of structure.

Fig.2 Soil studied.

Table 1 Soil mineralogical composition (%).

Minerals	Chemical formula	Mineralogical composition (%)
Calcite	Ca(CO3)	43.33
Quartz	SiO2	36.51
Dolomite	CaMg(CO3)2	16.18
Muscovite	H2KAl3Si3O12	3.98

X-ray fluorescence spectrometry (XRF) was used to determine the chemical composition of the soil. Table 2 shows the data obtained from this analysis.

Table 2 Soil chemical composition.

Components	Percentage %
SiO2	33.91
CaO	22.15
P.a.F	20.81
Al2O3	11.96
MgO	3.98
Fe2O3	3.57
K2O	1.74
Na2O	0.55
TiO2	0.49
CI	0.25
SO3	0.20
P2O5	0.11

2.2 Sample preparation

The material studied was obtained by mixing clay with date palm spathe fibers (Fig. 3). Five mass fractions (1%, 2%, 3%, 4% and 5%) of these fibers were selected. To obtain homogeneous composites prepared by hand, we followed the following steps: (1) immerse the fibers in water for five minutes before use, (2) mix the dry fibers with the soil, (3) then add the water solution in the required proportions. After demolding, the molds were placed in the laboratory to undergo a hardening process under controlled conditions. Pure samples (100% clay) of the same size were prepared to follow the evolution of thermal, mechanical and hydric properties. Samples were made in $250x250x30$ mm^3 molds for thermal characterization. To assess flexural strength, we used prismatic molds with dimensions of $40x40x100$ mm^3. For compressive strength, we used cubic molds measuring $40x40x40$ mm^3. Cylindrical samples measuring $100x100$ mm^2 were molded to determine the water absorption rate (Figure 3). After 3 days air-drying, the samples were dried at 50°C to remove moisture.

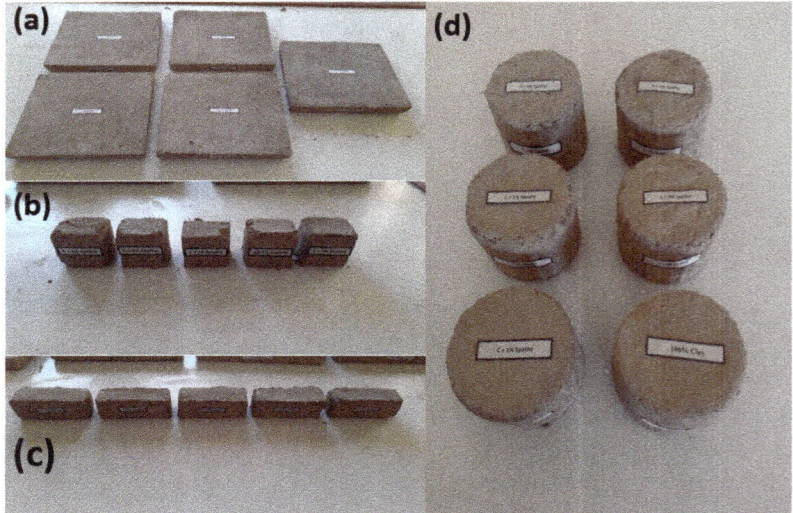

Fig.3 The various samples tested include (a) measurement of thermal conductivity, (b) evaluation of compressive strength, (c) measurement of flexural strength (d) determination of water absorption.

Table 3. Characteristics of the various samples studied.

Samples	Percentage of Fiber Spathe (%)	Number of samples	Average Density (Kg / m^3)
E1 : 100 %Clay	0	4	2985,18
E2 :1% Spathe	1	4	2827,56
E3 :2% Spathe	2	4	2758,85
E4 :3% Spathe	3	4	2613,59
E5 :4% Spathe	4	4	2540,70
E6 :5% Spathe	5	4	2319,25

3. Measurement methods

3.1 Thermal characterization method

The experimental tests were conducted within a highly insulated thermal house featuring interchangeable walls, as depicted in Figure 4. This approach has been employed by several researchers in previous studies [29,30]. The thermal house comprises four vertical walls, each equipped with a well-insulated square polystyrene opening measuring 210x210 mm^2. To provide controlled heating, a 100 W incandescent bulb was utilized as the heat source and connected to an internal thermal controller, allowing for the regulation of the desired temperature. During steady-state operation, the heat source maintained an internal temperature of approximately 50°C. Thermocouples were affixed to both the interior and exterior of the wall to capture the T_{int} and T_{ext} temperatures of the sample. The ambient temperature was rigorously controlled in a laboratory environment, maintained consistently at 20°C through the use of an air-conditioning unit. Following 8 hours of heating and the attainment of thermal equilibrium, the acquisition system recorded the three temperature values, T_{loc}, T_{int}, and T_{ext}, with an estimated error margin of 10% for this method [30].

Fig.4 PHYWE thermal insulation house.

To determine thermal conductivity, we take into account the conservation of heat flux that occurs by convection between the inner air and the inner wall surface, as well as by convection between the outer air and the outer wall surface, and by conduction through the sample is given by:

$$\varphi = h_{int}\left(T_{loc} - T_{int}\right) S = S \times \lambda \frac{T_{int} - T_{ext}}{e} = h_{ext}\left(T_{ext} - T_{amb}\right) S \qquad \text{Eq.1}$$

With:
h_{int}: internal convection coefficient.
h_{ext}: external convection coefficient.
T_{int}: temperature of the inside of walls.
T_{ext}: temperature of the outside of the wall.
T_{loc}: air temperature inside thermal house.
T_{amb}: ambient temperature.
e: wall thickness.

We determined the thermal conductivity value (λ) by applying equation (3) along with a coefficient value of 8.1 W/m^2·K, as indicated in prior references [31, 32]. In assessing the thermal conductivity of our construction materials, we followed the manufacturer's recommended protocol, utilizing a PHYWE thermal control stand and a set of thermocouples to monitor temperature fluctuations both inside and outside the thermal house [33]. Readings were recorded when the thermocouples stabilized at a consistent value, ensuring that the system had reached thermal equilibrium, typically 8 hours after initiating the measurement.

$$h_{int}\left(T_{loc} - T_{int}\right) = \frac{\lambda}{e}\left(T_{int} - T_{ext}\right) \qquad \text{Eq.2}$$

$$\lambda = \frac{\left(T_{loc} - T_{int}\right)}{\left(T_{int} - T_{ext}\right)} \times h_{int} \times e \qquad \text{Eq.3}$$

Thermal resistance can be calculated according to the following equation:

$$R = \frac{e}{\lambda} \qquad \text{Eq.4}$$

3.2 Mechanical performance tests

3.2.1 Dry and wet compressive strength test

Compressive strength tests are carried out on cubic specimens composed of a mixture of soil and fibers, at 28 days of age. Compression tests are carried out using a 30 KN capacity machine, in accordance with the requirements of Moroccan standard NM 10.1.538. Compressive strength can be calculated according to the following relationship:

$$R_C = \frac{F}{A_c} \qquad \text{Eq.5}$$

With;

R_C: compressive strength, expressed in MPa (N/mm^2);

F: is the maximum breaking load, expressed in N;

A_C: is the cross-sectional area of the specimen to which the compressive force is applied, calculated from the specimen's nominal size, expressed in mm^2.

Compressive strength should be expressed to the nearest 0.1 MPa (N/mm^2).

3.2.2 Dry bending strength test

Flexural strength was measured on prismatic samples of soil mixture and spathe fibers. Figure 5 below shows the set-up for the three-point bending experiment. Bending tests were carried out using a 30 KN capacity machine in accordance with the requirements of Moroccan standard NM 10.1.538. According to the bending relationship and knowing the load obtained at break, the bending strength can be calculated as shown in the relationship:

$$R_f = 1.5 \times \frac{F \times L}{b \times d^2} \qquad \text{Eq.6}$$

With;

R_F: is the compressive strength, expressed in MPa (N/mm^2);

F: is the maximum load applied to the sample, in newtons (N);

L: is the distance between the axes of the support rollers, in millimeters (mm);

b: is the width of the specimen, in millimeters (mm) ;

d: is specimen thickness, in millimeters (mm).

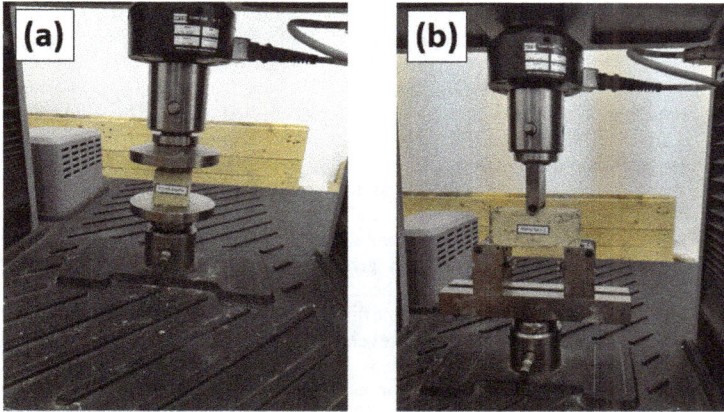

Fig.5 (a) Soil and fiber sample tested for strength; (b) C+5% spathe sample tested for flexural strength.

3.3 Immersion water absorption test

Immersion absorption was carried out in accordance with ASTM D570. Samples were immersed in distilled water at 27°C and sample weight changes were recorded every 0.5 h, 1.0 h, 2.0 h, 24 h, 48 h and 72 h respectively using a 0.01 mg resolution digital balance. The weighing process was carried out within 30 s, removing surface water from the sample using filter paper to avoid any errors. The water absorbed by the samples was determined using equation (7).

$$W(\%) = \frac{W_t - W_0}{W_0} \times 100 \qquad \text{Eq.7}$$

Where, W (%) is the average percentage of water absorption, W_0 and W_t are the initial and final sample weights after time t, respectively.

Fig.6 Measuring the degree of water absorption in the sample.

4. Results and discussion

4.1 Conductivity and thermal resistance of the samples studied

Thermal conductivity and thermal resistance are key parameters for assessing a building material's ability to conduct heat. Figure 7 illustrates the changes in thermal conductivity and thermal resistance concerning the incorporation of different fiber content in the various biocomposites developed. Notably, as fiber content increases, thermal conductivity decreases, and thermal resistance increases. Furthermore, it is evident that the sample enriched with 5% date palm spathe fibers exhibits superior insulation properties compared to the other samples. This enhancement is attributed to the reduction in thermal conductivity, which decreased from 0.514 W/(m·K) for the sample devoid of fibers to 0.274 W/(m·K) for the sample containing 5% of these fibers. This remarkable reduction amounts to an estimated 47% improvement in thermal insulation performanceThe results obtained for the 5% fiber content scenario can be rationalized in two ways. First, the gradual increase in the fiber quantity, characterized by their inherently low thermal conductivity, plays a significant role. Secondly, the heightened fiber content within the composites leads to a decrease in sample density, potentially generating air-filled pores within the material [34]. The latter can be attributed to the substantial cellulose content present in the fibers, which is associated with their insulating properties, evident through the high air permeability seen in the fibers as observed in SEM micrographs [35]. However, it is essential to acknowledge that the latter explanation is contingent on other factors such as fiber fragmentation and the uniformity of cellulose distribution within the fibers. In this context, date palm spathe fibers stand out due to their notably high cellulose content, which aligns with the observed lower thermal conductivity in samples containing these fibers.

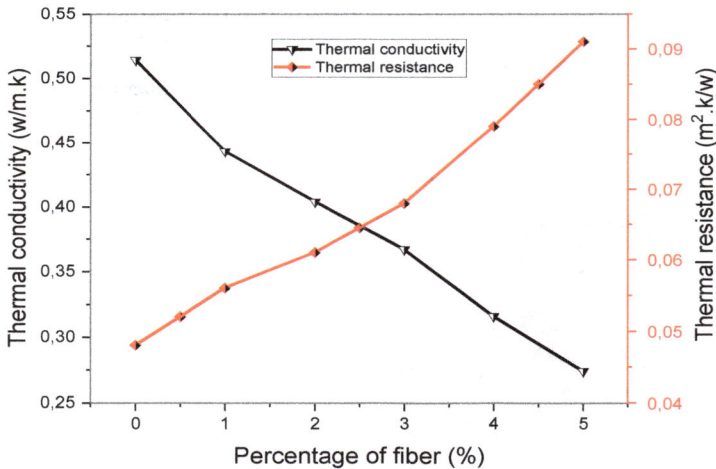

Fig. 7 Conductivity and thermal resistance as a function of sample fiber content.

4.2 Compressive strength

Figure 8 presents the outcomes of the compression tests. Notably, the compressive strength values of the stabilized soil samples remained within the specified range, with a maximum increase of 40% recorded at the optimal fiber content of 3%. These values did not surpass the upper limit of 5.6 MPa specified by the fiber or fall below the lower limit of 3.4 MPa set by the clay. Beyond this optimal threshold, the compressive strengths displayed a decline at higher fiber content levels, indicating that further increases in fiber content do not yield significant benefits in terms of enhancing block strength.

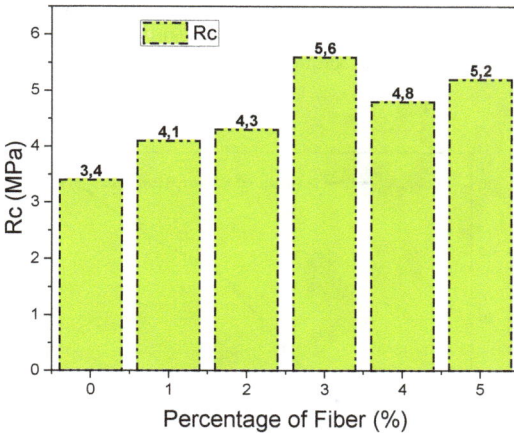

Fig.8 The effect of incorporating fibers into the mortar mix on compressive strength.

4.3 Flexural strength

In Figure 9, the outcomes of the flexural strength tests are depicted. The behavior of flexural strength in the stabilized soil blocks closely mirrored that of the compressive strength. The maximum enhancement, approximately 17% in comparison to the control samples, was observed at the optimal fiber content of 3%. Subsequently, the strength exhibited a reduction, which was then recuperated with higher fiber content, up to 3%.

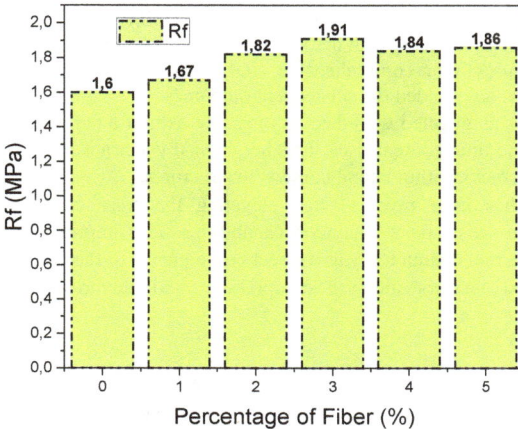

Fig.9 The effect of incorporating fibers into the mortar mix on flexural strength.

4.4 Analysis of hydric results

Figure 10 shows the evolution of water incorporation in a soil mixture reinforced with date palm spathe fibers. The increase in water incorporation is clearly perceptible and observable in correlation with the proportion of plant fibers. This observation is attributable to the presence of a large volume of voids produced by the addition of the fibers, as well as to the intrinsic nature of the latter.

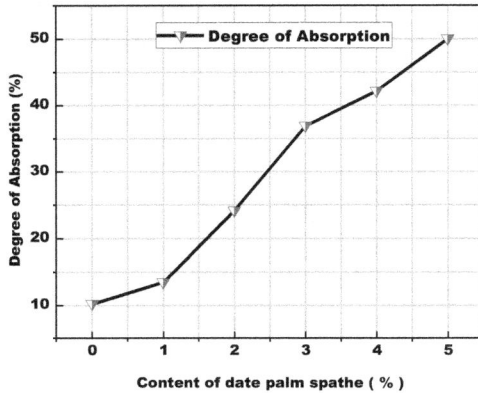

Fig.10 Variation in water absorption of a date palm soil and spathe mortar as a function of fiber mass percentage.

Conclusion

In this article, we present the findings of an experimental investigation focused on examining the thermo-physical and mechanical characteristics of soil that has been enhanced with date palm fibers. The primary objective of this study is to explore the viability of incorporating this enhanced material into construction practices, with the intention of reinforcing soil and mitigating heat loss within building structures. Our comprehensive analysis, encompassing both quantitative and qualitative assessments, has yielded the subsequent outcomes:

- Soil composites augmented with date palm spathe exhibit a noteworthy increase in water absorption as the fiber content rises. This heightened water absorption has the potential to result in microcracks within lateral composite structures.
- Empirical studies have revealed that elevating the fiber content contributes to an enhancement in the mortar's insulation capabilities. This improvement is attributed to a reduction in thermal conductivity as a direct consequence of the increased fiber content.
- Fiber-reinforced mortars increase compressive and flexural strength at low fiber percentages.

Reference

[1] ADEREE, Règlement Thermique de Construction au Maroc - Version simplifiée, 2014.
[2] United Nations Environment Programme. Global Status Report 2017: Towards a zero-emission, efficient, and resilient buildings and construction sector. 2017.

[3] L Lyu, J Lu, J Guo, et al, Sound absorption properties of multi-layer structural composite materials based on waste corn husk fibers, J. Eng Fiber Fabr 15(2020)1–8. https://doi.org/10.1177/1558925020910861

[4] M Bederina, L Marmoret, K Mezreb, et al,Effect of the addition of wood shavings on thermal conductivity of sand concretes: experimental study and modelling,J. Constr Build Mater 21(2007)662–668. https://doi.org/10.1016/j.conbuildmat.2005.12.008

[5] S Panyakaew and S Fotios, New thermal insulation boards made from coconut husk and bagasse, J. Energy Build 43(2011)1732–1739. https://doi.org/10.1016/j.enbuild.2011.03.015

[6] H.Binici, O.Aksogan, and T.Shah, Investigation of fibre reinforced mud brick as a building material,J. Construction and Building Materials. Letters 19(2005) 313–318. https://doi.org/10.1016/j.conbuildmat.2004.07.013

[7] S.Hejazi, et al, A simple review of soil reinforcement by using natural and synthetic fibers,J. Construction and Building Materials. Letters 30(2012)100–116. https://doi.org/10.1016/j.conbuildmat.2011.11.045

[8] M. Ben Mansour, A.Jelidi, A S Cherif and S. Ben Jabrallah, Optimizing thermal and mechanical performance of compressed earth blocks (CEB), J.Construction and Building Materials. Letters 104(2016) 44–51. https://doi.org/10.1016/j.conbuildmat.2015.12.024

[9] M Yusoff, M Salit, N Ismail, R Wirawan, Mechanical properties of short random oil palm fiber reinforced epoxy composites,J. Sains Malay 39(1)(2010)87–92.

[10] R.N Swamy. New reinforced concretes, Surry University Press; 1984. p. 200.

[11] I Salehan, Z Yaacob, Properties of laterite brick reinforced with oil palm empty fruit bunch fibers,J. Sci Technol 19(1)(2011)33–43.

[12] AR Estabragh, AT Bordbar, AA Javadi, Mechanical behavior of a clay soil reinforced with nylon fibers,J. Geotech Geol Eng29(5) (2011)899–908. https://doi.org/10.1007/s10706-011-9427-8

[13] J Prabakar, RS Sridhar, Effect of random inclusion of sisal fibre on strength behaviour of soil,J. Constr Build Mater 16(2)(2002)123–31. https://doi.org/10.1016/S0950-0618(02)00008-9

[14] S.S Namango. Development of cost-effective earthen building material for housing wall construction. In: Investigations into the properties of compressed earth blocks stabilized with sisal vegetable fibres. Cassava powder and cement compositions. A doctoral dissertation. Brandenburg Technical University; Cottbus (Germany): 2006.

[15] G Minke. Earth Construction Handbook. Boston: WIT Press; 2000.

[16] MCN Villamizar, VS Araque, CAR Reyes, RS Silva, Effect of the addition of coalash and cassava peels on the engineering properties of compressed earth blocks,J. Constr Build Mater 36(11)(2012)276–86. https://doi.org/10.1016/j.conbuildmat.2012.04.056

[17] V Rigassi. Compressed earth blocks. CRATerre-EAG, vol. 1. Germany: Manuel de production; 1995.

[18] RM Eko, ED Offa, TY Ngatcha, LS Minsili,Potential of salvaged steel fibers for reinforcement of unfired earth blocks,J. Constr Build Mater 35(10)(2012)340–6. https://doi.org/10.1016/j.conbuildmat.2011.11.050

[19] M. Ouakarrouch, , K El Azhary, Mansour, M, Laaroussi, N., Garoum, M,Thermal study of clay bricks reinforced by sisal-fibers used in construction in South of Morocco,J.Energy Reports6(2019)81–88. https://doi.org/10.1016/j.egyr.2019.11.045

[20] RDT Filho, K Scrivener, GL England, K Ghavami,Durability of alkali-sensitive sisal and coconut fibres in cement mortar composites,J.Cem Concr Compos 22(2)(2000)127–43. https://doi.org/10.1016/S0958-9465(99)00039-6

[21] J Khedari, S Charoenvai, J Hirunlabh, New insulating particleboards from durian peel and coconut coir,J.Build Environ 38(3)(2003)245–54. https://doi.org/10.1016/S0360-1323(02)00030-6

[22] S. Ismail, Z. Yaacob, Properties of laterite brick reinforced with oil palm empty fruit bunch fibres, Pertanika J. Sci. Technol. 19 (1) (2011) 33–43.

[23] S.N. Malkanthi, N. Balthazaar, A.A.D.A.J. Perera, Lime stabilization for compressed stabilized earth blocks with reduced clay and silt,J. Constr. Mater 12 (2020) https://doi.org/10.1016/j.cscm.2019.e00326

[24] B. Taallah, A. Guettala, The mechanical and physical properties of compressed earth block stabilized with lime and filled with untreated and alkali-treated date palm fibers,J. Constr. Build. Mater104 (2016) 52–62. https://doi.org/10.1016/j.conbuildmat.2015.12.007

[25] O. Benaimeche, A. Carpinteri, M. Mellas, C. Ronchei, D. Scorza, S. Vantadori, The influence of date palm mesh fibre reinforcement on flexural and fracture behaviour of a cement-based mortar,J. Compos. B Eng 152 (2018) 292–299. https://doi.org/10.1016/j.compositesb.2018.07.017

[26] R. Belakroum, A. Gherfi, M. Kadja, C. Maalouf, M. Lachi, N. El Wakil, T.H. Mai, Design and properties of a new sustainable construction material based on date palm fibers and lime,J. Constr. Build. Mater 184 (2018) 330–343. https://doi.org/10.1016/j.conbuildmat.2018.06.196

[27] A. Vatani Oskouei, M. Afzali, M. Madadipour, Experimental investigation on mud bricks reinforced with natural additives under compressive and tensile tests, J.Constr. Build. Mater. 142 (2017) 137–147. https://doi.org/10.1016/j.conbuildmat.2017.03.065

[28] D. Khoudja, B. Taallah, O. Izemmouren, S. Aggoun, O. Herihiri, A. Guettala, Mechanical and thermophysical properties of raw earth bricks incorporating date palm waste,J. Construct. Build. Mater270 (2021) 121824.https://doi.org/10.1016/j.conbuildmat.2020.121824

[29] M.A. Navacerrada, P. Fernández, C. Díaz and A. Pedrero, Thermal and acoustic properties of aluminium foams manufactured by the infiltration process,J.Applied Acoustics74(2013)496-501. https://doi.org/10.1016/j.apacoust.2012.10.006

[30] A. Gounni, M.T. Mabrouk, M. El Waznac, A. Kheiri, M. El Alami, A. El Bouari and O. Cherkaoui, Thermal and economic evaluation of new insulation materials for building envelope based on textile waste,J. Applied Thermal Engineering 149 (2019) 475-483. https://doi.org/10.1016/j.applthermaleng.2018.12.057

[31] M.A. Navacerrada, P. Fernandez, C. Díaz, A. Pedrero,Thermal and acoustic properties of aluminium foams manufactured by the infiltration process,J.Appl Acoust 74 (4)(2013)496–501. https://doi.org/10.1016/j.apacoust.2012.10.006

[32] A. Benallel, A. Tilioua, A. Mellaikhafi, A. Babaoui, M.A.A. Hamdi,Thermal characterization of insulating materials based on date palm particles and cardboard waste for use in thermal insulation building,in: AIP Conference Proceedings, AIP Publishing LLC, 2021, pp. 020024. https://doi.org/10.1063/5.0049442

[33] PHYWE, P2360300 PHYWE Series of Publications, Laboratory Experiments, Physics, PHYWE SYSTEME GMBH & Co. KG, Gottingen.

[34] H. Asan, Y.S. Sancaktar, Effects of Wall's thermophysical properties on time lag and decrement factor,J.Energy Build28(1998)159–166. https://doi.org/ 10.1016/S0378-7788(98)00007-3

[35] P. Brzyski, P. Kosiski, A. Skoratko, W. Motacki, Thermal properties of cellulose fiber as insulation material in a loose state, AIP Conference Proceedings 2133 (2019), 020006, https://doi.org/10.1063/1.5120136

Mediterranean Architectural Heritage - RIPAM10   Materials Research Forum LLC
Materials Research Proceedings 40 (2024) 55-63   https://doi.org/10.21741/9781644903117-6

Thermal Characterization of a New Bio-Composite Building Material based on Gypsum and Date Palm Fiber

Youssef KHRISSI[1,a,*], Amine TILIOUA[1,b,*], Najma LAAROUSSI[2,c]

[1]Research Team in Thermal and Applied Thermodynamics (2.T.A.), Mechanics, Energy Efficiency and Renewable Energies Laboratory (L.M.3.E.R.). Department of Physics, Faculty of Sciences and Techniques Errachidia. Moulay Ismael University of Meknès, B.P. 509, Boutalamine, Errachidia, Morocco

[2] Mohammed V University in Rabat, Material, Energy and Acoustics Team (MEAT), EST Sale, Morocco

[a]youssefkhrissi@gmail.com,[b]a.tilioua@umi.ac.ma,[c] lnajma@outlook.fr

Keywords: Building Materials, Gypsum-DPF, Hot Plate Method, Bulk Density, Thermal Conductivity

Abstract. In Morocco, the prevalent use of building materials with low thermal resistance has translated into substantial energy consumption. This has underscored the pressing need to promote the development and adoption of sustainable construction and insulation materials. The primary objective of our study is to enhance the thermal properties of plaster by incorporating date palm fibers (DPF) to create an exterior wall coating. To evaluate the thermal properties of the resulting Gypsum-DPF bio-composite material, we conducted several experimental measurements of thermophysical properties. These measurements encompassed the determination of bulk density and thermal conductivity, which were assessed using the steady-state hot plate method. Our findings reveal that the inclusion of date palm fiber in the material led to a noteworthy reduction in bulk density, amounting to approximately 17.16%. Furthermore, thermal conductivity decreased by approximately 26.24%. These outcomes underscore the potential and value of utilizing this bio-composite material in building construction to enhance thermal comfort and, critically, contribute to a reduction in greenhouse gas emissions, particularly CO_2 emissions.

1. Introduction

The construction industry plays a substantial role in global energy consumption, primarily driven by the escalating need for infrastructure. Following the transportation sector, it stands as the second-largest consumer of energy, representing 28% of the total energy usage. Notably, in Morocco, this sector contributes to 25% of the nation's overall energy consumption. In light of these statistics, Morocco has introduced a comprehensive energy efficiency plan specifically tailored to the construction sector. This strategic initiative sets out to conserve approximately 1.2 million tons of oil equivalent (MTep) in energy and simultaneously slash greenhouse gas emissions by approximately 4.5 million metric tons of CO_2 equivalent (MTeqco2) by the year 2020 [1]. Numerous endeavors have been undertaken to enhance energy efficiency within the construction industry. One key avenue for such improvement involves the enhancement of thermal properties in building materials. Morocco, endowed with abundant natural resources, presents a unique opportunity for this endeavor, particularly with its substantial reserves of gypsum. Although gypsum has a historical legacy as a construction material, its prevalent usage has historically been limited to decorative applications, primarily due to its fragility and subpar mechanical attributes. On another front, Morocco's bountiful date palm cultivation offers a valuable source of natural fibers boasting promising physical and mechanical characteristics. The central objective of this investigation is to explore the potential for elevating the worth of these eco-friendly materials by pioneering a novel composite material, which combines gypsum with reinforced date palm fibers.

Materials Research Forum LLC
https://doi.org/10.21741/9781644903117-6

It is worth noting that previous studies have already contributed to the characterization of diverse properties in gypsum composites strengthened with natural fibers, and these findings will be succinctly encapsulated below: Maaloufa et al [2] to enhance thermal insulation and mechanical characteristics, it is crucial to identify the ideal composition of cork and alpha fibers within a gypsum matrix. Notably, optimizing the thermal conductivity can be achieved, while also bolstering flexural strength through the incorporation of alpha fibers. However, it should be noted that the addition of cork may render the composite more susceptible to brittleness. F. Hernandez-Olivares and al [3] to enhance the strength of a gypsum composite, incorporating sisal fibers as reinforcement within the matrix is a viable approach. The introduction of randomly dispersed sisal fibers leads to a notable improvement in the overall mechanical properties of the composite. Mazhoud and al [4,5]. It has been observed that the utilization of hemp can effectively enhance the thermal and hydric properties of gypsum walls. Researchers have noted that hemp plasters exhibit significantly lower thermal conductivity in comparison to gypsum. Moreover, these hemp-based plasters demonstrate only a minor susceptibility to variations in temperature, highlighting their potential to contribute to reduced energy consumption in buildings [6]. Research efforts have extended to the exploration of composite materials incorporating date palm fibers, as exemplified by studies conducted by Benmansour and al [7]. They conducted experiments to evaluate the thermal and mechanical attributes of an innovative insulating material constructed using natural mortar reinforced with date palm fibers. Additionally, Kriker and al [8] investigated the mechanical characteristics of date palm fibers and their application in reinforcing concrete within hot and arid environmental conditions. In the study conducted by Gallala and al [9], they developed a gypsum composite incorporating date palm fibers (DPFW). Through a series of tests, it was determined that the inclusion of 17% DPFW, with a length of 20 mm, resulted in a composite exhibiting favorable thermal properties ($\lambda=0$, 52 w/m.k). Additionally, the mechanical characteristics of this composite outperformed those of non-fiber-reinforced samples. As a result of these findings, the authors advocate for the application of this biocomposite as an eco-friendly insulation material. In the experimental work by Boulaoued and al [10], an investigation into the thermal diffusivity and conductivity of palm fiber-reinforced plaster was conducted. The aim was to evaluate the feasibility of employing this innovative material for insulation, with the potential to decrease energy consumption in building structures. The outcomes of the study revealed a significant enhancement in the thermal characteristics of palm fiber-reinforced plaster. Additionally, it's worth noting that Almusawi et al. [11] highlighted the pivotal role of palm fiber reinforcement in enhancing the mechanical properties of plaster. In a study conducted by Naiiri et al. [12], an evaluation was carried out to determine the impact of doum palm fibers on gypsum mortar properties. Their research confirmed significant improvements in both mechanical and thermal attributes, as well as a reduction in damage, when doum palm fibers were integrated into plaster mortar. Djoudi et al. [13] also explored the influence of doum palm fibers on the thermal properties of plaster-gravel mixtures, revealing noteworthy enhancements in thermal conductivity, specific heat, and thermal diffusion.In a separate investigation, N. Fatma et al. [14] proposed the utilization of palm fibers in gypsum mortar to enhance density and thermal conductivity. Building on this body of research, the present study focuses on the development of a novel gypsum-based bio-composite material. This material incorporates varying mass fractions of date palm fiber (0%, 1%, 2%, 3%, 4%, and 5%) and it is designed for use in applications such as roof and interior wall cladding. The primary objectives of this innovative material are to meet thermal comfort standards, reduce energy consumption, and contribute to environmental sustainability through the utilization of waste resources.To gain insights into the critical factors influencing the performance of this material, an experimental characterization of its thermophysical properties was conducted. The key properties examined in this research encompass bulk density and thermal conductivity, and their evaluation was executed using the steady-state hot plate method. The resulting experimental findings are presented and comprehensively discussed in this paper.

2. Experimental materials and methods

2.1 Materials

2.1.1 Gypsum

Gypsum-based plaster is a construction material that finds common use in the building industry. A calcium sulfate material begins to harden and solidify when mixed with water. It releases heat when combined with water, unlike cement mortar, which is obtained by calcining natural gypsum (dehydrated calcium sulfate, CaSO4 2H2O) at a temperature of 120-150°C to obtain semi-hydrate with a certain amount of gypsum anhydride.

The chemical reaction in the solidification and hydration of hemihydrated calcium sulfate is:

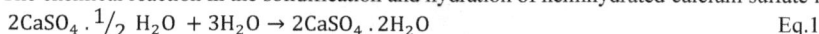

$$2CaSO_4 . {}^1\!/_2\ H_2O + 3H_2O \rightarrow 2CaSO_4 . 2H_2O \qquad \text{Eq.1}$$

In this work, the gypsum used is hemi-hydrated gypsum extracted from the town of Safi in Morocco. The X-ray fluorescence spectrometer (XRF) was used to determine the chemical composition of the gypsum. Data from this analysis are presented in Table 1.

Table 1 Chemical compositions of the plaster used.

Component	Percentage
SO_3	54.93
CaO	33.34
P.a.F	9.512
SiO_2	1.047
MgO	0.5414
Al_2O_3	0.3438
Fe_2O_3	0.1201
SrO	0.09384
K_2O	0.06964
Rb_2O	0.001740

2.1.2 Date palm fiber

Recently, there has been a noticeable worldwide trend towards the use of materials derived from renewable natural resources. An excellent example of this trend is the use of palm waste fibers, which are an important and valuable source of natural fibers. These fibers are extracted from various waste products derived from palm trees, a collection process that is an integral part of seasonal delimbing, a vital agricultural procedure. The potential applications for these residues are remarkably diverse and extensive. From a socio-economic point of view, particularly in the Middle East and North Africa, date palms are of significant importance. Date palm waste has found its way into the construction industry, where it is used for tasks such as acoustic improvement and thermal insulation [15]. In our specific case, we chose to incorporate these palm fibers into plaster, thus creating an innovative eco-material that exploits local resources. The palm fibers used in this project come from the male palms of the Draa Tafilalt oases in southeastern of Morocco.

A scanning electron microscope (SEM) image, as shown in Figure 1, reveals the microporous structure of a date palm fiber particle. This structural feature suggests the possibility of trapping air, which could improve the material's thermal insulation properties.

Mediterranean Architectural Heritage - RIPAM10
Materials Research Proceedings 40 (2024) 55-63

Materials Research Forum LLC
https://doi.org/10.21741/9781644903117-6

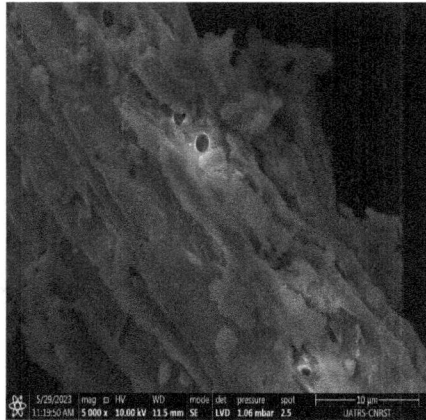

Fig.1 SEM image of a date palm fiber particle (CNRST-UATRS).

2.2 Sample preparation

The palm fibers gathered underwent a rigorous cleaning procedure, starting with a thorough washing using tap water to remove any dust or impurities. Subsequently, these fibers were manually separated into bundles. Afterward, the fibers were allowed to sit in the laboratory for a 24-hour period before being cut to the desired lengths, as depicted in Figure 2a, and then carefully stored in bags for future utilization. To enhance the compatibility of the palm fibers with the alkaline environment of the plaster, as illustrated in Figure 2b, and to eliminate weaker components such as lignin and hemicellulose that may not withstand it, a chemical treatment was implemented. This treatment not only reinforced the mechanical properties of the fibers but also optimized their bonding strength with the fiber-matrix interface. The specific treatment conditions were established based on insights from recent studies [16,17]. In this process, initially cleaned fibers were immersed in a 1% NaOH solution at 100°C for a duration of one hour. Following this, a thorough rinse was carried out to eliminate any residual NaOH on the fiber surface, and the fibers subsequently underwent a sodium chloride bleaching process. Subsequently, the fibers were meticulously washed once more and left to air dry at room temperature for a 24-hour period before deployment. The composite materials, identified as DPF, were prepared in steel parallelepiped molds measuring $150 \times 150 \times 30$ mm^3. To achieve the desired compositions, plaster was mixed with varying mass fractions of DPF (ranging from 1% to 5%). A water ratio of W/P = 0.6 was gradually introduced, with manual stirring to ensure the proper homogenization of the components, as illustrated in Figure 2c. The resulting mixture was then poured into the molds. In order to assess the impact of fiber addition on the thermophysical properties of the composite samples, a reference sample was generated using pure gypsum. All six samples underwent air-drying at room temperature for a 24-hour duration, with their outer surfaces smoothed to ensure consistent contact for subsequent experimental measurements. Following this; the samples underwent an oven-drying process at 60°C for two days to eliminate any remaining moisture. The drying process was intermittently paused until the sample mass remained constant, fluctuating by no more than ± 2 grams. Finally, the samples were sealed in plastic to preserve their dry state, as illustrated in Figure 2d.

Fig. 2 Different stages in sample preparation.

2.3 Experimental description of thermal properties measurement

2.3.1 Steady-state hot-plate method

Thermal conductivity of the composite materials under examination was determined using a steady-state hot-plate apparatus, as depicted in Figure 2. This experimental arrangement involves the controlled transmission of a constant heat flux, represented as ϕ, through a heating element positioned beneath the sample. The primary objective is to monitor temperature variations on the two upper surfaces of the sample as it interacts with the heat source [18]. The heating element itself assumes a parallelepiped shape, with dimensions measuring $150 \times 150 \times 0.2$ mm^3 and an electrical resistance of Re = 37.89 V. To ensure uniform temperature distribution on the unheated sides of the sample, the three heating elements are carefully situated between two aluminum blocks. Additionally, three K-type thermocouples, each with an uncertainty value of approximately +/- 0.6°C, are strategically positioned at the center of the sample faces. In this configuration, we are particularly interested in monitoring the temperature denoted T_0 at the center of the bottom face of the heating element,

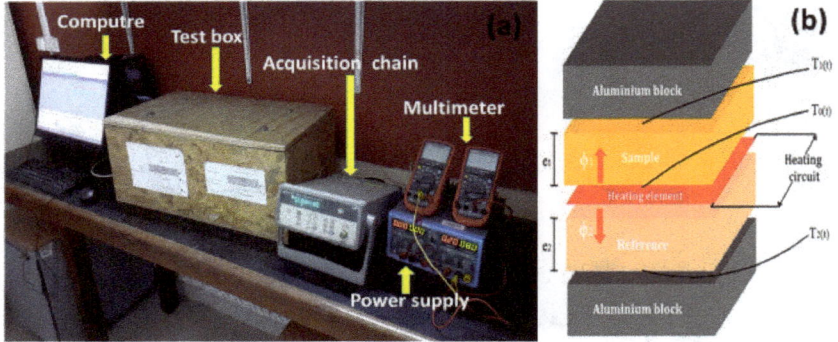

Fig. 3 Steady-state asymmetric hot-plate method: a) Experimental set-up, b) Principle of the method.

The temperatures we aim to track include T_1, positioned at the center of the unheated side of the sample, and T_2, situated at the center of the polyethylene foam, as delineated in Figure 3. It's important to highlight that the heat fluxes traveling through the sample, designated as ϕ_1, and the polyethylene foam, denoted as ϕ_2, are considered to be unidirectional, as outlined in equations (3) and (4). When the system reaches a steady state,

$$\phi = \frac{U^2}{RS} = \phi_1 + \phi_2 \qquad \text{Eq.2}$$

$$\phi_1 = \frac{\lambda}{e}(T_0 - T_1) \qquad \text{Eq.3}$$

$$\phi_2 = \frac{\lambda_i}{e_i}(T_0 - T_2) \qquad \text{Eq.4}$$

Where:

ϕ Represents the overall joule flux released by the heating element. In this context, $\lambda_i = 0.04$ $W.m^{-1}.K^{-1}$ and $e_i = 10mm$ are successively the thermal conductivity and the thickness of the insulating foam. Furthermore, **e** denotes the thickness of the sample, **R** and **S** denote the electrical resistance and surface area, respectively, of the heating element, through which an electrical current **I** flows under the influence of a voltage **U** applied to its terminals.

The thermal conductivity of the sample is calculated by the following equation:

$$\lambda = \frac{e}{(T_0 - T_1)}[\frac{U^2}{RS} - \frac{\lambda_i}{e_i}(T_0 - T_2)] \qquad \text{Eq.5}$$

4. Results and discussion

In this section, we will provide an overview of the outcomes derived from the thermophysical assessment of gypsum material with varying proportions of date palm fiber, specifically examining bulk density ρ_{app} (kg/m^3) and thermal conductivity λ (W/m.K). Each sample was rigorously examined to ensure the quality of the thermograms and verify compliance with experimental specifications. The process involved the systematic removal and replacement of samples between measurements to maintain the accuracy of the experimental set-up.

Materials Research Forum LLC
https://doi.org/10.21741/9781644903117-6

4.1 Bulk density

In the construction sector, the emphasis is on lightweight materials, which underlines the importance of bulk density assessment. This process involves measuring the weight and dimensions of a sample. Samples, in their dry state, the specimens were carefully weighed using an electric precision balance, and their dimensions were meticulously assessed with a high-precision caliper. Subsequently, the bulk density was determined using the following formula:

$$\rho_{app} = \frac{m_{app}}{v_{app}} \qquad \qquad \text{Eq.6}$$

Figure 4 illustrates the fluctuations in bulk density observed in gypsum composites with varying proportions of (DPF; 0% to 5%). In particular, the introduction of (DPF) into the gypsum matrix results in a significant decrease in bulk density, from 1386.13 Kg/m³ to 1148.25 Kg/m³. This is equivalent to a weight reduction of around 17.16%, particularly for the sample containing 5% fibers. The main factor in this reduction is the increase in material porosity resulting from the incorporation of date palm fibers. This result underlines the effectiveness of these materials for potential applications in interior wall and roof cladding, highlighting their value as bio-composite building materials characterized by a robust adhesive structure.

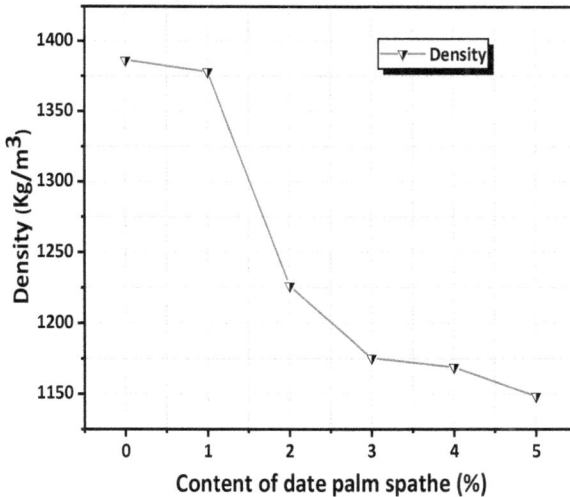

Fig. 4 Bulk density of different samples.

4.2 Thermal conductivity

The thermal conductivity of the samples tested was evaluated using the steady-state hot-plate method. The main input parameters, including sample thickness, voltage, electrical resistance and recorded temperatures (T_0, T_1 and T_2), as well as the resulting thermal conductivity measurements for the various samples, are presented concisely in Table 2. Notably, the maximum deviation observed in these measurements is around 4.1%, validating the reliability of the chosen method. An important observation in the data shows a steady decline in thermal conductivity as the proportion of date palm fiber (DPF) increases. Specifically, thermal conductivity decreases from 0.522 W/m.K, the value for pure gypsum, to 0.385 W/m.K for the composite that includes 5% DPF (P+5% DPF). This reduction in thermal conductivity can be attributed to two main factors. Firstly, it is linked to the intrinsically low thermal conductivity of the date palm fibers themselves. Secondly, it results from the formation of air-filled voids within the composite material due to the

Materials Research Forum LLC
https://doi.org/10.21741/9781644903117-6

inclusion of these fibers. In particular, the air has an extremely low thermal conductivity of 0.026 W/m.K, making a significant contribution to the overall reduction in the thermal conductivity of the composite material.

Table 2 Measurement input data and experimental results for thermal conductivity of various samples.

	e(m)	U (V)	R (Ω)	S (m²)	T$_0$(°C)	T$_1$ (°C)	T$_2$ (°C)	λ (W/m.K)
Pure plaster	0,0246	15,15	37,89	0,02250	40,96	31,271	26,718	0,522
P+1% DPF	0,0220	15,15	37,89	0,02250	39,479	30,409	24,499	0,490
P+2% DPF	0,0235	15,15	37,89	0,02250	42,727	33,41	25,587	0,485
P+3% DPF	0,0233	15,15	37,89	0,02250	38,558	27,941	22,179	0,430
P+4% DPF	0,0210	15,15	37,89	0,02250	45,704	35,663	28,348	0,400
P+5% DPF	0,0204	15,15	37,89	0,02250	38,562	27,974	23,062	0,385

Conclusion

The main objective of this study is to investigate the thermal characteristics of an innovative bio-composite material that combines gypsum and date palm fibers. We used experimental methods to evaluate the bulk density and thermal conductivity of gypsum samples with different mass fractions of date palm fibers (0%, 1%, 2%, 3%, 4% and 5%). The results of this study are as follows:

- The incorporation of 5% date palm fiber (DPF) into the composite material results in a substantial reduction of around 26.24% in thermal conductivity. In addition, the addition of DPF contributes to an overall reduction in the density of the building material, with a notable decrease of 17.16%.
- The adoption of innovative composite materials in construction makes it possible to improve thermal comfort without the need for heating or cooling systems, thus reducing greenhouse gas emissions. Lower energy consumption plays a crucial role in reducing CO_2 emissions.

References

[1] M. Ouakarrouch, N. Laaroussi and M. Garoum, Thermal characterization of a new bio-composite building material based on plaster and chicken feather waste,J. Renewable Energy and Environmental Sustainability 5 (2020) 2. https://doi.org/10.1051/rees/2019011

[2] A.KHABBAZI and J. KETTAR, Thermal and Mechanical Behavior of the Plaster Reinforced by Fiber Alpha or Granular Cork, J. Civil Engineering and Technology, (2017)1026-1040.

[3] F Hernández-Olivares., Ignacio Oteiza and Luis de Villanueva,Experimental analysis of the toughness and increase in modulus of rupture of gypsum hemihydrate reinforced with short sisal fibers,J.Composite structures 22.3(1992)123-137.https://doi.org/10.1016/0263-8223(92)90001-S

[4] B.Mazhoud, F.Collet, S.Pretot, J.Chamoin, Hygric and thermal properties of hemp-lime plasters,J. Build Environ, 96(2016)206–16. https://doi.org/10.1016/j.buildenv.2015.11.013.

Materials Research Forum LLC
https://doi.org/10.21741/9781644903117-6

[5] B.Mazhoud, F.Collet ,S. Pretot , C.Lanos,Development and hygric and thermal characterization of hemp-clay composite, J. Environ Civ Eng (2018)8189:0. https://doi.org/10.1080/19648189.2017.1327894

[6] S Amziane, L Arnaud. Bio-aggregate-based Building Materials. 2013.

[7] N Benmansour, B Agoudjil, Gherabli A, Kareche A, Boudenne A, Thermal and mechanical performance of natural mortar reinforced with date palm fibers for use as insulating materials in building,J. Energy Build 81(2014)98–104. https://doi.org/10.1016/j.enbuild.2014.05.032

[8] A Kriker, G Debicki, A Bali, Khenfer MM, Chabannet M, Mechanical properties of date palm fibres and concrete reinforced with date palm fibres in hot-dry climate,J. Cem Concr Compos 27(2005)554–564. https://doi.org/10.1016/j.cemconcomp.2004.09.015

[9] W. GALLALA, H.M.M. KHATER, production of low-cost biocomposite made of palm fibers waste and gypsum plaster, J.Int. Contam. Ambie36, 2(2020)475-483. https://doi.org/10.20937/rica.53541

[10] I. Boulaoued, I.Amara,A.Mhimid, Experimental Determination of Thermal Conductivity and Diffusivity of New Building Insulating Materials, J. heat and technology34,2(2016) 325-331. 10.18280/ijht.340224

[11] A.M. ALmusawi, T.S. Hussein, M.A. Shallal, Effect of Temperature and Sisal Fiber Content on the Properties of Plaster of Paris,J.Engineering & Technology7(2018)205-208.

[12] F.Naiiri, L.Allegue, M.Salem, R.Zitoune, M.Zidi, The effect of doum palm fibers on the mechanical and thermal properties of gypsum mortar, J .Composite Materials (2019)1–19. https://doi.org/10.1177/0021998319838319

[13] A. Djoudi, M. M. Khenfer, A. Bali, T. Bouziani, Effect of the addition of date palm fibers on thermal properties of plaster concrete: experimental study and modeling, J. Adhesion Science and Technology28,20(2014)2100-2111. https://doi.org/10.1080/01694243.2014.948363

[14] F. Naiiri, A.Lamis, S.Mehdi, The effect of doum palm fibers on the mechanical and thermal properties of gypsum mortar,J. Composite Materials 53.19 (2019)2641-2659. https://doi.org/10.1177/0021998319838319

[15] S. Bousshine, M. Ouakarrouch, A. Bybi, N. Laaroussi, M. Garoum, A. Tilioua, Acoustical and thermal characterization of sustainable materials derived from vegetable, agricultural, and animal fibers,J. Applied Acoustics 187 (2022) 108520. https://doi.org/10.1016/j.apacoust.2021.108520

[16] M. Rachedi, A. Kriker, Optimal composition of plaster mortar reinforced with palm fibers, J. Civil Eng Environ Sci 4(2018) 44-49. https://doi.org/10.17352/2455-488X.000027

[17] A. Achour, F. Ghomari, N. Belayachi, Properties of cementitious mortars reinforced with natural fibers, J.Adhesion Science and Technology 10 (2017)1–25. https://doi.org/10.1080/01694243.2017.1290572

[18] Y. Jannot, A. Degiovanni, Thermal properties measurement of materials, ISTE edn. (John Wiley & Sons, New Jersey, 2018)

Mediterranean Architectural Heritage - RIPAM10
Materials Research Proceedings 40 (2024) 64-78

Materials Research Forum LLC
https://doi.org/10.21741/9781644903117-7

Assessment of the Mechanical and Thermal Properties of Local Building Materials Stabilised with Gum Arabic in the Drâa-Tafilalet Region, South-East Morocco

Charaf Eddine EL MANSOURI[1,a,*], Amine TILIOUA[1,b], Fouad MOUSSAOUI[1,c]

[1]Research team in thermodynamics and applied thermodynamics (2.T.A.), Laboratory of Mechanics, Energy Efficiency and Renewable Energies (L.M.3. E.R.), Department of Physics, Faculty of Sciences and Techniques Errachidia, Moulay Ismaïl University of Meknès, B.P. 509, Boutalamine, Errachidia, Morocco

[a]charafeddineelmansoury@gmail.com, [b]tiliouamine@yahoo.fr, [c]bakkilimoussaoui@gmail.com

Keywords: Building Materials, Clay, Gum Arabic, Stabilization, Mechanical Characteristics, Thermal Conductivity, Durability

Abstract. Morocco enjoys a very remarkable earthen architectural heritage throughout the southeast of the country, earthen constructions which are characterized by its ability to absorb and reject moisture from the indoor air according to the fluctuations of the microclimate of the building guarantees a passive indoor comfort that would save energy. Unfortunately, earthen structures suffer from rapid degradation due to climatic changes (temperature, air humidity, water...). This study concerns mechanical, thermal characterization and durability of compressed earth blocks manufactured (CEB) with clay, gum arabic with different proportions. For this purpose, the mass percentages of 1%, 2%, 3%, 4% and 5% of gum arabic by contribution to the total mass are retained for this research work. cylindrical bricks of CEB are manufactured to carry out mechanical tests, and those of prismatic form are adapted for the determination of thermal conductivities with the method "house has high insulation". The use of gum arabic as a binder in construction has given satisfactory results. At a rate of 5% of gum arabic the bricks are associated with a compaction stress of 5.78 MPA for the compressive strength, allow us to obtain CEB with an acceptable mechanical strength and a better resistance to rainwater. In addition, the values of thermal conductivity measured, show that when the rate of gum arabic increases, the thermal conductivity rises. The thermal conductivities of all formulations vary between 0.72 and 1.05 W/(m.K). The durability test carried out on the stabilized and non-stabilized bricks, shows that the specimens not stabilized by gum arabic are totally degraded from 5 min of immersion, On the other hand those stabilized by gum arabic kept their shape more than 5 hours. This study proved the effectiveness of CEB stabilized by gum arabic for use as new sustainable construction materials in the region of Drâa-Tafilalet (southeast of Morocco).

1. Introduction

The demand for natural resources has increased globally due to spatial development and population growth, resulting in ecological imbalance and global warming. The construction sector alone consumes 40% of the planet's material and energy flows, making it one of the least sustainable sectors worldwide [1]. In Morocco, the construction sector is the largest energy consumer, accounting for over 33.6% of total energy consumption, followed by the transport sector at 38% [2]. This sector's impact on the environment is significant, necessitating the urgent need for sustainable housing solutions that meet both environmental and structural stability criteria.

Studies have shown that sustainable design, material production, construction, maintenance, and the reuse of construction materials can reduce the carbon footprint of buildings by an average of 25% [3]. One promising solution is the use of earthen soil, a natural and historical material, which can significantly reduce energy consumption and CO_2 emissions. Earthen materials offer excellent energy performance and are a cost-effective alternative to materials such as cement, lime, bitumen, and steel, which have high energy and environmental costs. It is estimated that one-third of the world's population resides in earth-constructed structures [5].

In Morocco, particularly in the Draa Tafilalet region, earthen constructions are still prevalent, with many heritage sites, such as Ksar d'Aït-Ben-Haddou, listed as UNESCO World Heritage sites [4]. Earthen construction materials align perfectly with environmental standards in arid climates with hot summers and cold winters. They possess numerous advantages, including excellent insulation properties that reduce the need for heating and cooling energy[5,6]. Additionally, earthen materials have the ability to store moisture and heat, contributing to a comfortable indoor environment. While the thermal conductivity of earth ranges from 0.3 to 1 W/(m.K) [7,8], which is not as low as conventional insulators, it still outperforms many conventional construction materials, making it advantageous in such environments.The low cost, availability, recyclability, and low embodied energy of these materials make construction particularly sustainable; the lifespan of earthen constructions is over 100 years. Therefore, earthen soil has been proven to be ecologically sustainable[9]. According to Minke[10], buildings constructed with earth can reduce global pollution by approximately 30% in 2009. Earth construction materials offer the potential to significantly reduce energy consumption during the production phase, with potential savings ranging from 80% to 90%. This efficiency tends to increase as production scale increases [8]. However, it is important to acknowledge that earth materials also have certain drawbacks, including low mechanical strength and high sensitivity to water. Humanity has always sought solutions to the shortcomings of earth material using various stabilization methods, including mechanical, chemical, and physical methods, which have led to the improvement and invention of different earth products. In our modern society, it seems important to ensure the safety of residents by providing minimum wet strengths. Thus, scientific studies have been conducted on the stabilization of raw earth. Among the different products of raw earth construction materials, rammed earth, or compressed earth blocks (CEB), is a recent version of adobe, which has the advantages of low shrinkage, high strength, low water sensitivity, and well-defined shape with straight edges. To improve its characteristics, various stabilizers are used, such as cement, lime, bitumen, natural or synthetic fibers, and biopolymer materials. The use of mineral binders in large proportions can question the ecological nature of the material [11]. In parallel, some traditional practices, especially in developing countries, and recent scientific studies have shown that the use of natural organic binders, namely biopolymers, could be a more environmentally virtuous alternative [12]. These organic products have great diversity and thus significant research potential, reflecting the variety of practices worldwide [13]. The accessibility and the effects of stabilization and reinforcement of biomaterials by environmentally friendly additives, such as hydrocolloid gums, also known as natural polysaccharides, have attracted considerable attention from researchers [14]. The use of biopolymer stabilizers can significantly improve the technical properties of soils and enhance their resistance to environmental soil conditions. Experimental tests have shown that soil strength tends to increase with an increase in biopolymer concentration [15], [16]. Research findings indicate that the incorporation of biopolymers in civil engineering projects can enhance the engineering properties of construction materials [17]. Specifically, the biopolymer b-1,3/1,6-glucan has been shown to effectively aggregate soil particles, leading to improved soil compression. Additionally, this biopolymer has demonstrated positive effects on compactness, Atterberg limits, and soil swelling index when applied to treated soil. However, it should be noted that the same biopolymer may have a negative impact on soil consolidation [18,19]. It can be observed that even with small amounts of biopolymer, the erosion reduction effect and improvement of inter-particle soil cohesion remain significant. The effect of xanthan gum on soil properties has been studied by various researchers, and it has been found that xanthan gum can significantly increase the compression and shear strength of soils, especially soils containing a significant amount of fine-grained aggregates [20,21,22]. Studies have shown that Persian gum is a biopolymer used to improve soils, and the results obtained are acceptable in terms of strength compared to xanthan and guar gums due to its property of reducing permeability and consequently high thermal stability for the treated soil. Our study aims to propose a low environmental impact method of stabilizing raw earth using local biopolymer materials through two approaches. First, we aim to limit the amount of mineral binders, namely cement and lime, and verify that the resulting solution meets sustainability and mechanical and thermal performance criteria suitable for the climatic and architectural conditions of the Draa Tafilalet region in Southeastern Morocco (fig. 1). Second, we aim to identify potential biopolymer binders that can effectively stabilize raw earth for construction. To do this, we will use a type of soil collected in Errachidia, from which we will determine its physico-chemical, mineralogical, and geotechnical characteristics to assess the impact of biopolymer binder (Arabic gum) stabilization on material efficiency. Therefore, starting from this article, we aim to propose a new local biobased material based on raw earth for contemporary construction.

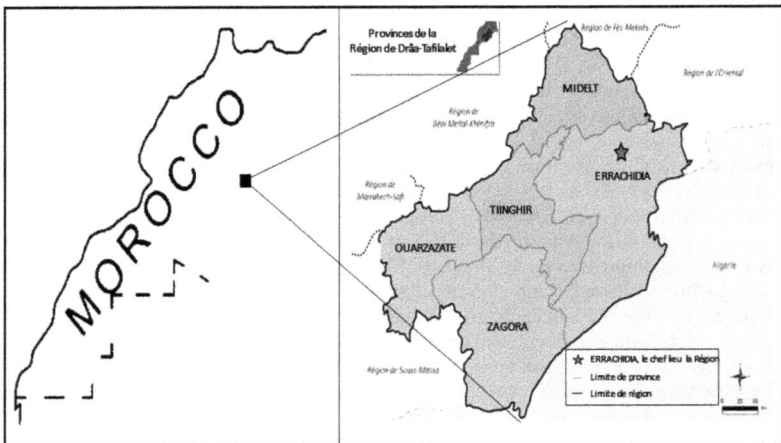

Fig.1: South-eastern Morocco (Errachidia Province)

2. Matrials and Methods

2.1 Materials

2.1.1 Soil

The soil used in our study comes specifically from the Ksar Ait Ben Omar Tinejdad region, located in Errachidia. This choice was made because of the availability and abundance of this soil in the region (fig.2). In selecting this soil, we took into account its composition and characteristics, which are essential for our research. In addition, the Ksar Ait Ben Omar Tinejdad region has particular geographical and climatic conditions that make it suitable for our study.

Mediterranean Architectural Heritage - RIPAM10   Materials Research Forum LLC
Materials Research Proceedings 40 (2024) 64-78   https://doi.org/10.21741/9781644903117-7

Fig.2 : Studied soil

2.1.2 Gum Arabic

Gum arabic, also known as "Gomme Sénégal", is a plant biomass obtained from the exudate of sap, solidified naturally or by incision, from the trunks and roots of trees in the acacia family. It is harvested mainly in Saharan Africa (Maghreb, Mali, Senegal, Chad, Egypt, Sudan, etc.). Gum arabic is commercially available in powder or crystal form. It is pale yellow to brownish yellow in color, odorless, soluble in water and insoluble in alcohol, but soluble in glycerol and propylene glycol with prolonged heating. The visible part of the crystals is "matt", with fine cracks that are difficult to see with the naked eye [26].

The gum used in our study was collected on the road between Risani and Tazarine (Morocco). This region is semi-Saharan (fig.3).

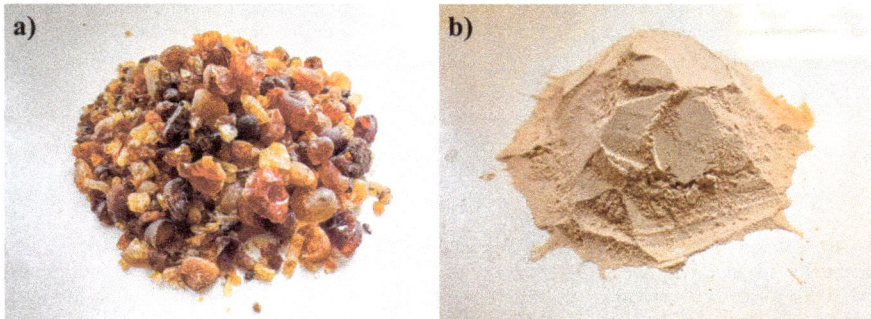

Fig.3: Gum arabic: a) form crystals, b)form powder

Materials Research Forum LLC
https://doi.org/10.21741/9781644903117-7

2.2 Methods

2.2.1 Exploring Soil Properties: Particle Size Analysis and Methylene Blue Values Examination

The particle size analysis of the soil samples in our study was conducted in accordance with established protocols outlined in French standards. The first standard utilized was NFP94056 (NF P94-056, 1996) [29], which is primarily concerned with particle size analysis through sieving for soil particles larger than 80 μm. To perform this analysis, the soil samples were initially soaked in water for 24 hours to ensure they reached a wet state. Subsequently, the samples underwent sieving, and the retained soil particles were then subjected to drying in an oven at 105°C until a constant mass was attained. The weight of the residue from each sieve was subsequently measured.

The second standard employed was NFP94057 (NF P94-057, 1992) [30], which complements the sieve-based particle size analysis by utilizing sedimentation for particles passing through an 80 μm sieve. This method entailed filling a test tube with distilled water to a precise volume of two liters. The suspended particles in the water settled at the bottom of the test tube based on their diameters. By utilizing a hydrometer, density measurements were regularly taken over time and at specific heights. These measurements allowed for the calculation of the proportions of particles of each diameter, which were recorded in a table and used to construct the grain size distribution curve. In addition to the particle size analysis, we also employed the methylene blue test to assess the clay fraction within the soil samples. The methylene blue test is commonly used in geotechnical engineering to determine the quantity and quality of clay present. Clay particles exhibit a strong affinity for methylene blue molecules, enabling the measurement of their adsorption capacity and, consequently, their water retention properties. The methylene blue values were determined following the guidelines outlined in the French standard NFP94068 (NF P94-068, 1998)[31].

2.2.2 Atterberg limits

Atterberg limits are employed to predict the behavior of materials, particularly soils that have been sieved to a size of 400 μm. This test involves varying the water content of the soil sample to assess its consistency. The procedure for this test adhered to the Moroccan standard NM 13.1.007 (NM 13.1.007, 1998) [32]. The test consisted of two stages. In the first stage, we identified the water content at which a groove formed in the soil sample placed in a bucket with specific characteristics closes when the bucket and its contents undergo repeated shocks. This closure indicates the plastic limit of the soil. In the second stage, we determined the water content at which the soil sample, when manually rolled, develops cracks. This particular water content corresponds to the liquid limit of the soil.

2.2.3 Specimen preparation

For the preparation of the test specimens, the mixture is prepared as follows: We collected the particles that passed through a 5 mm sieve for the clay and those that passed through a 2 mm sieve for the gum arabic. The mixture is prepared in a dry state, combining the clay and gum arabic, and then water is added. For the gum arabic, mass percentages of 1%, 2%, 3%, 4%, and 5% were chosen to prepare the mixture.

The mixture (clay + gum arabic) is prepared in a dry state. Subsequently, the entire mixture is manually kneaded with water. The mass of the dry mixture for making CEB test specimens is 13.50 kg for each dosage. The moisture content is monitored during the preparation of the test specimens. Once the mixture is well homogenized, it is placed in a plastic bag for about 4 hours for proper saturation. The press used for the production of CEBs is a manual press. The prepared material mixture is poured into the mold to obtain regular CEBs. After demolding, the bricks are dried in the laboratory of FST Errachidia to prevent rapid drying (see figure 7). In total, 18 bricks were manufactured (3 test specimens for each dosage).

Fig.7: Samples for testing a) mechanical b) thermal

Cylindrical test specimens (CEB) measuring 11 cm in diameter and 22 cm in height were used for mechanical tests (compression and tension), and test specimens measuring 24 x 24 x 4 cm were used for thermal conductivity measurements.

2.2.4 Thermal conductivity

We conducted an experimental study to evaluate the thermal conductivity of various bricks using the highly insulated house method (PHYWE, 2012) [33], a technique previously employed by researchers like Ben Zaid et al. (2020) [34] and Medina et al. (2017) [35]. The core concept of measuring thermal conductivity involves placing the sample to be assessed within a square opening measuring 210×210 mm2. The highly insulated house consists of four identical side walls enclosed by an insulated cover that includes a 5 cm thick polystyrene plate. Inside this setup, a small black box houses a 100 W bulb as a heat source, with precautions taken to minimize the impact of radiation (refer to Figure 8). The bulb is connected to a controller programmed to maintain indoor air temperature at 50 °C, while an air conditioner is employed to regulate external temperature at approximately 19 °C. We used K-type thermocouples to measure the temperature of both the inner and outer surfaces of all test specimens, with readings recorded at 1-minute intervals. Data acquisition was facilitated through a GL840 DATA-LOGGER equipped with a PC interface, and the accompanying data reading software allowed for data display and recording in Excel.

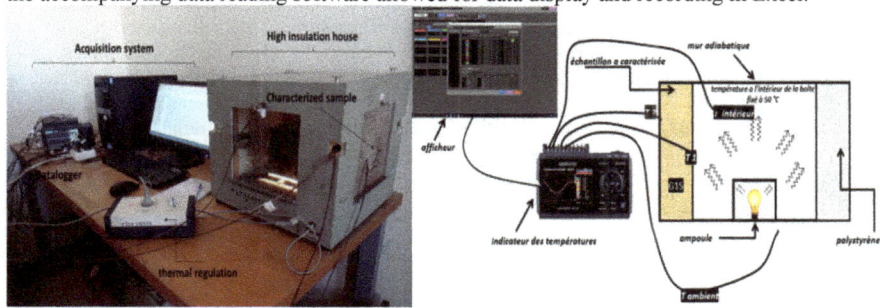

Fig.8: Thermal house connected to an acquisition system

The determination of thermal conductivity follows the procedure outlined below:

The convective heat transfer between the interior air and the surface of the inner wall of the sample is expressed by the equation:

$$\varphi = h_{int} * S * (T_{interior} - T_1) \qquad Eq.1$$

Here, h_{int} represents the convection coefficient of the interior air, S denotes the surface area of the sample wall, and $T_{interior}$ and T1 respectively represent the temperature inside the highly insulated house and the surface temperature of the interior sample wall.

The equation for φ is further defined as:

$$\varphi = \lambda * s * \frac{T_1 - T_2}{e} \qquad Eq.2$$

In this equation, λ represents the thermal conductivity, T_2 signifies the surface temperature of the outer wall of the sample, and (e) represents the thickness of the sample.

The calculation of thermal conductivity is based on the steady-state conservation of heat flux:

$$\lambda = h_{int} * e * \frac{T_{interior} - T_1}{T_1 - T_2} \qquad Eq.3$$

In the case of natural air movement in enclosed spaces, the reference value for h_int is 8.1 W/K m2 (PHYWE, 2012).

2.2.5 Simple compression tests

The uniaxial compression strength test of the prepared samples was carried out in accordance with the Moroccan standard for earth construction (Decree No. 2-12-666 of 17 Rejeb 1434, Decree on Seismic Regulations for Earth Construction and the Establishment of the National Committee for Earth Construction, 2013) [23]. The testing machine used was the FORM TEST SEIDNER D79400 from Germany, capable of exerting a maximum breaking load of 3000 kN. At least three samples for each soil and gum arabic mixture under the same conditions were tested, and an average value was recorded. The cylindrical specimen, measuring 100 mm in diameter and 200 mm in height, was dried in the laboratory at the Faculty of Sciences and Technology in Errachidia. It was then compressed between two flat plates (see Fig.9). The samples intended for the compression test were at least 28 days old at the time of testing and had smooth and regular end surfaces to ensure that the compression force was applied uniformly over the specimen's entire surface. Any sample with an irregular surface was coated with a thin layer of dental plaster, as specified by Walker et al. [25].

The measurement of compressive strength is given by the following formula:

$$\sigma c = \frac{F_r}{S} \qquad Eq.4$$

With: Fr (N): breaking force; S (mm^2): cross-sectional area of specimen; σ_c (MPa): compressive strength.

Mediterranean Architectural Heritage - RIPAM10   Materials Research Forum LLC
Materials Research Proceedings 40 (2024) 64-78   https://doi.org/10.21741/9781644903117-7

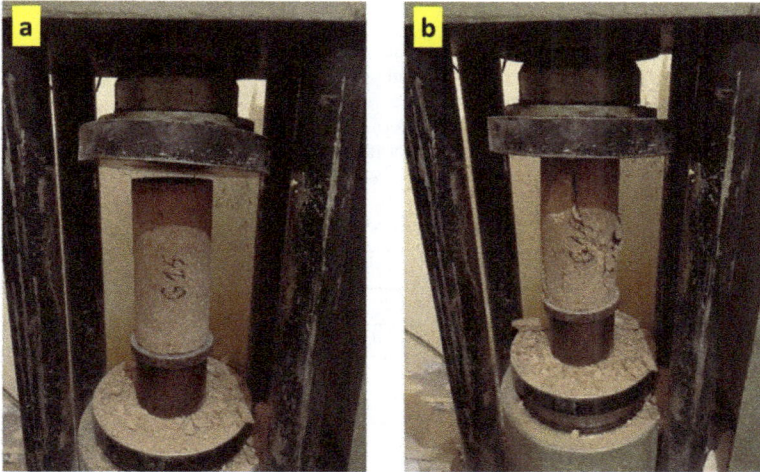

Fig.9: Mechanical test: a) before press, b) after press

2.2.6 Water erosion test

The Errachidia region receives 127 mm of rainfall during the rainy season [26]. Durability with respect to water is also a key aspect of construction materials quality.

The erosion test involves subjecting a sample of gum arabic-stabilized compressed earth block to immersion in a water basin for 24 hours. During this period, it was observed that the specimens not stabilized with gum arabic deteriorated completely (see Figure 11) after just 15 minutes of immersion. In contrast, those stabilized with gum arabic maintained their integrity in water for over 5 hours. For the specimens stabilized with gum arabic, the higher the gum arabic content, the greater the water resistance, and the better it retained its mass.

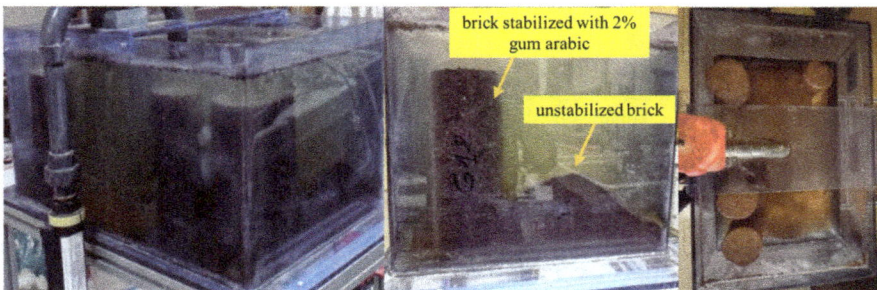

Fig.11: Cylindrical samples of clay-gum arabic mix immersed in water

In fact, in construction, the bricks only receive rainfall on one side. All other other parts are protected by the other bricks. We can say that stabilizing building materials building materials with gum arabic not only improves mechanical strength but also contributes to good resistance to rainwater.

3. Results and Discussion

3.1 Soil particle size and methylene blue value

Table 4 shows the grain sizes of our material, ranging from 20 mm to 0.002 mm, with an estimated 84% passing through the 20 μm to 2 mm sieve. Soil comprises 70% sand, 9% silt and 6% clay. The material in our study is made up of grains of various sizes, a large percentage of which are extremely fine. It is therefore necessary to perform Atterberg limit tests on the clay soil in order to classify it.

Table.4: Grain sizes of soil studied

Constituents	Diametres	Percentage%
Pebbles (%)	20 à 200 mm	4
Gravel (%)	2 à 20 mm	11
Coarse sand (%)	20 μm à 2 mm	27
Fine sand (%)	20 μm à 200 μm	43
Silt (%)	2 μm à 20 μm	9
Clay (%)	< À 2 μm	6
Methylene blue value		0,67

The methylene blue value of our soil is: 0.67

3.2 Atterberg limits

Atterberg limits and calcification of studied soils as a function of plasticity:
- Liquid limit W_l (%): 28
- Plasticity limit W_p (%): 21
- Plasticity index I_p (%): 7
- Soil plasticity: Low
- Swelling potential: Low
- Technical recommendations: CEB and Rammed earth

3.3 Drying kinetics of compressed earth blocks

We observed that most test specimens remain stable in their drying process from day 14 onwards. However, the test tube made with 5% gum arabic retained a higher moisture content until day 20. This observation can be explained by the fact that gum arabic retains water thanks to the sugars it contains.

In the Errachidia region, we benefit from sufficient sunshine to dry building materials. All we need to do is choose the right production period, preferably during the summer. This period offers favorable climatic conditions, with high temperatures and low humidity, which speeds up the drying process. By exploiting this natural resource, we can optimize specimen drying and guarantee reliable, consistent results for our studies.

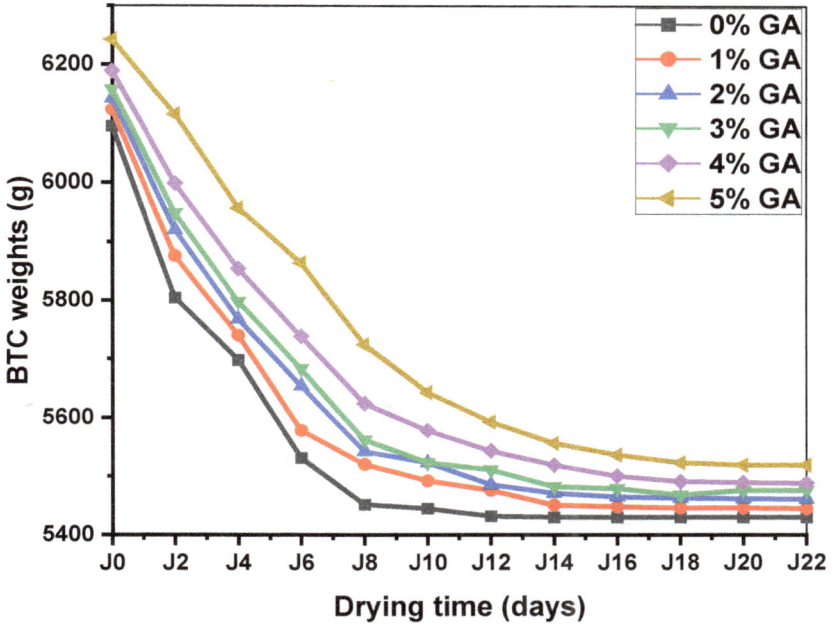

Fig.14: brick drying as a function of time

3.4 Thermal conductivity

Fig.12 illustrates the variation in thermal conductivity of the samples studied as a function of arabic gum content. Thermal conductivity values in all the formulations studied ranged from 0.72 to 1.05 $W.m^{-1}.K^{-1}$. The formulation with the highest thermal conductivity, around 1.05 $W.m^{-1}.K^{-1}$, is obtained with a composition of 95% clay and 5% gum arabic (G5).

Materials Research Forum LLC

https://doi.org/10.21741/9781644903117-7

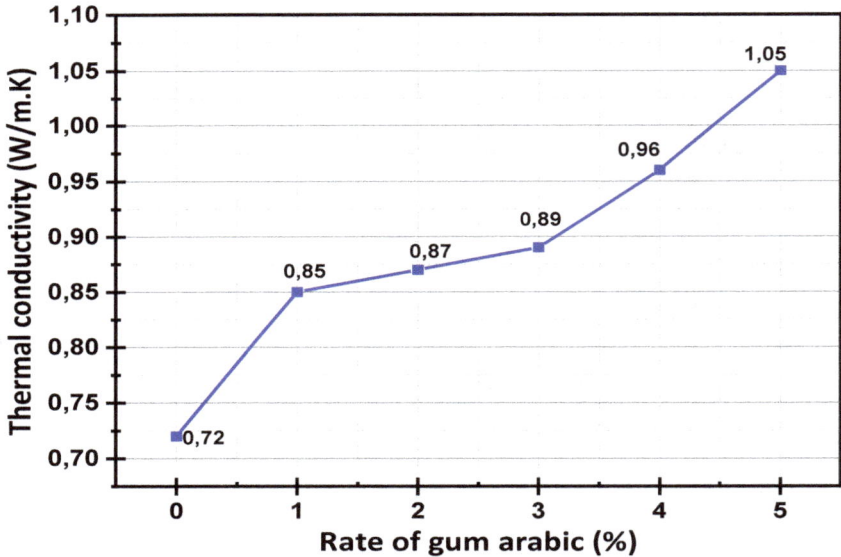

Fig.6: Variation in thermal conductivity of clay bricks as a function of Gum arabic

It was observed that for samples stabilized with gum arabic, an increase in the amount of gum arabic in the mixture is associated with an increase in the thermal conductivity value. This observation can be explained by the fact that a low proportion (1% to 3%) of gum arabic does not completely coat the grains of the materials, creating voids within the samples, leading to a decrease in thermal conductivity. However, as the proportion of gum arabic increases, all the pores are filled, leading to compaction of the material and, consequently, an increase in thermal conductivity.

3.5 Compressive strength

Tinejdad clay contains a high percentage of pores. This porosity adversely affects mechanical properties, particularly tensile strength. From the results obtained in (fig.13), we can see that Tinejdad clay stabilized with gum arabic gives satisfactory values for compressive strength. For example, clay stabilized with 1% gum arabic gives a 298% increase in simple compressive strength compared with unstabilized clay. At 3% gum arabic, we have a 437% increase, and at 5% gum arabic, strength rises by 803%. Our compressive strength results, which vary from 0.64 to 5.78 MPa, are in the same range as those obtained on Ndjamena clay in Chad, which range from 1.01 to 3.25 MPa [27], and on laterites from Burkina Faso, which range from 1.5 to 5 MPa [28].

Materials Research Forum LLC
https://doi.org/10.21741/9781644903117-7

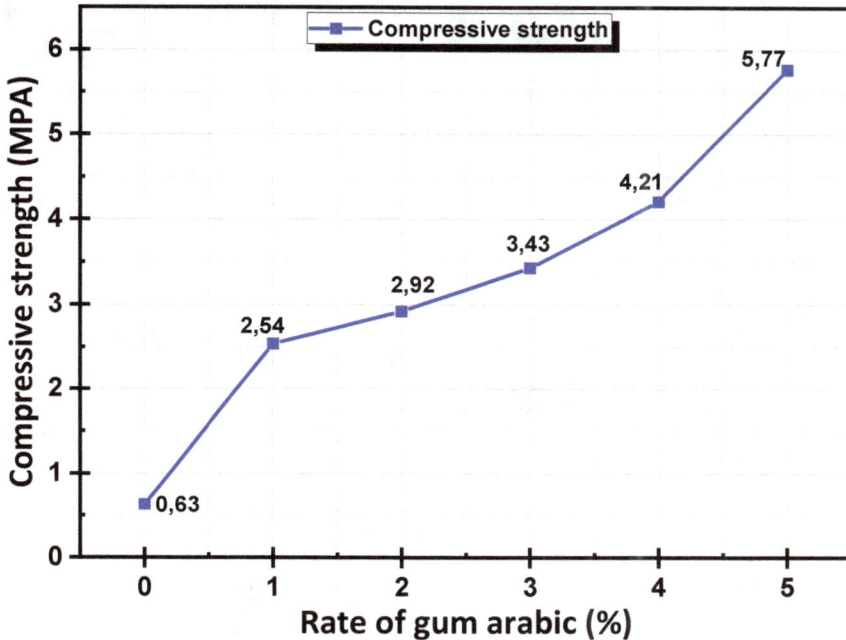

Fig.13: Variation in compressive strength of earth bricks compressed as a function of gum arabic content

The results obtained are very interesting, with a remarkable improvement in the mechanical strength of the materials. Maximum compressive strength exceeds 6 MPa.

The various characterizations made it possible to classify our clay as "Low plasticity clay". The results obtained in this work from mechanical tests are sufficient to justify its use in construction. Stabilization of the clay with gum arabic gives satisfactory results in compression. On the other hand, the sugar contained in gum arabic is responsible for the delayed setting and hardening of the composite. In fact, gum arabic can be used at a rate of 5%, where hardening is felt from the 20th day of age, or a drying method with low environmental impact can be used.

3.6 Water durability test

For gum arabic-stabilized specimens, the higher the gum arabic content, the greater the resistance to water and the greater the ability to retain its mass. In reality, in construction, bricks only receive rainfall on one side. All other parts are protected by the other bricks. We can say that stabilizing building materials with gum arabic not only improves mechanical strength but also contributes to good resistance to rainwater (fig.8).

- Bricks stabilized with biopolymers (gum arabic) can be considered suitable for use in exterior walls.

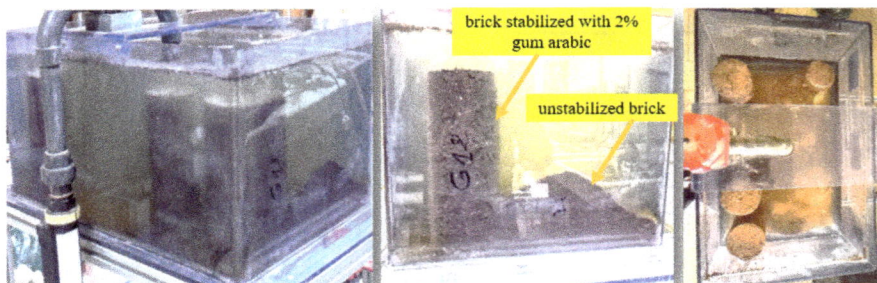

Fig.8: Cylindrical samples immersed in water

Conclusion

This research work has shed light on the significance of local materials in the field of construction. The thermal conductivity values obtained through the highly insulated construction method revealed that the various formulations examined in this study do not result in materials with high thermal insulation capacity. However, these materials possess a particularly valuable characteristic, namely, a high thermal inertia capacity. This property, while not making them top-tier thermal insulators, offers significant potential to enhance indoor comfort in construction environments.

Regarding mechanical strength, the results obtained can be described as modest, but they remain sufficient for common construction applications. It is worth noting that gum arabic-stabilized raw earth bricks are not suitable for use as building foundations. However, it is important to emphasize that these materials hold promising potential in sustainable construction. They could be advantageously employed for above-ground walls and partition construction, where their thermal inertia capacity and modest mechanical strength can bring significant advantages, both in terms of energy efficiency and building durability. This research thus paves the way for new opportunities in environmentally friendly construction.

References

[1] D. M. Roodman, N. Lenssen, A building revolution: How ecology and health concerns are transforming construction, Environment International, vol. 21, no. 3, p. 349, Jan. 1995, doi:10.1016/0160-4120(95)90083-7

[2] I. Merini, A. Molina-García, M. S. García-Cascales, M. Mahdaoui, M. Ahachad, Analysis and Comparison of Energy Efficiency Code Requirements for Buildings: A Morocco–Spain Case Study, Energies, vol. 13, no. 22, p. 5979, Nov. 2020. https://doi.org/10.3390/en13225979

[3] J. Kneifel, Life-cycle carbon and cost analysis of energy efficiency measures in new commercial buildings, Energy and Buildings, vol. 42, no. 3, pp. 333–340, Mar. 2010 . https://doi.org/10.1016/j.enbuild.2009.09.011

[4] U. C. du patrimoine UNESCO, unesco, UNESCO Centre du patrimoine mondial. Accessed: Jul. 18, 2022. [Online]. Available: https://whc.unesco.org/fr/list/444/

[5] A. Fabbri, N. Al Haffar, F. McGregor, Measurement of the relative air permeability of compacted earth in the hygroscopic regime of saturation, Comptes Rendus Mécanique, vol. 347, no. 12, pp. 912–919, Dec. 2019 . https://doi.org/10.1016/j.crme.2019.11.017

[6] T. Morton, Great Britain, Department of Trade and Industry, and Communities Scotland (Agency), Low cost earth brick construction: 2 Kirk Park, Dalguise : monitoring & evaluation : a Partners in innovation research project final report. Edinburgh: Communities Scotland?, 2005.

[7] G. A. Abanto, M. Karkri, G. Lefebvre, M. Horn, J. L. Solis, and M. M. Gómez, Thermal properties of adobe employed in Peruvian rural areas: Experimental results and numerical simulation of a traditional bio-composite material, Case Studies in Construction Materials, vol. 6, pp. 177–191, Jun. 2017 . https://doi.org/10.1016/j.cscm.2017.02.001

[8] M. Hall and D. Allinson, Analysis of the hygrothermal functional properties of stabilised rammed earth materials, Building and Environment, p. 8, 2009.

[9] B. Khadka and M. Shakya, Comparative compressive strength of stabilized and un-stabilized rammed earth, Mater Struct, vol. 49, no. 9, pp. 3945–3955, Sep. 2016 . https://doi.org/10.1617/s11527-015-0765-5

[10] G. Minke, Building with Earth: Design and Technology of a Sustainable Architecture, 2009, [Online]. Available: https://doi.org/10.1515/9783034612623

[11] H. Van Damme and H. Houben, Earth concrete. Stabilization revisited, Cement and Concrete Research, vol. 114, pp. 90–102, Dec. 2018 . https://doi.org/10.1016/j.cemconres.2017.02.035

[12] A. Vissac, A. Bourgès, D. Gandreau, R. Anger, L. Fontaine, argiles & biopolymères - les stabilisants naturels pour la construction en terre, p. 82.

[13] R. Anger and L. Fontaine, Interactions argiles/biopolymères : Patrimoine architectural en terre et stabilisants naturels d'origine animale et végétale, CRAterre-ENSAG/AE&CC/LRMH, Sep. 2013.

[14] S. Ahmad, M. Ahmad, K. Manzoor, R. Purwar, S. Ikram, A review on latest innovations in natural gums based hydrogels: Preparations & applications, International Journal of Biological Macromolecules, vol. 136, pp. 870–890, Sep. 2019 . https://doi.org/10.1016/j.ijbiomac.2019.06.113

[15] C. Chen, L. Wu, M. Perdjon, X. Huang, Y. Peng, The drying effect on xanthan gum biopolymer treated sandy soil shear strength, Construction and Building Materials, vol. 197, pp. 271–279, Feb. 2019 . https://doi.org/10.1016/j.conbuildmat.2018.11.120

[16] A. Soldo, M. Miletić, M. L. Auad, Biopolymers as a sustainable solution for the enhancement of soil mechanical properties, Sci Rep, vol. 10, no. 1, p. 267, Dec. 2020 . https://doi.org/10.1038/s41598-019-57135-x

[17] S. Muguda, Mechanical properties of biopolymer-stabilised soil-based construction materials, Géotechnique Letters, vol. 7, no. 4, pp. 309–314, Dec. 2017 . https://doi.org/10.1680/jgele.17.00081.

[18] C. Ilhan, G.C. Cho, Geotechnical behavior of a beta-1,3/1,6-glucan biopolymer-treated residual soil, Geomechanics and Engineering, vol. 7, no. 6, pp. 633–647, Dec. 2014 . https://doi.org/10.12989/GAE.2014.7.6.633

[19] W. J. Orts, R. E. Sojka, G. M. Glenn, Biopolymer additives to reduce erosion-induced soil losses during irrigation, Industrial Crops and Products, vol. 11, no. 1, pp. 19–29, Jan. 2000 . https://doi.org/10.1016/S0926-6690(99)00030-8

[20] I. Chang, J. Im, A. K. Prasidhi, G.C. Cho, Effects of Xanthan gum biopolymer on soil strengthening, Construction and Building Materials, vol. 74, pp. 65–72, Jan. 2015 . https://doi.org/10.1016/j.conbuildmat.2014.10.026

[21] S. K. Das, M. Mahamaya, I. Panda, K. Swain, Stabilization of Pond Ash using Biopolymer, Procedia Earth and Planetary Science, vol. 11, pp. 254–259, 2015 . https://doi.org/10.1016/j.proeps.2015.06.033

[22] H. Ghasemzadeh, F. Modiri, Application of novel Persian gum hydrocolloid in soil stabilization, Carbohydrate Polymers, vol. 246, p. 116639, Oct. 2020 . https://doi.org/10.1016/j.carbpol.2020.116639

[23] Décret n° 2-12-666 du 17 rejeb 1434, Décret le règlement parasismique pour les constructions en terre et instituant le Comité national des constructions en terre., 2013. . Maroc.

[24] E. Gartner, Industrially interesting approaches to "low-CO2" cements, Cem. Concr. Res. 34,1489–1498, 2004.

[25] D. Ciancio, P. Jaquin, P. Walker, Advances on the assessment of soil suitability for rammed earth. Construction and Building Materials. (2013). 42, 40–47.

[26] climate-data, 2021. [En ligne]. Available: https://fr.climatedata.org/afrique/maroc/errachidia/er-rachidia-37411/. [Accès le 17 09 2021]

[27] A. Ali, R. Benelmir, J.L. Tanguier, A. Saleh , Mechanical properties of gum arabic-stabilized Ndjamena clay, Afrique SCIENCE 13(5) (2017) 330 - 341, 2017.

[28] A. LAWANE, R. VINAI, A. PANTET, J. THOMASSIN, Characterization of laterite stone as building material in Burkina Faso, ournee Scientifique 2IE, 2011.

[29] NF P94-056, 1996. Sols : reconnaissance et essais - Analyse granulométrique - Méthode par tamisage à sec après lavage.

[30] NF P94-057, 1992. Sols : reconnaissance et essais - Analyse granulométrique des sols - Méthode par sédimentation.

[31] NF P94-068, 1998. Sols : reconnaissance et essais -Mesure de la capacité d'adsorption de bleu de méthylène d'un sol ou d'un matériau rocheux-Détermination de la valeur de bleu de méthylène d'un sol ou d'un matériau rocheux par l'essai à la tâche.

[32] NM 13.1.007, 1998. Essai d'identification - Détermination des limites d'Atterberg - limite de plasticité au rouleau - Limite de liquidité à la coupelle.

[33] PHYWE, 2012. P 3.6.03-00, Insulation House, Lab. Exp, PHYWE Series of Publications.

[34] Z. Ben Zaid, A. Tilioua, I. Lamaamar, O. Ansari, H. Souli, M.A. Hamdi Alaoui, An experimental study of the efficacy of integrating a phase change material into a clay-straw wall in the Drâa-Tafilalet Region (Errachidia Province), Morocco. Journal of Building Engineering 32, (2020). 101670. https://doi.org/10.1016/j.jobe.2020.101670

[35] N.F. Medina, D.F. Medina, F. Hernández-Olivares, M.A. Navacerrada, Mechanical and thermal properties of concrete incorporating rubber and fibres from tyre recycling, Construction and Building Materials 144, 563–573. (2017). https://doi.org/10.1016/j.conbuildmat.2017.03.196

Mediterranean Architectural Heritage - RIPAM10
Materials Research Proceedings 40 (2024) 79-89

Materials Research Forum LLC
https://doi.org/10.21741/9781644903117-8

Seismic Vulnerability Assessment of a Building Aggregate in the Historical Centre of Florence

Alessia LICO[1,a] *, Chiara MARIOTTI[1,b], Giulia MISSERI[1,c] and Luisa ROVERO[1,d]

[1]Materials and Structures Division, Department of Architecture, University of Florence, Piazza Brunelleschi 6, 50121 Florence, Italy

[a]alessia.lico@unifi.it, [b]chiara.mariotti2907@gmail.com, [c]giulia.misseri@unifi.it, [d]luisa.rovero@unifi.it

Keywords: Aggregate Buildings, Masonry, Internal Court, Vulnerability Index Method, FEM Analysis

Abstract. Safeguarding the built heritage represents an urgent challenge for the culture and identity of each country. In Italy, past seismic events have highlighted the vulnerability of historic urban centres, as aggregates of historic masonry buildings. In this work, the seismic vulnerability of the historic centre of Florence, a UNESCO heritage site since 1982, will be investigated in the context of the Vulnerability Index Method, an empirical approach for the vulnerability assessment at the territorial level, proposed by Benedetti and Petrini in 1984, adopted by the Italian Group of Defense from Earthquake in 1993 and integrated by Formisano in 2011 with the key factors linked to the influence of the aggregate layout in the seismic behaviour. In particular, an urban aggregate composed of fourteen masonry *in-line* buildings (two palaces in the corner and twelve serial intercluded buildings) is considered as a case study. Buildings show a long narrow plan and an *internal court* and have undergone many transformations throughout history. Historical and typological analysis and material and constructive investigations were carried out to aid in understanding the mechanical behaviour of these buildings. These preliminary analyses allowed us to highlight the specific features and vulnerabilities of the aggregate, such as the presence of an *internal court*, which was the object of a specific study carried out supported by non-linear FEM investigations. In particular, this study was aimed at understanding how the GNDT form of the Seismic Vulnerability Level II can describe the vulnerability induced by the internal court in the seismic behaviour of the typical historical buildings in the city centre of Florence. First, the parameters of the GNDT form, influenced by the internal court, have been identified. Some considerations are reported by evaluating the results relating to these parameters, obtained for the application of the case study. Subsequently, some possible proposals for integrating the GNDT form were formulated to include the local vulnerability induced by the internal court in the structural behaviour of the typical historical buildings of the UNESCO city centre of Florence.

Introduction

Italy is characterized by a remarkable seismic risk, also due to the high vulnerability that characterizes historic urban centres. Studying the seismic behaviour of historic buildings is not simple as the behaviour of the individual building is conditioned by the aggregate condition itself and therefore by the interactions between adjacent buildings [1]. Furthermore, complexity is linked to the intrinsic characteristics of historical buildings made up of material and structural elements with non-linear behaviour influenced by a large variety of both geometric and mechanical factors [2; 3; 4].

In the regional seismic classification (GRT resolution 421/2014) [5], the city of Florence is located in zone 3, nevertheless, the territory falls into the high-risk class, as it contains one of the most important assets in the world declared a UNESCO heritage site in 1982.

Mediterranean Architectural Heritage - RIPAM10
Materials Research Proceedings 40 (2024) 79-89

Materials Research Forum LLC
https://doi.org/10.21741/9781644903117-8

The study of seismic vulnerability of historic buildings has been the subject of research since the early 1970s. In particular, various methodologies found in the literature [6; 7; 8] have been categorized into three different typologies based on data quality and nature. These typologies include:

- Analytical Method;
- Empirical Method;
- Hybrid Method.

In this study, the Seismic Vulnerability assessment methodology employed derives from the Seismic Vulnerability Index Method [6; 7; 8], an empirical approach for vulnerability assessment at the territorial level. This method was proposed by Benedetti and Petrini in 1984 [9], adopted and modified by the Italian Group of Defense from Earthquake in 1993 [10] and updated by the Tuscany Region in 2003 [11].

The methodology involves formulating a Seismic Vulnerability Index by specifying the vulnerability class relative to a series of parameters to which a weight is associated. All parameters are grouped in the *GNDT form of the Seismic Vulnerability Level II*, preceded by the GNDT form of the Seismic Vulnerability Level I, to frame the object of investigation in the urban context [10]. A fundamental contribution to the Seismic Vulnerability Index Method [6; 7; 8] is due to Formisano and co-authors [12], who in 2011 introduced five key factors linked to the influence of the aggregate layout in seismic behavior.

In this study, this empirical and expeditious approach is studied in relation to the considerations that emerged from the preliminary analysis conducted on the case study. In particular, the preliminary investigations carried out on the urban historic city centre of Florence, also supported by FEM analysis, have highlighted the seismic vulnerability induced by the *internal court*, generally recurring in the typical construction typology of the city.

The objective of this study was to evaluate the capacity of the *GNDT form* concerning the vulnerability due to the *internal court*. This vulnerability is not covered directly in the GNDT form of the Seismic Vulnerability, but the evaluation of some parameters is influenced by the presence of the *internal court*.

The study carried out focused precisely on how to integrate the *GNDT form* to include the construction specificity of the case study analysed, taking advantage of the fact that the form can be adapted to construction contexts with their characteristics. With this objective, similar methodologies, such as the American normative [13], which considers the *internal court*, have also been considered.

Vulnerability Index Method

The methodology involves defining a Seismic Vulnerability Index by specifying the vulnerability class relative to a series of parameters collected in the *GNDT form* of the Seismic Vulnerability Level II, provided by the National Group for Earthquake Defense [10].

Each parameter is associated with a weight, typically ranging from 0.25 to 1.5, depending on the parameter's influence on seismic vulnerability. For each parameter, a judgment must be expressed through four ascending vulnerability classes (A, B, C, D), each corresponding to a score ranging from 0 to 45. The sum of the products of the different vulnerability classes and the weight of each parameter allows for the derivation of the relative Seismic Vulnerability Index, Iv^*. Normalizing this index within the range of 0 to 100, where 0 represents the absence of vulnerability and 100 represents maximum vulnerability, yields the Seismic Vulnerability Index.

Despite the limitations associated with the intrinsic subjectivity of the assessment, which relies on an expert judgment, this method does not define vulnerability based solely on the building's typology (as the macroseismic method does, which relies on the EMS-98 scale), allowing the vulnerability characteristics of the buildings under consideration to be determined in a specific way.

Furthermore, this methodology enables a large-scale qualitative assessment, facilitating an initial screening of constructed structures for the development of a management and action plan based on a priority scale for buildings considered to be at higher risk.

Figure 1. Urban plan with architectural emergencies and the identification of the aggregate.

Study Case

The aggregate object of the study is located near the Cathedral of Florence (Fig. 1) and is constituted of 14 buildings, two palaces occupy the head positions and 12 *serial intercluded buildings* are placed in between. The buildings were 13 in origin before a merge intervention of two adjacent buildings was carried out.

The aggregate shows an unusual shape, enclosed between two streets: *via dei Servi* and *via del Castellaccio*, which make the aggregate ideal for a line house typology, as a result of the transformations that occurred to the terraced house typology during the 19th Century [1; 14].

The serial buildings show a long narrow plan, accommodating the depth of the parcel, up to 40m. The depth-direction development of housing requires structurally bearing walls for both directions, orthogonal to the front streets as well as for the façades; the thickness of walls ranges around 45-50 cm.

The parcelling is not uniform: excluding the corner palaces, the first 4 buildings, from the left of Figure 2, show a width of 10 m, the 7 further houses show a width of 8,70 m, and the last two southmost parcels show a width of 9,57 m since they show a shorter depth compared to other buildings [15].

The presence of an *internal court* enables the sunlight to reach the internal rooms and is not in the centre of the plan. This position provides larger space for carrying out the main activities and gives increased brightness to the rooms facing the main street, via dei Servi.

Figure 2. View of the roofs. Plan with the identification of different units: 9 belonging to Arte della Lana (blue) and 4 belonging to Arte dei Mercanti (green).

Preliminary investigation

Historical Framework. The aggregate falls into the core zone of the UNESCO area of Florence and was built just outside the walls of 1172-1175.

Mediterranean Architectural Heritage - RIPAM10

Materials Research Proceedings 40 (2024) 79-89

Materials Research Forum LLC

https://doi.org/10.21741/9781644903117-8

Until the beginning of the 16th Century, the aggregate was occupied by an important *Tiratoio* (a building devoted to the production - tightening - of wood cloths), called *Tiratoio dell'Acquila*, who belonged to Noferi di Palla Strozzi, a rich banker of Florence. In 1417, he decided to pass the *Tiratoio* for 2/3 to the *Arte della Lana* and for 1/3 to the *Arte dei Mercanti*, two of the most prestigious and rich *Arti*, sort of professional societies and trade unions together [15].

Following the economic crisis that affected the textile industry, because of the deviation of the *Mugnone* river, the two Arti started a conversion of the area from production purposes to private housing, erecting 13 houses, 9 owned by the *Arte della Lana* and 4 owned of *Arte dei Mercanti* [16] (Fig. 2).

Typological Analysis. Three principal evolutionary phases of the aggregate can be identified: the first one (16th-17th Century) refers to the 16th Century unitary project of the *Arti* buildings; the second phase (18th-19th c.) refers to the first period of expansion, with the construction of elevations aligned and integrated with the masonry below; the third and last phase (from the 20th c.) refers to the smaller and back elevations. Despite the transformations that have taken place over the years, the original approach of the aggregate relating to the first evolutionary phase is still clear.

The typological reference model is the *court-terraced house*, which is single-family house. The study case shows variations of this typology connected to the attempt of increasing the available surface and implemented through saturation operations, i.e. development of the house in depth, with the annexation of further rooms up to the formation of a serial building with a *quintuple body* with double facing [17].

The typology is characterized by the presence of an *internal court* to guarantee adequate hygienic conditions of the rooms that cannot directly benefit from the lighting and ventilation from façades window.

Construction Technique. It was possible to define different types of masonry. In particular, the façade masonry, the masonry of longitudinal walls between buildings and the internal transverse walls, are all made of stone at least up to the second floor.

Superelevations (i.e. perimeter walls) at the third level are seldom implemented in stone masonry, while brickwork masonry is found as the unique technique on the third floor for internal partitions and renovation interventions.

There are also different types of floors: rooms on the ground floors and stairs are characterized by a vaulted roof; on the upper floors we can see wooden traditional floors, but also reinforced concrete slabs floors for the terraces referring to the recent elevations. For the roofs, the technique used is almost exclusively that of wood, with double or single-pitched roofs.

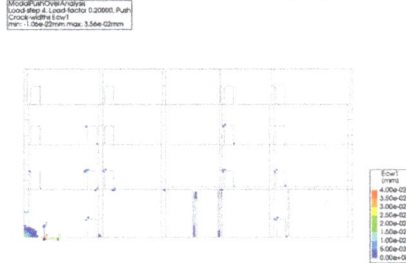

Figure 3. First configuration: the presence of the court is modelled thanks to the insertion of stone columns on the ground floor.

Mediterranean Architectural Heritage - RIPAM10

Materials Research Proceedings 40 (2024) 79-89

Materials Research Forum LLC

https://doi.org/10.21741/9781644903117-8

Figure 4. Second configuration: the presence of the court is modelled with walls also in the ground floor, in continuity with the walls in the higher levels.

Numerical analysis

The numerical investigation carried out on *finite element method* (FEM) models is aimed at highlighting the seismic vulnerability caused by the *internal court* of the building. The software used for model creation is *DIANA FEA*.

The modelling was carried out based on the information obtained during the preliminary investigation, particularly regarding wall thicknesses and mechanical properties associated with materials for each element. All models were fixed to the ground and fully constrained along the boundary walls in the transverse direction so that to resemble the containment effect of adjacent buildings. This constraint condition, although overestimating the rigidity provided by adjacent buildings, filtered the modes of vibration only in the longitudinal direction, i.e., the maximum length direction, orthogonal to the façades.

Initial modelling was created without considering the presence of floor slabs, an assumption made in favour of safety, given the deformability in the plane of the wooden floors. By neglecting the horizontal effect of slabs, a global response concentrated in the first vibration mode cannot be clearly identified.

On the other hand, it is possible to highlight the tendency of this kind of buildings to respond to seismic actions as a compound organism and through the activation of local damage mechanisms in the out-of-plane direction. Indeed, damage occurs due to vibration modes with lower frequencies, and involves small portions of the building and out-of-plane response of the masonry panels appears as predominant.

The second model of the building, which included the slabs, offers very different results. In particular, most of the participating mass is found in the first mode of vibration, so that a clear global behaviour can be identified. In this framework, expected damage is estimated at ground floor in the walls and at the corners of the internal opening due to a mainly in-plane response of the building. Also, the columns of the *internal court* undergo remarkable damage in either model.

Furthermore, two configurations with slabs were modelled: in both the slabs are perfectly clamped to the walls. In the first configuration, the presence of the court is modelled by the insertion of stone columns on the ground floor, resembling the current situation. In the second slab configuration, it is considered that the court has the same size, but it is defined by continuous walls rather than columns.

In the configuration with columns, at the same load step, the damage is more extensive, in particular, damage is also found at higher levels than in the first (Fig. 3; Fig. 4).

It is important to highlight that the assumption of perfectly anchored floor slabs to the walls does not describe the real condition of the building. Through the model, certain conditions are implicitly assumed in the best-case scenario. Consequently, the damage assessment emerging from the modelling is conservative but allows for defining the actual damage that will affect the real structure.

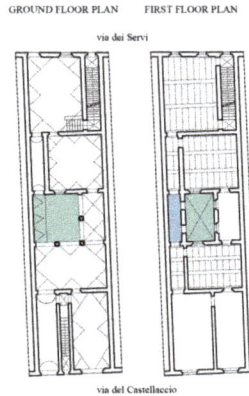

GROUND FLOOR PLAN   FIRST FLOOR PLAN

Figure 5. Reconstruction of the original configuration of the units: Ground floor (left) and First floor (right). Identification of the internal court (green) and of the additional cantilevered corridor on the first floor over the same internal court (blue).

Vulnerability associated to the presence of an inner court

Beginning with the *court-merchant house*, from which the *court-terraced house*, the reference typology for the buildings under study, derives, the expansion in the depth of the structure, along with subsequent added elements and adjunctions and the phenomenon of lot congestion, inevitably leads to the inclusion of an open space typically positioned centrally.

This space serves the purpose of illuminating and providing ventilation to the innermost areas. This element, referred to as a court, characterizes the majority of the historical architectural constructions in Florence, from smaller serial structures to grand noble palaces, where it reappear in larger dimensions.

In the buildings under examination, although the court belongs to the original layout, it inherently represents a significant point of discontinuity. Furthermore, the situation is exacerbated when considering the modifications the court has undergone during 19th Century practices, which altered its arrangement and functionality.

The presence of an *internal court* introduces irregularities at the floor level, which shows a hole that corresponds to the court and subsequent variations in the slab rigidity. Additional asymmetry and irregularity are observed in elevation. Specifically, on the ground floor, columns around the court must bear loads of the upper walls, highlighting structural discontinuity.

To this context, we must add the constructional weaknesses stemming from the evolution and modifications carried out over the years, beginning in the 19th Century. These include the construction of an additional cantilevered corridor on the first floor over the same *internal court* (Fig. 5).

These considerations make it impossible to associate the same stiffness to the cantilevered elements over the courtyard as to the other floor slabs. This extends the discontinuity and the vulnerability resulting not only from the *gap* created by the *internal court* but also to the portions of the floor slabs resulting from the gap itself, whose contribution to stiffness is negligible.

Table 1. Parameters for the determination of Seismic Vulnerability Index for masonry buildings. Identification of the analyzed parameters (red).

	PARAMETERS	CLASS				WEIGHT
		A	B	C	D	
1	Type and Organization of the structural system	0	5	20	45	1.5
2	Quality of the structural system	0	5	25	45	0.25
3	Conventional strength	0	5	25	45	1.5
4	Position of the building and type of foundation	0	5	25	45	0.75
5	Horizontal diaphragms	0	5	15	45	VAR
6	Plan irregularity	0	5	25	45	0.5
7	Height irregularity	0	5	25	45	VAR
8	Maximum distance between walls	0	5	25	45	0.25
9	Roof system	0	15	25	45	VAR
10	Non-structural elements	0	0	25	45	0.25
11	Physical condition	0	5	25	45	1

Vulnerability Index Method and internal court

In order to propose specific indicators capable of assessing vulnerabilities related to the court, this study presents a preliminary critical analysis conducted on those parameters from the *GNDT form* of the Seismic Vulnerability Level II [10], that are considered most useful to highlight the vulnerabilities connected to the presence of the *internal court* in the buildings (Table 1).

Parameter 3. Conventional Strength: Along with parameter *1. Type and Organization of the Resisting System*, this parameter holds the greatest weight within the *GNDT form*. The calculation considers the verification level, in general, the ground floor, assuming all floors are identical. However, this assumption does not hold when considering the presence of an *internal court* in the building. As demonstrated in the case study, the conventional strength calculated on the ground floor is different from the one calculated for the first floor, which is characterized by greater strength, even if only slightly, due to the presence of the walls corresponding to the columns below on the ground floor.

Parameter 5. Horizontal diaphragms. This parameter is based on two fundamental assumptions: the stiffness of horizontal diaphragms and the effectiveness of their connections with vertical elements. The evaluation is carried out on the general floor, and in cases of different types of horizontal elements, the condition defined by the worst-case scenario applies, provided it extends over a significant portion of the floor. The evaluation is done at a *global level* and is not suitable for interpreting the behaviour of the floor in the area of the court, in this specific case, where there is a change in geometry and certainly in stiffness. Furthermore, in this specific case, the nature of the connections changes, especially in the area of the cantilevered corridor (Fig. 5), supported by a beam.

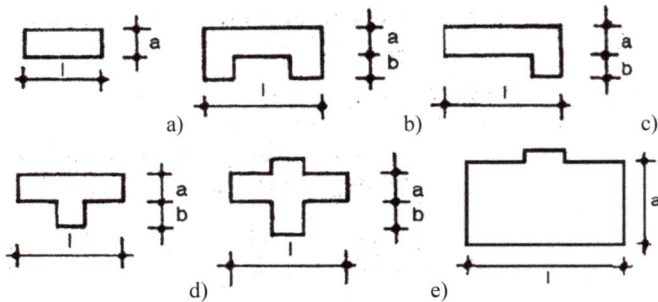

Figure 6. Plan irregularity. Schemes for calculating the plan irregularity β_2.

Parameter 6. Plan irregularity. This parameter depends on two factors: β_1 and β_2, representing the first one the ratio between the width and depth of the building and the second one the presence of plan irregularities. In addition to these, the *American seismic vulnerability form* [13] introduces an additional value to consider, which is the ratio between the area A_0 related to the court and the total area A_t, as well as the position of the court itself, central or lateral relative to the floor plan. As with the previous parameter, it is assumed that all floors are equal, and only the outer shape is considered. Moreover, articulations outside the outline <10% are considered inconsequential because, once again, the parameter focuses on global behaviour. The impact of changes in shape due to the presence of the court, when viewed globally over the total area, is not taken into account. The court, occupying a minimal area relative to the total area of the building, ends up being construed as a mere light well.

Parameter 7. Height irregularity. In masonry buildings, especially historical ones as in this case, the main cause of irregularities in elevation is precisely the presence of *portici* or *logge*. The parameter takes this into account, as well as changes in wall stiffness throughout the elevation of the building and any additions and reconstructions not contemporaneous with the original structure: once again, the behaviour is evaluated at a *global level*. For example, in this specific case, *logge* of modest dimensions, which affect less or at most 10% of the total area of the considered floor, lead to a downgrading concerning what could indeed be the level of vulnerability at the *local level*.

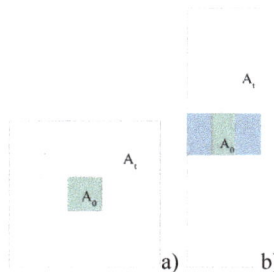

Figure 7. Plan irregularity. Distribution schemes to evaluate the transverse measure of the court in relation to the transverse measure of the plan. $A_0 = 20\ m^2$; $A_t = 314.64\ m^2$; $A_0/A_t = 6\%$. Identification of the cantilevered floor (blue).

Materials Research Forum LLC
https://doi.org/10.21741/9781644903117-8

Results

Several aspects emerge from the studied parameters that refer to global behaviour. The calculation of conventional strength performed on the model varies when comparing the ground floor to the first floor. When calculating the *coefficient C*, which represents the ratio between the ultimate shear T at the verification level and the weight P of the building portion above it, the value of the coefficient α for the ground floor is $\alpha = 0.15$, which is less than 0.4, where 0.4 is the reference value for seismic zones in the first category. For the first floor, $\alpha = 0.25$, also less than 0.4. Although this value also falls into Category D, it indicates a more adverse condition on the ground floor.

Regarding the plan irregularity, the ratio of the area of the *internal court* A_0 to the total area A_t is equal to 6%. For the calculation of the irregularity β_2 of the floor, which features an opening at the *internal court*, an approximate reference was made to the diagram in the manual (Fig. 6b), resulting in 9.8%, which is less than 10%, considered as the minimum value from the manual to indicate irregularity.

An assessment of the irregularity due to the presence of the *internal court* with respect to the floor slab could be made, not so much in relation to the A_0/A_t ratio, but to the ratio between the widths of the courtyard and the entire building. Taking into account the two models (Fig. 7), with $A_0 = 20$ m^2; $A_t = 314.64$ m^2, with an equal A_0/A_t ratio of 6%, considering the horizontal direction, model 7a) represents the better condition, in which the ratio between the width measurements is 0.2 (1:5), compared to case 7b) in which the width ratio is 0.3 (1:3), which becomes 1 (1:1) considering the further widening of the courtyard, considering that the slabs resting on beams do not offer stiffness comparable to the rest of the slabs, as in the case study.

In the height irregularity, we consider the portion affected by the *loggia*. This, concerning the total surface, is 5.7%, which is less than 10%, once again the minimum value to be taken into consideration.

Considering the issue at the *local level* and focusing the calculation on the individual cell, results naturally change. In the calculation of conventional resistance, while considering the ground floor, it remains in Category D with a value of $\alpha = 0.3$, less than 0.4. On the first floor, the value of $\alpha = 0.41$, which is greater than 0.4, falls into Category C. The same applies when considering the plan and height irregularities. In the first case, the A_0/A_t ratio is equal to 36%, and the β_2 coefficient is equal to 49.8%, which is greater than 10%, while the ratio of the area occupied by the *loggia* to the total surface is 34%, also greater than 10%.

Conclusion

Plan and elevation regularity are fundamental requirements for a better seismic response of a historical masonry building. Several factors can deviate a building from its regular configuration, and among these, the presence of an *internal court* has a fundamental role.

The structural response of the building is further worsened if the court is surrounded by a *loggia* which on the upper floors corresponds to closed spaces connecting adjacent rooms.

This construction system, in fact, implies an eccentric arrangement of resisting elements on the floor plan and a difference in the resisting area along the two principal directions. Furthermore, in elevation, a strong irregularity is represented by the different vertical structural elements around the court: columns on the ground floor and partition walls on the upper floors. The consequence of this irregularity also implies that 3 floors of partition walls weigh on a wooden beam supported by two stone columns.

This irregularity corresponds to the construction typology of the historic city centre of Florence. In this specific case, the court occupies the central core of the building which is also characterized by an extension in depth, much higher than the width.

This additional characteristic of the considered construction type results in the court's surface being small compared to the entire floor plan, but it practically occupies the central part of the floor plan's width, with negative effects on the seismic response concentrated mainly in the central part of the building. If the court is considered as a reduction in stiffness of the floor slab, the slabs adjacent to the court on the upper floors, resting on columns at the ground floor and not on continuous wall partitions, besides being the result of subsequent interventions compared to the original layout, cannot be assumed to have the same stiffness as the others.

In contrast to American normative that refer to *Diaphragm Discontinuity* [13], describing irregularities related to open portions of the floor structure with consequent changes in rigidity, the GNDT Seismic Vulnerability Index Method Level II [10], does not explicitly consider the case of the presence of a court, although many parameters consider irregularities induced by the presence of an internal court.

Considering the specific analysed case, the small size of the court compared to the total area means that even the criterion of [13], referring to the ratio of the area of the *internal court* A_0 to the total area A_t, does not capture its vulnerability.

In light of these considerations, it may be useful, in terms of results, to assess vulnerability by considering the transverse direction, the main weak axis if considering seismic action in the same direction. A possible modification to the parameter could be to specifically consider the ratio between width dimensions.

As revealed from the examination of parameters in the *GNDT form* (Table 1), considered most suitable and closely related to the interpretation of the vulnerabilities connected to the presence of the *internal court* in the buildings, the primary problem in the seismic vulnerability assessment referring to an *internal court* in the building, is related to the seismic vulnerability assessment of the building at a *global level*.

To investigate the onset of damage in correspondence to the internal courtyard of the building, a *localized* study would be necessary, considering the parameters of the *GNDT form* relating to the single cell where the presence of the court has the greatest impact.

In this study, a correlation between assessments of seismic vulnerability at the *global level* and at the individual element level is proposed.

References

[1] G. Caniggia, G. L. Maffei. Composizione architettonica e tipologia edilizia 1/2. Il progetto nell'edilizia di base. Marsilio editori, Venezia, 1979.

[2] N. Chieffo, (et al.). A vulnerability index-based approach for the historical centre of the city of Latronico (Potenza, Southern Italy), in Engineering Failure Analysis, n. 136. 2022. https://doi.org/10.1016/j.engfailanal.2022.106207

[3] A. De Maria, C. Donà. (a cura di). Manuale delle murature storiche. Analisi e valutazione del comportamento strutturale. Volume I. DEI Editore, Roma, 2011.

[4] A. Giuffrè. Letture sulla meccanica delle murature storiche. Edizioni Kappa, Roma, 1991.

[5] G. A. Centauro, C. Francini (a cura di). Progetto HECO (Heritage Colors). Metodologie Analisi Sintesi Apparati Valutazione d'impatto sul sito UNESCO Centro Storico di Firenze. DiDaPress, Firenze, 2017.

[6] A. Preciado (et al.). Earthquake risk assessment methods of unreinforced masonry structures: Hazard and vulnerability, in Earthquakes and Structures. 2015. https://doi.org/10.12989/eas.2015.9.4.719

[7] R. Maio. Seismic vulnerability assessment of old building aggregates. Master's thesis. University of Aveiro. 2013.

[8] G. M. Calvi (et al.). Development of Seismic Vulnerability Assessment Methodologies Over the past 30 years, in ISET Journal of Earthquake Technology, N. 472, Vol. 43. 2006.

Mediterranean Architectural Heritage - RIPAM10
Materials Research Proceedings 40 (2024) 79-89

Materials Research Forum LLC
https://doi.org/10.21741/9781644903117-8

[9] D. Benedetti, V. Petrini. Sulla vulnerabilità sismica di edifici in muratura: un metodo di valutazione. A method for evaluating the seismic vulnerability of masonry buildings, in L'industria delle Costruzioni, n.149. 1984.

[10] GNDT. Rischio sismico di edifici pubblici- Parte I. Proceeding of CNR- GNDT, Roma, 1993.

[11] M. Ferrini (et al.). Rilevamento della vulnerabilità sismica degli edifici in muratura. Manuale per la compilazione della Scheda GNDT/CNR di II Livello (versione modificata). Regione Toscana, 2003.

[12] A. Formisano (et al.). Un Metodo per la Valutazione su larga scala della Vulnerabilità Sismica degli aggregati storici. Anidis, Bari, 2011.

[13] National Institute of Building Sciences. NEHRP Recommended Seismic Provisions for new Buildings and other structures. FEMA, 2020.

[14] G. L. Maffei. La casa fiorentina nella storia della città dalle origini all'Ottocento. Marsilio editori, Venezia, 1990.

[15] G. Cataldi. Palazzetti a schiera in via dei Servi a Firenze, in Studi e documenti di architettura, n.14, Firenze, 1987.

[16] C. Tomasini Pietramellara. Tipologie edilizie, in Firenze, studi e ricerche sul centro antico. Nistri-Lischi, Pisa, 1974.

[17] C. Chiappi, G. Villa. Tipo/progetto/composizione architettonica. Alinea editore, Firenze, 1980.

[18] MIT. Circolare 21/01/2019. Istruzioni per l'applicazione dell'aggiornamento delle Norme Tecniche per le costruzioni, di cui al decreto ministeriale 17/01/2018. 2019.

Mediterranean Architectural Heritage - RIPAM10
Materials Research Proceedings 40 (2024) 90-96

Materials Research Forum LLC
https://doi.org/10.21741/9781644903117-9

Thermal, Mineralogical and Chemical Properties of Soil Building Blocks for Eco-Habitat Sustainable

A. AMMARI[1,2*], N. ZAKHAM[1], K. BOUASSRIA[3], H. BOUABID[1], M. CHERRAJ[1], S. CHARIF D'OUAZZANE[4], M. IBNOUSSINA[5]

[1]National School of Architecture of Marrakech

[2]Laboratory of Mechanics and Materials, Faculty of Science, Mohammed University, Rabat, Morocco

[3]LSIMO, Faculty of Science, Ibn Tofail University, Kenitra, Morocco

[4]LMTM, National School of Mineral Rabat, Morocco

[5]Université Cadi Ayyad, Lab. Géoenvironnements/Géorisques/Géotechnique, Marrakech, Morocco

* abdelmalek09@gmail.com

Keywords: Earth, Mineralogical, Chemical, Thermal Conductivity, Eco-Habitat

Abstract. The purpose of this study is to investigate the relationship between the thermal conductivity of Compressed Earth Block Stabilized (CEBs) by cement and the results of mineralogical and chemical examinations of the soil. The soil was taken from the Moroccan city of Fez. That; determination of the thermal conductivity of CEBs plays an important role when considering it's suitability for energy saving insulation. The measurement technique used in this study to determine thermal conductivity is hot ring method the thermal conductivity of the tested samples is strongly affected by the quantity of the cement added. The mineralogical and chemical analysis show the soilof Fez, mainly composed of the calcite, quartz and dolomite improved the behaviour of the material by the addition the optimum content of cement. The findings suggest that to manufacture lightweight samples with high thermal insulation properties, it is advisable to use clays that contain quartz. Ina ddition, quartz has high thermal conductivity.

1. Introduction

This paper recommends a more technical and scientific approach to sustainable housing, which acknowledges its multiple functions as a socio-cultural system and physical. In order to guarantee prosperous residential areas, it seeks to enhance and harmonize the environmental, material, cultural, and economic dimensions of eco-habitat sustainability. Sustainable housing is often considered primarily from an ecological point of view.

Since the earliest ages, the human has been keen to provide an environment suitable for housing. The human has developed his treatments for the surrounding environmental conditions through continuous experiments and accumulated experience in the practice of construction, has developed his treatments for the surrounding environmental conditions through continuous experiments and accumulated experience in the practice of construction. And we sable to recognize the characteristics of building materials and then use them to the maximum extent to meet his needs and requirements, because the internal environment of the building is very important for human comfort and this environment depends on several factors and the disruption of one of these factors upset the balance of this environment, and the temperature is at the top of the factors that must be controlled within the ranges of human comfort.

The present study investigates the impact of the initial raw material mixture mineralogical and chemical composition on the dry state thermal conductivity of eco-habitat sustainable CEB samples [1]. Furthermore, a study was conducted to examine the impact of varying proportions of cement addition on the thermal conductivity of material composed of soil reinforced by compression at a pressure of two MPa [2]. The need of developing soil stability with materials is low cost and environment friendly is necessary, with simple rates of stability and limited and according to the standards, we have used cement for stabilization because they are safe, effective and affordable soil stabilizers.

2. Materials and methods

2.1 Material

A leftover natural soil that came from the Fes city, Morocco site centre that F mentioned in this work. The materials used were soil and cement. The soil came from the Moroccan region of Fez, which is well-known for its earthenware industry and traditional stability pottery production, and the cement type used was CPJ 35 Portland. The compressed earth block samples, we have mixed soil used and stabilised with content of cement, with a mechanical compressive strength of 2 MPa and placing the mixture into moulds that is compressed by machine, and then left it to dry away from the ambient temperature for a 28 days to allow the cement to reach its maximum strength. In this study, we used cement dosing to manufacture soil block, with also a few ratios doses of cement (0%;4%; 7% and 10% of cement). Every sample is cylinder-shaped, measuring 12 cm in height and 8 cm in diameter. Every sample was split into two halves. Prior to testing, the samples were dried. This took place over the course of 72 hours in an electric oven set to 60 °C until the mass of the samples stabilized. Utilizing a hot ring method apparatus, the thermal conductivity was calculated. The device consists of a hot ring probe that uses a kapton slab to measure the thermal conductivity.Positioned between two test pieces of the CEB samples to be characterized or submerged in the medium, the hot-ring device works on the slab of kapton principle. After that, a heat flow is created by injecting current in the ring. This movement of heat eliminates the specimen's CEBs over a predetermined period of time before reaching the ring's centre, where the measurement is made. The heating period of 400 s, the measurement time of 500 s, and the heating power of 2 W were the same for every sample in our experiment.

2.2 Methods

2.2.1 Measurement of thermal conductivity

The specimens' thermal conductivity was determined in this work using the process of hot rings. the process of hot rings predicts the CEB specimens' thermal conductivity by using the temperature evolution. The thermal conductivity is determined using the hot ring probe, also known as the hot-ring apparatus slab of kapton. The hot-ring apparatus's slab kapton is positioned in the middle of the CEB specimens between two components, and current is subsequently pumped into the ring to create a heat flow. The heat flow within the CEB sample diminishes and eventually reaches the centre of the ring, where a thermocouple is positioned for temperature measurement (fig. 2).The element that needs to be described is thought of as an isotropic, homogenous, infinite medium. Thermal measurements were performed on the dry specimens. The drying process was done in an electric laboratory oven at 60 °C until the specimens' mass stabilized (i.e., there was no water present). The drying period for each test was 72 hours. In our experience, the hot ring method's thermal parameters were as follows: heating power of 2 W, measurement time of 500 s, heating time of 400 s, and probe resistance of 2.5 Ω. These parameters remained consistent for every specimen during the experiment. We used a multimeter (hot ring method) (Fig.1)[1].

Fig. 1. Hot ring method

3. Results and discussion

3.1 Chemical and mineralogy analysis

3.1.1 Analysis of X-ray fluorescence [2]
The table 1 show that the chemical composition of clay in clay F.

Table.1: Thechemical elements of clay F

Composition	clay of Fez (%)(*)
Cao	**31,8**
SiO$_2$	**22,3**
Fe$_2$O	1,72
K$_2$O	0,536
SO$_3$	1,29
Al$_2$O$_3$	5,96
TiO$_2$	0,226
P$_2$O$_5$	0,753
MnO	0,0503
Na$_2$O	0,323
SrO	0,0125
Loss on ignition at975°C)	32,9

3.1.2 X-ray diffraction analysis of clay F
The soil F (Fig. 2) stabilised at 0% and 4% cement with X-ray analysis revealed the formation of high peaks of calcite; quartz; and dolomite. The calcite, quartz, and silicate of aluminum that make up the majority of the block of the same earth of F (Fig. 3), enhance the material enhanced thermal conductivity and behavior [2].

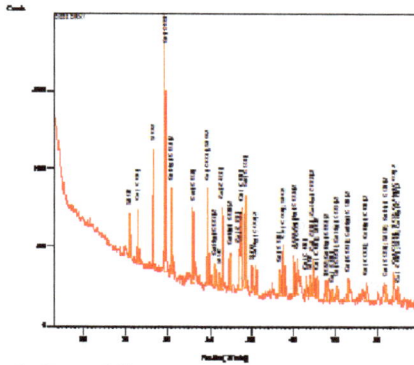

Fig. 2: X-ray diffractogram of studied clay of the soil F[2]

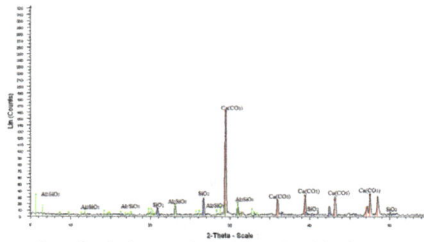

Fig. 3: X-ray diffractogram of studied clay stabilized with 4% of cement for the soil F block[2]

3.2 Thermal results and discussion

The hot-ring technique is a temporary approach for establishing a material thermal conductivity. Samples are heated to a higher temperature when the hot-ring apparatus's kapton slab begins to heat them.

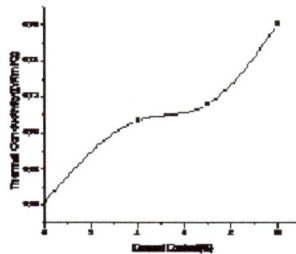

Fig.4 Impact of cement dosage in thermal conductivity

Figure 4 shows that stabilizing samples with cement increases their thermal properties; increased cement content in the blocks results the increased thermal conductivity

Mediterranean Architectural Heritage - RIPAM10   Materials Research Forum LLC
Materials Research Proceedings 40 (2024) 90-96   https://doi.org/10.21741/9781644903117-9

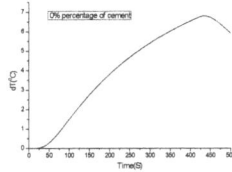

Fig.5. The temperature variation curve, dT(0C), in the CEBs at 0% cement dose versus time (s) .

The temperature evolution in sample 1 with 0% cement content is depicted in Figure 5. Sample 1's density is 1654,6 (kg/m3), Specific heat is 1023,4 (J/kg.k) and thermal of conductivity is 0,662 (W/m.K).

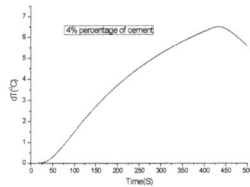

Fig.6. The curve of the temperature variation dT(^{0}C) in the CEBs at 4% dose of cement versus time (s).

The temperature evolution in sample 2 with a 4% cement dosage is depicted in Figure 6. The density is 1722,6 (kg/m3), Specific heat is 1029,4 (J/kg.k) and thermal of conductivity is 0,707 (W/m.K).

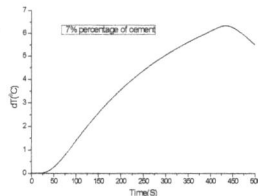

Fig.7. The curve of the temperature variation dT(^{0}C) in the CEBs at 7% dose of cement versus time (s).

Figure.7, shows that the evolution of the temperature in sample 3 with 7% dosage of cement, density of sample 3 =1749,9 (kg/m3) ,thermal conductivity =0,716 (W/m.K), specific heat= 1035,8 (J/kg.k).

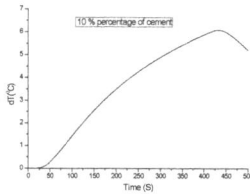

Fig.8. The curve of the temperature variation dT(^{0}C) in the CEBs at 10% dose of cement versus time (s).

The temperature evolution in sample 4 with a 10% cement dosage is depicted in Figure 8. The density is 1822.6 (kg/m3), specific heat is 1040.7 (J/kg.k) and thermal of conductivity is 0.761 (W/m.K).

The heat transfer is affected when the hot-ring measuring method is applied to samples of CEBs, from the figures 5 to 8, as can be observed, the temperature variation curve for the CEBs specimens with varying cement stabilizer contents. Thermal conductivity vary with the temperature difference, in this experiment we used the same parameters for all the samples, which there were as follows the heating time= 400s, the measure time=500s and the heating power = 2 W, for the reason all that is necessary to understand is how cement dosage affects thermal conductivity. According to the behaviour of temperature in the curves 5,6,70 and 8, the high temperatures continue the increase in to 450°C, after that start it to decrease, because in the experimental parameter the heating time was 400s for each sample.

The main mineral in sand and gravel is quartz, which has a significantly higher thermal conductivity than other minerals. Consequently, variations in the mineral composition lead to variations in thermal conductivity. The quartz content of the material used in Lei Zhang et al. study was significantly higher than ours, and their research revealed a more noticeable quartz decrease for the same porosity increase, causes a more noticeable decrease in thermal conductivity [3]. The percentage of $CaCO_3$ rises with the addition of cement; it reacts with H_2O to produce $Ca(OH)_2$ + CO_2, the CO_2 gas that is produced during the breakdown. This probably explains why, in 0% of the cement content, the real part of the conductivity decreased [4]. Zhaoyu Wang et al. [5] found an excellent linear correlation between thermal conductivity and average $CaCO_3$ content. According to research by A. Ammari et al., in order to fix inert clay particles, the amount of lime needed between 1% and 3% by weight of soil must be greater than the fixation limit. Furthermore, a portion of the fine sand makes up the quartz, which is not a binder [2,6,7,9]. As a result, the carbonation of the lime cements the quartz grains in the clay Fez. It is commonly known that the presence of SiO_2 results in significantly higher thermal conductivity and an enhanced heat transfer rate. Quartz increases heat conductivity as a result.

It should be highlighted that adding a small amount of cement does not stabilize the sand and gravel fraction (80%) of soil F. This consequences in a slight variation in the thermal of conductivity, ranging from 0% to 4%. The thermal conductivity value grows to 0.75 W/m.K. above 4% of cement. Additionally, the cement has a good coating over the soil F.

Dolomite and calcite decrease the thermal conductivity of the block when process elements are kept constant. This is due to porosity is greatly increased when dolomite and calcite break down at specific temperatures. higher than 700°C. However, dolomite and calcite improve heat conductivity. The crystalline phases of aluminium silicate, calcium and magnesium silicate, and magnesium silicate form in these compositions at temperatures higher than 860°C [8].

Moreover, Eq. 1 provides a summary of the dolomite element powder reaction in cement.

$$6CaMg(CO_3)_2 + 2Al(OH)_4^- + 8OH^- + 5H_2OMg_6 \longrightarrow Al_2CO_3(OH)_{16} 5H_2O + 6CaCO_3 + 5CO_3^{2-}. \text{ Eq. 1}$$

Conclusions

The conclusions of this study are as follows:

-The soil of the Fez (F) earth formed by CaO with a high SiO2 (22.3%) mainly because there is a peak of quartz and kaolinite.

- The proportion of the calcite and the small amount of aluminum silicate present in the earthen F test pieces are both increased by the addition of cement.

- The determination of thermal conductivity and the enhancement of soil building blocks for the sustainability of eco-habitats are determined by the mineral and chemical compositions, such as dolomite, quartz, and to a certain extent dolomite.

-Increase thermal conductivity of samples as temperature rises and cement content increases due to cement's ability to interact with soil contents.

- The thermal conductivity is significantly increases due of presence higher percentage of SiO_2

-Increasing the thermal conductivity of the earth's building blocks for eco-habitat sustainability is dependent on the mineralogical and chemical compositions, such as quartz and dolomite.

References

[1] N. Zakham, A. Ammari, Y. El Rhaffari, M. Cherraj,H. Bouabid, K. Gueraoui, A. Samaouali , The Effect of Cement Content ontheThermo-Mechanical Performance of Compressed Earth Block, International Review of Civil Engineering (I.RE.C.E.), Vol. 11,N.6 ISSN 2036 - 9913 November 2020. https://doi.org/10.15866/irece.v11i6.17438

[2] Ammari A., Bouassria K., Cherraj M., Bouabid H., Charif D'ouazzane S., "Combined effect of mineralogy and granular texture on the technico-economic optimum of the adobe and compressed earth blocks", Case Studies in Construction Materials 7 (2017) 240–248. https://doi.org/10.1016/j.cscm.2017.08.004

[3] Lei Zhang, Arild Gustavsen, Bjørn Petter Jelle, Liu Yang, Tao Gao, Yu Wang,. Thermal conductivity of cement stabilized earth blocks., Construction and Building Materials 151 (2017) 504–511. https://doi.org/10.1016/j.conbuildmat.2017.06.047

[4] Štefan Csáki, Tomáš Húlanb, Ján Ondruška, Igor Štubňab, Viera Trnovcováb, František Lukáča,c, Patrik Dobron, Electrical conductivity and thermal analyses studies of phase evolution in the illite – CaCO3 system, Applied Clay Science 178 (2019) 105140. https://doi.org/10.1016/j.clay.2019.105140

[5] Zhaoyu Wang, Nan Zhang, Jinhua Ding , Qi Li, Junhao Xu , Thermal conductivity of sands treated with microbially induced calcite precipitation (MICP) and model prediction, International Journal of Heat and Mass Transfer

[6] Ammari A., Bouassria K., Zakham N., Cherraj M., Bouabid H., Charif D'ouazzane S., "Durability of the earth mortar: Physico-chemical and mineralogical characterization for the reduction of the capillary rise " MATEC Web of Conferences 149, 01024 (2018). https://doi.org/10.1051/matecconf/201714901024

[7] Bouassria K., Ammari A., Tayyibi A., Bouabid H., Zerouaoui J., Cherraj M., Charif D'ouazzane S., "The effect of lime on alumino-silicate and cement on the behavior of compressed earth blocks"., J. Mater. Environ. Sci. 6 (12) (2015) 3430-3435(JMES)

[8] J. Garcı ́a-Ten, M.J. Orts , A. Saburit, G. Silva, Thermal conductivity of traditionalceramics Part II: Influence of mineralogical composition, Ceramics International 36 (2010) 2017–2024. https://doi.org/10.1016/j.ceramint.2010.05.013

[9] Ammari A. , Bouassria K, Tayyibi A, Bouabid H, Cherraj M., Charif D'ouazzane S., Ibnoussina M., "Promotion de la technique du bloc de terre comprimée dans le secteur de la construction par l'amélioration de son comportement mécanique", Journal of Materials and Environmental Science(JMES).

Mediterranean Architectural Heritage - RIPAM10
Materials Research Proceedings 40 (2024) 97-108

Materials Research Forum LLC
https://doi.org/10.21741/9781644903117-10

Physico-Chemical and Mechanical Characteristics of Traditional Marrakech Lime

N. AQELMOUN[1,a,*], A. BELABID[1,b], H. AKHZOUZ[2,c], Han ELMINOR[3,d],
Has ELMINOR[1,e]

[1] Department of Mechanics, Materials and Civil Engineering, National School of Applied Sciences, Ibn Zohr University, Agadir 80000, Morocco

[2] Departement of Civil Engineering, Euromed University of Fez, Fez 30000, Morocco

[3] Polytechnic school, Universiaspolis University of Agadir, Agadir 80000, Morocco

[a]nissrine.aqelmoun@edu.uiz.ac.ma, [b]abderrahim.belabid@edu.uiz.ac.ma,
[c]h.akhzouz@ueuromed.org, [d]hanane.elminor@e-polytechnique.ma, [e] h.elminor@uiz.ac.ma

Keywords: Marrakech's Traditional Lime, Lime Traditional Concrete, Physio-Chemical Characteristics, Mineralogy, Mechanical Characteristics

Abstract. To tackle the problem of construction-related pollution, especially the greenhouse effects mainly caused by carbon dioxide emissions, governments have begun to encourage the use of traditional building materials and techniques characterized by their low carbon impact. Among these alternatives is Marrakech lime, produced using traditional processes. Historically, it has been used to produce a special mortar known as Tadlakt, an ancestral Amazigh skill used to waterproof parts in contact with water. In order to identify other ways of adding value to Marrakech lime, a literature search was launched. Physico-chemical and mechanical characterization was also carried out on samples of traditional lime concrete sold in the Moroccan market under the name "Jer de Gram". Morphological and mineralogical analyses were examined using scanning electron microscopy (SEM), energy-dispersive X-ray spectroscopy (EDS), and X-ray diffraction (XRD). The results showed that the lime examined is slightly hydraulic of varied forms, it is mainly composed of calcite, belite, calcium oxides, silicon, and carbons. Crushing tests show a low compressive strength of traditional lime concrete, not exceeding 1 MPa in the best cases, an unexpected result that contradicts the literature. The reason lies in the sandy nature of the aggregates and the traditional lime production process, which does not allow for effective quality control of the final product.

Introduction

Protecting the environment and preserving natural resources has become a major global concern. The world is increasingly aware of the ecological issues we face today, specifically CO_2 emissions and their impact on climate change. The construction sector has been identified as the most responsible for carbon dioxide emissions, due to cement, one of the basic building materials. The most recent studies on air pollution indicate that cement is behind an estimated 10% increase in CO_2 emissions (observed between 2015 and 2016) [1]. With this in mind, and intending to limit climate fluctuations, the European Union has shown global leadership, planning to reduce greenhouse gas emissions by 40% by 2030 and to cut energy consumption by 20% by 2020 and 50% by 2050, in addition to implementing action plans that encourage the use of natural resources as well as improving the energy efficiency of buildings [2].

From antiquity to the first decades of the 20th century, lime mortars played a crucial role in the history of architecture and engineering [3]. However, the knowledge of how to prepare and apply these mortars, which was passed down from parents to sons and from masters to apprentices, thus ensuring good durability and performance, as can be seen in surviving ancient buildings, has been abandoned, and this lack of artisanal knowledge has created difficulties for their practical use [4]. In this context, in-depth scientific studies have been carried out on the nature, quality, and proportions of components, such as the effect of environmental conditions, the evolution of carbonation and change in pore structure, the influence of mortar preparation and application process, and the impact of lime production conditions [4].

The methodological approach adopted in this article is mixed (qualitative and quantitative), combining an intensive bibliographical analysis of the scientific literature related to hydraulic lime, with experimental work in the laboratory, integrating both theory and empirical data.

In this paper, we will focus on the Marrakech traditional lime. First of all, a literature review was launched. Then, using scanning electron microscopy (SEM), energy dispersive X-ray spectroscopy (EDS), and X-ray diffraction (XRD), morphological and mineralogical investigations were performed. Realizations were also applied to samples of the traditional lime concrete known as "Jer de Gram", sold in the Moroccan market, to determine the material's physico-chemical and mechanical properties.

Previous literature

Hydraulic lime

The use of lime has grown considerably in recent years. The "lime revival", a set of rules for good lime practice first developed in 1970, gained momentum after the conservation of the west front of Wells Cathedral under Robert and Eve Baker, and Crowland Abbey and Exeter Cathedral after 1975 [5]. Nevertheless, according to Copsey [5], this "revival" requires an uncompromising reassessment to improve the handling of hydraulic lime while respecting the main foundations of conservation. Referring to the latter, and despite scientific studies encouraging the use of hydraulic lime in this sense, Veiga [4], wondered why the employment of this material was still restricted. He identified three major factors: the need to acquire more scientific knowledge about local conditions and the heterogeneity of raw materials; the importance of the human factor, including the training of technicians and operators in the practical application of the material; and finally, the difficulty of managing the time required for applying and drying lime.

Hydraulic lime is extracted from the calcination of limestone and siliceous rock. During combustion at around 1200°C, silica and calcium oxide (C_aO) are combined to create calcium silicates. The decarbonization of $CaCO_3$ and the extinction of C_aO take place in the same way as for calcium lime. The result is a binder, called belite, composed of $C_a(OH)_2$ and C_2S (dicalcium silicates) [6]. The ratio (C_aO/S_iO_2) influences the lime's hydraulicity, which, in simple terms, affects the amount of belite present in the lime. Three different categories of hydraulic lime can be distinguished: NHL_2, $NHL_{3.5}$, and NHL_5, which are differentiated according to their minimum compressive strength after 28 days, as defined in European standard EN 459-1. NHL_2 has a strength of 2 MPa, $NHL_{3.5}$ has a strength of 3.5 MPa and NHL_5 has a value of 5 MPa. Consequently, the type of lime used in construction basically depends on the strength you need once it has had time to harden [6]. The hydraulic lime cycle is mentioned below in Figure 1:

Mediterranean Architectural Heritage - RIPAM10   Materials Research Forum LLC
Materials Research Proceedings 40 (2024) 97-108   https://doi.org/10.21741/9781644903117-10

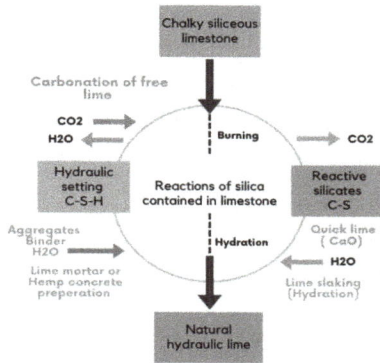

Figure 1: Hydraulic lime cycle [6]

Marrakech traditional lime

Lime was for long an essential conventional material in the construction of Moroccan structures. It's known for its wide range of uses, from building and rehabilitating architectural heritage to husbandry and crafts. For over eight centuries, Marrakech has been producing the lime plaster known as" Tadelakt", which boasts waterproofing and antibacterial rates, from lime made using traditional styles. The original occupants used this mortar admixture specifically to cover areas in contact with water. It consists substantially of lime, water, black cleaner, and several colorings [7].

The Tadelakt method almost disappeared from Morocco in the early 1970s, in favor of new pre-mixed ingredients that allowed for a faster application process. Renewed interest in traditional architecture, particularly in Marrakech, was a major factor in its return in the 1980s. Tadelakt has become a material appreciated all over the world for its waterproofing capabilities, durability, beauty, and variety of colors, as well as its increasingly appreciated ecological components [8]. Local architecture has always been based on this unique blend as shown in the picture bellow (Figure 2).

Figure 2: A Marrakech bathroom covered with Tadlakt [8]

Many studies have examined the remarkable properties of hydraulic lime, but few have focused on the elements and characteristics of traditional Marrakech lime. Accordingly, Amrani et al [8]

Mediterranean Architectural Heritage - RIPAM10   Materials Research Forum LLC
Materials Research Proceedings 40 (2024) 97-108   https://doi.org/10.21741/9781644903117-10

studied the slaked lime of Marrakech and succeeded in extracting its chemical and mineralogical characteristics, which are presented in the tables below:

Table 1 : Mineralogical composition of   Marrakech lime [8]

Mineralogical composition (%)	
Calcite	5.7 ± 0.1
Tricalcium aluminate	1.0 ± 0.3
Belite	13.3 ± 0.4
Katoite	5.7 ± 0.3
Portlandite	35.5 ± 0.6
Dolomite	2.1 ± 0.2
Quartz	9.5 ± 0.2
Periclase	0.8 ± 0.1
Palygorskite	1.0 ± 0.1
Brucite	1.0 ± 0.1
Amorphous and minor phases	24.0 ± 1.0
on dry sample (105°C)	

Table 2: Chemical composition of   Marrakech lime [8]

Chemical composition (%)	
CaO	65.77 ± 0.5
SiO_2	23.32 ± 0.5
Al_2O_3	4.01 ± 0.5
MgO	2.54 ± 0.5
Fe_2O_3	1.97 ± 0.5
SO_3	0.14 ± 0.5
K_2O	1.08 ± 0.5
Na_2O	0.59 ± 0.5
TiO_2	0.35 ± 0.5
SrO	0.06 ± 0.5
MnO	0.04 ± 0.5
P_2O_5	0.09 ± 0.5
As_2O_3	0.07 ± 0.5

Mediterranean Architectural Heritage - RIPAM10   Materials Research Forum LLC
Materials Research Proceedings 40 (2024) 97-108   https://doi.org/10.21741/9781644903117-10

Chemical characteristics of Marrakech lime

For this study, we used hydraulic lime, which is produced from marly limestone traditionally fired in the Marrakech region. A significant portion of the unfired marly limestone has been elaborated by this process, which was then combined with the siliciclastic fraction to form the aggregates of the mixture [7].

Scanning Electron Microscopy (SEM)

Microstructure analysis of Marrakech lime was carried out using a JEOL scanning electron microscope, to produce images of the sample surface as shown in Figure 3.

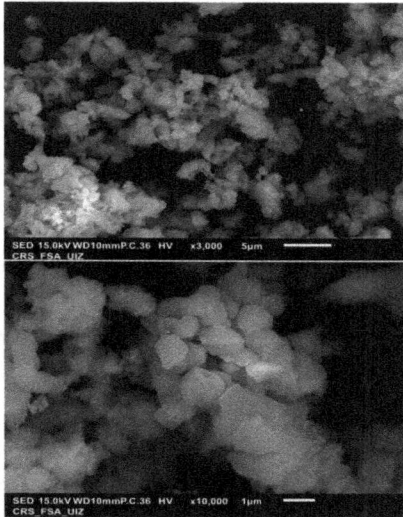

Figure 3: Scanning electron microscope (SEM) photos of Marrakech lime

These images show that Marrakech lime has a highly varied morphology, given its multiple mineralogical components. Minerals formed at low temperatures (around 1200°) are generally poorly crystallized, with small sizes of less than 5 μm [9]. The rhombohedral and orthorhombic crystals in the center of the SEM image represent calcite due either to the unfired portion of the parent rock (traditional firing does not allow decarbonization of the entire mixture) or to carbonation of the portlandite by moisture. The hexagonal layered structures visible in the SEM images represent portlandite, the main component of hydrated lime. Crystals in ovoid or globular form represent belite (C_2S). Further studies are needed to identify the other minority phases (Tricalcium aluminate, Katoite, Periclase, Palygorskite, and Brucite [8]), which appear as interstitial phases between the majority phases [10].

Energy-Dispersive X-ray Spectroscopy analysis (EDS)

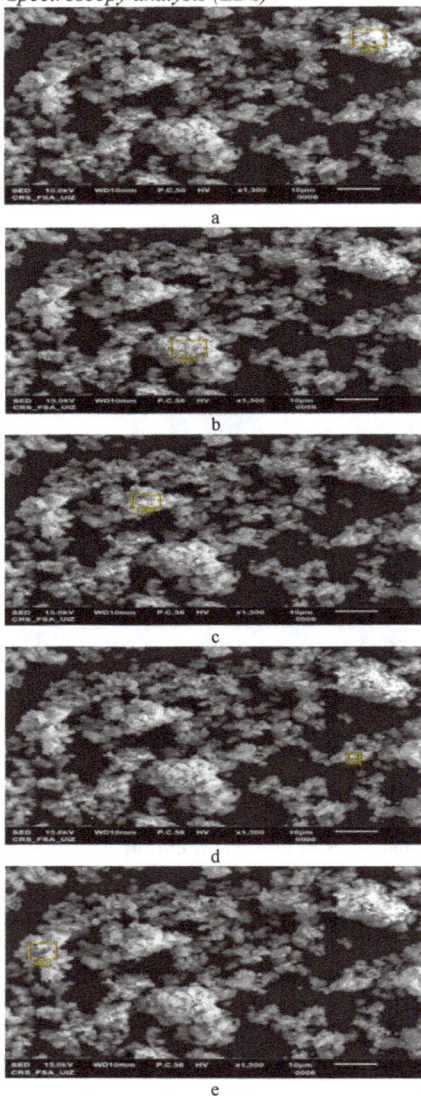

Figure 4: Areas scanned on the Marrakech lime sample concerned by EDS analysis (a: EDS001, b: EDS002, c: EDS003, d: EDS004, e: EDS005)

As shown in Figure 4, we used the energy dispersive spectrometry (EDS) technique to determine the phases present in the sample. Figure 5 below shows the EDS spectra of the scanned areas.

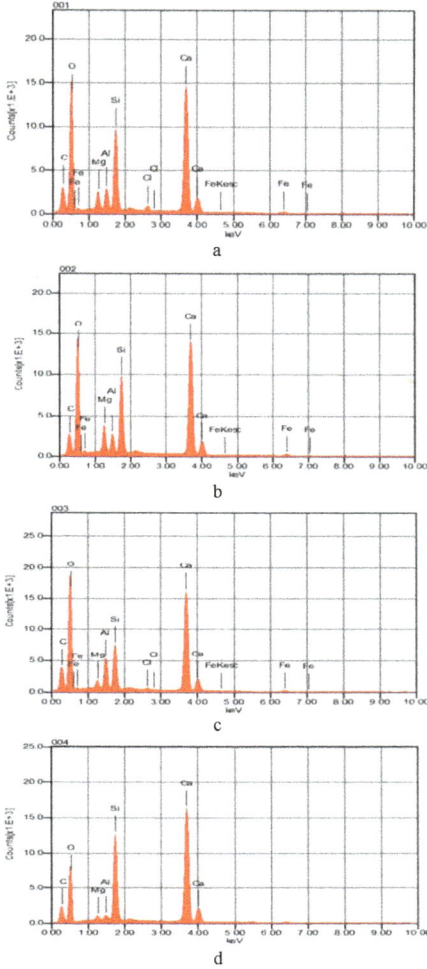

a

b

c

d

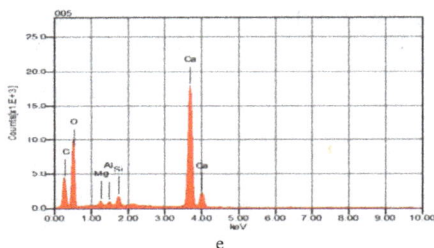

Figure 5: Spectrum of Marrakech lime in swept areas (a: S001, b: S002, c: S003, d: S004, e: S005)

According to the EDS analysis carried out, Marrakech lime consists mainly of calcium, silicon, and carbon oxides, with a minority chemical components such as magnesium, aluminum dioxides, and traces of chlorine and iron dioxide. Hydrogen and oxygen, the fundamental structural elements of portlandite, were not found in the analysis. Table 3 summarizes the results obtained.

Table 3: The quantification of Marrakech lime

	EDS1 mass %	EDS2 mass %	EDS3 mass %	EDS4 mass %	EDS5 mass %
C	8.46	7.32	9.38	6.35	9.38
O					
CaO	57.77	56.28	59.91	62.93	83.13
SiO_2	21.92	22.75	16.22	28.05	3.98
MgO	3.99	6.34	2.38	1.25	1.84
Al_2O_3	4.99	4.79	9.97	1.42	1.67
CI	0.94		0.53		
FeO		2.53			

X-ray diffraction analysis of Marralech lime

XRD analysis of the material was conducted using a Bruker D8 Eco X-ray diffractometer from the Agadir Faculty of Science. This diffractometer is connected to a computer by EVA operating software. It was used to record all the diffraction lines, identifying the different phases based on peak intensity analysis. Each mineral is characterized by these peaks. The secondary peaks are compared if two minerals have very similar first-order peaks. For XRD testing, CuKa radiation having a wavelength of l.54 A at a voltage of 30 KV, a scan speed of 3/min, and a current of 30 mA, was used.

Figure 6 shows the XRD diagram for Marrakech lime.

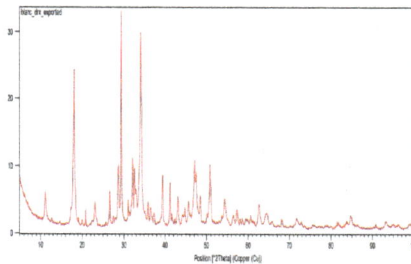

Figure 6: X-ray diffraction pattern of   Marrakech slaked lime powder (intensity as a function of peak positions)

The mineral distribution was extracted from the analysis of emission spectra according to the norms indicated by Brindly and Brown [11] using X'pert High Score software. Consequently, the sample consists of the majority phases of portlandite, calcite, quartz, and belite. As the study performed failed to identify the minority phases, further investigations are recommended.

Physico-mechanical characteristics of Marrakech lime

Physical properties
In order to derive the physical properties of the material in question, a particle size analysis was carried out for the fraction of soil larger than 80 µm, accompanied by a sedimentation test for fine particles following French standard NF P94-056, NF P94-057[12], [13]. Figure 7 shows the weight distribution of grain sizes in the lime used.

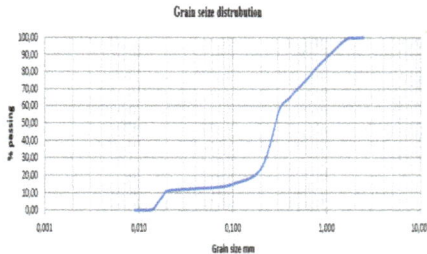

Figure 7: Diagram showing the weight distribution of grain sizes in the lime used.

It has been observed that 24% of the particles have a diameter of less than 200 µm by weight, and this fraction constitutes the binder phase of the mix [8]. Thus, the mixture is made up of three parts binder to three parts aggregate by weight.

Mechanical properties
- Sample preparation

First, we prepared the mixture of Marrakech hydraulic lime and water, adopting a Water/Lime ratio equal to 0.60 [14]. We then proceeded to fill the cylindrical test tubes, tightening them manually with a pricking bar. To get rid of air bubbles, each layer of concrete was struck 25 times. After 24 hours, we moved on to the demolding phase, keeping the specimens for 90 days under controlled laboratory conditions (T=22±3 and Rh= 65% ± 5).

Mediterranean Architectural Heritage - RIPAM10   Materials Research Forum LLC
Materials Research Proceedings 40 (2024) 97-108   https://doi.org/10.21741/9781644903117-10

- Testing equipment

To better characterize the material in its hardened state, we opted for a uniaxial compression test to derive ductility parameters, following French standards NF P94-420 and NF P94-425[15],[16]. The experimental setup (Figure 8) comprises a press with a maximum load capacity of 110 KN and a displacement sensor with an accuracy of ±0.05 mm. A displacement speed of 0.1 mm/min was selected for the test. The machine applies a progressively increasing compression force until the sample yields. It is connected to a computer that processes the results obtained and plots the stress-strain curve (Figure 9), facilitating the study of the mechanical behavior of the material in question. All the work was carried out in the laboratory of Agadir National School of Applied Sciences (NSAS).

Figure 8: Picture of the compression machine taken in the NSAS laboratory

- Results and discussion

To evaluate the resistance of Marrakech lime concrete, we were able to determine the ultimate stresses and maximum strains shown in Table 4 from the stress-strain diagram illustrated below:

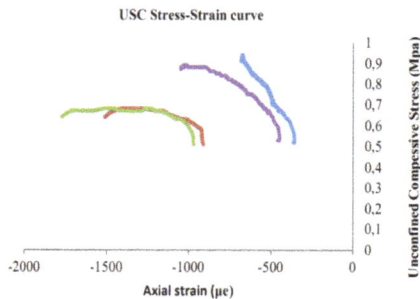

Figure 9: Stress-strain diagram generated by the computer for four specimens

Table 4: Stress-strain results for the four samples generated by the computer

	ρ g/cm³	Ultim. Strength σ_u Mpa	Fracture Strength σ_f MPa	Strain at maximum stress ε_u µe	Ultimate strain ε_f µe
Specimen 1	1.217	0.93	0.88	-671	-681
Specimen 2	1.143	0.89	0.82	-1017	-1033
Specimen 3	1.110	0.68	0.64	-1129.6	-1798.8
Specimen 4	1.108	0.68	0.6	-1278.8	-1516.6
AVG	1.144	0.795	0.74	-1024.1	-1257.35

The ultimate stresses obtained vary between 0.68 and 0.93 MPa, low values due mainly to the sandy nature of the aggregates, as well as to the traditional hydraulic lime production procedure, which does not effectively control the quality of the final product. Maximum deformations range from 681 to 1798.8 µe. These rather low values correspond well to the low values for stress at break. Given that the device was unable to record deformations relative to stresses under 0.5MPa as shown in Figure 9, we cannot conclude the elastic characteristics of the hydraulic lime concrete used.

Conclusion and recommendations:
This article provides a concise and accurate review of the literature on hydraulic lime, its characteristics, field of use, and utility. It's a rich material, used since antiquity, with ecological and architectural benefits.

Analyses carried out on samples from the Marrakech region have demonstrated the major constituents of this material. It consists mainly of portlandite, calcite, quartz and belite. Calcite was found to have a high percentage due to the presence of unfired marl, in addition to the phenomenon of carbonation due to the CO_2 present in the air. The iron, aluminum, and magnesium oxides detected by EDS analysis are linked to the presence of Katoite, Dolomite, Periclase, Palygoskite, and Brucite minerals, minority phases characterizing Marrakech lime [8].

Marrakech lime was also tested from a mechanical point of view, by assessing its strength in the hardened state, opting for a uniaxial compression test. It was confirmed that lime concrete is not resistant to compression, given the low values obtained, which did not exceed 1MPa. This opens up a wide avenue for researchers to consider ways of strengthening this useful material and contributing to its conservation.

Recommended ideas for future studies include:
- Improve lime performance by incorporating reinforcing fibers.
- Set up units for the production of traditional hydraulic lime in Morocco.
- Draw up a guide to the use of lime to encourage its use in construction.
- Implement reliable methods for reformulating lime concretes.
- Apply Carbon cure technology to improve the performance of lime concretes.

References

[1] A. Aziz *et al.*, « Effect of slaked lime on the geopolymers synthesis of natural pozzolan from Moroccan Middle Atlas », *Journal of the Australian Ceramic Society*, vol. 56, p. 67, mai 2019 . https://doi.org/10.1007/s41779-019-00361-3

[2] B. Abu-Jdayil, A.-H. Mourad, W. Hittini, M. Hassan, et S. Hameedi, « Traditional, state-of-the-art and renewable thermal building insulation materials: An overview », *Construction and Building Materials*, vol. 214, p. 709-735, juill. 2019 .
https://doi.org/10.1016/j.conbuildmat.2019.04.102

[3] M. del M. Barbero-Barrera, L. Maldonado-Ramos, K. Van Balen, A. García-Santos, et F. J. Neila-González, « Lime render layers: An overview of their properties », *Journal of Cultural Heritage*, vol. 15, n° 3, p. 326-330, mai 2014 . https://doi.org/10.1016/j.culher.2013.07.004

[4] R. Veiga, « Air lime mortars: What else do we need to know to apply them in conservation and rehabilitation interventions? A review », *Construction and Building Materials*, vol. 157, p. 132-140, déc. 2017 . https://doi.org/10.1016/j.conbuildmat.2017.09.080

[5] N. D. Copsey, « A CRITICAL REVIEW OF HISTORIC LITERATURE CONCERNING TRADITIONAL LIME AND EARTH-LIME MORTARS ».

[6] M. Chabannes, E. Garcia-Diaz, L. Clerc, J.-C. Bénézet, et F. Becquart, *Lime Hemp and Rice Husk-Based Concretes for Building Envelopes*. in SpringerBriefs in Molecular Science. Cham: Springer International Publishing, 2018. doi: 10.1007/978-3-319-67660-9

[7] A. Belabid, H. Elminor, et H. Akhzouz, « Study of parameters influencing the setting of hydraulic lime concrete: an overview », vol. 14, n° 3, 2023.

[8] A. E. Amrani *et al.*, « From the stone to the lime for Tadelakt: Marrakesh traditional plaster », 2018.

[9] « W. Yanmou, D. Junan, and S. Muzhen, "Sub theme 1.3," in ´ Proceedings of the 8th International Congress on the Chemistry of Cement (ICCC '86), vol. 2, pp. 363–371, Rio de Janeiro, Brazil, 1986. »

[10] I. Maki, « Morphology of the so-called prismatic phase in Portland cement clinker », *Cement and Concrete Research*, vol. 4, n° 1, p. 87-97, janv. 1974 . https://doi.org/10.1016/0008-8846(74)90068-4

[11] G. Brown, *Crystal Structures of Clay Minerals and their X-Ray Identification*. The Mineralogical Society of Great Britain and Ireland, 1982.

[12] « (NF P94-056, 1996) Standard AFNOR: NF P94-056, reconnaissance et essais – Analyse granulométrique - Méthode par tamisage à sec après lavage, 1996. »

[13] « (NF P94-057, 1992) Standard AFNOR: NF P94-057, Sols : reconnaissance et essais - Analyse granulométrique des sols - Méthode par sédimentation, 1992. »

[14] M. M. Barbero-Barrera, N. Flores Medina, et C. Guardia-Martín, « Influence of the addition of waste graphite powder on the physical and microstructural performance of hydraulic lime pastes », *Construction and Building Materials*, vol. 149, p. 599-611, sept. 2017 .
https://doi.org/10.1016/j.conbuildmat.2017.05.156

[15] « (NF P94-420, 2000) Standard AFNOR: NF P94-420, Détermination de la résistance à la compression uniaxiale, 2000. »

[16] « (NF P94-425, 2002) Standard AFNOR: NF P94-425, Méthodes d'essai pour roches - Détermination du module d'Young et du coefficient de Poisson, 2002. »

Mediterranean Architectural Heritage - RIPAM10
Materials Research Proceedings 40 (2024) 109-118

Materials Research Forum LLC
https://doi.org/10.21741/9781644903117-11

Experimental Characterization of the Thermal and Mechanical Properties of Earth Bricks Stabilized by Alkaline Solution and Reinforced with Natural Fibers: A Comparative Study

Mohamed CHAR[1,a,*], Amine TILIOUA[1,b], Fouad MOUSSAOUI[1,c], Youssef KHRISSI[1,d]

[1] Research Team in Thermal and Applied Thermodynamics (2.T.A.), Mechanics, Energy Efficiency and Renewable Energies Laboratory (L.M.3.E.R.), Department of Physics, Faculty of Sciences and Techniques Errachidia, Moulay Ismaïl University of Meknès, B.P. 509, Boutalamine, Errachidia, Morocco

[a]char.mohamed94@gmail.com, [b]tiliouamine@yahoo.fr, [c]bakkilimoussaoui@gmail.com, [d]youssefkhrissi@gmail.com

Keywords: Earth Bricks, Natural Fibers, Alkaline Solution, Thermal Properties, Mechanical Properties, Building Materials

Abstract. The use of earth bricks is a sustainable and environmentally friendly alternative to traditional building materials. However, these bricks can be vulnerable to erosion and extreme climatic conditions, which may limit their use in arid and semi-arid regions. In this experimental study, the aim was to improve the thermal and mechanical properties of earth bricks by stabilizing them with a mixture of alkaline solution and reinforcing them with natural fibers (maize, reeds, and olive). To this end, we designed earth bricks with different fiber percentages (from 0% to 8% of soil weight) and a fixed percentage of alkaline solution of 1.5%. After 28 days, the bricks were subjected to an experimental study to assess their thermal and mechanical cleanliness. The results showed that bricks stabilized with a fiber percentage of 2% and 3% had the best mechanical properties. They also showed an increase in thermal resistance as the percentage of fibers used increased. In addition, these bricks had higher compressive and tensile strengths than unstabilized bricks. This experimental study demonstrated that stabilizing earth bricks with a mixture of alkaline solution and fibers significantly improved their thermal properties.

1. Introduction

in the field of construction and energy efficiency, it's essential to find sustainable, environmentally friendly building materials. Vernacular building materials such as earth have many environmental advantages: local availability, low embodied energy, low CO2 emissions, and biodegradability. They also offer social and economic benefits. However, The building industry is responsible for a third of all emissions, uses 40% of materials, and generates 40% of waste[1]. Using earth bricks in construction can cut carbon dioxide emissions by 80% compared to using fired bricks[2]. Furthermore, earth bricks have a lower structural strength than modern building materials such as reinforced concrete. They can be more prone to cracking and spalling, particularly under extreme climatic conditions such as earthquakes or heavy rain. Earth bricks are also sensitive to moisture. If not properly protected against water, they can deteriorate rapidly.

Stabilizing earth bricks can improve their structural strength and durability. This can be achieved by adding materials such as straw, renewable palm fibers, plant fibers, cement, lime or other admixtures to the raw earth during brick manufacture to strengthen the bricks and make them more resistant to weathering and mechanical stress. A study by A. Thennarasan Latha et al. examined the impact of adding soda-treated sisal fibers and cement to stabilized compressed earth blocks (CSEB). The findings revealed that the compressive strength of CSEB increases with the addition of cement up to 10%, but decreases beyond that. Additionally, the incorporation of 1%

30 mm sisal fibers with 10% cement resulted in the highest compressive strength, and the flexural strength of CSEB also improved with the inclusion of sisal fibers [3]. Another study by K. Bougtaib et al. investigated the effects of adding doum palm fibers and lime to hand-made compressed earth bricks (CEB). The results indicated that 10% lime and 1% fiber were the optimal proportions for maximizing the performance of earth bricks [5]. Furthermore, A. Jesudass, V. Gayathri, and R. Geethan et al. explored the use of natural fibers in earth blocks as a sustainable solution for enhancing thermal performance and durability [4]. E.B. Ojo et al. assessed the reinforcing effects of various fibers (sisal, eucalyptus pulp, polypropylene) and alkaline activators (NaOH and Na silicate) on the thermal and physical properties of earth bricks. The study revealed that a 2% sisal content provided the best strength-density balance, and polypropylene conferred strain-hardening behavior [6]. K. AlShuhail, A. Aldawoud, and J. Syarif et al. investigated how the addition of natural additives such as date palm fibers enhances the mechanical and physical performance of compressed earth bricks, making them more resistant to moisture [7]..(R. Mateus et al.) carried out an Environmental Life Cycle Assessment of earth building materials.This analysis found that vernacular earth building materials have a low environmental impact thanks to their local availability and low-energy manufacturing processes. They can help reduce the impact of buildings [8]. C.R. Ganesh, J. Sumalatha, K.S. Sreekeshava et al., the impact of adding ground granulated blast furnace slag (GGBS) and polypropylene geofibers on the properties of compressed stabilized earth blocks (CSEB) was investigated. The results showed that increasing GGBS enhanced the dry and wet compressive strength of CSEBs, while slightly reducing water absorption and block density. The optimal combination was found to be 30% GGBS, 5% lime, and 0.2% geofibers [9]. I. Bouchefra et al. examined the effects of treated and untreated doum fibers as an eco-friendly material. The study revealed that as the fiber content increased, the density of the fiber-reinforced composite decreased, with a reduction of up to 15.2% observed with 2% raw fiber. The compressive strength also decreased with higher fiber content, showing a decrease of 25% with treated fibers and 35% with 2% raw fibers. Additionally, the addition of fibers led to a slight decrease in thermal conductivity, with a reduction of 27% observed with 2% treated fiber. Increasing the compaction pressure from 4.75 to 9.7 MPa resulted in a 50% increase in compressive strength and an 11% increase in thermal conductivity. Furthermore, increasing the fiber length from 0.4 to 2.5 cm led to a 20% reduction in compressive strength. It was also noted that a cement content of 10% resulted in increased compressive strength by 7.5%, while a lime content of 9% provided maximum strength [10]. M. Saidi et al., the influence of stabilization on the thermal conductivity and moisture absorption capacity of compressed earth bricks was examined. The results indicated that the thermal conductivity of the earth bricks increased with higher cement and lime content [11].

The primary objective of this comprehensive comparative study is to evaluate both the thermal and mechanical characteristics of earth bricks that have been stabilized with an alkaline solution and reinforced with a range of renewable natural fibers. These fibers include reed, corn, and olive, and the study aims to analyze how these additions impact the overall properties of the earth bricks, with a specific focus on their thermal conductivity, density, and mechanical strength.

2. Materials and methods

2.1 Materials

2.1.1 Soil

The soil used in our study comes from the Errachidia region in south-eastern Morocco. This soil is a very fine material and is widely used in construction work due to its ideal properties for certain types of structures (fig.1) . Errachidia clay is renowned for its ability to absorb water, making it highly resistant to temperature fluctuations. This property is particularly useful in regions where temperatures can vary considerably over the course of a day or season.

Fig.1 : The soil of Errachidia

2.1.2 Alkaline Solution

For the synthesis of geopolymers between soil and alkaline solution, we used commercial alkaline solutions of sodium aluminosilicate (AlNa12SiO5) and sodium hydroxide (NaOH) (fig.2) . These alkaline solutions, also known as activator solutions, are used to form geopolymers, which are inorganic polymers formed by a chemical reaction between a siliceous material and an alkaline activator. Geopolymers have a dense crystalline structure and are known for their resistance to deformation and degradation, making them highly durable building materials.

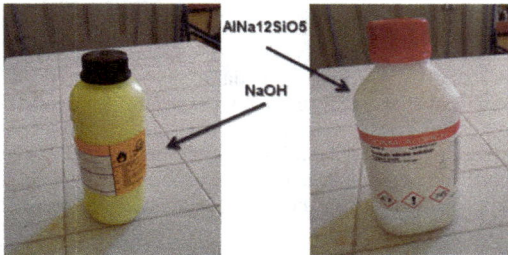

Fig.2. Alkaline solutions of sodium aluminosilicate (AlNa12SiO5) and sodium hydroxide (NaOH).

In this study, we utilized renewable natural fibers such as reeds, corn fibers, and olive fibers. The reeds were harvested from the oases of Errachidia, near the river. The corn and olive fibers were obtained from corn and olive waste, which were collected and cleaned before being dried for use (fig.3).

Fig.3 Fibers of (a): reeds, (b): corn, (c): olive.

111

Mediterranean Architectural Heritage - RIPAM10   Materials Research Forum LLC
Materials Research Proceedings 40 (2024) 109-118   https://doi.org/10.21741/9781644903117-11

2.2 The experimental procedure

Sample preparation involved mixing soil and fibers in varying percentages ranging from 0% to 8%. Next, we added the alkaline solution and water, with a fixed optimum percentage of alkaline solution of 1.5% by weight of water. Mixing was carried out manually, then the bricks were designed using specific molds. The dimensions of the molds are $24\times24\times4$ cm^3 (for thermal testing), and cylindrical molds with diameter D= 10 cm and height H= 20 cm (for mechanical testing). Sample preparation involved mixing soil and fibers in varying percentages ranging from 0% to 8%. Next, we added the alkaline solution and water, with a fixed optimum percentage of alkaline solution of 1.5% by weight of water. Mixing was carried out manually, then the bricks were designed using specific molds. The dimensions of the molds are $24\times24\times4$ cm^3 (for thermal testing), and cylindrical molds with diameter D= 10 cm and height H= 20 cm (for mechanical testing).

Fig.4. Bricks prepared after demolition.

After demolding, the bricks were carefully removed from the molds and placed in our laboratory for a natural drying process. This step was essential to allow the bricks to harden and stabilize before being subjected to further testing. During this 28-day drying period, the bricks were kept in a controlled environment. This was to avoid any distortion or deformation of the bricks during the drying process. Natural drying allowed the bricks to gradually lose their water content, helping to strengthen their structure and improve their mechanical strength. This drying stage is crucial to guaranteeing the quality and durability of the bricks, minimizing the risk of subsequent cracking or deformation. (fig.4) .

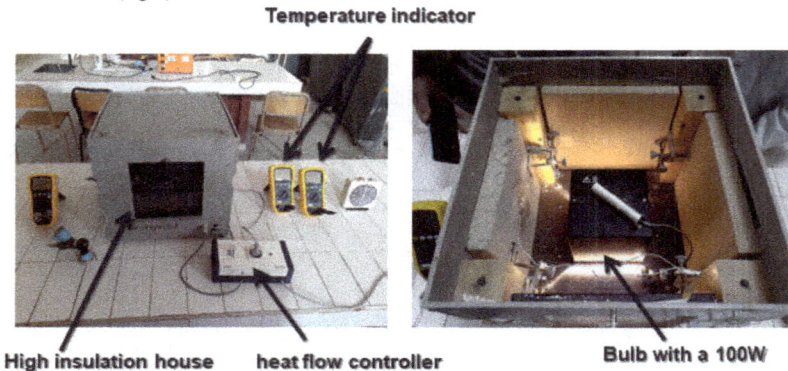

Fig.5 : Thermal house

Mediterranean Architectural Heritage - RIPAM10
Materials Research Proceedings 40 (2024) 109-118

Materials Research Forum LLC
https://doi.org/10.21741/9781644903117-11

In this study, we utilized a thermal conductivity measurement device known as the thermal house (fig.5) with interchangeable walls measuring 40x40x40 cm³. This apparatus has been extensively detailed in various publications for determining the thermal conductivity of building materials in controlled laboratory conditions, with an uncertainty of under 10% [13-12]. The thermal enclosure consists of four side walls, each featuring a square aperture measuring 21x21 cm², which can be sealed using a 5 cm thick polystyrene sheet known for its exceptional insulation properties. To produce heat and minimize the influence of radiation, a 100 W incandescent lamp enclosed in a small black box was utilized. Furthermore, a thermal regulator was employed to sustain an internal temperature of 50°C, while an air conditioner was used to maintain the room temperature at 20°C.

During the trial, the specimen to be analyzed, measuring 24x24x4 cm³, was positioned on one of the side walls of the thermal enclosure, with the remaining walls sealed using polystyrene panels to prevent heat dissipation. Thermocouples with an uncertainty of +/-0.7°C were employed to monitor temperature changes inside and outside the thermal enclosure. These temperature fluctuations were recorded using a data acquisition system. Once a stable condition was reached (approximately 6 hours after the commencement of each measurement), the temperature data were utilized to estimate the thermal conductivity.

In computing the thermal conductivity, we took into account the conservation of heat flow arising from convection between the internal air and the inner wall surface, as well as convection between the external air and the outer wall surface. Additionally, conduction through the sample was taken into account:

$$\varphi = h_{int}(T_{loc} - T_{int})S = S \times \lambda \frac{T_{int} - T_{ext}}{e} = h_{ext}(T_{ext} - T_{amb})S \qquad (Eq.1)$$

With :
- h_{int}: internal convection coefficient.
- h_{ext}: external convection coefficient.
- T_{int}: temperature of the inside of walls.
- T_{ext}: temperature of the outside of the wall.
- T_{loc}: air temperature inside thermal house.
- T_{amb}: ambient temperature.
- e: wall thickness.

We calculated the thermal conductivity (λ) by applying Equation (Eq.1) with a heat transfer coefficient of 8.1 W/m²K [14,13]. The assessment of our samples' thermal conductivity adhered to the prescribed methodology outlined by the manufacturer (PHYWE). This procedure entailed the utilization of a PHYWE thermal control stand, along with a set of thermocouples for monitoring temperature fluctuations both inside and outside the thermal housing [15]. We recorded the readings once the thermocouples had stabilized, ensuring that the system was close to equilibrium (approximately 8 hours after the start of the measurement)

$$h_{int}(T_{loc} - T_{int}) = \frac{\lambda}{e}(T_{int} - T_{ext}) \qquad (Eq.2)$$

$$\lambda = \frac{(T_{loc} - T_{int})}{(T_{int} - T_{ext})} \times h_{int} \times e \qquad (Eq.3)$$

with:
- e = 0,04 m.

Mediterranean Architectural Heritage - RIPAM10   Materials Research Forum LLC
Materials Research Proceedings 40 (2024) 109-118   https://doi.org/10.21741/9781644903117-11

To assess the mechanical compressive and tensile strength of the samples, we used a FORM TEST SEIDNER hydraulic press (fig. 6). This device applies a controlled mechanical force to the samples to measure their strength.

The force applied to the samples during compression and tensile tests is measured using a load cell integrated into the hydraulic press. This load cell converts the applied force into a numerical value, which is then used to calculate the mechanical strength of the samples.

Mechanical strength is calculated using the following relationship:

$$\text{Mechanical strenght} = \frac{Applied\ force}{Sample\ surface} = \frac{F}{r^2 * \pi} \qquad \text{(Eq.4)}$$

Fig.6: SEIDNER FORM TEST 500KN hydraulic press

3. Results and discussion

3.1 The effect of natural fiber content and alkaline solution

3.1.1 Conductivity and thermal resistance

The findings of the study indicate that the thermal properties of earth bricks are affected by the presence of reed, corn, and olive fibers, as well as the type of alkaline solution used for stabilization. The study maintained the percentage of alkaline solution at 1.5% relative to the weight of water.

Figure 7 illustrates the changes in thermal conductivity and thermal resistance with varying fiber content. It was observed that the thermal conductivity of alkaline-stabilized earth bricks reinforced with natural fibers decreases as the fiber content increases. For corn fiber contents ranging from 0% to 8%, the thermal conductivity of the blocks was measured at 0.702, 0.7, 0.642, 0.6, 0.58, 0.42, 0.41, 0.38, and 0.32 W/mK. Similarly, the thermal conductivity of the blocks reinforced with reed and olive fibers reached values of 0.566 W/mK and 0.501 W/mK, respectively, for a fiber content of 8%. Additionally, it was noted that the thermal resistance of the earth bricks increased in proportion to the increase in fiber content. This is due to the fibers creating air spaces within the bricks, which act as thermal insulation. These air pockets impede the transfer of heat through the material, resulting in an increase in thermal resistance [16,17,18].

This study's results align with previous research demonstrating the impact of natural fibers on the thermal properties of construction materials [19,20]. The findings suggest that the incorporation of natural fibers in earth bricks can be a viable strategy for enhancing their thermal performance, potentially leading to more energy-efficient and sustainable construction practices.

Mediterranean Architectural Heritage - RIPAM10   Materials Research Forum LLC
Materials Research Proceedings 40 (2024) 109-118   https://doi.org/10.21741/9781644903117-11

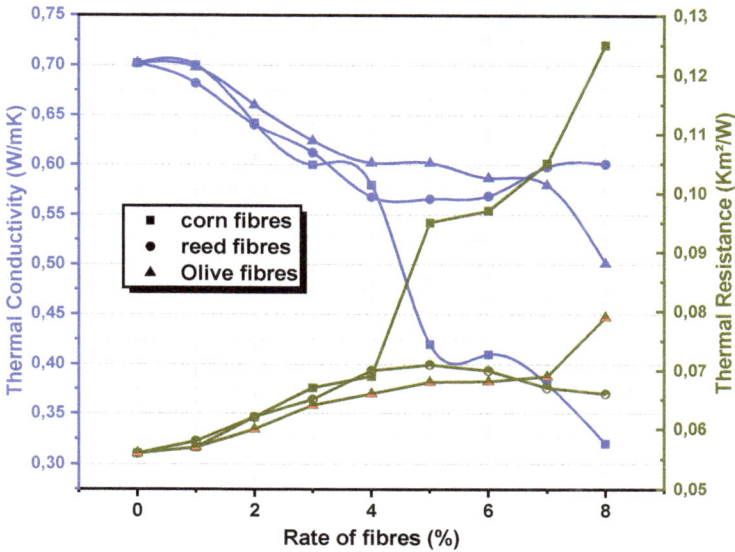

Fig.7. The variation of thermal conductivity and resistance of the clay bricks studied.

The incorporation of reed, corn, and olive fibers in alkaline-stabilized earth bricks leads to a reduction in thermal conductivity and an increase in thermal resistance. This highlights the potential of using natural fibers as reinforcement in earth construction to enhance thermal insulation properties and improve energy efficiency in buildings.

3.1.2 Compressive and tensile strength

In our mechanical testing, we examined both unstabilized and unreinforced bricks, and found that they exhibited compressive and tensile strengths of 4.2 MPa and 0.82 MPa, respectively (see fig.8). Upon the addition of corn fiber and alkaline solution, the compressive strength increased to a peak of 5.78 MPa with 2% corn fiber, after which it began to decrease. Meanwhile, the tensile strength rose to 1.6 MPa with 1% fiber content. When reed fiber was introduced, the compressive strength reached 6.1 MPa with 3% fiber, and the tensile strength increased to 1.8 MPa. Finally, olive fibers led to a maximum compressive strength of 6.4 MPa with 2% fiber content, and a maximum tensile strength of 1.8 MPa with 3% fiber content.

These results are consistent with previous studies that have demonstrated the potential of natural fibers to enhance the mechanical properties of construction materials [21,22,23]. The findings suggest that the incorporation of specific natural fibers, such as corn, reed, and olive fibers, in earth bricks can significantly improve their compressive and tensile strengths, which is crucial for the development of more durable and resilient construction materials.

Mediterranean Architectural Heritage - RIPAM10   Materials Research Forum LLC
Materials Research Proceedings 40 (2024) 109-118   https://doi.org/10.21741/9781644903117-11

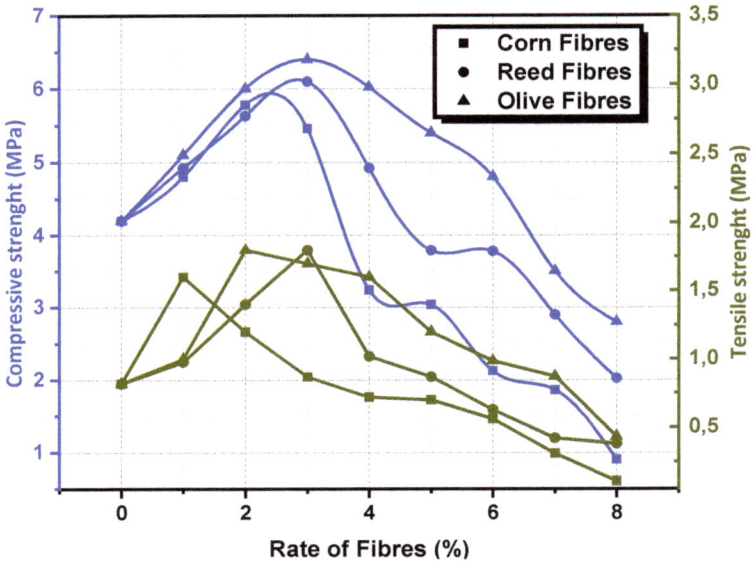

Fig.8: the variations of compressive and tensile strength of samples

Above a certain optimum percentage (between 2% and 3%, depending on the type of fiber), the addition of more fiber can lead to a reduction in the mechanical strength of the composite material. This can be due to several factors. Firstly, too many fibers can overload the material, compromising its cohesion and overall strength. In addition, too high a fiber content can lead to poor dispersion of the fibers in the matrix, reducing their effectiveness in terms of reinforcement.

However, it is important to note that mechanical performance also depends on the type of fibers used. In the case of olive fibers, they appear to offer the best mechanical performance for this composite material. Olive fibers have unique characteristics that make them particularly suitable for reinforcing the material matrix.

Conclusion

The objective of this experimental investigation was to evaluate the impact of incorporating renewable natural fibers such as reeds, corn, and olives, along with an alkaline solution comprising sodium aluminosilicate (AlNa12SiO5) and sodium hydroxide (NaOH), on the thermal and mechanical characteristics of earth bricks. The findings demonstrated that the utilization of these additives in stabilizing earth bricks significantly enhanced their performance. The introduction of fibers resulted in a decrease in thermal conductivity and an enhancement in the thermal resistance of the bricks, thereby imparting improved insulating properties. Moreover, the fibers contributed to an enhancement in the mechanical compressive and tensile strength of the bricks. Optimal outcomes were achieved with a fiber dosage of approximately 2-3% by mass and a consistent percentage of alkaline solution at 1.5%. Beyond this dosage, an excess of fibers tended to diminish the mechanical properties of the bricks. Among the various fibers examined, olive fibers exhibited the most favorable performance in this study.

These results underscore the substantial potential of stabilized earth bricks for environmentally friendly and sustainable construction, particularly in semi-arid regions where they can endure extreme climatic conditions. Their production from local materials renders them a promising substitute for traditional materials. Further research will be essential to refine the performance of these bricks and investigate their applicability in diverse construction scenarios.

References

[1] D.A. Ness, K.e. Xing, Toward a Resource-Effiient Built Environment: A Literature Review and Conceptual Model, J. Ind. Ecol. 21 (3) (2017) 572–592, https://doi.org/10.1111/jiec.12586

[2] J.E. Oti, J.M. Kinuthia, J. Bai, Engineering properties of unfied clay masonry bricks, Eng. Geol. 107 (3-4) (2009) 130–139. Dio https://doi.org/10.1016/j.enggeo.2009.05.002.

[3] A. Thennarasan Latha, B. Murugesan and B. Skariah Thomas, Compressed earth block reinforced with sisal fiber and stabilized with cement: Manual compaction procedure and influence of addition on mechanical properties, Materials Today: Proceedings, https://doi.org/10.1016/j.matpr.2023.04.373

[4] A. Jesudass, V. Gayathri, R. Geethan, M. Gobirajan, M. Venkatesh, Earthen blocks with natural fibres - A review, Materials Today: Proceedings 45 (2021) 6979–6986 DIO : https://doi.org/10.1016/j.matpr.2021.01.434

[5] K. Bougtaib, Y. Jamil, S. Nasla, K. Gueraoui and M. Cherraj, Compressed earth blocks reinforced with fibers (doum palm) and stabilized with lime: manual compaction procedure and influence of addition on mechanical properties and durability, JP Journal of Heat and Mass Transfer 26 (2022), 157-177. http://dx.doi.org/10.17654/0973576322018

[6] E.B. Ojo, K.O. Bello, K. Mustapha, R.S. Teixeira, S.F. Santos, H.S Jr, Effects of fibre reinforcements on properties of extruded alkali activated earthen building materials, Construction and Building Materials 227 (2019) 116778. Dio: https://doi.org/10.1016/j.conbuildmat.2019.116778 .

[7] K. AlShuhail, A. Aldawoud, J. Syarif, I. Abu Abdoun, Enhancing the performance of compressed soil bricks with natural additives: Wood chips and date palm fibers, Construction and Building Materials 295 (2021) 12361. Dio : https://doi.org/10.1016/j.conbuildmat.2021.123611

[8] R. Mateus, J. Fernandes, E.R Teixeira, Environmental Life Cycle Analysis of Earthen Building Materials, Elsevier Inc (2019).

[9] C.R. Ganesh, J. Sumalatha, K.S. Sreekeshava et al., Experimental study on strength behaviour of geofibre reinforced stabilized mud blocks using industrial by-products, Materials Today: Proceedings, https://doi.org/10.1016/j.matpr.2023.04.045

[10] I. Bouchefra, F. EL Bichri, H. Chehouani, B. Benhamou, Mechanical and thermophysical properties of compressed earth brick reinforced by raw and treated doum fibers, Construction and Building Materials 318 (2022) 126031. Dio: https://doi.org/10.1016/j.conbuildmat.2021.12603

[11] M. Saidi, A.S. Cherif, B. Zeghmati, E. Sediki, Stabilization effects on the thermal conductivity and sorption behavior of earth bricks, Construction and Building Materials 167 (2018) 566–577. Dio :https://doi.org/10.1016/j.conbuildmat.2018.02.063

[12] F.J.G. Silva, A. Baptista, G. Pinto, R.D.S.G. Campilho , M.C.S. Ribeiro, Characterization of hybrid pultruded structural products based on preforms, Compos B Eng 2018;140: 16–26.

[13] M.A. Navacerrada, P. Fernandez, C. Díaz, A. Pedrero, Thermal and acoustic properties of aluminium foams manufactured by the infiltration process, Appl Acoust 2013;74 (4):496–501

[14] A. Benallel, A. Tilioua, A. Mellaikhafi, A. Babaoui, M.A.A. Hamdi, Thermal characterization of insulating materials based on date palm particles and cardboard waste for use in thermal insulation building, in: AIP Conference Proceedings, AIP Publishing LLC, 2021, pp. 020024

[15] PHYWE, P2360300 PHYWE Series of Publications, Laboratory Experiments, Physics, PHYWE SYSTEME GMBH & Co. KG, Gottingen.

[16] H. Chaib, A. Kriker, A. Mekhermeche, Thermal Study of Earth Bricks Reinforced by Date palm Fibers, Energy Proc. 74 (2015) 919–925, https://doi.org/10.1016/j.egypro.2015.07.827

[17] B. Tayeh, M. Hadzima-Nyarko, M.Y.R. Riad, R.D.A. Hafez, Behavior of ultrahigh-performance concrete with hybrid synthetic fiber waste exposed to elevated temperatures, Buildings 13 (1) (2023), https://doi.org/10.3390/buildings13010129

[18] P.O. Awoyera, A.D. Akinrinade, A.G. de Sousa Galdino, F. Althoey, M.S. Kirgiz, B.A. Tayeh, Thermal insulation and mechanical characteristics of cement mortar reinforced with mineral wool and rice straw fibers, J. Build. Eng. 53 (2022), https://doi.org/10.1016/j.jobe.2022.104568

[19] S. Hakkoum, A. Kriker, A. Mekhermeche, Thermal characteristics of Model houses Manufactured by date palm fiber reinforced earth bricks in desert regions of Ouargla Algeria, Energy Proc. 119 (2017) 662–669, https://doi.org/10.1016/j.egypro.2017.07.093

[20] B. Taallah, A. Guettala, The mechanical and physical properties of compressed earth block stabilized with lime and filled with untreated and alkali-treated date palm fibers, Constr. Build. Mater. 104 (2016) 52–62, https://doi.org/10.1016/j.conbuildmat.2015.12.007

[21] S.M. Marandi, M.H. Bagheripour, R. Rahgozar, H. Zare, Strength and ductility of randomly distributed palm fibers reinforced silty-sand soils, Am. J. Appl. Sci. 5 (3) (2008) 209–220, https://doi.org/10.3844/ajassp.2008.209.220

[22] K. Alshuhail, A. Aldawoud, J. Syarif, I. Abu, Enhancing the performance of compressed soil bricks with natural additives: Wood chips and date palm fibers, Constr. Build. Mater. 295 (2021) 123611, https://doi.org/10.1016/j.conbuildmat.2021.123611

[23] B. Taallah, A. Guettala, S. Guettala, A. Kriker, Mechanical properties and hygroscopicity behavior of compressed earth block filled by date palm fibers, Constr. Build. Mater. 59 (2014) 161–168, https://doi.org/10.1016/j.conbuildmat.2014.02.058

Mediterranean Architectural Heritage - RIPAM10   Materials Research Forum LLC
Materials Research Proceedings 40 (2024) 119-126   https://doi.org/10.21741/9781644903117-12

The Role of Local Population in Safeguarding Heritage Case of Chefchaouen

Samia NAKKOUCH

National Institute of Planning and Urbanism (INAU), Morocco

samia.nakkouch@inau.ac.ma

Keywords: Participatory Initiatives, Inhabitants' Involvement, Heritage Through Practice, Chefchaouen, Morocco

Abstract. Local community engagement in heritage safeguarding has been a growing trend during the last two decades. With international models and charts that introduced the participatory paradigm into public restoration and interventions, community-based conservation approaches have started to emerge. The involvement of local inhabitants into restoration processes has proven effective contribution into sustaining public interventions and developing behaviors that mitigate the risks of deterioration of architectural heritage. The case of Chefchaouen has demonstrated multiples initiatives taken by inhabitants and local council towards chaouni medina aged from de 15th century. Made of stone, earth, wood and lime, the participatory dynamic around safeguarding the fabric has contributed to improve public efforts of restoration and rehabilitation. The contribution has gone beyond raising awareness to tangible actions and engagement. The article purpose is to analyse the modalities and role of inhabitants involvement into sustaining heritage. It presents the results of an empiric research based on qualitative data and analysis, and develops the dynamics around the medina of Chefchaouen, exploring the local initiatives and their effects on safeguarding heritage, with acknoledging the local configurations behind such dynamics.

Introduction

Safeguarding urban heritage has known during the last decade an increasing interest especially after the launch of the royal rehabilitation program of traditional cities (medinas) in December 2011. By definition, medinas are traditional fabrics that composed of historical landmarks, traditional houses, streets and patterns, built with local knowhow and traditional materials, reflecting a valuable genius loci [1].

As a living heritage, it encounters daily uses and transformations, which make its preservation challenging at many levels.The public efforts deployed to preserve and value the medinas as an important economic and cultural asset for local development are multiple. From architectural charters, to safeguarding plans, including requalification and upgrading programs, the repertory is large [2, 3].

In parallel, modalities of state control and action have slightly shifted toward involving inhabitants and civil society in general. Seeking legitimacy and effectiveness at a first priority, this renewal with a participatory paradigm have acknowledged the later potential of optimizing the public interventions. The injunction of inhabitants as emerging actors has been confirmed by local experiences through the world in addition to global development models, proving that effective protection of heritage depends on the involvement and support of the community continuing use and maintenance [4.5].

As social practices and anthropic factors have as much impact as other types of risks on the degradation of heritage, it has from the opposite perspective, the involvement of inhabitants as permanent actors on heritage may influence safeguarding process [6].

The potential highlighted of inhabitants involvement beside institutional actors and elected representatives are mainly articulated about a mutual understanding of social practice and constructive knowledge in the medina taken into account by the stakeholders for efficient and harmonized interventions. The expected effects emphasis the valorization of physical and historical identity and image of the medina, through the promotion of cultural heritage [7, 8].

The case of Chefchaouen had demonstrated a local dynamic around medina heritage, combining inhabitants and associations initiatives to a political leadership. The medina of Chefchaouen, a living traditional fabric from the 15th century, has known for more than a decade, multiples initiatives. The article focuses on understanding the local dynamic of collective management of heritage on mitigating the different risks of its deterioration.

The methods of collecting and analyzing data are mainly qualitative, including official documents and reports, interviews of different stakeholders and participatory observations[1].

Chefchouen Medina: a strong image of the traditional fabric

Worldwide known as the blue pearl of north morocco, the medina is located in the mountainous city of Chefchaouen, in the region of Tanger Tetouan El Hoceima. Chefchaouen City is located at 112km from the city of Tangier.

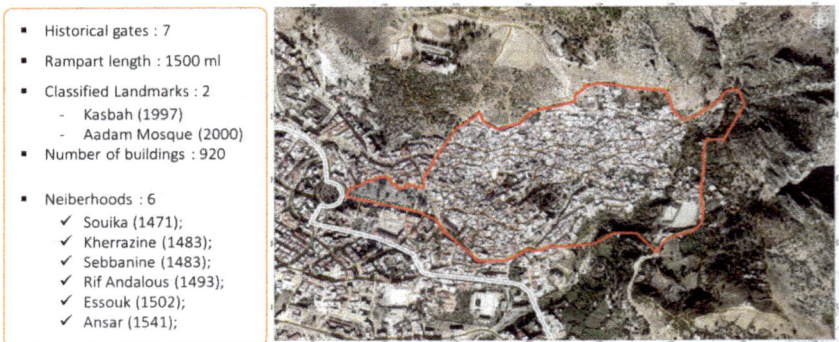

- Historical gates : 7
- Rampart length : 1500 ml
- Classified Landmarks : 2
    - Kasbah (1997)
    - Aadam Mosque (2000)
- Number of buildings : 920

- Neiberhoods : 6
    - ✓ Souika (1471);
    - ✓ Kherrazine (1483);
    - ✓ Sebbanine (1483);
    - ✓ Rif Andalous (1493);
    - ✓ Essouk (1502);
    - ✓ Ansar (1541);

Figure 1: Delimitation and characteristics of Chefchaouen Medina's fabric (Author)

The medina represents the core traditional fabric buit since de 15th century. It is implemented in an area exceeding 20 ha, and counts more 18 000 inhabitants, which nearly represents 42% of the whole city population. The medina is the most populated fabric in the city, with a density of 900 inahbitants /ha (Figure 2).

Although the medina is still in process of classification, a delay due to extreme densification and hybridization of heritage character, it maintains authentic and strong physical and historical image with its narrow alleys, monuments, gates and ramparts, and especially the shades of local materials (Table1).

---

[1] The period of investigation is 2019-2022.

Table 1 : Building materials of Chefchaouen medina's heritage (Author)

Building materials		Use
Stone	Calcareous and siliceous	Structural components
Terracota	Bricks and tiles	Roofs, part of walls, arches Masonry support elements Ceilings, awnings
Lime	Mortar / Plaster	Rendering exterior walls Plastering interiors Finishing roofpaving Protection of water infiltration
Timber Wood	Wooden beams (cedar, soha, red fir)	Roofs, ceilings, doors, lintels
Earth	Masonries, mortar	Rammed earth walls

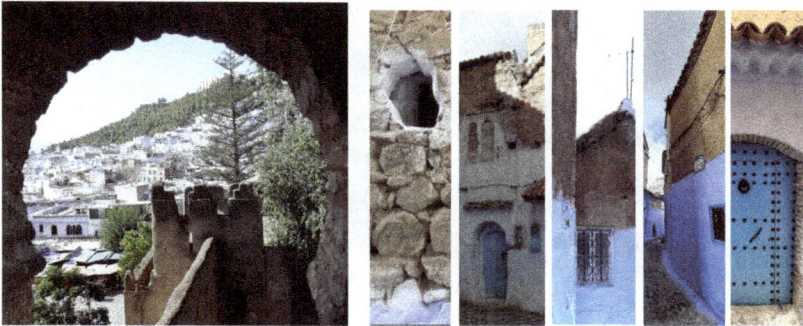

Figure 2 : Overview of the Medina of   Figure 2 : Compilation of materials and shades
Chefchaouen from the Kasbah (Author)   in Chefchoauen Medina (Author)

Between the ochre shades of stones, terracotta and earth mortal, in addition to the white color of lime plaster, and the indigo tints obtained by mixing lime to a woad plant extract called nila, the medina's fabric has an homogenous image (Figure 2, 3).

Its authenticity lays on the continuum of shads and subtle functional details such timber roofs, architrave doors and windows with different types of arches, awnings with curved tiles…etc[2]. Details that are extended to public realm with the counter-arches, fountains, traditional pavement, a whole hierarchized structure offering different experiences from private to public space [9].

For the safeguarding and valorization of such heritage, a large rehabilitation program has been implemented since 2011 mobilizing more than 115 Mdhs in rehabilitating the built environment, the landmarks, and enhance culture animation (Figures 4, 5). The interventions on built environment have cost more than half of the budget (66dhs) including facades, alleys, public spaces facilities maintenance and rehabilitation [10,11].

---

[2] Architectonic components are detailed in the architectural charter of the Medina of Chefchouen, published as a reference and guidelines to follow for the development of the city since 2006;

Mediterranean Architectural Heritage - RIPAM10   Materials Research Forum LLC
Materials Research Proceedings 40 (2024) 119-126   https://doi.org/10.21741/9781644903117-12

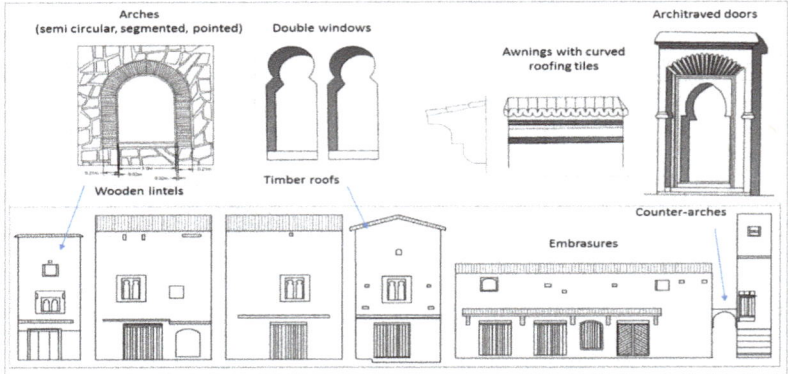

*Figure 4 : Constructive details taken into account in the restoration work
in Chefchaouen Medina (Author)*

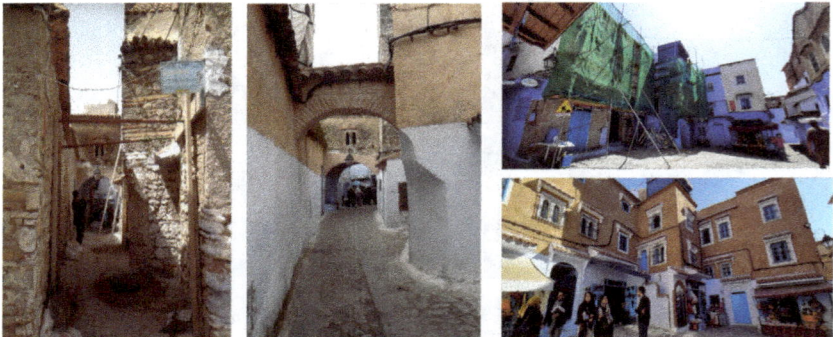

*Figure 3: Restoration work conducted in Chefchaouen medina by public actors
(AUT report, 2022)*

Approaching heritage through practice: participatory initiatives

Beside public actors and programs to rehabilitate the medina's heritage, local initiatives are conducted by inhabitants periodically, such as Al Awacher.

Awacher is annual event in chefchaouen that mobilises inhabitants in a collective maintenance operation, with the support of neighborhood associations and municipal services. Local population proceed to the liming of houses, streets, public paving…etc. and embellishment of their built environment with flower pots, street art and handicrafts.

As deep rooted tradition in chaouni's culture since the andalousian period, it has revived since 2010 from a social practice of families and women embellishing their home, to an extended public manifestation organized during "Chefchaouen spring" festival, covering a larger scale, and involving different age and gender groups of inhabitants, beside the municipality and neighborhoods associations' assistance (Figure6). This initiative has been promoting the local culture and reinforcing restoration efforts with the participatory and collaborative approach. The municipality provides the materials and the lime preparation to obtain a unified shade, while the associations assist inhabitants in the maintenance work [12, 13].

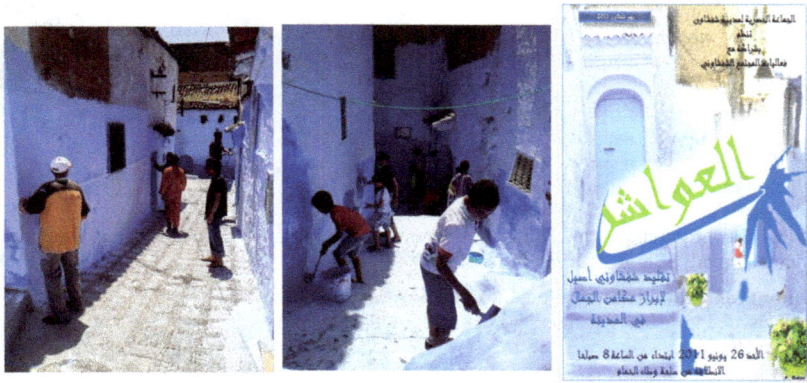

Figure 5 : Inhabitants 'maintenance work in Chefchaouen medina, poster of the initiative (Municipality archive. 2019)

Another initiative that involved inhabitants and neighborhood associations is the "participatory budget" for the rehabilitation of the medina that has been adopted as regular collective managing instrument[3]. As an instrument of co-decision, it consists on making available 2% of the investment budget of the municipality to inhabitants to decide interventions to implement in the medina, according to inhabitants needs, priorities and aspirations [14].

The process of the participatory budget progresses in several phases:

1. Diagnosis, propositions: the neighborhoods associations are involved to diagnose and make propositions of interventions.
2. Feasibility, prioritization: the driving group details their feasibility with the assistance of municipal services and representatives. The criteria of sorting is the respect of the budget,
3. Public vote: the sorted propositions are submitted to public vote (Figure 7);
4. Implementation: the terms of reference are prepared by the municipality to implement the project in the medina.

---

[3] Experimented first as part of an international cooperation project in 2015, the purpose of the project was to promote citizen participation in the local management. The second edition in 2017 was an independent initiative of the municipality to adopt the mechanism as a permanent instrument the involve inhabitants in the rehabilitation of the Medina. They municipal council dedicated a budget of 200.000.00dhs for the project. The associations that participated represented the six neighbourhoods of the medina. Some adaptations of the process were introduced, such as the gathering spot for the public vote in Outa Hmam square for more visibility. The period of voting was extended to one week instead of one day, during the first edition;

Inhabitants are involved to cogitate, diagnose and express their perspective of medina's heritage through the ideas proposed. The process allowed them to understand more closely the municipality work through the sorting and feasibility study. They participated also in the monitoring of the implementation.

« *We voted the proposition of protecting alleys and add raillings in some dangerous sections in the medina. It was needed. Each year during raining season, we endure accessibility difficulties with rain flows, Sometimes accidents and injuries occur. The edition that followed, we voted the same project for many other sections, in addition to benches to install for more comfort in public space* »[4].

Figure 6 : Participatory budget public vote (Municipality archive, 2019)

The local configuration allowed the expression of such initiatives and dynamics involving the inhabitants directly in safeguarding processes[5].

Inhabitants as ambassadors of local know-how and cultural heritage

The effects of the local dynamics around safeguarding medina's heritage are considered from various levels.

  a) The remedial of inadequate practices on built heritage

The direct involvement of inhabitants provided a shared space of knowledge and expertise that contributed in the remedial of many inadequate practices by inhabitants. With 55% of inhabitants as renters, the local traditions were in the beginning maintained only by the natives and active associations. Since the imitative got promoted into a large maintenance operation, the construction culture became more accessible.

"We've noticed that some inhabitants started to use synthetic paintings and cement for plastering their facades instead of the lime based mortar. Those are not local technics. In Chaouen we use lime as an essential material for our homes and streets. We've managed to raise awareness among inhabitants unfamiliar with our culture about the importance to use lime to prevent water infiltration, to regulate heat, and to refresh air in the alleys and interiors. We also use natural tints, extracted from "nila" a woad plant that gives as the indigo shades. The next year, we've prepared collectively the lime mortar, and it was very instructive for neighborhoods associations and inhabitants"[6]

---

[4] Interview with inhabitant in Chefchaouen medina, Mars 2019;

[5] The local dynamic around safeguarding the medina heritage is related to many factors. The living traditions like Al Awacher provided a key configuration to promote the culture heritage in a large and more organized initiative. Local initiatives have been more structured since 1990 through an active associative network in the medina, with key figures advocating the collective management culture. One of them is the current mayor the city that has switched from associative activism into politics leading the municipality strategy to openness toward inhabitants and civil society in general. This political endorsement of local initiatives and dynamics has allowed the promotion of cultural heritage nationally and internationally;

[6] Interview with a municipal agent in charge of the coordination of Awacher initiative, Mars 2019;

b) Mutual understanding of needs and risks around heritage

The initiatives have brought inhabitants closer to public management and its functioning. The experience of participatory budgeting has offered a space to discuss both sides' constraints and needs, which leaded to mutual understandings of risks and challenges.

The environmental challenges prevailed during this experience. The proposed project of protecting some sections in the medina from rain flows was highlighted in the public vote twice. As the hydrological risks are recurring in Chefchaouen, especially in rainfall episode, inhabitants helped through this initiative to identify the damaged sections in the neighborhoods around Ras Al Ma spring. The identification of this intervention as priority is based on uses of space and a practical knowledge of inhabitants, which allows to take into account their perspective and priorites, and thus, provide more accuracy to public intervention.

Conclusion

The case study of chefchaouen has demonstrated an endogenous dynamic around safeguarding heritage based on inhabitants initiatives endorsed by a political leadership.

The rehabilitation and maintenance work conducted by inhabitants in parallel with public interventions has allowed mutual understanding of heritage character and challenges, which contributed to value it and to mitigate the impact of different risks. Thus, the coherence in heritage interventions are more controllable.

This approach of heritage through practice has also permitted to access to knowledge systems on built heritage, which have a key role in promoting the culture heritage. Through the strengthening of the collective identity and intergenerational bridges, the community is empowered to fulfill the role of permanent ambassador of local heritage and know-how (Figure 8).

Having permanent figures and ambassadors of the cultural heritage is essential to sustain local dynamics by insuring long term community involvement in safeguarding their heritage.

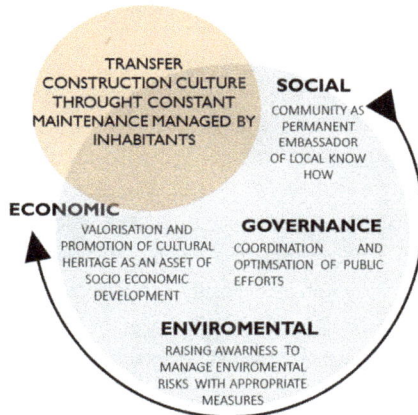

Figure 7: Effects of inhabitants' involvement in safeguarding heritage (Author)

References

[1] L. Dispasquale, Understanding Chefchaouen : traditional knowledge for a sustainable habitat, Etudes Euro-méditerannéennes, Firenze university press, 2020. https://doi.org/10.36253/978-88-5518-178-5

[2] GIZ, Réseau marocain des anciennes médinas : Etat des lieux des médinas des villes membres, 2018, https://issuu.com/programme-comun-giz/docs/etat_des_lieux_des_m__dinas

[3] GIZ, Réseau marocain des anciennes médinas : État des lieux de la gestion des médinas, 2018, https://issuu.com/programme-comun-giz/docs/catalogue_remam__mise____jour__web

[4] M.H. Bacqué, R. Henry, Y Sintomer, Gestion de proximité et démocratie participative : une perspective comparative », La Découverte, Paris, 2005. https://doi.org/10.3917/dec.bacqu.2005.01

[5] ICOMOS, Charter on the built vernacular heritage, 1999, http://www.icomos.org/charters/vernacular_e.pdf

[6] P. Frey (dir), Learning from vernacular: pour une architecture vernaculaire, édition Actes sud, 2010

[7] I Asquith, Lessons from the vernacular: integrated approaches and new methods of housing research, in Asquith I. Velliga M. « Vernacular architecture in the twenty first century: theory, education and practice, Taylor and Francis, Oxon, p, 128-144, 2006

[8] H. Fathy, Construire avec le peuple, édition Actes Sud, 1999

[9] DA, Charte architecturale de la médina de Chefchaouen, 2006

[10] AUT, Programme de réhabilitation et de renforcement de la médina de Chefchaouen, 2009

[11] AUT, Report of monitoring rehabilitation work in medina, 2022

[12] Chefchaouen municipality, report of "the spring of Chechaouen" festival, 2019

[13] Chefchaouen municipality, report of Al awacher, 2019

[14] Chefchaouen municipality, report of the participatory budgeting, 2019

Mediterranean Architectural Heritage - RIPAM10 Materials Research Forum LLC
Materials Research Proceedings 40 (2024) 127-133 https://doi.org/10.21741/9781644903117-13

Preservation of Local Architectural Heritage in Slovenian Istria Against Coastal Flooding

Polona OBLAK

Grudnova ulica 3, 6330 Piran, Slovenia

Polona.oblak@emuni.si

Keywords: Architectural Heritage, Istria, Coastal Flooding, Stone Construction, Sečovlje Salt Pans

Abstract. This article examines the problem of preserving the local architectural heritage of the town of Piran against of flooding. It presents an overview of the materials and techniques used in traditional construction in Piran and in the Sečovlje salt pans as well as it constitutes a review of the state of the flooding situation in these two places. The gap between what is done on the ground and the needs of the place have their source in the poor management of such emergency situations and the possible erroneous choice of techniques and materials used for the restauration of the infrastructure or buildings. The main conclusion is that, beyond crisis management, flood risks need to be considered in a holistic way, respecting the ancient knowledge or techniques and materials of construction, and including different stakeholders in order to better anticipate, organize and preserve the local and architectural heritage for future generations.

Introduction

Piran, also known as the pearl of Slovenia, is a municipality and town in the extreme south of the country situated in the Slovenian coastal region of Istria. This ancient city lies on a narrow peninsula bordered by the Adriatic Sea, Croatia and Italy. Thanks to its historical past, the medieval town of Piran boasts a rich nautical, architectural, cultural, historical, linguistic, gastronomic and natural heritage. The list is long but not exhaustive: Venetian and Gothic-inspired buildings, ancient Greek, Roman and Byzantine remains, medieval ramparts and paintings by famous artists such as Carpaccio and Tintoretto. Piran benefits also of an extraordinary linguistic diversity ranging from Greek and Latin to Istro-Venetian, Italian and Slovenian language, as well as a Mediterranean gastronomy enriched by the salt from the Piran salt pans in Sečovlje, now national heritage site and natural park.

Over the centuries, this town in Slovenian Istria has always been exposed to and influenced by many people thanks to its strategic position. Istria was populated as far back as the Stone Age. The first fortified hills were built in the middle of the 2nd century B.C. The first civilization to live there was the Histres, mentioned as far back as the 5th century B.C. The Romans attacked the Histres around the 3rd century B.C., conquering them definitively around 178-177 B.C. Then came the Goths, Byzantines and Longobards, followed by the Franks in 788 [1]. The first Slavs probably appeared in the 7th century. In the mid-ninth century, Istria was annexed to the Holy Roman Empire and came under Italian rule. After 1209, Istria became a margrave of the Patriarch of Aquileia, and finally, in 1283, Piran was the last independent town in Istria to come under the rule of the Republic of Venice, although Istrian towns periodically came under the influence of Venice as early as 970 [2]. With the Serenissima, Piran began to flourish. It was during this period, marked by the intense exploitation of local production as a survival strategy[1], that Piran salt pans were first mentioned.

---

[1]The Serenissima became interested in exploiting local resources when sea routes became dangerous, or even inaccessible, due to the presence of pirates [3].

Mediterranean Architectural Heritage - RIPAM10   Materials Research Forum LLC
Materials Research Proceedings 40 (2024) 127-133   https://doi.org/10.21741/9781644903117-13

With the fall of the Republic of Venice in 1797, the Venetian territories, including Piran and Istria, passed to Austria. In 1806, French troops occupied Venetian Istria. As a result, it became part of the Illyrian provinces from 1809 to 1813. In 1814, the region came under Habsburg control and remained so until the end of the First World War in 1918, when it was annexed to Italy. After the Second World War, the region was part of Zone B of the Free Territory of Trieste for a few years. In 1954, it became part of the former Yugoslavia, and in 1991, when Slovenia gained its independence, Piran officially became a municipality and city of the Republic of Slovenia [2].

This paper aims to present the traditional building materials used in the coastal city of Piran and its Sečovlje salt pans and wishes to show that these construction materials and methods play a crucial role in helping to prevent natural risks, especially flood damage.

The first part of the article explains that the traditional stone architecture found in Piran, as well as in the rest of the Mediterranean, is considered local heritage and is renowned for its durability and resistance to earthquakes, weather (floods, waves, strong winds), sea air and humidity. This is due to the characteristics of the stone used, as well as the choice of traditional building techniques employed. The second part of the article is an overview of the flooding situation in Piran and its salt pans, while the conclusion suggests some possible solutions to better anticipate, organize and preserve the local architectural heritage.

Materials and techniques

The building materials used in Piran came mostly from Piran and its inland territory (Fizine and Kanegra quarries, Karst plateau), but also from the quarries of Pazin and Aquileia or other places. However, as the researcher Tommasini [4] explains, the finest stones, such as those from the Rovigno quarry, which were white and red, were not used locally but were taken to the factories in Venice. However, Piran (but also Koper), could boast two or three kilns for tiles, flagstones, bricks and lime.

The choice of location and building materials did indeed reflect the economic power of the owner. Thus, the best houses were those around the harbour[2] and in Piazza Vecchia [6]. Wealthy families, nobles and the clergy could afford more prestigious, expensive and resistant materials to build their sumptuous residences, such as white stone[3], mortar, rendering, marble, lime and sand[4], brick and wood. Modest families, on the other hand, built smaller, simpler homes. They used cheap or free[5] mixed materials and dry-stone construction. The outer walls of their dwellings were thicker, while the inner walls were thinner and plastered with wooden [4].

Piran adopted mostly a vertical urbanism of individual houses, separated from each other by narrow cobbled streets. These dwellings covered a small area, but were multi-storey and housed a single family.

Rather than fighting nature, the ancient "Piranesi" lived in harmony with it, using the omnipresent natural elements to their advantage: sea, water, wind and sun. Thus, the first floors, which in the autumn months and with heavy rains risked being flooded, were never inhabited. With its own cool, damp microclimate, this part of the house was used as a storehouse for water, barrels of wine, oil, fruit, vegetables, fishing nets and animals (chickens, donkeys). An internal

---

[2] Piran's harbour, which corresponds to today's Tartini Square, was covered in 1894 for hygiene reasons [5]. Researcher Apollonio [6] points out that in the 19th century, the architectural appearance of the town of Piran was different. The most striking detail is undoubtedly the absence of the shoreline and the sea that bathes the houses.

[3] White stone from Pazin was softer than the one from Kanegra and therefore less waterproof. For this reason, it was appreciated and used for decorative elements on buildings. Kanegra's white stone was harder, and as water penetrates it only slightly, it was used in the construction of buildings.

[4] Sand was purchased in a centre at the mouth of the Soča, while the quicklime used to coat walls was bought from local concrete producers [7].

[5] The author explains that the "modest" houses of Piran used the stones they took from the surrounding land to build their homes.

Mediterranean Architectural Heritage - RIPAM10   Materials Research Forum LLC
Materials Research Proceedings 40 (2024) 127-133   https://doi.org/10.21741/9781644903117-13

(and rarely external) staircase in wood or stone led to the warmer upper floors, where the living quarters were located: kitchen and bedrooms. Most houses also had an "altane" or terrace, ideal for drying corn and laundry. The roof was usually two-sloped or low-pitched, covered with curved tiles and rarely with stone shingles [8]. In winter, when they needed warmth, families slept in the bedrooms upstairs, while in summer, in the cool of the first floor.

It wasn't just the choice of building materials that enabled Piran's inhabitants to live in harmony with the elements of water, wind, sea air and sun, but also the architectural construction of the town itself.

Indeed, the narrow cobbled lanes inside the city of Piran, similar to those in Venice, were built in the shape of a turtle shell, so that water and dirt could drain away into the side sections. Similarly, these city's sloping lanes and numerous gateways allowed warm air to circulate and rise to the top, creating draughts, shade and coolness. Today, you can still admire this heritage, revealing the traces of the past, and walk through the labyrinth of small, winding streets, hidden between Piran's beautiful buildings.

The Piran salt pans are also an example of local architectural heritage designed to withstand the challenges of nature and the elements.

In the past, Piran's economy was based on fishing, maritime trade, agriculture and, above all, salt. Salt, also known as white gold, was highly prized and, in the days of the Venetian Republic, was the currency par excellence. Although Cumin [9] believes that the activity of salt harvesting was already present in the territory of Istria in prehistoric times, other documents lead us to believe that the beginning of the history of the Istrian salt pans was around the year 543 [10], when they were mentioned in an ecclesiastical document. However, the researcher Nicolich [11] believes that they existed already in the 6th or 7th centuries.

In any case, the first document attesting to the presence and activity of the Piran salt pans is the Statute of Piran dating from 1274 [12]. The Piran salt pans were built following the example and know-how of the Pag salt pans, and the municipality of Piran had 3 sites or salt pans: those in the Sečovlje Valley, those in Strunjan and those in Fasano-Santa Lucia, which no longer exist today. Near Piran, salt was also harvested in the Isola, Koper, Trieste and Muggia salt pans [3].

According to research by Professor Nicolich [11], Piran had 1,200 crystallization basins by the time the town politically entered the sphere of economic interests of the Venetian Republic. Around 1574, Piran had 2,680 basins: 327 basins in the Fasano-Santa Lucia salt pans, 167 in the Strunjan salt pans and 2,186 basins in the Sečovlje Valley salt pans [13].

During the years 1945-1954 the displacement of a large part of the indigenous population took with it the knowledge, know-how and experience of the salt production and other professions. This is important because of the loss of knowledge about salt production, the tradition of seasonal work, the seasonal migration and living in the salt pans.

The years '54-'60 of the twentieth century were marked by the destruction of dwellings in the salt works and the abandonment of salt production activity, not only due to the post-war political situation, but also because of the lack of manpower and the large-scale tourism development planned for the region, particularly in Portorose. It was only with Slovenia's independence that the municipality of Piran began to take a greater interest in protecting the salt pans as a cultural heritage and characteristic feature of Slovenian Istria. In addition to being recognized as a protected nature reserve in 1989, the Sečovlje peninsula and its salt pans have also been a cultural, ethnographic and technical monument since 2001 [3].

Situated next to the basins and along the navigable channels in order to be functional and easily accessible by boats, the typical salt pans dwellings or "salari"[6] were built using traditional Istrian

---

[6] These temporary dwellings were also present in the salt pans of Strunjan, Fasano-Santa Lucia, Koper and Muggia, but to a limited extent. Indeed, because of the proximity of the town, the workforce did not need to itch during the

materials. These included grey stone or sandstone, known as the least expensive, Kanegra limestone, which was the most resistant, and Pazin limestone, also known as white stone or Istrian stone, which was the most beautiful but porous and therefore not very resistant to the destructive action of the sea air [7].

The structure of these dwellings was made entirely of stone, often without plaster, and built on a floating base frame of wooden[7] planks laid on a compact soil of limestone sand and clay. Load-bearing walls were either made of grey stone or square blocks of limestone (white stone). Exterior walls were plastered with lime, but never painted to prevent the sea air from destroying the layer of colour.

In the salt pans, many dwellings were built with mixed materials. In this case, the lower part was built of limestone and white stone, while the upper part was made of grey stone [7].

The last part of the structure ended with fir-wood transverse joists, which were fixed at the ends by metal[8] beam brackets and embedded in the load-bearing wall. The roof was pitched and covered with curved tiles. The rooms were divided by thin plastered wooden walls that reached only as far as the rafters. There was no attic, and small windows with solid shutters, facing at least two different directions, enabled to observe weather changes and shelter salt in good time [14].

Two materials were used to build and consolidate the paths and dikes lining the basins: limestone sand and clay. They were laid in three vertical layers, with the outer two always made of sand and the middle one of beaten clay, which ensured that the construction was watertight. Naturally, the channels and dikes, as well as the silt, water and impurities in the basins, had to be cleaned regularly.

Each "salaro" consisted of an upper living area and a storage area on the first floor. Some "salari" had an adjoining external storage area. The ground-floor warehouse used to store salt and tools had two doors. The door that opened onto the basins adjacent to the "salaro" was also the one through which salt was brought into the warehouse, while the door that opened onto the canal was the one through which salt was loaded into the boats that docked along the channels.

Some dwellings, probably the oldest, had an external stone staircase. In the rest of the dwellings, an internal wooden staircase linked the first floor to the second floor, where the kitchen and one or two bedrooms were located [14].

"Salari" being temporary dwellings, they were inhabited by the families renting the basins during the period from late April/May to late September/early October. The heads of the families waited until the end of winter to leave for the salt pans to clean and repair the dwellings, depots, connecting channels, paths adjacent to the basins, dikes and evaporation basins that had suffered damage from the sea air and bad weather [7].

Despite the importance and high activity of the salt pans in the past, today only four of the 440[9] "salari" that existed in the first half of the 19th century bear witness to life in the salt pans in the Sečovlje Valley, having been restored and turned into a museum. Some of the ponds have also been restored, and salt production in Piran continues today. The restoration work at the salt pans has brought together a number of local players who have collaborated to bring this project, led by the maritime museum in Piran Pomorski muzej Sergej Mašera, to a successful conclusion.

---

season, as was the case in the Sečovlje Valley salt pans [9]. In the first half of the 19th century, there were 17 "salari" in the Strunjan salt pans and 35 in the one in Fasano-Santa Lucia [14].

[7] The planks were arranged in a frame to distribute the load, and even if they were made of wood, they would not rot, as bacteria, deprived of oxygen, could not grow on them.

The use of metal was reduced to a minimum in the salt marshes, as it was not resistant to salt water and sea air.

[8] The use of metal was reduced to a minimum in the salt marshes, as it was not resistant to salt water and sea air.

[9] In 1984, 118 "salari" were counted in the salt pans of the Sečovlje Valley. They were all abandoned, while some of them were well preserved, others less.

Mediterranean Architectural Heritage - RIPAM10
Materials Research Proceedings 40 (2024) 127-133

Materials Research Forum LLC
https://doi.org/10.21741/9781644903117-13

Indeed, in Piran, as in all coastal towns in this region, it has been essential to adapt the way of life to the surrounding nature and its elements: sea, water, wind and sun. In fact, it is not uncommon for high tides to cause coastal flooding in Piran. This happens almost regularly in October, November and December. According to various researchers [15; 16], flooding is the result of a combination of factors such as south wind and wave action, low atmospheric pressure, meteorological fronts and the orographic configuration of the Adriatic Sea, i.e. its land-locked northern part. Other authors also add the importance of the influence of the autumn full moon on tides and sea level rise [17].

When the high tide is accompanied by a strong southerly wind and heavy rain, the material damage is considerable. In this region, flooding due to high tide was first mentioned in 1343 in a document found in the archives of the town of Pirano [18]. To date, the highest sea level ever recorded reached 395 cm in 1969.

On the Slovenian coast, Piran is the town most at risk due to its position as it is the most exposed to high tides and southeast winds. As a result, many buildings that are part of the town's architectural and cultural heritage are located directly on the shore, facing the sea and the wind.

During high tides, a large part of the old town, home to almost a fifth of the population, is flooded [16]. The areas that are most frequently flooded in Piran are those from Punta to Street Gregorčičeva ulica, the 1.maj Square, the Shoreline Prešernovo nabrežje, the Tartini Square, the Shoreline Cankarjevo nabrežje, a part of the Street Župančičeva ulica and partially the Street Dantejeva ulica and the stretch from Riviera beach to Fornače.

Like the town of Piran, the Sečovlje Valley and salt pans are also located in a high-risk zone. Covering an area of 650 hectares and rising above sea level, the Sečovlje salt pans are among the most threatened wetlands in the Mediterranean. During the autumn months and the annual floods, they are largely inundated. And in the event of extreme flooding, they risk being almost entirely underwater.

Other potential negative effects of rising sea levels include accelerated coastal erosion, the destruction of basins, saltwater intrusion into salaries and subsidence. For example, in October 1896, as a consequence of the difficult management of the Dragonja river's waters following the poor state of cultivated land and because of the deforestation in the upper reaches of the valley, the salt pans were destroyed by a flood that wreaked havoc with the private management of this important and ancient city trade [6].

Conclusion

For decades now, floods have been occurring at the same time, under the same weather conditions and causing the same damages. It is clear, therefore, that the damage caused by marine flooding can be prevented. To achieve this, a combined approach that integrates prevention, protection of architectural heritage, preservation and application of ancestral building materials and techniques, use of appropriate building materials and techniques and proper management of the emergency situations is required.

Firstly, it would be necessary to conscientiously monitor the meteorological situation (in particular wind direction and speed, but also atmospheric pressure), while comparing it with the predicted tidal height [15].

In the city of Piran, channels and manhole covers should be regularly cleaned to ensure that they are clean and that water can flow freely into the watercourses [19].

It might be convenient to install one way valves/systems in order to prevent sea water flowing through drain pipe lines in the opposite direction (from the sea to Piran town surfaces).

Similarly, when restoring streets and buildings, it is essential to keep traditional construction and ancient building materials (street pavers, sand, stone) and do not replace it with other (modern) materials, such as concrete, or other types of stone other than the local ones that are already in use. The second one are more expensive and do not meet the needs of the place.

Mediterranean Architectural Heritage - RIPAM10   Materials Research Forum LLC
Materials Research Proceedings 40 (2024) 127-133   https://doi.org/10.21741/9781644903117-13

In addition, it would be necessary to install flood barriers all along the shoreline. In the flood of October 27, 2023, new flood barriers were tested in Piran's harbour area. The objective to prevent flooding in Tartini Square and Street Ulica Svobode was achieved, but to the detriment of Street Župančičeva ulica, which was flooded more than it would have been without the barrier. In the same context, it would also be interesting to see the best practices of other coastal cities and countries. The city of Venice, for example, has invested in the installation of a "mose" floating dike system, which has proved to be effective.

It would also be very important for buildings in the old town, especially those classified as architectural and cultural heritage, to be properly maintained against humidity and sea air. In this case, collaboration between the owners of the buildings, the municipality, the National Institute for the Protection and Restoration of Piran's Natural and Cultural Heritage, architects and experts in the field, as well as the professionals who carry out the restoration work, is essential.

Finally, storeowners and property owners in flood-prone areas must continue to use flood barriers or flood bags. And more important, ground floors should no longer be converted into apartments, as this choice has consequences - the property will be flooded and destroyed by sea water and humidity.

In the Sečovlje salt pans is therefore essential not only to restore, rehabilitate, clean and maintain these salt pans which are classified as a national heritage site and nature reserve. Here, too, the collaboration of many stakeholders is essential, going from the Maritime Museum Pomorski muzej Sergej Mašera in Pirano and the National Institute for the Protection and Restoration of Natural and Cultural Heritage in Pirano, to architects, experts and schools.

Above all, children in schools and future generations need to be educated and made aware not to fight nature but to live with it, while respecting and protecting ancestral know-how and the architectural, cultural, historical, linguistic, gastronomic and natural heritage that surrounds them.

References

[1] A. Pucer, Istrske štorije: legende, miti, pripovedi in zapisi, Umetniško ustvarjanje Istral, Padna, 2013.

[2] D. Kladnik, P. Pipan, Bay of Piran or Bay of Savudrija? An example of problematic treatment of geographical names, Acta geographica Slovenica, 48 (1), 2008, pp. 57-91. https://doi.org/10.3986/AGS48103

[3] O. Selva, Ecoturismo e turismo culturale in Slovenia: il Parco naturale delle Saline di Sicciole, EUT Edizioni Università di Trieste, 2007.

[4] G. F. De Tommasini, Commentari storici-geografici della provincia dell'Istria, libri otto, con appendice. Archeografo triestino, vol. 4, 1837.

[5] N. Terčon, Kako je piransko pristanišče postajalo kulturna dediščina, Pomorski muzej "Sergej Mašera" Piran, regionalno srečanje dekd 2017, Mediadom Pyrhani, Piran, 2017.

[6] A. Apollonio, Autogoverno comunale nell'Istria asburgica il caso di Pirano: seconda fase 1888-1908, Atti, 26(1), 1996, pp. 15-70.

[7] Pomorski Muzej "Sergej Mašera", Sečoveljske soline včeraj, danes, jutri / Le saline di Sicciole ieri oggi, domani, Piran, Katalog k razstavi, 1988.

[8] R. Starec, Aspetti della casa rurale istriana. Rilevazioni sul territorio e fonti d'archivio, Atti, vol. 27(1), 1997, pp. 345-379.

[9] G. Cumin, Le saline istriane, Bollettino della R. Società Geografica Italiana, (7)2, 1937, pp. 373-392.

[10] G. Zalin, Il sale nell'economia delle marine istriane. Produzione, commercio e congiuntura tra Cinquecento e Seicento, Atti del Convegno di Bari 1979, Napoli, 1981, pp. 239-267.

[11] E. Nicolich, Cenni storico-statistici sulle saline di Pirano, Appolonio, 1882.

[12] M. Pahor, Statut občine Piran iz leta 1274. Zgodovinski časopis, 29(1-2), 1975, pp. 77-88.

[13] J. C. Hocquet, Il sale e il potere. Dall'anno 1000 alla Rivoluzione francese, Genova, ECIG, 1990.

[14] A. Danielis, Le vecchie saline di Pirano, L'Archeografo Triestino (3)6, Volume del Centenario, Parte II, 1930-1931, pp. 409-417.

[15] F. Bernot, Vzroki in pogostost poplav ob slovenski obali, in Gams: Naravne nesreče kot naša ogroženost, SAZU, Ljubljana, 1983, pp. 50-53.

[16] N. Kolega, Slovenian coast sea floods risk, Acta geographica Slovenica, (46)2, 2006, pp. 143–169. https://doi.org/10.3986/AGS46201

[17] B. Komac, K. Natek, M. Zorn, Geografski vidiki poplav v Sloveniji, Geografija Slovenije št. 20, Ljubljana, 2008. https://doi.org/10.3986/9789612545451

[18] G. Zupančič, M. Centa, L. Gosar, Modeliranje valovanja v slovenskem morju, aktualni projekti s področja upravljanja in urejanja voda, 26. Mišičev vodarski dan, 2015.

[19] I. Gams, Naravne nesreče v Sloveniji kot naša ogroženost, Založba ZRC, 1983.

[20] D. Plut, Pregled negativnih vplivov na življenjsko okolje s pomočjo matrice in bodoči prostorski razvoj Koprskega Primorja, Geographica Slovenica 9, 1978, pp. 41-48.

Mediterranean Architectural Heritage - RIPAM10　　　　　　　　Materials Research Forum LLC
Materials Research Proceedings 40 (2024) 134-140　　　　https://doi.org/10.21741/9781644903117-14

Integrated Digital Documentation for Conservation, the Case Study of the Torre deli Upezzinghi Called Caprona, in Vicopisano (PI) Italy

Giovanni PANCANI

Department of Architecture, University of Florence, Italy

giovanni.pancani@unifi.it

Keywords: Tower of Caprona, Heritage Management, Digital Survey, Aerial Photogrammetry, Documentation, SFM

Abstract. Located on the Rocky Promontory overlooking the town of Caprona, the Upezzinghi Tower is a 19th-century reconstruction of a watchtower that once served the ancient castle, which existed in the mid-11th century and was destroyed by the Florentines in 1433. The hill on which it stands has been gradually eroded due to stone quarrying, significantly altering the landscape around Caprona. Until the mid of the last century, the rocky promontory was still substantially intact, and the remains of the medieval fortress could be identified beneath the tower. However, at its base, the remains of the medieval tower's foundation are still visible. The structure is currently in an advanced state of architectural decay, and the extraction of stone material has been so aggressive that the quarry's limit has come within about 50 cm of the tower's profile. The small square-shaped building appears to be smaller than the one demolished in 1433 since measurements at the base of the current tower have confirmed one side to be approximately 4.50 meters, while the remains of the medieval tower had a side of about 5.00 meters. Digital surveying has been carried out for the preservation and conservation of the tower, which is in urgent need of restoration. TLS (Terrestrial Laser Scanning) and UAS (Unmanned Aerial Systems) tools were used with multiple acquisitions that were subsequently compared and calibrated, using the laser scanner point cloud as a reference. The maximum misalignment error of the TLS point cloud was within a maximum range of 0.015 meters. The delivery of the survey results, considering the modest size of the structure, was performed at a 1:20 scale.

Introduction

On the rocky spur overlooking the town of Caprona, you can see the "Upezzinghi Tower," which is a 19th-century scaled-down replica of the tower from the ancient castle that existed in the mid-11th century and was dismantled by Florence in 1433 [1]. The extraction of stone from the quarries in Caprona has gradually transformed the landscape of the town. In fact, at the beginning of the 20th century [2], the rocky promontory was still largely intact, and it was possible to see the remains of the medieval fortress around the small tower (fig. 1). Unfortunately, at the present state, there are not many significant traces of the entire fortified complex, except for a foundational base at the foot of the 19th-century tower, which likely could belong to the medieval tower. The medieval tower was constructed to control the narrow strip of land situated between the Arno River and the southern extension of Monte Pisano and to facilitate communication with the surrounding fortified structures, such as Rocca della Verruca and the towers of Uliveto. The Upezzinghi Tower has been the subject of a multidisciplinary study involving the Department of Architecture (DIDA) at the University of Florence, the Department of Earth Sciences, and the Department of Energy, Systems, Territory, and Construction at the University of Pisa. This collaborative effort has already conducted a territorial-scale survey [3].

In this article, focusing on the study conducted by the Department of Architecture at the University of Florence, we will discuss the data acquisition methods, the technologies employed, and the results obtained concerning the digital 3D architectural survey work.

Mediterranean Architectural Heritage - RIPAM10
Materials Research Proceedings 40 (2024) 134-140

Materials Research Forum LLC
https://doi.org/10.21741/9781644903117-14

Fig. 1, A picture of the Caprona tower at the beginning of the 20th century where the remains of the medieval fortress can be clearly seen.

Methodology

For the TLS (Terrestrial Laser Scanning) survey, a Zoller+Frogich Image 5016 laser scanner was utilized. This instrument is a Phase Difference-based tool with a range of 360 meters, a vertical field of view of 300 degrees, and a horizontal field of view of 360 degrees. It includes an integrated camera with HDR technology in a dual configuration and integrated LED flash. The measurement noise does not exceed 0.2mm at 25m with 80% reflectance, without the use of optimizations. The specific shape of the rocky promontory on which the tower stands, with one side bordered by a 90-meter cliff and the others presenting steep slopes, required careful handling and precise leveling of the instrument (fig. 2). The ten scans were performed along a polygonal path on three sides around the tower, while the fourth side was reached by extending the polygonal network for capturing the interior of the tower [4].

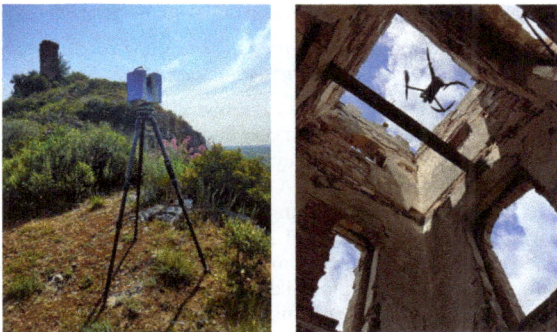

Fig. 2, On the right, the cliff where the Caprona tower stands, and the scanner used is the Z+Inager 5016.

Fig. 3, On the left, the interior of the tower with the Dji Mini 3 Pro drone at work.

Mediterranean Architectural Heritage - RIPAM10   Materials Research Forum LLC
Materials Research Proceedings 40 (2024) 134-140   https://doi.org/10.21741/9781644903117-14

The scans were recorded and processed using Recap-pro software (fig. 4). The 3D point cloud model was then exported in e57 format, followed by verification and certification using Cyclone software. Certification and testing of the point cloud were necessary because each survey requires validation of its reliability and accuracy before it can be made available to scholars who will use it for their research needs [5]. Verification of compliance with minimum error and reliability requirements involved extracting a series of horizontal and vertical cross-sections from the point cloud, both during partial scans and from the overall point cloud of all scans. In the identified sections, it was verified that the section lines (fig. 5), representing the profiles obtained by cutting through the point cloud, from different scans had a matching alignment. Even when not perfectly aligned, the section line thickness did not exceed 15 millimeters. Point cloud certification ensures that the results are based on reliable and credible data for subsequent analysis and considerations [6].

Fig. 4, On the left, the point cloud of the survey of Caprona Tower, sectioned but with a view of the space (Section Elevation), is visible along with the horizontal cutting plane.

Fig. 5, On the right, the sliced point cloud with a thickness of 1 cm, the section lines are clearly visible, which are evaluated for survey certification.

In addition to the laser scanner survey, a highly accurate Structure from Motion (SFM) photogrammetric survey was conducted with a very high shooting density, approximately 10 pixels per square centimeter. For ground-based surveying, a Sony alpha 900 camera with a 24-megapixel CMOS sensor and a Sony 24-70mm f/2.8 ZA SSM Carl Zeiss Vario-Sonnar T* lens was used. Aerial acquisitions (fig. 3)were carried out using a DJI Mavic Mini 3 Pro Unmanned Aerial System (UAS) equipped with a camera featuring a 1/1.3-inch CMOS sensor with 12 effective megapixels and a flexible ISO range. The acquired photographs were processed using Reality Capture software. The results from the two point clouds, one from the 3D laser scanner survey and one from the photogrammetric survey, were optimized and compared using Cloud Compare software [7].

The meshes obtained from the laser scanner and photogrammetry were exported from Reality Capture and compared using Geomatic Wrap. The results revealed a maximum positive distance of 0.100 meters and a maximum negative distance of -0.100 meters. The areas at the top of the tower showed the most significant differences, partly due to the scanner's difficulty in reaching all points. The average distance between the meshes was 0.001 meters, with a maximum positive value of 0.056 meters and a maximum negative value of -0.056 meters. The standard deviation was 0.062 meters, and the RMS estimate was 0.062 meters (fig. 6).

Fig. 6, The comparison between the point clouds where critical points are evident and those where the laser scanner survey point cloud closely corresponds to the photogrammetric survey point cloud.

To achieve a high-quality 3D photomodel, essential for producing quality photoplanes, certain principles were considered during both the photographic acquisition and subsequent data processing phases. To create a photoplane at a 1:20 scale and print it at 300 DPI, the following conditions needed to be met. In order to achieve a 300 pixels per inch print definition without quality loss, the images had to have a resolution that adhered to the following parameters. In the chosen scale for this survey, which is 1:20, one real meter is represented by 5 cm of print. Therefore, for simplicity, we will assess pixel density in centimeters rather than inches. Namely, 1 inch equals 2.54 centimeters, which means 300 pixels divided by 2.54 cm equals 118.11 pixels per centimeter. Thus, the print needed to meet the 118 pixels per centimeter parameter, rounded to 120 pixels. Consequently, in a 1:20 scale print, one square meter of real surface had to be represented by at least 120 pixels in 5 centimeters, or 600 pixels. Therefore, the images for creating the photomodel had to be captured with a resolution of at least 6 pixels per centimeter. Considering that during acquisition, dimensions of the surfaces to be photographed are estimated empirically and that native resolution may be lost during photomodel processing, it's highly advisable to acquire images with a resolution greater than the necessary minimum. This implies that the images should have a resolution of approximately 10 pixels per centimeter [8].

Fig. 7, On the left, shows the floor plan with the wireframe and the photomap of the ground level, at a height of 1.20 meters.

Fig. 8, On the right, displays the floor plan with the wireframe and the photomap of the level at 4.50 meters

In the deliverable, the certified point cloud, exported from the Recap software in e57 format, was imported into AutoCAD, where section planes were identified. The representation of the structure was organized following the classical scheme, with a plan view relative to the rocky outcrop on which the small tower stands, a ground level plan, one at a height of 4.50 meters, another at a height of 7.50 meters, and finally, a top-down view above the battlements (fig. 7, 8, 9, 10). Additionally, four internal section-elevation views and the four elevations were created. The elevation of the remains of the base, likely belonging to the previous medieval tower, was also depicted. These graphic representations provided a detailed description of the deteriorated state of the tower's masonry surfaces and were overlaid with high-resolution photomaps, with a resolution of at least 300 DPI [9]. All deliverables were presented at a scale of 1:20.

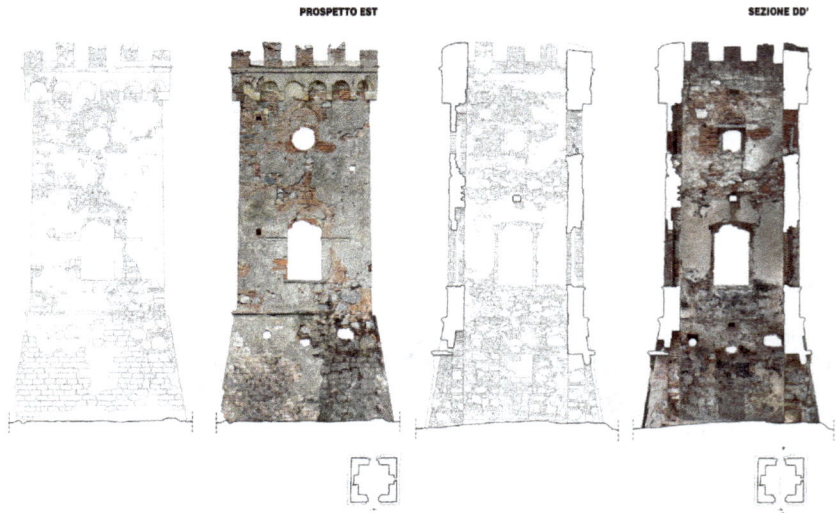

PROSPETTO EST   SEZIONE DD'

Fig. 9, The East elevation, on the right, and the elevation section D-D', were first rendered with the wireframe, and then with the photomap on the right.

Considerations

This research has significant implications for the documentation and preservation of the historical and cultural heritage of Caprona. The detailed 3D representation of the Upezzinghi Tower will allow for an in-depth study of the structure and its changes over time. Initial assessments from the survey data suggest that the current tower is smaller than its medieval counterpart, appearing to have been constructed more as a memory marker of the past. The external dimensions of the current square-plan tower have a base side measuring approximately 4.50 meters, while the dimensions at the narrowing of the buttress are 3.40 meters, with a height of 10.00 meters at the level of the merlons, which are about 0.50 meters. When compared to the length of the found foundational base, which is over 5.00 meters, it can be presumed that the medieval tower was considerably larger and, most importantly, taller.

The comparison of data acquired from different technologies provides a unique opportunity to assess the accuracy of data acquisition methodologies. This knowledge can be applied to other historical sites with similar purposes. However, to enhance the representation quality of this survey, it is believed that, in compliance with European regulations regarding public works management, it would be appropriate to enrich the documentation with a Historical Building

Mediterranean Architectural Heritage - RIPAM10
Materials Research Proceedings 40 (2024) 134-140

Materials Research Forum LLC
https://doi.org/10.21741/9781644903117-14

Information Modeling (HBIM) process. For this purpose, a project could be developed to model and parameterize the data obtained from this work [10].

Finally, despite being aware of the destructive effects of the adjacent stone quarrying, small archaeological tests could be conducted in the areas surrounding the remaining foundational base near the tower. This would provide insight into what the unwise quarrying activity has spared. Additionally, further data can be extracted from the study of the mortar present in the current tower and the mortar in the foundational base [11].

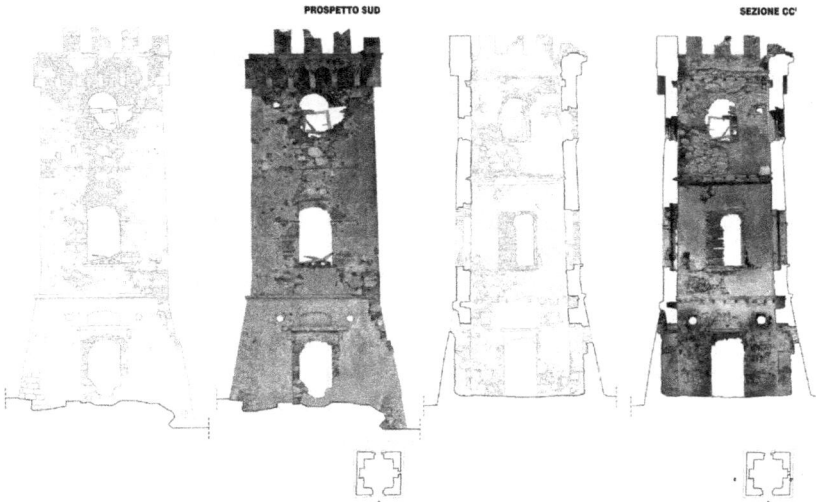

PROSPETTO SUD   SEZIONE CC'

Fig. 10, The Sud elevation, on the right, and the elevation section C-C', were first rendered with the wireframe, and then with the photomap on the right.

Conclusions

The ongoing multidisciplinary analysis and 3D reconstruction of the Upezzinghi Tower in Caprona have provided a detailed understanding of the structure and its surrounding environment. The integration of data from laser scanning, photogrammetry, and drones has yielded results that are sufficiently accurate and detailed. This information has allowed the research team to gain knowledge of the real state of decay in which the tower is situated, and the dimensional information has enabled the formulation of historical typological considerations. Finally, the high-resolution photoplans deliver an updated and accurate depiction of the site. This research represents a well-established methodological approach developed for the documentation and preservation of historical architectural and cultural heritage, which can serve as a valuable tool for the conservation of such an emblematic and significant site for the entire Pisan territory. In this regard, we would like to highlight the studies conducted by Pascal Cotte, who advances the intriguing idea that the Caprona tower, along with the Verruca Fortress, is part of the background in Leonardo's Mona Lisa [12].

Mediterranean Architectural Heritage - RIPAM10   Materials Research Forum LLC
Materials Research Proceedings 40 (2024) 134-140   https://doi.org/10.21741/9781644903117-14

References

[1] E Repetti. (1833) Dizionario geografico, fisico, storico della Toscana contenente la descrizione di tutti i luoghi del Granducato. Firenze, Repetti, Vol. I, pp. 366-367.

[2] A Alberti. (2014) Monasteri e castelli sul monte pisano. Insediamenti medievali in un'area di confine (X-XII secolo), in: Salvatori E. (a cura di) Studi di Storia degli Insediamenti, Pisa, Pacini Editore, pp. 149-163.

[3] D Billia, V Croceb., G Montalbanoc., P Rechichid. (2023), La Torre degli Upezzinghi a Caprona:Caprona: analisi storico-archivistica e rilievo digitale per la documentazione dell'evoluzione temporale, in Bevilacqua M. G. , Ulivieri (a cura di), Defensive Architecture of the Mediterranean, Vol. XIII / (Eds.), Pisa University Press (CIDIC) / edUPV.

[4] F Condorelli. & F Rinaudo. (2018), *Cultural heritage reconstruction from historical photographs and Videos*, ISPRS Int. Arch. Photogramm. Remote Sens. Spatial Inf. Sci., XLII–2, 259-265.

[5] A Arrighetti., F Fratini., G Minutoli., G Pancani., (2022), *Historical Seismic Events and Their Traces on Medieval Religious Buildings*, in D'Amico S., Venuti V. (a cura di), *Handbook of Cultural Heritage Analysis*, pp. 2182-2209, Srpinger Nature Switzerland (ISBN: 978-3-030-60015-0).

[6] G. Pancani (2017), *Il centro storico di Poppi, analisi a livello urbano per la valutazione del rischio sismico*, "DisegnareCon", Vol 10, No 18 (2017), PDF 9.1-9.10. (ISSN 1828-5961).

[7] R De Marco.;, S Parrinello. (2021). Digital surveying and 3D modelling structural shape pipelines for instability monitoring, in, historical buildings: a strategy of versatile mesh models for ruined and endangered heritage, ACTA IMEKO, vol. 10, pp. 84-97, ISSN:2221-870X.

[8] G. Pancani, M Bigongiari. (22 June 2019), *The integrated survey of the Pergmum by Nicola Pisano in the cathedral of Pisa*, in a cura di Kremers H., *Digital Cultural Heritage*, Springer, Cham, Basel, (January 2020), pp. 373-388, DOI: 10.PA1007/978-3-030-15200-0, eBook ISBN: 978-3-030-15200-0.

[9] S Parrinello.; R De Marco.; A Miceli (2020). *Documentation strategies for a non- invasive structural and decay analysis of medieval civil towers: an application on the Clock Tower in Pavia*. RESTAURO ARCHEOLOGICO, vol. 28, pp. 18-43, ISSN:2465-2377.

[10] A Dell'Amico., S Parrinello. (2021). *From Survey to Parametric Models: HBIM Systems for Enrichment of Cultural Heritage Management*. In: Cecilia Bolognesi, Daniele Villa. 978-3- 030-49278-6, pp. 89-107, Cham: Springer.

[11] F. Fratini (2023), The Rocca Vecchia fortress in the Gorgona island (Tuscany, Italy): building materials and conservation issues, in Bevilacqua M. G. , Ulivieri (a cura di), Defensive Architecture of the Mediterranean, Vol. XIII / (Eds.), Pisa University Press (CIDIC) / edUPV.

[12] P. Cotte (2019) *Monna Lisa dévoilée: les vrais visages de la Joconde*. Parigi, éditions Télémauqe.

Mediterranean Architectural Heritage - RIPAM10
Materials Research Proceedings 40 (2024) 141-148

Materials Research Forum LLC
https://doi.org/10.21741/9781644903117-15

Analyzing the Literature on Seismic Resilience in Rammed Earth Construction: A Cartographic Approach

Yassine RAZZOUK[1] *, Khadija BABA[1], and Sana SIMOU[1]

[1]Civil Engineering and Environment Laboratory (LGCE), Mohammadia School of Engineering, Mohammed V University Rabat, Morocco

yassine.razzouk@um5s.net.ma

Keywords: Historic Buildings, Cultural Heritage, Seismic Resilience, Rammed Earth Construction, Literature Map, Bibliometric Analysis

Abstract. The study explores the seismic impact on Rammed Earth Constructions through an analysis of various bibliographic factors. These factors encompass publication volume, authorship, geographical origin, institutional affiliations, and relevant scholarly journals. Employing a rigorous examination of bibliographic data retrieved from reputable databases such as Scopus, the research identifies a discernible uptick in pertinent publications since 2014. Moreover, it discloses prominent figures within this academic domain, delineating their geographical origins, institutional affiliations, and contributions to influential journals. Additionally, the investigation scrutinizes prevalent keywords in search queries and recurrent themes in research undertakings. The citation analysis is directed towards identifying noteworthy authors and seminal documents that hold substantive significance within this scholarly discourse. The principal aim of this inquiry is to discern primary areas of interest by analyzing co-citations among authors. Biased assessments have been systematically excluded, and the linguistic framework employed adheres to an objective and value-neutral stance. Technical terminology is elucidated upon initial usage, and conventional academic sections are seamlessly integrated into the narrative.

Introduction

Rammed earth housing, an enduring architectural tradition utilizing earth-based materials, encapsulates a rich cultural legacy and a nuanced understanding of local resources [1], [2]. This study scrutinizes the global landscape of structures, with a specific focus on rammed earth constructions, to elucidate their seismic resilience. The examination extends to evaluating the seismic performance of these structures in the aftermath of various earthquakes that have significantly impacted regions worldwide [3], [4]. Recent attention has been directed towards the seismic vulnerability of rammed earth constructions, prompting a comprehensive examination of their resilience. This research delves into the seismic history of rammed earth homes worldwide, highlighting instances where these structures demonstrated remarkable resistance to seismic forces. The vulnerability of rammed earth constructions to seismic forces underscores the imperative to implement robust seismic regulations. Recognizing this necessity, numerous countries have established seismic building codes and standards to safeguard both heritage and inhabitants. This article navigates through global seismic regulations, emphasizing their significance in mitigating the seismic risks associated with rammed earth constructions. Morocco, in particular, stands out for implementing seismic regulations tailored to earth-based constructions in 2011, reflecting a forward-thinking initiative to preserve architectural heritage while ensuring community safety [5]. This exploration aims not only to unveil the seismic challenges faced by rammed earth constructions but also to underscore the importance of regulatory frameworks in safeguarding these invaluable structures. By delving into global seismic experiences and spotlighting Morocco's proactive measures [6], [7], this article contributes to the ongoing dialogue

Mediterranean Architectural Heritage - RIPAM10   Materials Research Forum LLC
Materials Research Proceedings 40 (2024) 141-148   https://doi.org/10.21741/9781644903117-15

surrounding the seismic resilience of rammed earth constructions in our ever-evolving civil engineering landscape.

Furthermore, this study extends its inquiry by utilizing a cartographic analysis of the literature. By mapping out the existing body of knowledge on seismic resilience in traditional earth constructions, we provide a visual representation of the global progress and barriers in this field [8]. This analytical approach allows for a comprehensive understanding of the current state of research, identifying key trends, gaps, and areas for future exploration. As we navigate through the varied seismic experiences worldwide, it becomes evident that the challenges faced by traditional earth constructions are diverse and context-specific [9]. The nuanced nature of these challenges underscores the importance of a tailored and informed approach to seismic resilience. By amalgamating insights from diverse sources, our cartographic analysis aims to facilitate a holistic view, fostering informed discussions and guiding future research endeavors.

In the context of our exploration, the proactive measures taken by Morocco in implementing seismic regulations for earth-based constructions serve as a beacon of best practices [5]. In conclusion, our article not only contributes to the scholarly understanding of seismic resilience in traditional earth constructions but also advocates for a proactive and context-aware approach to safeguarding these structures. By combining historical perspectives, contemporary case studies, and a cartographic analysis of the literature, we hope to inspire ongoing efforts aimed at balancing heritage preservation and community safety in the dynamic landscape of seismic risk.

Methodology

To achieve the research objectives, a bibliometric analysis employing a quantitative approach was initially conducted. This methodology has gained popularity among researchers due to its utility in identifying emerging trends, reviewing outcomes in the specified research field, and establishing global collaboration networks among authors, institutions, and documents. The bibliometric approach enables the identification of research gaps, emerging themes, and the correlation of results in the specific scientific domain combined with relevant research areas [10].

Initially, publications were gathered using a specific database, widely recognized as one of the most comprehensive, and covering a greater number of academic documents than others. Query strings related to the study's topics were utilized with Boolean operators "AND" and "OR" to circumscribe the search. The initial search focused on articles published on traditional constructions. For the given query, 2965 documents were found in the dataset. While the substantial number of documents emphasizes the importance of the subject, it falls short in pinpointing the impact of the specific variable on the specific domain.

To address this, the variable "Seismic" was introduced, resulting in the modification of the query. The revised query encompassed the main keywords related to the study's topics and introduced the variable "seismic," broadening the scope of the research. The refined search, conducted until 2023, aimed to understand the seismic resilience of traditional houses. In summary, this research focused on the main keywords "rammed earth construction," "traditional construction," and "seismic."

This comprehensive selection incorporated a significant number of published studies dedicated to the specific domain that simultaneously address rammed earth construction and the impact of seismic activity on them. The use of asterisks allowed for the inclusion of all possible variants of the given words, accommodating differences after the asterisk.

Mediterranean Architectural Heritage - RIPAM10   Materials Research Forum LLC
Materials Research Proceedings 40 (2024) 141-148   https://doi.org/10.21741/9781644903117-15

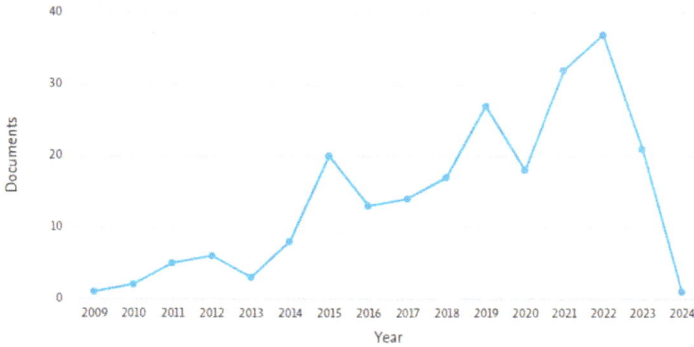

Fig. 1 Documents by year

Our findings reveal a gradual increase in publications from 2009 to 2013, followed by a second phase characterized by moderate growth in the number of publications between 2013 and 2020. Subsequently, there was a significant surge in research activity from 2020 onwards.

To achieve the research objectives, a bibliometric analysis employing a quantitative approach was conducted, with a focus on documents published until 2023. This analysis considered various criteria, including the typology of publication (conference paper, article, conference review, book chapter, and review), language of publication, distribution of publications by territory and institution, citation analysis, and keyword co-occurrence. The search, conducted in the Scopus database, resulted in a number of documents, all representing publications in their final stages [11]. These documents spanned various scientific disciplines, highlighting the interdisciplinary nature of the research. The typology of documents included various categories such as "articles," "conference papers," "conference reviews," "book chapters," and "reviews." Significantly, some documents featured bilingual content, emphasizing the global perspective of the research.

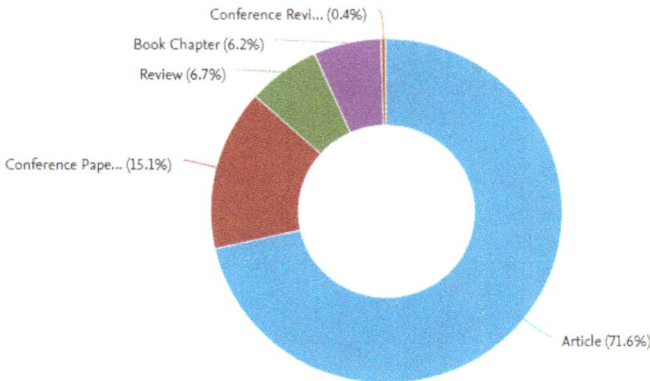

Fig. 2 Documents by type

In summary, this bibliometric analysis aimed to provide a comprehensive overview of the research landscape related to a specific intersection of topics. The methodology employed is

Mediterranean Architectural Heritage - RIPAM10   Materials Research Forum LLC
Materials Research Proceedings 40 (2024) 141-148   https://doi.org/10.21741/9781644903117-15

applicable across various domains, enabling researchers to adapt and apply similar approaches in different fields of study.

Following the initial data analysis conducted in Scopus, a more in-depth examination was carried out using VOSviewer. While several bibliometric tools such as BibExcel, CiteSpace, Sci2, or HistCite are available, VOSviewer was employed to analyze the data extracted from the Scopus database. VOSviewer, as a free software tool, is utilized for constructing and visualizing bibliometric networks that depict relationships between stakeholders such as authors, publishers, and institutions. These networks are established based on factors such as citations, keyword co-occurrence, and co-authorship within the scientific literature [12].

Results

To begin, Figures 3 and 4 illustrate the publication trends for the key territories contributing to research on the combined subjects. In the case of the combined research between "Traditional Building" AND "Seismic" (Fig. 3), the results indicate that Italy stands out as the primary contributor with the highest number of articles, accounting for nearly 10% of global publications until 2023. Following Italy, China contributed 123 articles, and Portugal contributed 63 articles. The United Kingdom, Japan, the United States, Turkey, Greece, India, and Iran are among the top 10 countries that have contributed. Morocco, on the other hand, has 2 articles on these subjects. This distribution highlights the global dissemination of research on the topic, with various countries making significant contributions to the knowledge base.

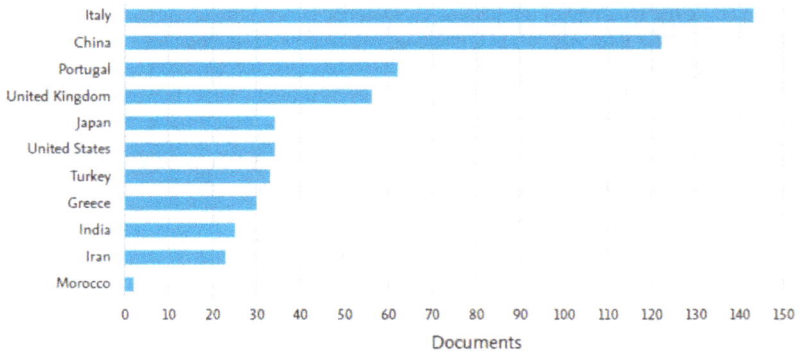

Fig. 3 Documents by country or territory:"Traditional Building" AND "Seismic"

Figure 4, on the other hand, depicts the results of the combined search for "Rammed Earth Construction" AND "Seismic", where Portugal emerges as the leading country contributing with 53 articles. In the second position is China with 35 articles, followed by France with 27 articles in the third position. Morocco has 3 articles.

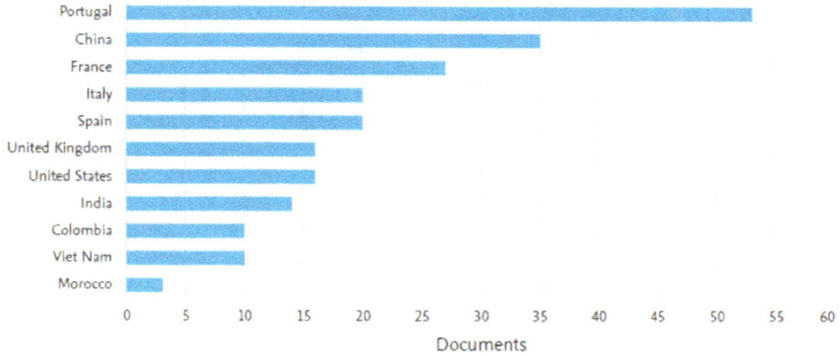

Fig 4. Documents by country or territory:"Rammed Earth Construction" AND "Seismic"

To determine the most frequently used expressions in the selected articles, a keyword analysis was conducted (Fig. 5). The aim of this study was to identify key concepts and research topics that have garnered significant attention in the discipline. The frequency and distribution of keywords were examined to gain insights into current research interests and priorities. Five unique clusters were identified after the keyword analysis, representing the main areas of interest for further studies. These domains cover important topics such as "Seismic Performance", with 55 links distributed across 5 clusters, "Rammed Earth", with 130 links in 1 cluster, "Compressive Strength", with 99 links, "Finite Element Method", with 70 links spread across 3 clusters.

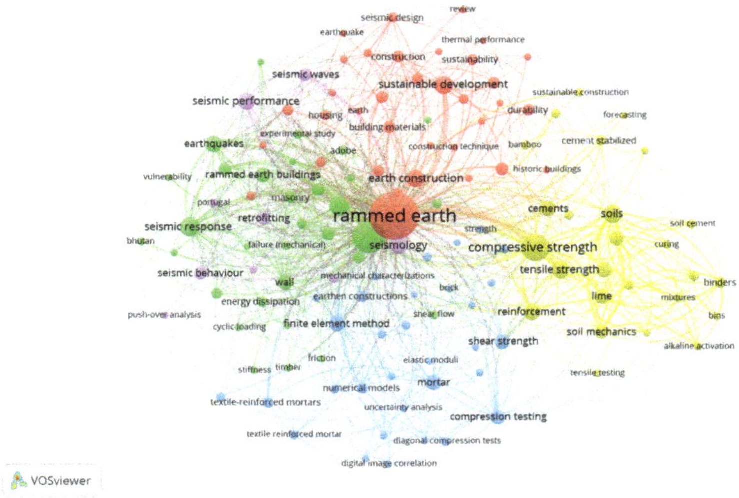

Fig 5. Keyword co-occurrence map, 5 clusters and 2784 links

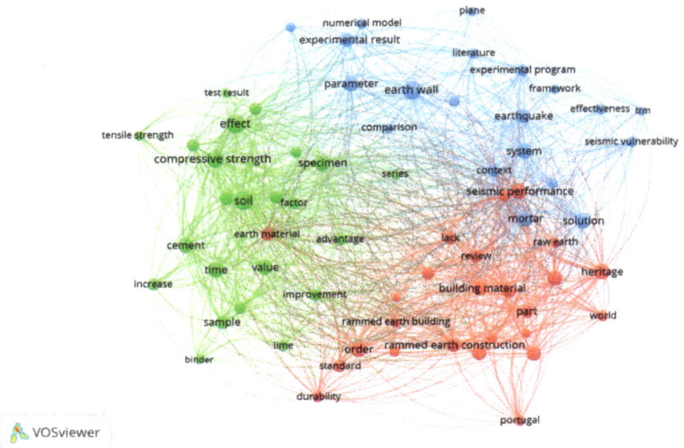

Fig. 6 Title and abstract fields

By conducting an analysis of recurring keywords in titles and abstracts (Fig. 6), we observe the repetition of the same keywords, supporting our initial analysis. Researchers with a high number of citations are often considered authorities in their field. Identifying experts in a specific subject and determining whether a particular country dominates that field can be a valuable endeavor, offering insights that can guide research. A co-publication analysis (Fig. 7) was carried out to identify the most influential authors. To assess the impact of individual documents and authors on the field, the number of citations received by each was examined. Through co-citation analysis, clusters of related studies and prominent research topics within the literature have been identified.

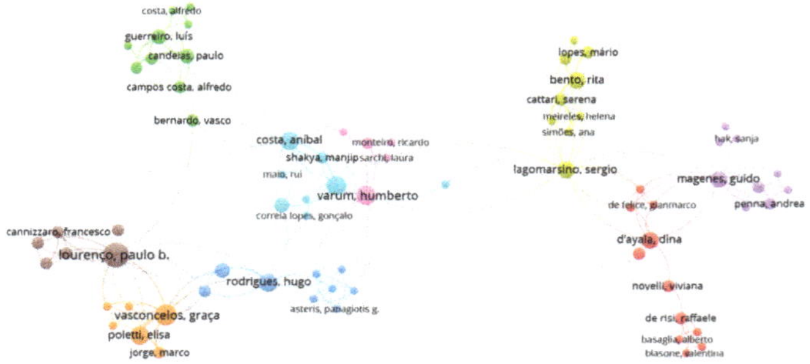

Fig. 7 Co-authorship map, 9 clusters and 241 links

On the illustrated graphical representation, it is evident that there are 9 distinct clusters of authors interconnected by a total of 241 links. Among these clusters, notable prominence is observed within two primary figures: Lourenço P.B., who is connected with 14 other authors and has contributed to 10 documents [13], and Varum H., who has authored 6 documents while maintaining connections with 16 other authors [14]. Identifying experts and assessing the dominance of countries (Fig. 8) in a specific subject can help target collaborations and investments in research areas where specific knowledge gaps exist.

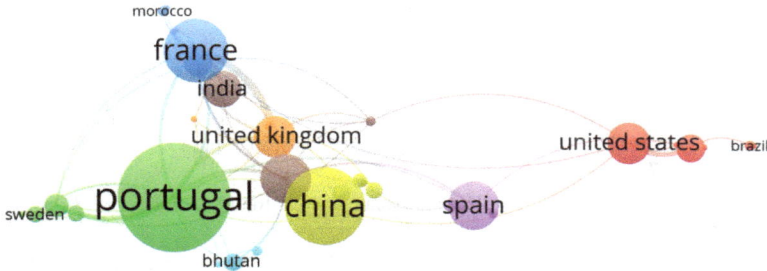

Fig. 8 Country co-occurrence map, 8 clusters and 90 links

The country-wise analysis unveils the presence of 8 clusters connected by a total of 90 links (Fig. 8). It is noteworthy that these clusters exhibit significant variations in terms of socio-economic conditions and respective regulatory frameworks. The countries displaying the highest interest in rammed earth constructions and seismic resilience are Portugal, China, and France. Morocco finds itself linked to France, contributing together to the origin of 3 documents.

Conclusion

In conclusion, this in-depth study on rammed earth constructions and their seismic resilience has provided an enlightening perspective on recent research trends in the field of civil engineering. By integrating bibliometric analyses, we have identified significant developments, established connections among influential researchers, and highlighted the evolution of study areas in this specific domain. This research underscores the growing importance of seismic resilience in the context of rammed earth constructions on a global scale. The compiled information provides a solid foundation to guide practitioners and researchers in understanding and enhancing the performance, sustainability, and operational resilience of rammed earth buildings.

By focusing on topics such as seismic regulations and proactive initiatives in certain countries, notably Morocco, this study offers valuable insights to steer future research and design practices. It also demonstrates the ongoing need to develop robust seismic standards to preserve architectural heritage while ensuring the safety of communities. Ultimately, this research contributes to the ongoing discourse on the seismic resilience of rammed earth constructions in a constantly evolving world. It paves the way for future advancements in the design, construction, and preservation of these structures, emphasizing the importance of reconciling tradition and innovation to address contemporary challenges.

References

[1] S. Simou, K. Baba, N. Akkouri, M. Lamrani, M. Tajayout, and A. Nounah, "Mechanical Characterization of the Adobe Material of the Archaeological Site of Chellah," in *Recent Thoughts in Geoenvironmental Engineering*, H. Ameen, M. Jamiolkowski, M. Manassero, and H. Shehata, Eds., in Sustainable Civil Infrastructures, Cham: Springer International Publishing, 2020, pp. 118–130 . https://doi.org/10.1007/978-3-030-34199-2_8

[2] S. Simou, K. Baba, N. Akkouri, M. Lamrani, M. Tajayout, and A. Nounah, "Mechanical characterization and reinforcement of the adobe material of the Chellah archaeological site," *E3S Web Conf.*, vol. 150, p. 03022, 2020 . https://doi.org/10.1051/e3sconf/202015003022

[3] D. V. Oliveira, A. Romanazzi, R. A. Silva, A. Barontini, and N. Mendes, "Seismic Behaviour and Strengthening of Rammed Earth Constructions," in *Structural Analysis of Historical Constructions*, vol. 47, Y. Endo and T. Hanazato, Eds., in RILEM Bookseries, vol. 47.

, Cham: Springer Nature Switzerland, 2024, pp. 1214–1225 . https://doi.org/10.1007/978-3-031-39603-8_98.

[4] F. J. Baeza, L. Estevan, and S. Ivorra, "Seismic Retrofitting of Heritage Structures, Actual Techniques and Future Challenges for Earth and Masonry Constructions," in *Structural Analysis of Historical Constructions*, vol. 47, Y. Endo and T. Hanazato, Eds., in RILEM Bookseries, vol. 47, Cham: Springer Nature Switzerland, 2024, pp. 1088–1101 . https://doi.org/10.1007/978-3-031-39603-8_87

[5] M. of Housing and U. Policy, The Moroccan Seismic Regulation for Earth Constructions 2011. 2011.

[6] Y. Razzouk, M. Ahatri, K. Baba, and A. E. Majid, "Optimizing Seismic Bracing Systems for Reinforced Concrete Buildings Based on Height and Seismic Zone," *cea*, vol. 11, no. 6, pp. 3430–3450, Nov. 2023 . https://doi.org/10.13189/cea.2023.110615

[7] Y. Razzouk, M. Ahatri, K. Baba, and A. El Majid, "Optimal Bracing Type of Reinforced Concrete Buildings with Soil-Structure Interaction Taken into Consideration," *Civ Eng J*, vol. 9, no. 6, pp. 1371–1388, Jun. 2023. https://doi.org/10.28991/CEJ-2023-09-06-06

[8] F. Orsini and P. Marrone, "Approaches for a low-carbon production of building materials: A review," *Journal of Cleaner Production*, vol. 241, p. 118380, Dec. 2019 . https://doi.org/10.1016/j.jclepro.2019.118380

[9] Q. Yang, M. Zhou, and K. Liu, "Seismic Performance and Vulnerability Analysis of Traditional Chinese Timber Architecture Considering Initial Damage," in *Structural Analysis of Historical Constructions*, vol. 47, Y. Endo and T. Hanazato, Eds., in RILEM Bookseries, vol. 47, Cham: Springer Nature Switzerland, 2024, pp. 1114–1124 . https://doi.org/10.1007/978-3-031-39603-8_89

[10] W. Rong and A. Bahauddin, "A Bibliometric Review of the Development and Challenges of Vernacular Architecture within the Urbanisation Context," *Buildings*, vol. 13, no. 8, p. 2043, Aug. 2023 . https://doi.org/10.3390/buildings13082043

[11] L. F. Cabeza, M. Chàfer, and É. Mata, "Comparative Analysis of Web of Science and Scopus on the Energy Efficiency and Climate Impact of Buildings," *Energies*, vol. 13, no. 2, p. 409, Jan. 2020 . https://doi.org/10.3390/en13020409

[12] G. Demir, P. Chatterjee, and D. Pamucar, "Sensitivity analysis in multi-criteria decision making: A state-of-the-art research perspective using bibliometric analysis," *Expert Systems with Applications*, vol. 237, p. 121660, Mar. 2024 . https://doi.org/10.1016/j.eswa.2023.121660

[13] R. A. Silva, D. V. Oliveira, L. Schueremans, P. B. Lourenço, and T. Miranda, "Modelling the Structural Behaviour of Rammed Earth Components," presented at the The Twelfth International Conference on Computational Structures Technology, Naples, Italy, p. 112 . https://doi.org/10.4203/ccp.106.112

[14] H. Varum, A. Costa, J. Fonseca, and A. Furtado, "Behaviour Characterization and Rehabilitation of Adobe Construction," *Procedia Engineering*, vol. 114, pp. 714–721, 2015 . https://doi.org/10.1016/j.proeng.2015.08.015

Mediterranean Architectural Heritage - RIPAM10
Materials Research Proceedings 40 (2024) 149-159

Materials Research Forum LLC
https://doi.org/10.21741/9781644903117-16

Comparative Analysis of the Effect of Thermal Insulation on the Energy Requirements of a Tertiary Building in Meknes

A. BOUCHARK[1]*, M.C. EL BOUBAKRAOUI[1], A. EL BAKKALI[1] and M. HADDAD[1]

[1] Laboratoire de Spectrométrie des Matériaux et des Archéomatériaux (LASMAR), URL-CNRST N°7, Faculty of Sciences - Moulay Ismail University of Meknes, Morocco

*Abouchark@gmail.com

Keywords: Thermal Insulation, Insulation Materials, Energy Efficiency, Tertiary Building, TRNSYS Software

Abstract. Thermal insulation materials are essential for minimizing heat loss in winter and heat gain in summer in buildings, irrespective of the presence or absence of air conditioning systems. The right choice of insulation materials paves the way for considerable savings in buildings' energy requirements, while rationalizing the use of air-conditioning systems. This is all the more important in Morocco, where the building sector is one of the biggest consumers of energy. Consequently, improving the energy efficiency of buildings is an imperative, especially in the current context characterized by the gradual depletion of fossil resources and ever-rising energy costs. Our study focuses on the practical impact of integrating different insulation materials, including phase-change materials (PCMs), hemp concrete and polystyrene, into the structure of a tertiary building in Meknes, Morocco. The results of this research highlight that the incorporation of effective thermal insulation in the building's various construction elements results in substantial reductions in energy requirements, both in terms of heating and cooling. It should be noted that this study was carried out using energy simulations with TRNSYS software.

Introduction

Today, energy efficiency has become a global issue, as energy demand increases exponentially in sectors such as building, transport and industry. Indeed, the majority of energy policies target the building sector first and foremost as an energy-intensive entity in order to achieve energy consumption reduction targets [1].

In Morocco, the building sector is the second largest consumer of energy, prompting the Moroccan government to adopt thermal regulations (RTCM) in 2014 to integrate energy efficiency measures into the Moroccan construction context [2].

The building envelope is one of the fundamental elements on which energy efficiency measures are based, and is considered an effective passive strategy for improving the thermal performance and energy requirements of both tertiary and residential buildings [3-5]. For this purpose, a great deal of research has been carried out, focusing on thermal insulation [6-10].

A numerical simulation study to measure the effect of several insulation materials on the energy demand of an educational building, revealed that extruded polystyrene and polyurethane were the best choices for low energy consumption [11].

Another study evaluates the thermal behavior of six thermal insulants introduced separately into a wall using a simulation with ANSYS/FLUENT16 software, with the results showing that polystyrene is the most suitable insulator [12].

Phase-change materials (PCM) have also been used as insulation elements [13-17]. An experimental and numerical study was carried out to assess the thermal performance of a wall panel incorporating PCMs. According to the study, the use of the PCM layer reduced the thermal load by 15% [18].

Other research has also focused on the use of bio-sourced materials for thermal insulation [19-21], such as hemp and hemp concrete, were subjected to a dynamic simulation analysis using TRNSYS software to assess their performance in a building in Meknes, Morocco. The results showed a 36.78% reduction in annual heating and cooling requirements when the building's roof and external walls were insulated with hemp [22].

In the present study, we are interested in the choice of a thermal insulation material adapted to the specific climatic conditions in Meknes city, while respecting the recommendations of the Moroccan thermal regulations (RTCM), in order to optimize the energy efficiency of a primary school annex in this region. We carried out a numerical simulation using TRNSYS software to compare the energy performance of three types of insulation: polystyrene, hemp concrete and phase-change materials. These materials were chosen for their distinct thermophysical properties, offering a variety of potential solutions to meet the specific energy requirements of the Meknes region, Morocco's third climate zone [2]. We aim to rigorously validate our results according to RTCM norms, thereby reinforcing the reliability and relevance of our study in the Moroccan regulatory context.

Description of the studied building
Our study focused on an extension being built to a primary school in Meknes, comprising three classrooms. The basic dimensions of the building are 20.45m long, 9.20m wide and 3.20m high. These parameters prioritize the space in which pupils will learn.

As part of our analysis of this school annex, we took detailed measurements to assess the building's thermal performance. Specific features include the presence of 12 single-glazed windows measuring 1.6 x 1.2 m^2, as well as three wooden doors measuring 2.1 x 1 m^2. Our analysis took these details into account to assess the significant impact of these components on our school's energy consumption. At the same time, we also noted the presence of 18 light bulbs of 40W each, which added a further dimension to our determination of the building's energy load. In addition, we identified a permanent occupancy of 93 people for 10 hours a day from 8h00 to 18h00, as this school works in groups, the first group in the morning from 8h00 to 13h00 and the second group in the afternoon from 13h00 to 18h00, an essential parameter for assessing heating and cooling requirements, and defining precise strategies for optimizing the energy efficiency of this school annex currently under construction.

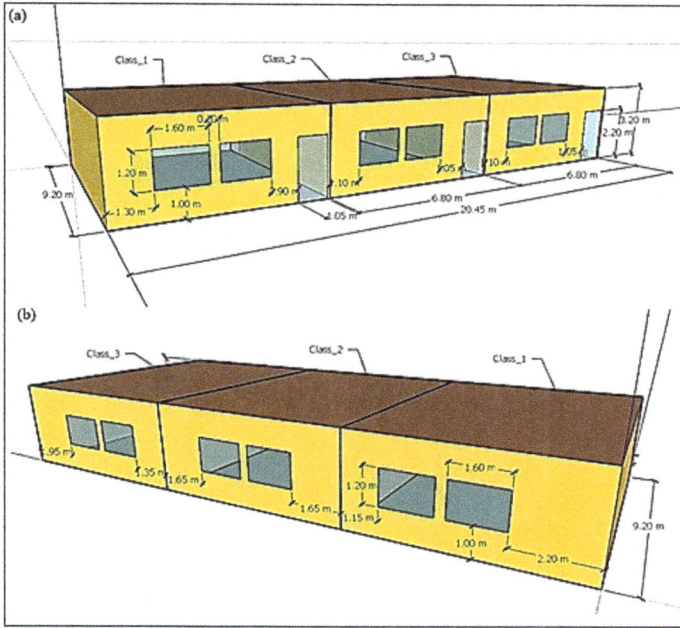

Fig. 1: 3D school plan: (a) Front facade, (b) Rear facade

The thermophysical characteristics of walls, roofs and low floors are of crucial importance in the thermal design of buildings. The essential data relating to these properties are detailed in the tables below:

Table 1: Exterior wall compositions

Materials	Mortar	Brick (8holes)	Air gap	Brick (8holes)	Mortar
Thickness (cm)	1	10	10	10	1
Thermal conductivity (W/m K)	1.4	1.15	0.556	1.15	1.4
Density (kg/m^3)	2000	743	1	743	2000

Table 2: Interior wall compositions

Materials	Mortar	Brick (8holes)	Mortar
Thickness (cm)	1	10	1
Thermal conductivity (W/m K)	1.4	1.15	1.4
Density (kg/m^3)	2000	743	2000

Table 3: Roof compositions

Materials	Plaster	Concrete slabs	Slab	Mortar	Tiles
Thickness (cm)	1	16	7	10	5
Thermal conductivity (W/m K)	0.351	1.23	1.65	1	1
Density (kg/m³)	1500	1300	2150	1700	2500

Table 4: Floor on ground compositions

Materials	Exposed aggregate concrete	Form	Slab	Sand	Gravel
Thickness (cm)	2	6	12	3	20
Thermal conductivity (W/m K)	1	0.39	1.7	0.39	0.47
Density (kg/m³)	2000	600	2400	750	570

Building orientation is of crucial importance in determining energy loads. The way our building is positioned in relation to the sun directly influences its energy requirements. For the orientation of our school's facades (Fig. 1), we have:

Table 5: Orientation of the school facades

Façade	Orientation
Front façade	South
Rear façade	North
Left lateral	West
Right lateral	East

Description of the studied climate
The city of Meknes, located in northern Morocco at an altitude of 531m, has a subtropical climate with hot, sunny summers and mild winters. Although it is slightly cooler (on average, it's around two degrees cooler in winter and even three quarters cooler in summer). It is also rainier, due to its more northerly location [23].
- In winter, temperature variations between day and night are significant: the average daytime temperature is 14°C, while the minimum night-time temperature is 5°C.
- In summer, the average daytime temperature is 35°C, with little difference from the night-time temperature. Similar to many regions of Morocco.

Mediterranean Architectural Heritage - RIPAM10
Materials Research Proceedings 40 (2024) 149-159

Materials Research Forum LLC
https://doi.org/10.21741/9781644903117-16

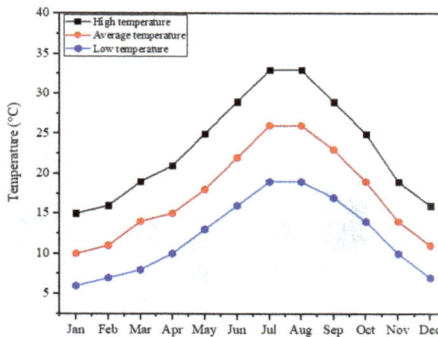

Fig. 2: Temperature chart of Meknes city.

Building simulation

In this stage of our study to evaluate and optimize the energy efficiency of the annex of our primary school in Meknes, we resorted to advanced energy simulations. These simulations were carried out using TRNSYS software. This energy simulation methodology offers an in-depth and accurate perspective on the building's thermal performance, enabling a detailed assessment of the impact of the various architectural elements and energy systems considered. The three-dimensional representation of the structure was created using SketchUp software. Simulation of the school's thermal behavior is carried out using multi-zone transient modeling with a time interval of one hour. The TRNBuild tool is used to input the data required for the multi-zone building simulation, describing the specific characteristics of the envelope (building materials, thickness of each layer and thermophysical parameters), window details and the school's occupancy schedule.

Our study is based on the annual working calendar, with the exception of the period from mid-July to August, when schools are closed for the school vacations. These months were not included in our analysis to ensure an accurate and relevant representation of working conditions throughout the year, avoiding variations linked to school closures during the summer. In addition, we took into consideration the thermal comfort set-point temperatures established by the RTCM, i.e. 26°C in summer and 20°C in winter. These values were incorporated into our model to ensure a realistic assessment of energy requirements. In addition, we also took into account the high occupancy rate, which is a crucial element for a comprehensive analysis of energy loads. It is important to note that we carried out the simulation with the single glazing for the windows, with a transmission coefficient of 5.5 W/m² K and a solar factor of 0.871.

Results and discussion

Firstly, a fundamental step in our analysis is to calculate the energy requirements of our school under baseline conditions, using conventional parameters. This initial assessment will then be compared with the recommendations of the RTCM to validate the consistency of our study. Once this validation has been carried out, we plan to introduce insulating materials to improve the energy performance of the building. The results obtained by calculating the energy requirements of our building and comparing them with RTCM standards are clearly presented in Fig 3.

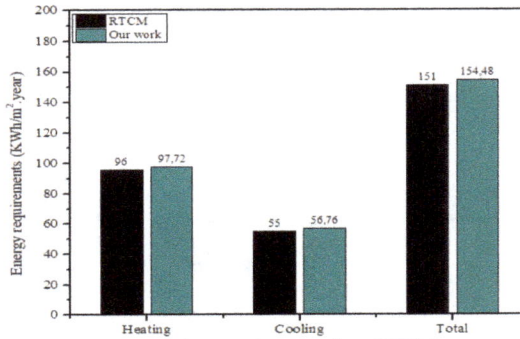

Fig. 3: Validation of our work to RTCM.

A comparative analysis of the annual energy requirements of our building, as calculated in our work compared with the references set out in the RTCM, reveals significant results. For cooling requirements, our simulation produced a figure of 56.76 kWh/m² year, slightly higher than the 55 kWh/m² year defined by the RTCM. On the other hand, for heating, our assessment indicated a requirement of 97.72 kWh/m² year, compared with the RTCM reference of 96 kWh/m² year. The sum total of our calculations is 154.48 kWh/m² year, slightly exceeding the RTCM figure of 151 kWh/m² year. These results indicate a remarkable concordance between our analysis and the regulatory standards, validating the reliability of our approach in determining the energy requirements of our building.

Having validated our results in accordance with the RTCM, we are now committed to the next phase, which focuses on the introduction of insulation materials to improve the energy efficiency of our building. We have chosen to compare three types of insulation: polystyrene, hemp concrete and phase change materials, to determine which best meets our requirements in our conditions, particularly in Meknes, located as Morocco's third climatic zone. The focus will be on energy efficiency, assessing the ability of each insulation to meet the specific cooling and heating needs in the city's distinct climatic context. This targeted approach aims to optimize the energy performance of our building by selecting the insulation best suited to our particular climatic environment.

Before comparing the insulation, materials selected, it is essential to know the thermophysical characteristics of each material. This preliminary step will enable us to rigorously assess the performance of each insulation material, whether polymeric, biobased or phase change materials. The data are shown in Table 6.

Table 6: insulation characteristics

Materials	Expended Polystyrene	Hemp Concrete	PCM (BioPCM M27/Q23)
insulation thickness in exterior walls and Roof (cm)	4	4	4
insulation thickness in Floor on ground (cm)	2	2	2
Thermal conductivity (W/m K)	0.037	0.082	0.2
Density (kg/m³)	30	317	860

After a detailed study of the thermophysical characteristics of each insulating material, we will proceed to the numerical simulation of our building by introducing 4 cm of each insulating material in the external walls and the roof, while 2 cm of insulating material will be added to the floor on ground. The aim of this simulation is to assess the relative performance of each insulating material in our specific context. The detailed results of this numerical analysis, illustrated in Fig 4.

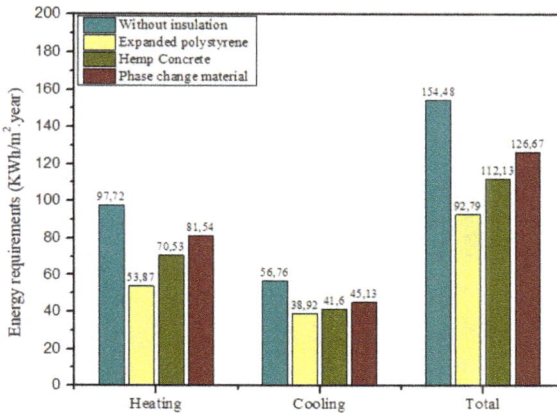

Fig. 4: energy loads with and without the three insulators.

The results presented in Figure 4 provide an in-depth analysis of the impact of different insulation methods on the annual energy requirements of our building in Meknes. Without insulation, the total energy requirement is 154.48 kWh/m² year, with a heating component of 97.72 kWh/m² year and a cooling component of 56.76 kWh/m² year. The introduction of polystyrene results in a significant reduction, bringing total energy requirements down to 92.79 kWh/m² year, an impressive reduction of almost 40.13%. Heating showed a substantial reduction of 45.02%, from 97.72 to 53.87 kWh/m² year, while cooling fell by 31.42%, from 56.76 to 38.92 kWh/m² year. For hemp concrete, a total reduction of 27.43% was observed, with significant reductions of 28.01% for heating (70.53 kWh/m² year) and 26.67% for cooling (41.60 kWh/m² year). Finally, the introduction of the PCM results in a total reduction of 18.03%, with decreases of 16.57% for heating (81.54 kWh/m² year) and 20.42% for cooling (45.13 kWh/m² year).

To ensure performance equivalent to that achieved with polystyrene, we undertook a series of simulations varying insulation thicknesses in different parts of the building, including external walls, the floor on ground, and roof (Table 7). This approach enabled us to explore various scenarios for each insulation material, seeking to identify the optimum thicknesses for achieving performance levels comparable to those of polystyrene. The results obtained from these simulations were carefully analyzed to inform our final choice of insulation material for our school in Meknes. The thicknesses selected for each material were adjusted to ensure optimum thermal performance, enabling us to make an informed decision in favor of the insulation that best met our specific energy and climatic criteria.

Table 7: various insulation thicknesses

Materials	Expended polystyrene	Hemp Concrete	PCM (BioPCM M27/Q23)
insulation thickness in exterior walls (cm)	4	8	12
insulation thickness in roof (cm)	4	8	18
insulation thickness in Floor on ground (cm)	2	2	10

The results obtained by varying the thicknesses of insulating materials such as hemp concrete and PCM are illustrated in detail in Fig 5.

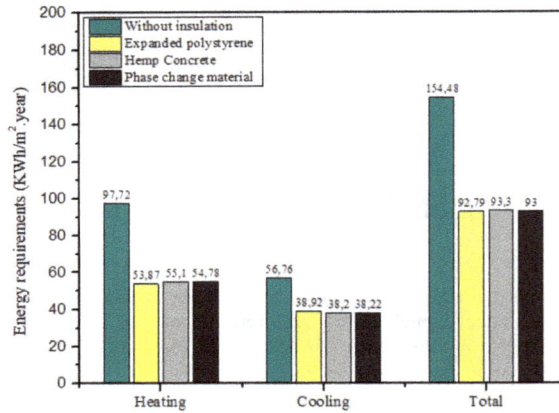

Fig. 5: energy load with and without the three insulators (with different thicknesses)

These results highlight the significant impact of thickness variation on the reduction in energy requirements, underlining the outstanding performance of both hemp concrete and PCM. For hemp concrete, judicious thickness adjustment resulted in considerable reductions, with total energy requirements falling to 93.3 kWh/m² year, reflecting a significant reduction of 39.59%. The reduction in heating requirements is 43.65%, from 97.72 to 55.1 kWh/m² year, while the reduction in cooling requirements is 32.70%, from 56.76 to 38.2 kWh/m² year. For the PCM, thickness adjustments have also led to significant improvements, with total energy requirements reduced to 93 kWh/m² year, representing a significant reduction of 39.84%. The reduction in heating requirements is 43.93%, from 97.72 to 54.78 kWh/m² year, and the reduction in cooling requirements is 32.52%, from 56.76 to 38.22 kWh/m² year.

These results show that, despite the significant increase in insulation thickness, hemp concrete and PCM achieve energy performances almost equivalent to those of polystyrene in its initial configuration. In this case, polystyrene stands out as the optimum insulator, maintaining equivalent or even superior energy performance despite variations in the thickness of the other insulators chosen.

According to our study carried out in the climatic context of Meknes, polystyrene stands out as the best performing insulation for our specific application. Simulation results using TRNSYS software highlight its remarkable efficiency, largely attributed to its low thermal conductivity. However, it's essential to note that this doesn't mean that other insulants, such as hemp concrete and PCM, don't meet our requirements. On the contrary, these insulants demonstrate a significant capacity to reduce energy requirements. In particular, hemp concrete offers outstanding performance, with a reduction of almost 27.43% in cooling requirements and 18.3% in heating requirements compared to the situation without insulation. As for PCM, although its performance is slightly lower, it offers a distinct advantage thanks to its high energy storage capacity. This unique feature could be a key element in specific applications where thermal storage is an important consideration. Thus, although polystyrene emerges as the optimal choice in our current context, it would be interesting to consider further studies to improve other materials, particularly phase-change materials. The latter could benefit from further investigation in order to fully exploit their potential as innovative solutions for energy efficiency, while taking into account the specific characteristics of the Meknes climate.

Application: Real case

Following this study, polystyrene appears to be the optimal insulation choice for our building, representing an effective passive strategy for significantly improving its thermal performance and reducing its energy requirements. The simulation results clearly demonstrate the considerable advantages offered by polystyrene in our specific Meknes context.

Based on these conclusions, we decided to adopt polystyrene in our specific case, namely in our school building. This decision was based on concrete data and simulation results, guaranteeing a significant improvement in the thermal performance of our building. Here are a few images of the installation of polystyrene insulation in our school building envelope (fig 6):

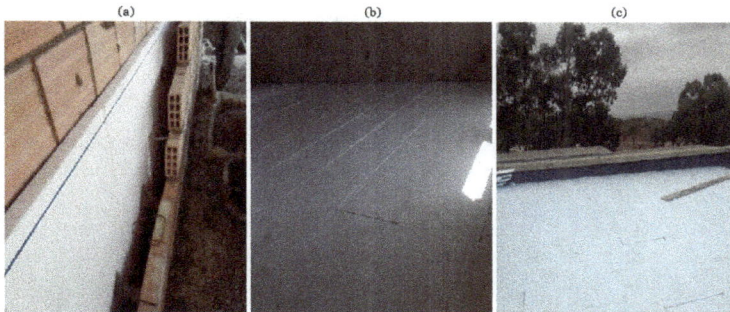

Fig. 6: Installing polystyrene as insulation in the building envelope: (a) external wall, (b) floor on ground, (c) roof.

Conclusion

In conclusion, our study involved a comparative analysis of the effect of thermal insulation on the energy requirements of a primary school annex in Meknes. We carefully compared energy requirements without insulation with those resulting from the integration of three different insulating materials: polymers, bio-sourced materials and phase change materials. The results, obtained from a numerical simulation using TRNSYS software, offer significant insight into the efficiency of each insulator in the specific climatic context of Meknes.

The overall percentage reduction in energy requirements is encouraging, with polystyrene leading the way with a substantial reduction of around 40.13 %, followed by hemp concrete with 27.43 % and PCM with 18.03 %. These results guided our strategic choice to adopt polystyrene as the insulation in our real case, particularly in our school building. This decision was based on polystyrene's ability to offer the most significant reductions in energy costs, reinforcing its status as the preferred choice for improving the energy efficiency of our building in Meknes.

It is important to note that other scenarios aimed at further reducing these energy requirements form part of the objectives of future studies, with the aim of continually optimizing the energy performance of the building. A concrete example of such scenarios could be the integration of solutions such as more efficient glazing to minimize thermal losses, thus contributing to a further reduction in the building energy requirements.

However, this conclusion does not close the debate. It opens the door to future studies, inviting further exploration of the possibilities offered by other insulating materials. The constant evolution of technologies and knowledge in the field of energy efficiency suggests that further innovations could be envisaged to further optimize the thermal performance of buildings in similar climates.

References

[1] Ministère de l'Aménagement du territoire national, de l'Urbanisme, de l'Habitat et de la Politique de la ville, General construction regulations, 2014, https://www.mhpv.gov.ma/wp-content/uploads/2018/03/RT-fr-1.pdf. Accessed 10 November 2023.

[2] Agence Nationale pour le Developpement des Energies Renouvelables et de l'Efficacite Energetique, Thermal_Construction_Regulations_in_Morocco, https://www.amee.ma/sites/default/files/inline-files/Reglement_thermique_de_construction_au_Maroc.pdf. Accessed 10 November 2023.

[3] N. Ghabra, L. Rodrigues, P. Oldfield, The impact of the building envelope on the energy efficiency of residential tall buildings in Saudi Arabia, Int. J. Low-Carbon Technol. 12 (4) (2017) 411-419. https://doi.org/10.1093/ijlct/ctx005

[4] F. Ascione, N. Bianco, G. Maria Mauro, DF. Napolitano, Building envelope design: multi-objective optimization to minimize energy consumption, global cost and thermal discomfort. Application to different Italian climatic zones, Energy. 174 (2019) 359-374. https://doi.org/10.1016/j.energy.2019.02.182

[5] X. Meng et al., Optimization of the wall thermal insulation characteristics based on the intermittent heating operation, *Case Studies in Construction Materials*. 9 (2018) e00188. https://doi.org/10.1016/j.cscm.2018.e00188

[6] D. Kumar, M. Alam, P. X. W. Zou, J. G. Sanjayan, and R. A. Memon, Comparative analysis of building insulation material properties and performance, Renewable Sustainable Energy Revi. 131 (2020) 110038. https://doi.org/10.1016/j.rser.2020.110038

[7] L. Aditya, et al., A review on insulation materials for energy conservation in buildings, Renew Sustain Energy Rev 73 (2017) 1352-1365. https://doi.org/10.1016/j.rser.2017.02.034.

[8] B. Abu-Jdayil, A. H. Mourad, W. Hittini, M. Hassan, and S. Hameedi, Traditional, state-of-the-art and renewable thermal building insulation materials: an overview, Construct Build Mater 214 (2019) 709-735. https://doi.org/10.1016/j.conbuildmat.2019.04.102

[9] L. D. Hung Anh and Z. Pásztory, An overview of factors influencing thermal conductivity of building insulation materials, *J. Build. Eng.*, 44 (2021) 102604. https://doi.org/10.1016/j.jobe.2021.102604

[10] F. Asdrubali, F. D'Alessandro, S. Schiavoni, A review of unconventional sustainable building insulation materials, SM&T. 4 (2015) 1-17. https://doi.org/10.1016/j.susmat.2015.05.002

[11] M. Morsy, M. Fahmy, H. Abd Elshakour, and A. M. Belal, Effect of Thermal Insulation on Building Thermal Comfort and Energy Consumption in Egypt, J. Adv. Res. App. Mech. 43 (2018) 8–19. www.akademiabaru.com/aram.html

Mediterranean Architectural Heritage - RIPAM10
Materials Research Proceedings 40 (2024) 149-159

Materials Research Forum LLC
https://doi.org/10.21741/9781644903117-16

[12] H.Q. Hussein, A.F. Khalaf, A.K. Jasim, A Numerical Study To Investigate Several Thermal Insulators To Choose The Most Suitable Ones For Thermal Insulation In Buildings, Int. J. Mech. Eng. 7 (2022) 921-929.

[13] P. K. S. Rathore, N. K. Gupta, D. Yadav, S. K. Shukla, and S. Kaul, Thermal performance of the building envelope integrated with phase change material for thermal energy storage: an updated review, *Sustain Cities Soc,*. 79 (2022) 103690. https://doi.org/10.1016/j.scs.2022.103690

[14] R. A. Kishore, M. V. A. Bianchi, C. Booten, J. Vidal, and R. Jackson, Enhancing building energy performance by effectively using phase change material and dynamic insulation in walls. *Appl Energy*, 283 (2021) 116306. https://doi.org/10.1016/j.apenergy.2020.116306

[15] A. R. El-Sayed, A. Talaat, and M. Kohail, The effect of using phase-changing materials on non-residential air-conditioning cooling load in hot climate areas, *A S E J*, 14, no. 6 (2023) 102-109. https://doi.org/10.1016/j.asej.2022.102109

[16] Q. Al-Yasiri and M. Szabó, Building envelope-combined phase change material and thermal insulation for energy-effective buildings during harsh summer: Simulation-based analysis, *Energy Sustain Dev*. 72 (2023) 326–339. https://doi.org/10.1016/j.esd.2023.01.003

[17] Z. Zhang, N. Zhang, Y. Yuan, P. E. Phelan, and S. Attia, Thermal performance of a dynamic insulation-phase change material system and its application in multilayer hollow walls, *J Energy Storage*, 62 (2023) 106912. https://doi.org/10.1016/J.EST.2023.106912

[18] A. Fateh, F. Klinker, M. Brütting, H. Weinläder, and F. Devia, Numerical and experimental investigation of an insulation layer with phase change materials (PCMs), *Energy Build*, 153 (2017) 231–240. https://doi.org/10.1016/j.enbuild.2017.08.007

[19] M. Charai, M. Salhi, O. Horma, A. Mezrhab, M. Karkri, and S. Amraqui, Thermal and mechanical characterization of adobes bio-sourced with Pennisetum setaceum fibers and an application for modern buildings, *Constr Build Mater*. 326 (2022) 126809. https://doi.org/10.1016/j.conbuildmat.2022.126809

[20] F. Balo and L. Sagbansua, Investigating the ecological efficiency of widely utilized bio-sourced insulation materials in the building lifecycle, Proceedings of Engineering to Thrive 2022. https://scholar.uwindsor.ca/cgi/viewcontent.cgi?article=1003&context=wtel

[21] A. Benallel, A. Tilioua, and M. Garoum, Development of thermal insulation panels bio-composite containing cardboard and date palm fibers, *J Clean Prod*. 434 (2024) 139995. https://doi.org/10.1016/J.JCLEPRO.2023.139995

[22] M. Dlimi, R. Agounoun, I. Kadiri, R. Saadani, and M. Rahmoune, Thermal performance assessment of double hollow brick walls filled with hemp concrete insulation material through computational fluid dynamics analysis and dynamic thermal simulations, e-prime Adv. Electri. Eng. Electro. Energy, 3 (2023) 100124. https://doi.org/10.1016/j.prime.2023.100124

[23] Weather Spark, Climate, weather, average temperature for Meknes (Morocco). https://fr.weatherspark.com/y/34043/M%C3%A9t%C3%A9o-moyenne-%C3%A0-Mekn%C3%A8s-Maroc-tout-au-long-de-l'ann%C3%A9e. Accessed 20 November 2023.

Mediterranean Architectural Heritage - RIPAM10
Materials Research Proceedings 40 (2024) 160-166

Materials Research Forum LLC
https://doi.org/10.21741/9781644903117-17

Wall Paintings from The Roman City of Volubilis in Morocco: XRF, Raman and FTIR-ATR Analyses

Imane FIKRI[1], Mohamed EL AMRAOUI[1], Mustapha HADDAD[1,*],
Ahmed Saleh ETTAHIRI[2], Christophe FALGUERES[3],
Ludovic BELLOT-GURLET[4], Taibi LAMHASNI[1,2], Saadia AIT LYAZIDI[1],
Lahcen BEJJIT[1]

[1] Laboratoire de Spectrométrie des Matériaux et Archéomatériaux (LASMAR, URL-CNRST N°7), Université Moulay Ismail, Faculté des Sciences, Meknès, Maroc

[2] Institut National des Sciences de l'Archéologie et du Patrimoine (INSAP), Rabat, Maroc

[3] Muséum National d'Histoire Naturelle, UMR7194, Paris, France

[4] MONARIS, UMR 8233, Sorbonne Université, Paris, France

* m.haddad@umi.ac.ma

Keywords: Volubilis, Roman Period-Morocco, Wall-Paintings, Pigments, Vibrational and XRF Spectroscopies

Abstract. The work is an in-depth investigation of painting remains from the roman city of Volubilis in Morocco, classified World Heritage. Raman and ATR-FTIR structural and XRF elemental spectroscopies were crossed to decrypt the pigments adopted by roman craftsmen in the south Mediterranean region. Red-ochre alone or in admixture with cinnabar was used in brown-red paintings, while yellow ochre, green earth and Egyptian blue pigments were used to achieve yellow, green and blue ones. All pigments highlighted had been commonly used in the roman world, among which some ones continue until the medieval period in Morocco. In addition to documenting built heritage in Morocco, the results provide a helpful background for archaeologists interested in Roman sites around the Mediterranean space.

Introduction

The scientific community has devoted increasing efforts to the investigation of roman wall paintings owing to the widespread presence of the Roman empire [1–6]. In the case of Roman sites in Morocco, archaeometric investigations are very rare; to our knowledge the unique study on painting remains is the one carried out by Gliozzo and al. [7]. Painting analysis provides background supporting conservators, restorators and archeologists.

The Volubilis Roman site in Morocco dates back to the 3rd C. BC [8]; founded initially as a headquarters to control the north of Africa, in the 8th C it became the starting point of Islamisation by Idriss 1st who took refuge in Walila (ancient Volubilis) [9]. Exhibiting exceptional remains (forum, baths, capitol, shopping streets, basilica, etc.), the city of Volubilis was listed as World Heritage by UNESCO in two times: a first part in 1997 and the all 42 hectares in 2008. It is one of the most attracting architectural Roman sites in the south Mediterranean region.

The present on-going experimental work aims exploring a set of eighty wall painting fragments sampled during recent archaeological excavations of Volubilis remains. To achieve the purpose of characterizing painted plasters, different analytical techniques have been combined: elementary compositions were determined by X-ray fluorescence spectrometry (XRF), while coloring phases were identified by crossing micro-Raman and infrared vibrational spectrometries with optical reflectance and X-ray diffraction (XRD) analyses [10–12]. This investigation, which is the first fairly depth one on Roman paintings, is part of a large research program taking aim of the

Mediterranean Architectural Heritage - RIPAM10   Materials Research Forum LLC
Materials Research Proceedings 40 (2024) 160-166   https://doi.org/10.21741/9781644903117-17

establishment of a scientific documentation on ancient architectural heritage of Morocco [13]. The results we present here relate to a sub-set of five representative painting fragments; only XRF, Raman and ATR-FTIR results are reported.

Materials and methods

Materials

Eighty wall paintings fragments sampled in the site of Volubilis have been studied; we will show the results relative to only five representative ones. All samples are stored at the reserve of the INSAP in Rabat.

The five samples are denoted VER6, VER9, VEJ, VEV7 and VEB1 (fig.1); they are selected to show all the colours present in the corpus: brown-red, yellow, green and blue. Their sizes vary from 1 cm x 1,5 cm to 3,5 cm x 4,5 cm, with thicknesses ranging between 1 and 3 cm.

| VER6 | VER9 | VEJ | VEV7 | VEB1 |

Fig. 1. Photos of the representative samples

Techniques

As a first step, the painting fragments were examined visually by means of a stereomicroscope ZEISS SteREO Discovery V8. The colours were measured as a*, b* and L* chromatic coordinates, using a portable Konica Minolta CM-700d colorimeter piloted by SpectraMagicTMNX; the illuminant is the D65 one and the observer set at 10 °.

The XRF elemental analyses were carried using the handheld portable analyzer S1 Titan from Brucker, based on Rh target/50 kV X-ray tube. The instrument was used in the conditions of its original factory energy calibration, and the acquisition time adopted was 2×30 s.

Two spectrometers with two different laser excitation wavelengths were used in Raman analysis; the first one is a Renishaw RM1000 spectrometer equipped with a laser (He-Ne) emitting at 632.8 nm and coupled to a Leica DMLM microscope with 4 objectives (x5, x20, x50, x100). The second one is a HR800 Jobin-Yvon Horiba spectrometer equipped with an ionized Argon Laser providing 458 and 514 nm excitations, and coupled to an Olympus microscope equipped with lenses (x5, x10, x50 and x100). The equipment was adjusted to obtain a power ranging from 70 μW to 4 mW.

The ATR-FTIR spectra were collected, in the range of 4000–400 cm^{-1} at a spectral resolution of 2 cm^{-1}, by means of a Vertex 70 spectrometer (Bruker) equipped with a diamond crystal using a single reflection ATR-Golden Gate accessory (Specac) and managed by the OPUS software. Less than 1 mg of sample powder was pressed on the surface of the crystal.

Mineral phases were identified using Raman and Infrared databases [14,15], and also by means of comparison with published works cited later.

Results and discussion

Brown-red

Samples VER6 et VER9 show brown-red colour with different hues, the corresponding chromatic coordinates a* and b* are positive (fig. 2), with mean values of sample VER6 (a* ≈ 24 and b* ≈ 23) higher than those of sample VER9 (a* ≈ 17 and b* ≈ 14); the lightness ones are medium increasing from L*≈46 for VER9 to L*≈58 for VER6. These chromatic coordinates are practically similar to those reported for red painted plasters from Roman sites [11,16].

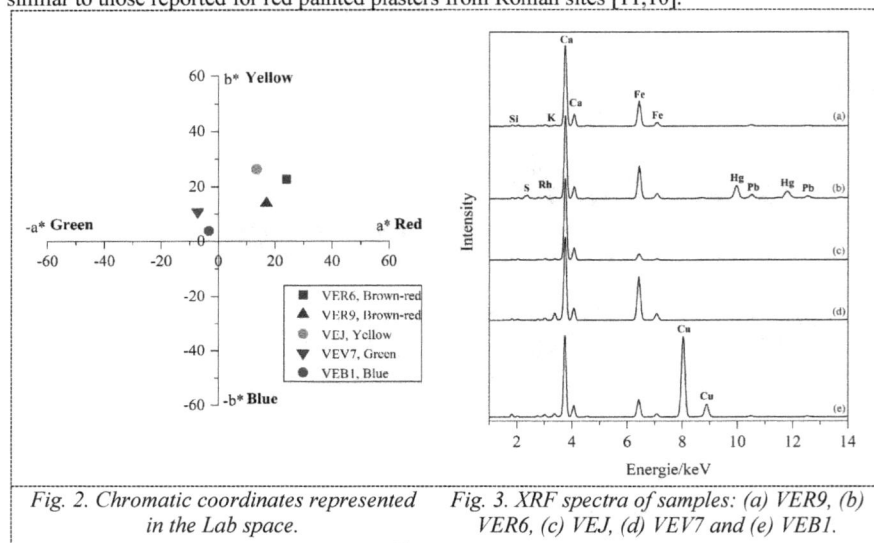

Fig. 2. Chromatic coordinates represented in the Lab space.

Fig. 3. XRF spectra of samples: (a) VER9, (b) VER6, (c) VEJ, (d) VEV7 and (e) VEB1.

The XRF spectra collected on brown-red areas of VER9 and VER6 samples (fig. 3-a,b) reveal iron (Fe) as a main element with, in the case of the sample VER6, some amount of mercury (Hg) and sulfur (S); in both fragments Calcium (Ca) is the main element. Traces elements such as Si, K and Pb were also detected. To identify the coloring phases, Raman and Infrared spectra were measured. The Raman lines observed on fragment VER9 and located at 222, 242, 290, 405, 608 and 1317 cm^{-1} (fig. 4A-a) are characteristic of hematite (Fe_2O_3) [14], while bands emerging at 250, 283 and 340 cm^{-1} in the case fragment VER6 (fig. 4A-b) are those of Cinnabar (HgS) [14]. The Infrared measurements further the vibrational analyses; hence in both samples the presence of hematite is confirmed by the bands observed at 465 and 527 cm^{-1} (fig. 4B-a,b) [15]. In concordance with the XRF spectra, calcite is also emerging at 217, 285, 711, 870 and 1401 cm^{-1} vibrational frequencies [15]. The non-appearance of cinnabar on these infrared spectra may be explained by the facts that this inorganic red pigment does not absorb in the region 400-4000 cm^{-1} at one hand [17,18], and probably also because of its high refractive index [19].

Other components such as clays in the form of kaolinite, emerging at 1008 and 1030 cm^{-1}, as well as Quartz (SiO_2) in small quantities appearing at 797, 778 cm^{-1} and 695 cm^{-1} [15] are also observed.

Mediterranean Architectural Heritage - RIPAM10   Materials Research Forum LLC
Materials Research Proceedings 40 (2024) 160-166   https://doi.org/10.21741/9781644903117-17

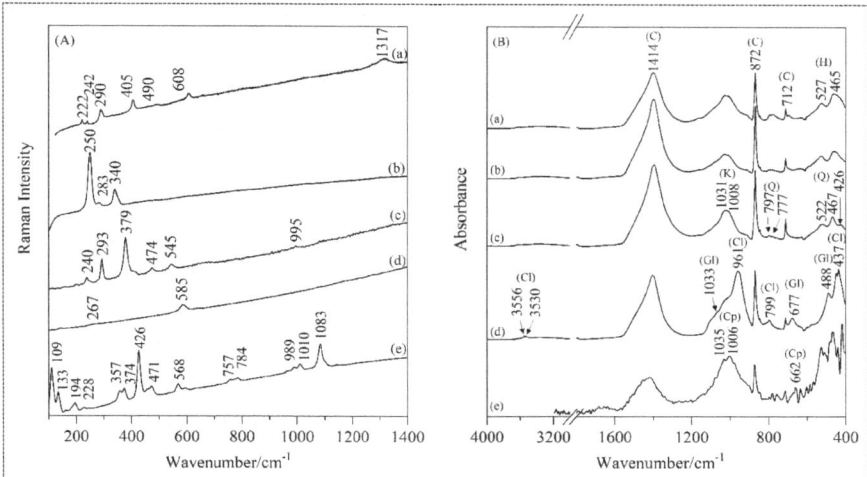

Fig. 4. (A) Raman and (B) ATR-FTIR spectra of samples: (a) VER9, (b) VER6, (c) VEJ, (d) VEV7 and (e) VEB1. (H: hematite, Cn: Cinnabar, Gl: Glauconite, Cl: Celadonite, Cp: Cuprorivaite, C: Calcite, Q: Quartz and K : kaolinite).

The correlation of all the previous results, indicating a coexistence of hematite with clays (possibly in the form of kaolinite [$Al_2Si_2O_5(OH)_4$]), leads to consider red ochre as the main pigment used to prepare the red-brown color. In the case of sample VER6, this pigment is mixed with cinnabar. The silica or quartz observed may be associated with iron oxide or contamination by the mortar layer, they may also relate to impurities present on the surface of the fragments due to their underground burial. Calcite is likely origination from the preparatory layer during the painting process or even while the powder sampling.

The red ochre was widely used by Roman artists in ancient mural paintings [5,20]. In some cases, it used to be mixed with cinnabar, a very expensive mineral whose use dates back to the antiquity [19]; primary cinnabar mines were in Spain [16]. The practice of pigment Admixtures was fairly common in Roman and medieval times as an economic necessity, especially for wall paintings [20,21].

Yellow colour
The chromatic coordinates mean values obtained on the yellow area of fragment VEJ are: a*≈ 13,5 for a*, 26,5 for b* and 70 for the lightness L* (fig. 2). They are characteristic of yellow shades and similar to those already reported in the case of other sites [1,16]. The corresponding XRF elemental composition (fig. 3,c) is similar to those revealed by the brown red fragments (fig. 3-a,b). However, the relative low content of iron (Fe) leads to the assumption that the yellow pigment used is an iron-based one such as goethite. This assumption is confirmed by the structural Raman analysis recorded on the yellow area, showing lines at spectra at 240, 293, 379, 474 and 545 cm^{-1} (fig. 4A-c) characteristic of goethite ((α-FeO(OH)) [14]. The complementary infrared analysis of VEJ doesn't show any goethite signal probably a very low amount of this pigment was used to achieve the yellow color. Similarly to the case of brown-red areas, calcite and quartz are also identified on the infrared spectrum. Consequently, the main pigment used is yellow ochre in the form of goethite which was widely used in Roman times [5,20].
Green colour

Regarding the green area of fragment VEV7, the chromatic coordinate values are $\approx$ -7,21 for a* and $\approx$ 2,7 for b* quietly consistent with the green shade. The measured lightness L*$\approx$ 54 is also in concordance with the observed clarity of the green painting layer (fig. 2). The elementary analysis of this green colour revealed by XRF measurements (fig. 3-d) shows a high content of iron (Fe) with some amount of K and Si, in addition to the commonly observed high content of calcium (Ca). This indicates the use of an iron-based green pigment. The Raman spectra recorded on this green area (fig. 4A-d) exhibits a low intensity band around 267 cm^{-1} along with another signal at 585 cm^{-1}, attributable respectively to glauconite [(K,Na)(Fe^{3+}, Al, Mg)$_2$(Si,Al)$_4$O$_{10}$(OH)$_2$] and to celadonite [K[(Al, Fe^{3+}), (Fe^{2+},Mg)](AlSi$_3$, Si$_4$)O$_{10}$(OH)$_2$] which are minerals forming the green earth pigment and responsible of the observed green color [22,23].

The high intensity bands at 437 and 961 cm^{-1} observed on the ATR-FTIR spectrum, measured on a micro-sample taken on the same green zone (fig. 4B-d), are those of celadonite; the lower intensity and less resolved ones located around 488, 677 and 1033 cm^{-1} correspond to glauconite [15,22]. Calcite characteristic bands are also observed. Green earth pigments are composed of several minerals whose proportions vary depending on the ore deposit; the most abundant ones are glauconite and celadonite [23]. Consequently, the green colours were obtained by the use of green earth pigments that have been identified in painted plasters dating back to the Roman period [5,20]. During Antiquity, the Romans especially employed these green earth pigments when painting in fresco where the pigments are impregnated on a freshly spread moist lime plaster, technique well known to fix the colours on the wall [23].

Blue colour
The chromatic coordinates mean values corresponding to the blue area of fragment VEB1 are: a* $\approx$ -3,7, b* $\approx$ 2,7 and L*$\approx$ 67. The corresponding point on the Lab diagram seems very close to the achromatic region because of the low value of the yellow/blue coordinate b* (fig. 2). The XRF elementary composition measured on this blue paint revealed a high content of copper (Cu) along with other relatively low amount elements such as iron (Fe), silicon (Si) and lead (Pb) (fig. 3,e), indicating thus the use of a copper-based blue pigment. The structural Raman analysis carried out on the same zones (fig. 4A-e) revealed signals at 109, 426 and 1083 cm^{-1} characteristic of Egyptian blue ((cuprorivaite, CaCuSi$_4$O$_{10}$), besides other bands of different intensities located at 133, 194, 228, 357, 374, 471, 568, 592, 757, 784, 989 and 1010 cm^{-1} [16,24]. The Infrared measurements of micro-sample from this blue pictorial layer corroborate the use of Egyptian blue by its bands observed at 662, 1006 and 1035 cm^{-1} (fig. 4B-e) [5,25]. This pigment has also been identified in wall paintings from the Roman site of Thamusida in Morocco [7]; it was commonly used in several Roman sites in Spain and Italy [2,26].

Conclusion
In the course of the establishment of a scientific documentation on architectural heritage sites in Morocco, a set of eighty wall painting fragments from the Roman city of Volubilis have been investigated by means of different analyses combining elemental XRF and structural Raman with ATR-FTIR techniques, supported by other ones.

Here, are presented the results relative to a sub-set of five representative painting samples exhibiting a large palette of colours ranging between red-brown, yellow, green and blue. All pigments used have identified. Hence, brown-red paintings had been achieved by using either red-ochre alone or in admixture with cinnabar, yellow ochre was used for yellow, green earth for green and Egyptian blue for blue. This palette represents the typical one commonly used in the roman world.

On the insight of the on-going CH scientific documentation results, it seems that in Morocco, the use of red ochre and cinnabar had been maintained until the medieval period for brown-red shades, while malachite replaced green earth for green colours and azurite or lazurite replaced Egyptian blue for blue ones.

Going deeper into the investigation, stratigraphic analysis and cross-section examinations are in progress; they are focusing on the painting techniques adopted throughout Roman and medieval periods in Morocco.

References

[1] M. Edreira, Spectroscopic analysis of roman wall paintings from Casa del Mitreo in Emerita Augusta, Mérida, Spain, Talanta. 59 (2003) 1117–1139. https://doi.org/10.1016/S0039-9140(03)00020-1

[2] I. Aliatis, D. Bersani, E. Campani, A. Casoli, P.P. Lottici, S. Mantovan, I.-G. Marino, Pigments used in Roman wall paintings in the Vesuvian area, J. Raman Spectrosc. 41 (2010) 1537–1542. https://doi.org/10.1002/jrs.2701

[3] A. Duran, J.L. Perez-Rodriguez, M.C. Jimenez de Haro, M.L. Franquelo, M.D. Robador, Analytical study of Roman and Arabic wall paintings in the Patio De Banderas of Reales Alcazares' Palace using non-destructive XRD/XRF and complementary techniques, J. Archaeol. Sci. 38 (2011) 2366–2377. https://doi.org/10.1016/j.jas.2011.04.021

[4] M. Bakiler, B. Kırmızı, Ö. Ormancı Öztürk, Ö. Boso Hanyalı, E. Dağ, E. Çağlar, G. Köroğlu, Material characterization of the Late Roman wall painting samples from Sinop Balatlar Church Complex in the black sea region of Turkey, Microchem. J. 126 (2016) 263–273. https://doi.org/10.1016/j.microc.2015.11.050

[5] V. Guglielmi, V. Comite, M. Andreoli, F. Demartin, C.A. Lombardi, P. Fermo, Pigments on Roman Wall Painting and Stucco Fragments from the Monte d'Oro Area (Rome): A Multi-Technique Approach, Appl. Sci. 10 (2020) 7121. https://doi.org/10.3390/app10207121

[6] S. Bracci, E. Cantisani, C. Conti, D. Magrini, S. Vettori, P. Tomassini, M. Marano, Enriching the knowledge of Ostia Antica painted fragments: a multi-methodological approach, Spectrochim. Acta. A. Mol. Biomol. Spectrosc. 265 (2022) 120260. https://doi.org/10.1016/j.saa.2021.120260

[7] E. Gliozzo, F. Cavari, D. Damiani, I. Memmi, Pigments and plasters from the Roman settlement of thamusida (Rabat, Morocco): Pigments and plaster from the Roman settlement of Thamusida (Rabat, Morocco), Archaeometry. 54 (2012) 278–293. https://doi.org/10.1111/j.1475-4754.2011.00617.x

[8] J.-L. Panetier, H. Limane, Volubilis. Une cité du Maroc antique, Maisonneuve & Larose, Paris : Casablanca, 2002.

[9] N. Brahmi, Volubilis : approche religieuse d'une cité de Mauretanie Tingitane (milieu Ier-fin IIIème siècles apr. J.-C.), These de doctorat, Le Mans, 2008.

[10] I. Aliatis, D. Bersani, E. Campani, A. Casoli, P.P. Lottici, S. Mantovan, I.-G. Marino, F. Ospitali, Green pigments of the Pompeian artists' palette, Spectrochim. Acta. A. Mol. Biomol. Spectrosc. 73 (2009) 532–538. https://doi.org/10.1016/j.saa.2008.11.009

[11] D. Miriello, A. Bloise, G.M. Crisci, R. De Luca, B. De Nigris, A. Martellone, M. Osanna, R. Pace, A. Pecci, N. Ruggieri, Non-Destructive Multi-Analytical Approach to Study the Pigments of Wall Painting Fragments Reused in Mortars from the Archaeological Site of Pompeii (Italy), Minerals. 8 (2018) 134. https://doi.org/10.3390/min8040134

[12] A. Duran, J.L. Perez-Rodriguez, Revealing Andalusian wall paintings from the 15th century by mainly using infrared spectroscopy and colorimetry, Vib. Spectrosc. 111 (2020) 103153. https://doi.org/10.1016/j.vibspec.2020.103153

[13] I. Fikri, Caractérisation des matières colorantes dans des enduits muraux issus de fouilles archéologiques sur des sites antiques et islamiques du Maroc : développement de protocoles analytiques, Thèse de Doctorat, Université Moulay Ismail-Meknès & Muséum National d'Histoire Naturelle -Paris, 2021.

[14] Infrared and Raman Users Group Spectral Database, (n.d.). http://www.irug.org/search-spectral-database?reset=Reset.

[15] N.V. Chukanov, Infrared spectra of mineral species, Springer Netherlands, Dordrecht, 2014. https://doi.org/10.1007/978-94-007-7128-4

[16] I. Garofano, J.L. Perez-Rodriguez, M.D. Robador, A. Duran, An innovative combination of non-invasive UV–Visible-FORS, XRD and XRF techniques to study Roman wall paintings from Seville, Spain, J. Cult. Herit. 22 (2016) 1028–1039. https://doi.org/10.1016/j.culher.2016.07.002

[17] S. Vahur, A. Teearu, P. Peets, L. Joosu, I. Leito, ATR-FT-IR spectral collection of conservation materials in the extended region of 4000-80 cm–1, Anal. Bioanal. Chem. 408 (2016) 3373–3379. https://doi.org/10.1007/s00216-016-9411-5

[18] M.L. Franquelo, A. Duran, L.K. Herrera, M.C. Jimenez de Haro, J.L. Perez-Rodriguez, Comparison between micro-Raman and micro-FTIR spectroscopy techniques for the characterization of pigments from Southern Spain Cultural Heritage, J. Mol. Struct. 924–926 (2009) 404–412. https://doi.org/10.1016/j.molstruc.2008.11.041

[19] R.J. Gettens, R.L. Feller, W.T. Chase, Vermilion and Cinnabar, Stud. Conserv. 17 (1972) 45–69. https://doi.org/10.1179/sic.1972.006

[20] R. Siddall, Mineral Pigments in Archaeology: Their Analysis and the Range of Available Materials, Minerals. 8 (2018) 201. https://doi.org/10.3390/min8050201

[21] M. Gutman, M. Lesar-Kikelj, A. Mladenovič, V. Čobal-Sedmak, A. Križnar, S. Kramar, Raman microspectroscopic analysis of pigments of the Gothic wall painting from the Dominican Monastery in Ptuj (Slovenia), J. Raman Spectrosc. 45 (2014) 1103–1109. https://doi.org/10.1002/jrs.4628

[22] F. Ospitali, D. Bersani, G. Di Lonardo, P.P. Lottici, 'Green earths': vibrational and elemental characterization of glauconites, celadonites and historical pigments, J. Raman Spectrosc. 39 (2008) 1066–1073. https://doi.org/10.1002/jrs.1983

[23] A. Fanost, A. Gimat, L. de Viguerie, P. Martinetto, A.-C. Giot, M. Clémancey, G. Blondin, F. Gaslain, H. Glanville, P. Walter, G. Mériguet, A.-L. Rollet, M. Jaber, Revisiting the identification of commercial and historical green earth pigments, Colloids Surf. Physicochem. Eng. Asp. 584 (2020) 124035. https://doi.org/10.1016/j.colsurfa.2019.124035

[24] A. Cosentino, FORS Spectral Database of Historical Pigments in Different Binders, E-Conserv. J. (2014) 54–65. https://doi.org/10.18236/econs2.201410

[25] V. Crupi, B. Fazio, G. Fiocco, G. Galli, M.F. La Russa, M. Licchelli, D. Majolino, M. Malagodi, M. Ricca, S.A. Ruffolo, V. Venuti, Multi-analytical study of Roman frescoes from Villa dei Quintili (Rome, Italy), J. Archaeol. Sci. Rep. 21 (2018) 422–432. https://doi.org/10.1016/j.jasrep.2018.08.028

[26] E.J. Cerrato, D. Cosano, D. Esquivel, R. Otero, C. Jimémez-Sanchidrián, J.R. Ruiz, A multi-analytical study of a wall painting in the Satyr domus in Córdoba, Spain, Spectrochim. Acta. A. Mol. Biomol. Spectrosc. 232 (2020) 118148. https://doi.org/10.1016/j.saa.2020.118148

Mediterranean Architectural Heritage - RIPAM10   Materials Research Forum LLC
Materials Research Proceedings 40 (2024) 167-178   https://doi.org/10.21741/9781644903117-18

Spectrometric Characterization of Moroccan Architectural Glazed Tiles

Mohamed EL AMRAOUI[1,*], Mustapha HADDAD[1], Lahcen BEJJIT[1],
Saadia AIT LYAZIDI[1], Abdelouahed BEN-NCER[2]

[1] Laboratoire de Spectrométrie des Matériaux et Archéomatériaux (LASMAR) URL-CNRST N°7, Faculté des Sciences, Université Moulay Ismaïl, B.P. 11201 – Zitoune, Meknès (Maroc)

[2] Institut National des Sciences de l'Archéologie et du Patrimoine (INSAP), Rabat (Maroc)

*elamraouimohamed43@gmail.com

Keywords: Glazed Tiles, Spectrometric Characterization, Moroccan Historical Sites and Monuments, Colouring Phases, Crystalline Phases

Abstract The present work relates to a multi-analytic characterization of glazed tiles consisting of green monochrome glazed ceramics used in Moroccan architecture to protect ceilings, walls and roofs from rainwater. These tiles originate from five sites and date back to different historical periods: Bou-Inania Madrasa in Meknes (14th century), Prison of Qara in Meknes (18th century), Dar El-Beida Palace in Meknes (18th century) and Al-Hibous Cemetery of Mdaghra in Errachidia (19th century). Different analysis techniques were used in view to go back to the ancient technological processes adopted (materials, coloring pigments, firing temperatures, etc..). Optical absorption spectrometry revealed two different types of chromogenic ions in green glazes, chromium Cr^{3+} in the case of the tiles from Dar El-Beida Palace and Prison of Qara, and copper Cu^{2+} in the case of the tiles from Bou-Inania Madrasa and Al-Hibous Cemetery. Raman microspectroscopy identified different coloring phases with two types of green glazes, escolaite (Cr_2O_3) in the case of the glazes of the Prison of Qara and copper phthalocyanine mixed with a chromium-based pigment in the case of the glazes of the Dar El-Beida Palace. However, the origin of the green color in the glazes from Bou-Inania Medersa in Meknes and Al-Hibous cemetery of Errachidia may be due to the dissolution of copper in the vitreous glazes. X-ray diffraction, supported by Raman microspectrometry, revealed the mineralogical compositions of the terracotta tiles. Quartz and calcite are the main phases, while hematite and "high temperature" phases (anorthite, gehlenite and diopside) appear as minority ones. These identified phases permit to estimate the firing temperature of the tiles at around 950 °C in an oxidizing atmosphere. The chromatic coordinates of all glazes, represented in the Lab CIE color space, made it possible to discriminate objectively all green colors. The present investigation of glazes from different historical sites allowed the exploration of the coloring materials, revealed differences in the adopted technological protocols and permitted the establishment of a color reference database to follow glazes degradation and to help while replacing missing or degraded tile pieces.

1. Introduction

Several cities in Morocco conceal a very rich and diversified historical architectural heritage. Meknes city in particular, founded in the 11th century by the Almoravids as a military settlement and becoming a capital under Sultan Moulay Ismaïl (1672–1727), has several monuments and sites in good state : Madrasa Bou-Inania, Madrasa Filalia, Palace Dar-El Beïda, Palace Al Mansour, Sultan Moulay Ismaïl Mausoleum, Quara Prison, portals Bab Al Mansour Laalaj, Bab Al Khmis, Bab Berdaine, ramparts, fountains, Kasbahs. This architectural heritage is recognized worldwide, for example, the old medina of Meknes (Lahdim) has been classified since 1996, as an UNESCO World Heritage Site. Several types of building materials were served as part of this architectural heritage. Among these materials, Moroccan glazed ceramic called (Zelliges or zellij), roofing glazed tiles called (Quermoud), wall paintings, plasters, gypsum, carved wood and painted wood are used in architectural decoration and the protection of monuments. For most monuments in Meknes, studies on Moroccan glazed ceramics (zellige) as decorative and protective materials, have been carried out [1-7]. But studies of roofing glazed tiles remain absent. Thus, investigations of especially Moroccan glazed roofing tiles "Quermoud" are necessary, because several of them, in most ancient monuments, show degradation, destruction, cracking, flaking and detachment of their glazes. Also, they have lost their impressive image due to weather conditions (rain, wind, sunshine) and pollution.

Green glazed roofing ceramic tile is an archictural element that covers the roofs of Palaces, Mosques, Medersas. A glazed ceramic is a composite material composed by a ceramic support called also terracotta or ceramic body on which a glaze is deposited (enamel, glass) that serves both to waterproof and to decorate it. The main components of a glaze are: silica as vitrifier and network former, fluxes or network modifiers (lead, natron,...), stabilizing oxides, colouring elements (copper, cobalt, iron, manganese,...) and opacifying elements (tin oxide, ...). These glazed roofing tiles are often monochrome green colour.

The presented work is a part of a spectrometric studies project of Moroccan historical architectural glazed ceramics and especially glazed roofing ceramic tiles started many years ago [1-7]. The main objective is to identify the colouring materials and ancient handcrafting techniques adopted to manufacture green glazed roofing tiles of different sites selected. Results presented, in this work, will contribute to create a database of Moroccan architectural glazed roofing tiles, which will be useful for conservation and restoration. Thus, Raman micro-spectrometry, optical reflectance spectroscopy, colorimetry and X-ray diffraction, have been used to characterize samples from sites cited above.

2. Materials and Techniques

2.1 Materials
Samples of green glazed roofing tiles, from four monuments and sites (figure 1), were selected and referenced as: MEK-GGRTMBIM from Madrasa Bou Inania (14th century), MEK-GGRTPDEBM from Dar-El Beïda Palace (18th century), MEK-GGRTQPM from Quara Prison (18th century) in Meknes and MEK-GGRTGHME from Graveyard Hibous Mdaghra (19th century) in Errachidia.

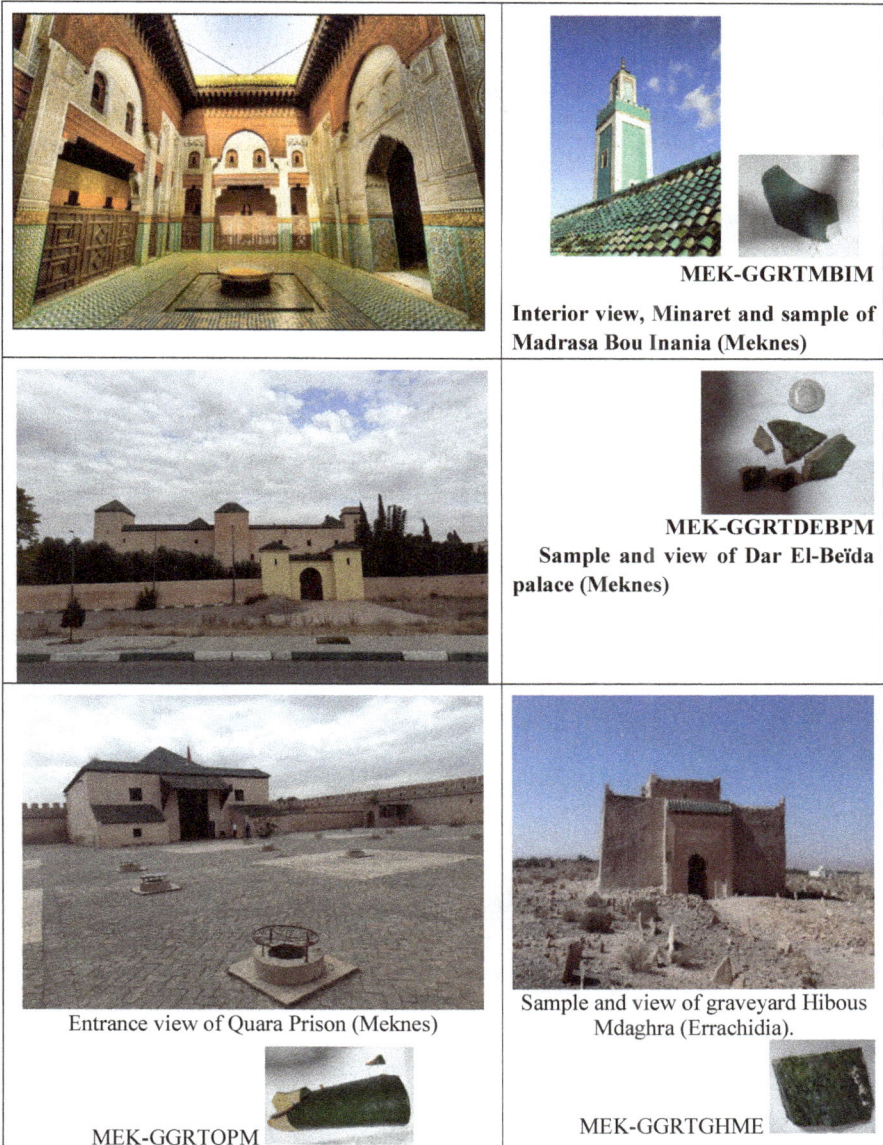

MEK-GGRTMBIM

Interior view, Minaret and sample of Madrasa Bou Inania (Meknes)

MEK-GGRTDEBPM
Sample and view of Dar El-Beïda palace (Meknes)

Entrance view of Quara Prison (Meknes)

MEK-GGRTQPM

Sample and view of graveyard Hibous Mdaghra (Errachidia).

MEK-GGRTGHME

Figure 1: Samples and view of sites

Mediterranean Architectural Heritage - RIPAM10
Materials Research Proceedings 40 (2024) 167-178

Materials Research Forum LLC
https://doi.org/10.21741/9781644903117-18

2.2 Characterization Techniques

Raman analyses

Raman spectra were recorded using a Renishaw RM1000 spectrometer equipped with a CCD detector and an external Leica DMLM confocal microscope with 5×, 20×, 50× and 100× objectives. The excitation source is a He-Ne laser of wavelength 632.8 nm. The laser beam is focused directly on the sample without any preparation.

Colorimetric and visible measurements

The portable Konica Minolta CM700d spectrophotometer working in the visible wavelength range 400–700 nm with a D65 illuminant, is used for optical and colorimetric measurements. The instrument is provided with its own white reference (100% reflective) and a zero calibration box (0% reference). Data are collected as L*, a* and b* coordinates.

Mineral phase analyses

Powder mineralogical analyses of ceramic bodies were performed using a BRUKER D8 ADVANCE x-ray diffractometer. It is equipped with an x-ray tube with a copper anode (Cu-$K_{\alpha1}$, λ=1.5406 Å) operating at 40 kV and 30 mA. The position 2θ varies from 5 to 70° with a step of 0.02°. The mineral phases were identified using the Rruff database [8].

3. Results and discussion

3.1 Colorimetric coordinates

Table 1 presents the chromatic coordinates of the four studied glazes in (L, a*, b *) space and figure 2 show their presentation in (a*, b*) diagram. According to the chromatic (a*, b*) space, we can notice that the chromatic coordinates of the green glazes are located in the dial placed between the green axis (- a*) and the yellow axis (b*) i.e. the green-yellow dial. The chromatic coordinates of samples MEK-GGRTMBIM, MEK-GGRTGHME and MEK-GGRTQPM present a significant contribution to green colour than yellow, unlike that of MEK-GGRTDEBM sample which presents a significant contribution of yellow than green colour.

Table 1: chromatic coordinates of glazes of samples in (L, a, b*) space*

Samples	L*	a*	b*
MEK-GGRTMBIM	46.34	-16.23	5.33
MEK-GGRTDEBM	43.91	-4.96	14.49
MEK-GGRTQPM	37.44	-12.58	12.66
MEK-GGRTGHME	33.78	-9.27	8.11

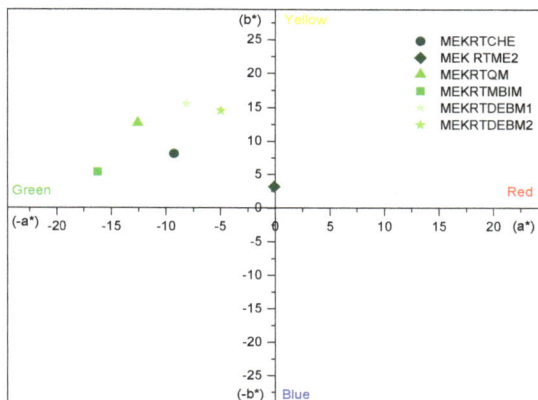

Figure 2: chromatic coordinates of green glazes studied in (a*, b*) space

3.2 Optical reflectance spectrometry

Figure 3 presents the optical reflectance spectra of green glazes of samples MEK-GGRTMBIM and MEK-GGRTGHME. The spectra show the presence of a broad band centred at 520 nm which can be attributed to copper–ions Cu^{2+}. These ions are responsible of the green colour of glazes of ceramic tiles [6; 9].

The optical reflectance spectra of green glazes of MEK-GGRTDEBM and MEK-GGRTQPM samples are presented in figure 4. They show the presence of two bands centred at 410 and 540 nm. These bands are characteristic of d-d electronic transitions of chromium ions Cr^{3+} in an octahedral site in a vitreous matrix. It is well known that the chromium ions are responsible of the green colour of glazes of ceramic tiles [6; 9].

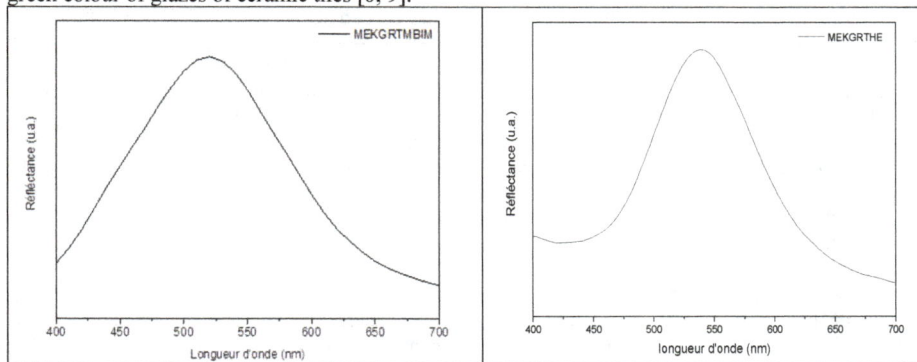

Figure 3: optical reflectance spectra of green glazes of MEK-GGRTMBIM and MEK-GGRTGHME

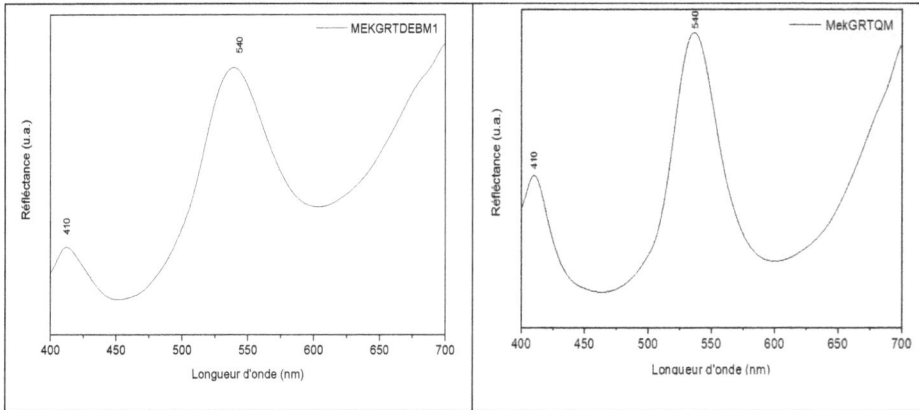

Figure 4: optical reflectance spectra of green glazes of MEK-GGRTDEBM and MEK-GGRTPQM

3.3 Raman spectrometry of glazes

Figure 5 presents the Raman spectra recorded on different areas of the green glaze of the sample MEK-GGRTMBIM. These spectra generally show the same bands: 130; 324; 460; 633; 776 and 990 cm^{-1}. Bands at 460 and 990 cm^{-1} are characteristic of silicate glasses [5; 6; 10; 11; 12]. The first band located at 460 cm^{-1} corresponds to the bending modes of the SiO$_4$ tetrahedral, especially, to the displacement of the bridging oxygen along the line bisecting the Si-O-Si angle. The second band located at 990 cm^{-1} corresponds to the Si-O stretching vibrations in the silicates glasses [5; 6; 7; 10; 11; 12]. This band is generally located at 1000 cm^{-1} for silicate glasses and it is shifted towards 990 cm^{-1} which is a Raman signature of lead glazes [6; 7; 10; 15; 16]. This stretching block of Si-O is more intense than that of bending Si-O-Si which indicates strong depolymerisation of the silicate network consistent with a lead-rich glaze composition [15]. The doublet located at 633 and 776 cm^{-1} is a Raman signature of cassiterite a tin oxide (SnO$_2$) [5; 6; 7; 10; 11-13] used as an opacifier of glazes [5; 6; 7; 10; 11-13]. The bands located around 130 and 324 cm^{-1} can be attributed to lead antimonite also called Naples yellow (Pb$_2$Sb$_2$O$_7$) [10;13; 17]. It should be noted that tin oxide and Naples yellow were also used as glaze opacifiers [6; 13].

Although, no colouring phase responsible for the green colour of the glaze is not detected in the spectrum, the green colour of silicates glazes could be due to the dissolution of copper ions Cu^{2+} in a silicate vitreous matrix called transparent glazes [6; 7; 13; 18]. We note that the ions Cu^{2+} are detected by optical absorption spectrometry (figure 3). Therefore, the colour of this green glaze of sample MEK-GGRTMBIM is a copper-based pigment.

Raman Analyse of MEK-GGRTGHME sample presented in figure 6 shows the same bands than MEK-GGRTMBIM except the band at 633 cm^{-1}. Therefore, we can conclude that a copper-based pigment is responsible of the green colour in this glaze since it is generally dissolved in transparent lead glazes [6; 7 ; 13; 18].

Raman spectra recorded on the green glaze of MEK-GGRTDEBM (Figure 7) present several bands: 236, 260, 303, 486, 595, 682, 750, 781, 835, 954, 1011, 1042, 1110, 1143, 1164, 1199, 1217, 1308, 1343, 1433, 1452 and 1531 cm^{-1}. These bands are the Raman signatures of copper phthalocyanine, it is a synthetic organic pigment [19, 20]. As the optical absorption spectroscopy indicated that the chromogenic ions responsible for the green color of the glaze are Cr^{3+} ion, we suggest that the chromium-based pigment is mixed with the identified organic pigment to give the green colour of this glaze. The absence of the Raman signatures of the chromium-based pigment can be explained by the fluorescence phenomenon due to the organic character of copper phthalocyanine [20]. The existence of copper phthalocyanine in these glazes suggest that the roofing tiles were undergone restorations in the past [21].

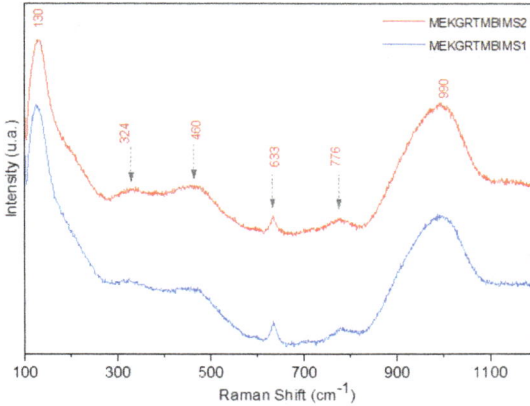

Figure 5: Raman spectra of green glaze of MEK-GGRTMBIM sample.

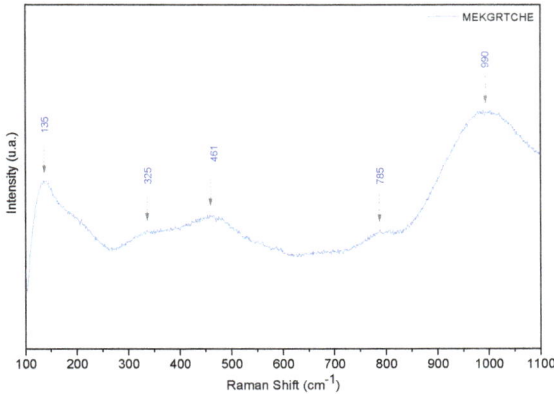

Figure 6: Raman spectrum of glaze of MEK-GGRTGHME sample

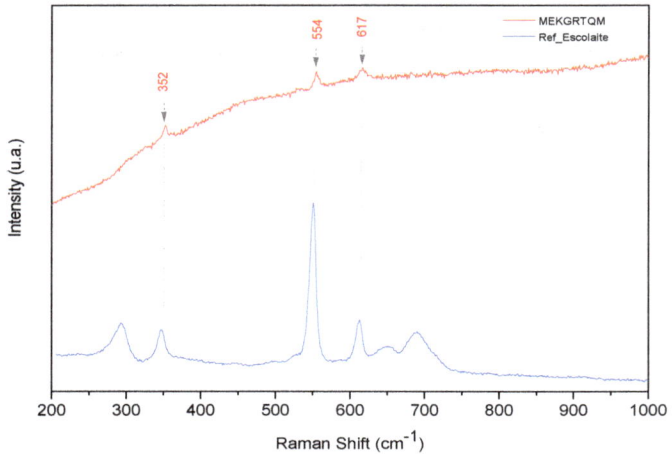

Figure 7: Raman spectra of green glaze of MEK-GGRTPQM and reference of Eskolaite

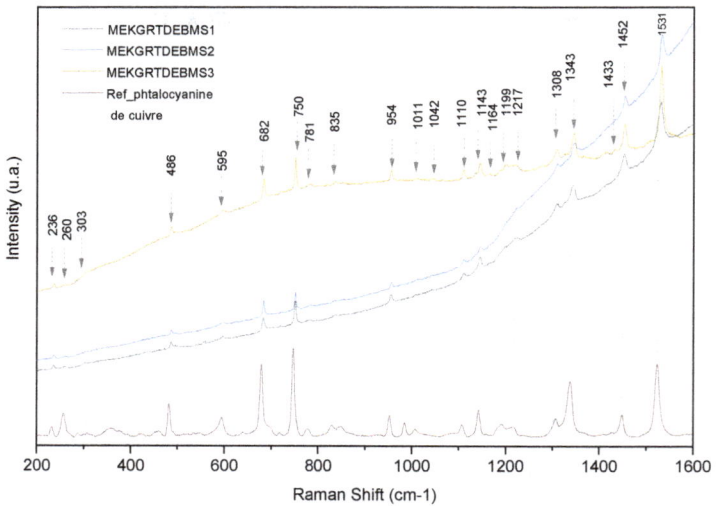

Figure 8: Raman spectra of green glaze of MEK-GGRTDEBM and reference of copper phtalocyanine

3.4 Mineralogical composition of ceramic bodies

The X-ray diffraction patterns recorded on powder of ceramic bodies of MEK-GGRTMBIM, MEK-GGRTDBM, MEK-GGRTQPM and MEK-GGRTGHME samples are shown in figures 9 to 12. They highlight quartz as the most abundant crystalline phase. Anorthite, $(CaAl_2Si_2O_8)$ gehlenite $(2CaO \cdot Al_2O_3 \cdot 2SiO_2$ and diopside $(MgCaSi_2O_6)$, so-called high temperature phases, are also identified. The presence of the newly formed minerals (Anorthite, gehlenite and diopside) in the ceramic bodies of samples let know that calcic clays have been used and their firing temperatures were around 950°C [6; 7; 22; 23]. Calcite and hematite have been detected but in low amount. The presence of hematite indicates that the firing of the tiles was carried out in an oxidizing atmosphere during production [6; 9; 13; 24; 25]. The XRD pattern of the ceramic body of sample MEK-GGRTGHME presents the wollastonite phase in addition in comparison to other samples studied. Amount of calcite is low, because of it contributes to the neo-formed phases (Anorthite, gehlenite and diopside).

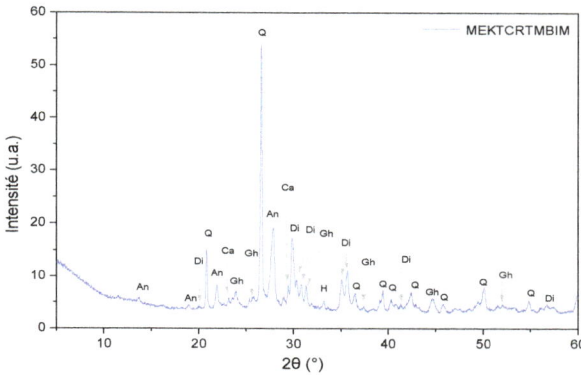

Figure 9: XRD pattern of ceramic body of sample MEK-GGRTMBIM.
[Quartz (Q) ; Gehlenite (Gh) ; Anorthite (An) ; Diopside (Di) ; Calcite (Ca); Hematie (H)]

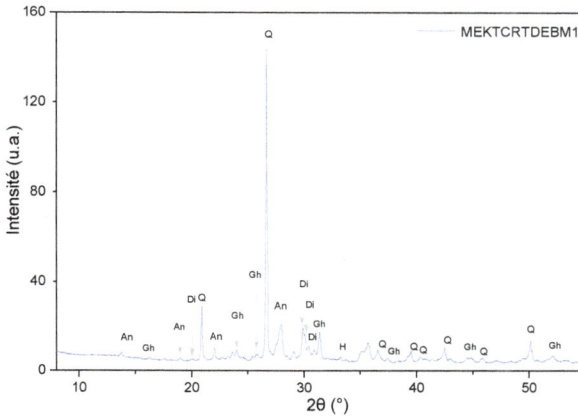

Figure 10: XRD pattern of ceramic body of sample MEK-GGRTDBM
[Quartz (Q); Gehlenite (Gh) ; Anorthite (An) ; Diopside (Di) ; Calcite (Ca); Hematie (H)]

Mediterranean Architectural Heritage - RIPAM10

Materials Research Forum LLC

Materials Research Proceedings 40 (2024) 167-178

https://doi.org/10.21741/9781644903117-18

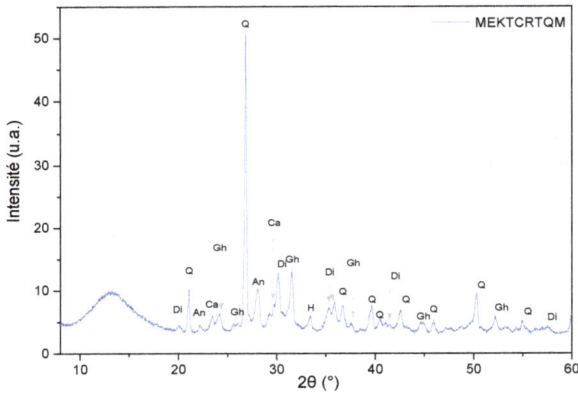

Figure 11: XRD pattern of ceramic body of MEK-GGRTQPM sample
[Quartz (Q) ; Gehlenite (Gh) ; Anorthite (An) ; Diopside (Di) ; Calcite (Ca); Hematie (H)]

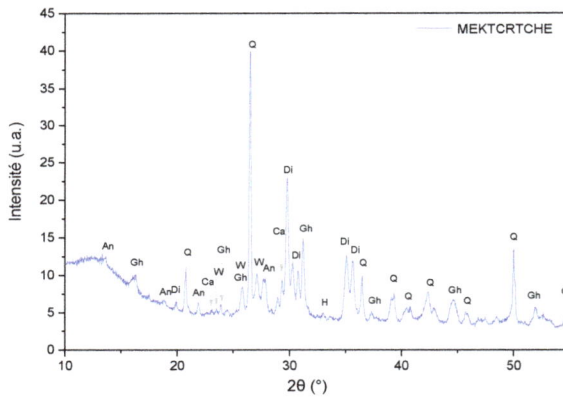

Figure 12: XRD pattern of ceramic body of MEK-GGRTGHME sample
[Quartz (Q) ; Gehlenite (Gh) ; Anorthite (An) ; Diopside (Di) ; Calcite (Ca); Hematie (H);
wollastonite(W)]]

Conclusion

In this work, we have characterized the green glazes of Moroccan glazed roofing tiles. We have obtained access to some technological parameters such as, chromophore elements, coloring phases, firing temperature and conditions of firing, involved in the manufacture of glazed roofing tiles coming from the four sites.

Chromametry has shown difference in chromatic coordinates (L, a*, b*) of green glazes. Optical reflectance spectroscopy has shown that the green colour of glazes is mainly related to the properties of light absorption by metal ions:

- Trivalent chromium ions Cr^{3+} for green glazes of the roofing tiles of Dar El-Beida Palace and Quara Prison in Meknes
- Copper ions Cu^{2+} for green glazes of samples coming from Medersa Bou Inania in Meknes and graveyard of Hibous Mdaghra in Errachidia.

Raman microspectroscopy has identified the colouring phases, namely escolaite (α-Cr_2O_3) for glaze from Quara prison, copper phthalocyanine mixed with a chromium-based pigment for Dar El-Beida Palace. For glazed tiles from Madrasa Bou Inania (Meknes) and Hibous Mdaghra (Errachidia) , the green color is due to dissolution of copper in a vitreous matrix in lead glazes opacified with tin oxide and/or Naples yellow. Ca-rich clays fired at about 950°C in an oxidizing atmosphere have used to make ceramic bodies composed mainly quartz. Other phases as anorthite, gehlenite, diopside, hematite and calcite have been also detected. The comparative study has highlighted difference in manufacturing technology between green glazes of glazed roofing tiles coming from the four sites.

References

[1] M. El Amraoui, A. Azzou, M. Haddad, L. Bejjit, S. Aït Lyazidi, Y. El Amraoui, "Zelliges of Dar-El Beïda Palace (18th-century) in Meknes (Morocco): Optical Absorption and Raman Spectrometry study" Spectroscopy Letters, Volume 40, issue 5, (2007), pp. 777-783. https://doi.org/10.1080/00387010701521892

[2] A. Zucchiatti, A. Azzou, M. El Amraoui, M. Haddad, L. Bejjit, S. Aït Lyazidi, "PIXE analysis of Moroccan architectural glazed ceramics (XIV-XVIII centuries)", International Journal of PIXE, Volume: 19, Issues: 3-4, (2009), pp. 175-187. https://doi.org/10.1142/S0129083509001862

[3] M El Amraoui., A Azzou., M Haddad., L Bejjit., Y El Amraoui., S Aït Lyazidi. "Elemental and crystalline analysis of architectural ceramics of Dar-El Beïda Palace (18th century) in Meknes (Morocco)" Minbar Al Jamiaa (7); Actes de la 1ère Rencontre Internationale sur le Patrimoine Architectural Méditerranéen (RIPAM 1) Edition de l'Université Moulay Ismaïl, (2007), p.188-194

[4] M. El Amraoui, A. Azzou, A. Zucchiatti, M. Haddad, L. Bejjit, *"Analyse par PIXE de Zelliges marocains du XIVème siècle"*, Actes de la 2ème Rencontre Internationale sur le Patrimoine Architectural Méditerraneen (RIPAM 2) édité par M. Haddad et M. Ibnoussina, Minbar Al Jamiaa, édition de l'Université Moulay Ismaïl, (20), (2010), pp. 212-221.

[5] A. Azzou, M Haddad., S Ait Lyazidi., L Bejjit., A Ben Amara., M. Schvoerer, 2007, " Microspectrométrie Raman appliquée à la céramique ancienne de Fès (Maroc, 14ème siècle)", MINBAR AL JAMIAA n°7, Actes de la RIPAM1-(2005), Meknes, Maroc.

[6] M.El Amraoui, 2011," Etudes par des techniques physiques de céramiques glaçurée de l'architecture marocaine (14ème - 20ème s), de mosaïques romaines de Volubilis et de matériels archéologiques provenant de fouille", Thèse de Doctorat, Faculté des sciences, Université Moulay Ismaïl, Meknès, Maroc, 219 p.

[7] A. Azzou, 2005, " céramiques glaçurée architecturales marocaines: caractéristiques physicochimiques et techniques de fabrication des zelliges anciennes (14ème et 17ème siècle) ", Thèse de Doctorat, Faculté des sciences, Université Moulay Ismaïl, Meknès, Maroc, 150 p.

[8] RRUFF project database. http://rruff.info

[9] T Cavaleri., A Giovagnoli. et M. Nervo, 2013, " Pigments and mixtures identification by Visible Reflectance Spectroscopy ", Procedia Chemistry, 8 : p.45–54. https://doi.org/10.1016/j.proche.2013.03.007

[10] A. Tournié, Analyse Raman sur site de verres et vitraux anciens : modélisation, procédure, lixiviation et caractérisation, Thèse de Doctorat, Universite Pierre et Marie Curie, 163p.

[11], Ph. Colomban, 2003, "Polymerisation Degree and Raman Identification of Ancient Glasses used for Jewellery, Ceramics Enamels and Mosaics", Non-Crystalline Solids, 323 :p. 180-187. https://doi.org/10.1016/S0022-3093(03)00303-X

Materials Research Forum LLC
https://doi.org/10.21741/9781644903117-18

[12] Ph. Colomban, V Milande. et L Le Bihan., 2004, " On site Raman analysis of Iznik pottery Glazes and pigments ", J. Raman spectroscopy , 35 :p.527-535. https://doi.org/10.1002/jrs.1163

[13] Ph. Colomban, G. Sagon, A Louhichi., H Binous. N Ayed., 2001 " Identification par Microscopie Raman des tessons et pigments de glaçures de céramiques de l'Ifriqiya (Dougga, XI-XVIIIèmes siècles)" Revue d'Archéométrie, 25 p.101-112. https://doi.org/10.3406/arsci.2001.1005

[14] Ph Colomban., F. Treppoz, 2001, "Identification and differentiation of ancient and modern European porcelains by Raman macro- and micro-spectroscopy", Raman Spectroscopy, 32, 93–102. https://doi.org/10.1002/jrs.678

[15] L.F. Vieira Ferreira, R. Varela Gomes, M.F.C. Pereira, L.F. Santos et Ferreira Machado I., 2016, "Islamic ceramics in Portugal found at Silves Castle (8th to 13th c.): An archaeometric characterization", Archaeological Science: Reports, 8 : p.434–443. https://doi.org/10.1016/j.jasrep.2016.06.051

[16] L.F. Vieira Ferreira, E. De Sousa, M.F.C Pereira., Guerra S. et Ferreira Machado I., 2019, "An archaeometric study of the Phoenician ceramics found at the São Jorge Castle's hill in Lisbon", Ceramics International. https://doi.org/10.1016/j.ceramint.2019.11.267

[17] Ph. Colomban, B. Kırmızı, B Zhao., J.B Clais., Yang Y. et Droguet V., 2020, " Non-Invasive On-Site Raman Study of Pigments and Glassy Matrix of 17th–18th Century Painted Enamelled Chinese Metal Wares: Comparison with French Enamelling Technology",. Coatings 2020, 10, 471. https://doi.org/10.3390/coatings10050471

[18] S.S Pawełkowicz., D Rohanová. et P Svora., 2017, " Gothic green glazed tile from Malbork Castle: multi-analytical study", Heritage science, 5:27. https://doi.org/10.1186/s40494-017-0141-6

[19] M.C Caggiani., A Cosentino. et A. Mangone, 2016, "Pigments Checker version 3.0, a handy set for conservation scientists: A free online Raman spectra database", Microchemical Journal, 129:p.123–132. https://doi.org/10.1016/j.microc.2016.06.020

[20] T. Aguayo, E. Clavijo, A. villagrán , F Espinosa., F. E. sAgüés et Campos-Allette M., 2010, "Raman vibrational study of pigments with patrimonial interest for the chilean cultural heritage", J. Chil. Chem, 55, N° 3. https://doi.org/10.4067/S0717-97072010000300016

[21] A. Coccato, D Bersani., A. Coudray, J Sanyova., L. Moens, Vandenabeele P., 2016, 'Raman spectroscopy of green minerals and reaction products with an application in Cultural Heritage research' J. Raman spectroscopy. https://doi.org/10.1002/jrs.4956

[22] A. Zoppi, C Lofrumento., E.M Castellucci., M.G Migliorini., 2002, " Micro-Raman technique for phase analysis on archaeological ceramics", J. Raman spectroscopy, p.16–21.

[23] H. AZZOUZ, R. ALOUANI and S. TLIG "Mineralogical characterization of ceramic tiles prepared by a mixture of Cretaceous and Mio-Pliocene clays from Tunisia: factory and laboratory products" Journal of the Ceramic Society of Japan 119 [2] 93-100 2011. https://doi.org/10.2109/jcersj2.119.93

[24] G. Raja Annamalai1, R. Ravisankar, A. Chandrasekaran "Analytical investigation of archaeological pottery fragments excavated from Porunthal, Tamil Nadu, India," Cerâmica 66 (2020) 347-353. https://doi.org/10.1590/0366-69132020663792811

[25] M. Sendova, V. Zhelyaskov, M. Scalera, M. Ramsey, J. Raman Spectrosc. 36 (2005) 829. https://doi.org/10.1002/jrs.1371

Mediterranean Architectural Heritage - RIPAM10
Materials Research Proceedings 40 (2024) 179-185

Materials Research Forum LLC
https://doi.org/10.21741/9781644903117-19

From the Great Synagogue of Algiers to Jamma Lihoud, Architectural Monography of a Centuries-Old Building

Naouel NESSARK

University of Bordeaux de Montaigne Bordeaux France

University of Biskra Algeria

nawel_n@ymail.com

Abstract This article proposes a monographic study of the great synagogue of Algiers. An important architectural and symbolic construction, which is not only representative of the changes experienced by the Jewish community and their places of worship after 1830, but also of the contradictions of the colonial administration toward them. The monumental character, the use of many elements of the local architecture, and the Moorish style, have made of it a singular building in this middle of the Algerian nineteenth century imbued with the Parisian inspiration on architecture. Designed in an Arabian style, on a plan close to the plan of the traditional North African synagogues, the building was converted into a mosque after independence, without major consequences on its formal appearance. The communication proposes a detailed analysis of the spatial context, of this conversion.

Introduction

The Grand Synagogue of Algiers has symbolic significance for several reasons. It is very representative of the changes experienced by the Jewish community and their places of worship after 1830, but also of the contradictions of the colonial administration vis-à-vis them. Its construction was planned in 1837 to compensate for the demolished buildings, but the project was materialized many years after. The location of its was a longstanding issue, as were funding issues. Apart from the architectural analysis of the building, the article focuses on the historical context and social influences that surrounded the design of the project. It is based on the analysis of its shape, its spatial organization and its aesthetics. Our objective is to explain, the connections that have been made between innovation displayed by the authorities and traditional pattern of pre-colonial North African synagogues. The aim is not only to shed light on the architectural history of the colonial period, but also after independence. In fact, the question of conversion and the changes brought to the building by becoming a mosque is the second part of this article.

History of Construction of the Great Synagogue

Despite the renewed interest observed in recent years in the architecture of the colonial era in Algeria, the great synagogue is absent from scientific productions. Indeed, except for the work of Dominique Jarrassé, no research is conducted on this building. Based on the meticulous research in archives and architectural analyses, this first part retraces the history of implantation and the first architectural proposals.

Establishment of the Great Synagogue.

The first proposition was a piece of land near the Place de Chartres in 1840; it was subsequently transferred to the land occupied by the lions' barracks. This change is explained by the refusal of local representatives to see a building intended for the natives on a newly created main square and artery. Opposition was quickly expressed even for the new site of the lions' barracks, one of the council members felt it was regrettable that such a beautiful site was destined to host a synagogue. The proximity to Bab Azzoun Street made the prospect of building such a visible synagogue unthinkable. The construction project was then temporarily abandoned. In 1844, another piece of land was proposed, it is that of the old Moorish house located on the street of the

Mediterranean Architectural Heritage - RIPAM10
Materials Research Proceedings 40 (2024) 179-185

Materials Research Forum LLC
https://doi.org/10.21741/9781644903117-19

revolution. The house was used as a temporary barracks. But the minister proposed a new location because the military could not get the site back. Land occupied by vacant federal buildings, is then proposed near the impasse Orali . The smallness of this land quickly eliminated it. The same year a commission composed of the notables of the Jewish community of Algiers is constituted to find the ideal ground, *"four hypothesis are envisaged including that of transforming the mosque of La Pecherie.* The option of reconverting a mosque that corresponded to churches at that time, is mentioned for the first time for synagogues. However, to alienate the Muslim community once again, by converting the largest mosque in the city, and providing such a visible location for a synagogue was not feasible for the authorities. The choice finally came across the site

Figure 1*: Main façade of the synagogue of Algiers*

of an outdated mosque, located near the new Place Randon. This piece of land, proposed by the military, was quickly agreed upon for several reasons; both in the Jewish quarter, far from the main axes and sufficiently visible in the urban fabric by constituting the wall of a public square, something that was unthinkable until then. The choice to build a synagogue on the site of a mosque symbolically materializes the change of status of Jews vis-à-vis Muslims. The slowness of the choice of land of the Great synagogue tells us perfectly about the difficulty of the Jews to make admit the construction of their temples. However, it was supposed to be a compensation for all the submitted expropriations and demolitions.

Early Architectural Proposals

Giauchain designed four projects not accepted in 1838. The first proposal, with a total area of 441 m², could hold up to 1300 people, consisted of a prayer room for men, a platform for women and swimming pools. The second project in an area of 351 m², can accommodate 1000 people. The third project, covering almost 472 m², is designed for more than 1 400 people. Finally, the fourth and final project is more in line with the architect's expectations. Designed for more than 1100 people, it spreads over 371 m², occupying almost the same area as the second project. These projects were quickly abandoned in 1839, because this site was not retained. The architect, who had designed these projects in a neoclassical style, had proposed another not accepted, on the site of the old barracks in a rather neo-Gothic style. In 1844, Giauchain drew another project on the Rue de la Révolution in the marine district. The project was designed in an Egyptian Revival style with papyriform columns with open corolla, and winged sculptures under the law tables. This project is certainly the first to send Jews of Middle Eastern origin. The project fell far short of the expectations of the Jewish community and its aspirations to have a building worthy of all its losses. Algerian liberal architects very probably mandated by the newly installed consistory proposed new projects around 1847. Léon and Rochet had drawn up a sketch for nearly 1,600 people, 1,000 men and 600 women, on the plot located at the Orali impasse. The authorities had not accepted their project.

In 1848, Ravoisié had in turn drawn a proposal on the same ground. The inclination towards an Arabian style for the synagogue is expressed for the first time in this project. The project takes up the characteristic features of the North African synagogues with a plan centered around the Tevah and in the back the wardrobe of the Torah. It still adds a space for women in gallery upstairs. This

Mediterranean Architectural Heritage - RIPAM10
Materials Research Proceedings 40 (2024) 179-185

Materials Research Forum LLC
https://doi.org/10.21741/9781644903117-19

project will undoubtedly have an influence on the final project of the great synagogue and on the architectural choices of the final project, designed towards the end of 1840.

The final project of the synagogue

The architect of the civil buildings Viala du Sorbier designed the final project of the great synagogue, by the end of 1840. The shape was a square overhanging a central octagonal dome and four small domes in the corners, could contain 900 men and 200 women. The influence of Ravoisié's project and the architect's familiarity with Islamic architecture after the rehabilitation of the mosques of Tlemcen, according to Jarrassé[1], may justify this choice. As for the doors and windows, the architect had opted for horseshoe arches resting on twisted columns and bordered by green and white ceramic. Under the dome and its cornices, openings reminiscent of the shape of the tables of laws were lined with stained-glass windows to sift the light inside.

The interior of the synagogue

The building is a square with an irregular octagon in the center with four long sides opened by poly-lobed arches. The entrance is oriented towards the East and not the Holy Ark, which faces it directly. It must be emphasized that the orientation towards Jerusalem is only a recommendation and not an obligation that would invalidate the edifice. No doubt, the desire to open it on the square had taken precedence in the mind of the designer. Moreover, during the 19th century many synagogues built in France were not oriented to the East. The Great Synagogue of Victory is oriented towards the North and the Sacred Heart church according to Dominique Jarrassé. The internal spatial distribution is inspired by that of the traditional North African synagogues centered around the *Tevah*. The most remarkable element in this synagogue is the refinement of its interior decoration, which is steeped in Moorish references.

Figure 2. View of the Holy Ark
Source: IAU Archive

A stalactite arch surmounted the holy arch; the eight ribs of the vault and the windows that illuminate it were adorned with stucco embroidery and chiseled plaster. The stone carvings were the work of Jean-Émile Latour. The same artist was in charge of harmonizing the decorations of Saint-Philippe Cathedral. There are also similarities in the interior decorative elements, particularly in the domes, edges and borders. It seems rather logical that the two most important religious buildings built in the same period, have a reciprocal influence in terms of decorative elements. Paul Alfred Magdonel made carvings on wood[2]. In another register, the synagogue contained important objects and relics, among which there are "sepharim", one of which dates back to the 15th century, objects belonging to rabbis Barchichat and Duran, were also kept in the tabernacle. Marble and steel commemorative plaques were hung around the pulpit. They bore the names of the Jewish soldiers who died for France, just opposite were those of all the benefactors of the community who worked for the building of the temple[3].

---

[1] Dominique, Jarassé, Orientalism, Colonialism, and Jewish Identity in the Synagogues of North Africa under French domination, *Art Judaica*. 2011. P. 1-22.

[2] Claudine Piaton; Juliette, Hueber Boussad, Aiche, and Thierry, Lochard,Algiers - City and Architecture 1830-1940. Algiers. 2016. P. 98

[3] Paviot Marcel, October 24th 1951, a brief history of our shrines "the Jewish temple Grand Rabbi Bloch", in the echo of Algiers Archives of the Diocesan Studies Centre – glycines Algiers. 270-96, AAJ. 01 (4).

Figure 3: *The deterioration of the Great Synagogue in 1961*

Source : https://www.morial.fr/communautes-et-traditions-3/synagogues/949-la-profanation-de-la-grande-synagogue-d-alger.html *consulté en décembre*

Although many similarities are remarkable with the traditional synagogues of North Africa, the beginning of the process of francisation of Jewish places and their places of worship is also evident. The complexity of this project, combined with administrative and especially financial barriers, delayed it for nearly two decades. It was inaugurated on September 19, 1865. The synagogue was, but its capacity remained lower than all the demolished ones. To compensate and offer decent places of worship, many other projects for the construction or developments of synagogues were begun.

The Great synagogue of Algiers, by its monumental character, is one of the most important built in Algeria. It is also, unique by the re-appropriation of certain elements of

Figure 4: *Interior Ornament*
Source: Author

Mediterranean Architectural Heritage - RIPAM10   Materials Research Forum LLC
Materials Research Proceedings 40 (2024) 179-185   https://doi.org/10.21741/9781644903117-19

traditional North African synagogues in particular, the centered aspect of the plan. Generally, on facades, the Moorish language continued to be widely used. The windows often had a shape reminiscent of the forms of tables of law. They were often paired. The predominance of the twisted columns in particular to mark the Holy Ark is also a constant element that probably refers to the temple of Jerusalem. Although it remained an external dressing, the choice of orientalism for the Algerian synagogues, according to Dominique Jarrassé, is the continuous mental projection of the Jews on the side of the colonized. Semitism through Orientalism works in the same way in the mentality of administrators and architects as in metropolitan France. As simple as it was, there was confusion of indigenous culture; the synagogue that represents them was assimilated by its architecture to the mosque.... However, vis-à-vis the Arabs, they are francized and their synagogues usurp forms that do not belong to them4. This use of elements of Orientalist architectural language has remained superficial, an exterior dressing that contrasts with the radical changes made to the interior. The latter is deeply marked by metropolitan influences. The monumentality, the basilicale plan, the arrangement of the benches, the introduction of the organ as well as other musical instruments, and the stands for women are all elements borrowed from other architectures

The Great Synagogue as a mosque
The only synagogue converted to a mosque is the old Randon Square, which became the Ibn Fares Mosque just after independence. Located on Arbadji Abed Street, the building gave its name to the whole neighborhood which became Djamaa Lihoud or Jewish mosque. After independence, the legal manuscripts and various silver religious objects belonging to this synagogue were classified as national heritage and the synagogue converted into a mosque.

Pre independence deterioration
The first chapter of this process began before independence when it suffered a sacking attributed to the FLN in the w ake of the demonstrations of December 11, 1960, but which is probably the work of supporters of French Algeria. The deterioration was substantial on both the furniture and the building itself. A symbolic ceremony was organized the next day to bury the destroyed objects near the Rabbis of Algiers in the cemetery of Saint-Eugene. The building probably did not regain its previous influence until its reconversion in 1962.

Conversion into a mosque
Originally built on the site of a mosque, the synagogue was quickly claimed and reco vered. In this conversion, the building has not undergone any significant modification except the addition of an octagonal minaret. The configuration of the synagogue and its Moorish architecture greatly facilitated its conversion. The women's prayer space has retained this function; this one is located at the height of 5.05 m. An intermediate wooden floor, at the height of 2.94 m, is added below to increase the capacity of the prayer room for men of about 240 m². The height under the dome of the building is about 18 m. A library and a Maksoura or imam's office are located on the western façade; these spaces are 25 m² each. The mihrab, meanwhile, is leaned against the old main gate of the synagogue. The old rabbinical school, underlying the building on the South-West side, which had long been abandoned, was recently converted into a Koranic school. During the 70s and 90s, the mosque had undergone renovations to seal the domes with bituminous felt. Other fit-up and maintenance work was carried out during the same period. A new restoration operation is planned shortly, especially since the recent closures due to the pandemic have accelerated its degradation.

---

4 Jarrassé Dominique, 1997, une *histoire des synagogues françaises entre orient et occident*, éditions actes sud. Paris. P. 254.

Mediterranean Architectural Heritage - RIPAM10
Materials Research Proceedings 40 (2024) 179-185

Materials Research Forum LLC
https://doi.org/10.21741/9781644903117-19

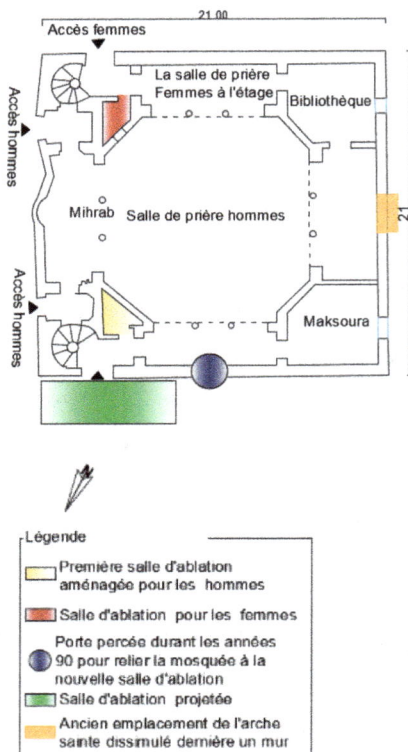

Figure 5: *Schematic reconstruction of the mosque plan Source: author*

Current Building Condition

Today cracks are visible, the coatings are detached inside and outside, and traces of moisture are visible even at the level of the dome... On another register, since the 1990s, it was recommended to carry out a historical study and more in-depth surveys to restore the missing elements of the history of the building and for a better architectural knowledge of it. It should be emphasized that during our investigations we learned that traces of its former synagogal function could remain behind the South-West wall, which initially housed the niche of the holy arch. We have no way of verifying the veracity of that information. The completion of surveys and the search for these traces and their preservation can only enhance the architectural and heritage value of this building.

Conclusion

The Great synagogue built during the nineteenth century in a style combining a desire to break with pre-colonial synagogue. But the dichotomy of visibility/discretion has governed all its construction process. This situation was hardly different in architectural terms. The external aspect of the synagogue was heavily influenced by Moorish influences. The Spanish origins in particular, of the organizers of the Jewish community of Algiers, can explain in part, the option of the architects for this Moorish dress. They remain, however, very characteristic by their monumentality and by the many changes brought about inside; introduction of the organ, adoption of the basilical plan and especially creation of a space for women. Beyond ideological questions, this building halfway between mosque and church is very representative of the situation of Jews at the time, in search of European modernity, but attached to their cultural and social origins. Today, synagogues and our entire architectural and urban heritage are in a state of perilous disrepair.

References

Archives
Archives of the Diocesan Studies Centre glycines Algiers 270-96, AAJ.01 (3).
Archives of IAU Israelite alliance universal obout the great synagogue
Others

[1] Claudine Piaton; Juliette, Hueber Boussad, Aiche, and Thierry, Lochard,Algiers - City and Architecture 1830-1940. Algiers. 2016.

[2] Dominique, Jarassé, Orientalism, Colonialism, and Jewish Identity in the Synagogues of North Africa under French domination, *Art Judaica*. 2011.

[3] Dominique, Jarrassé, «The Synagogue of Peace in Strasbourg (1958): The Weight of the Legacy». In *Religious architecture in the 20th century, what heritage.* Presse universitaire, Rennes 2007.

[4] Jean, Laloum and Jean Luc Allouche , the Jews of Algeria, éditions du scribes, Paris, 1987.

[5] Notra, Lafi, «Being Jewish in Ottoman Algeria», in *Juifs d'Algérie*, Flammarion, Paris, 2012.

[6] Valerie, Assan, « Synagogues in Colonial Algeria of the 19th century», in *Archives Juives. Vol 37)* 2004/1. https://agorha.inha.fr/inhaprod/ark:/54721/00281482. Accessed June 2020 https://athar.persee.fr/issue/feldj_1112-0649_1937_hos_1_1_1?sectionId=feldj_1112-0649_1937_hos_1_1_119. Accessed June 2021

Mediterranean Architectural Heritage - RIPAM10

Materials Research Proceedings 40 (2024) 186-197

Materials Research Forum LLC

https://doi.org/10.21741/9781644903117-20

Physical and Thermal Properties of Raw Earthern Bricks from Ksar Ait Benhaddou

Mahdi LECHHEB[1,a*], Mohammed CHRACHMY[1,b], Achraf HARROU[2,c], Hassan OUALLAL[1,d], Meryem BEN BAAZIZ[1,e], M'barek AZDOUZ[1,f], Mohamed AZROUR[1,g], Polona OBLAK[3,h], Meriam EL OUAHABI[4,i], El Khadir GHARIBI[2,j]

[1]Materials Sciences for Energy and Sustainable Development Team, Department of Chemistry, FST Errachidia, Moulay Ismail University, Errachidia, Morocco

[2]Laboratory of Applied Chemistry and Environment Solid Mineral Chemistry Team, Faculty of Sciences, Mohammed Premier University, Oujda 60000, Morocco

[3]Euro-Mediterranean University (EMUNI), Slovenia

[4] UR Argile, Geochemistry and Sedimentary Environment (AGEs), Department of Geology, Quartier Agora, University of Liège, Bâtiment, B18, Allée du six Aout, 14, Sart-Tilman, B-4000 Liège, Belgium

[a]m.lechheb@edu.umi.ac.ma, [b]chrachmy@gmail.com, [c]harrou201@gmail.com , [d]h.ouallal@umi.ac.ma, [e]m.benbaaziz@edu.umi.ac.ma, [f]m_azdouz@hotmail.com, [g]m.azrour@umi.ac.ma, [h]polona.oblak@emuni.si, [i]meriam.elouahabi@uliege.be, [j]gharibi_elkhadir@yahoo.fr

Keywords: Raw Earth, Adjuvants, Compressive Strength, Thermal Conductivity, Recycling

Abstract. The deterioration of historical earthen architecture in the Drâa-Tafilalet region of Morocco is becoming a significant concern. These structures are susceptible to various natural and human-induced factors, leading to their deterioration. Our research focuses on the recycling and valorization of debris from the deteriorated walls of Ksar Ait Benhaddou to potentially facilitate restoration. Various techniques were employed to characterize the debris from Ksar Ait Benhaddou, including geotechnical analysis (Atterberg limits, grain size), physicochemical assessment (X-ray diffraction, Infrared spectroscopy, X-ray fluorescence), and microscopic examination (Scanning electron microscopy). Standardized brick specimens were created from debris waste paste mixed with 22% water by mass. Additional specimens were prepared by incorporating stabilizers (portland cement or lime) or natural plant fibers (wheat straw) into the debris, with a water/solid ratio of 22%. The prepared specimens underwent aging for different periods (0 to 4 days). The study investigated the impact of aging duration and three additives on mechanical properties, material thermal conductivity, and hydrate formation. The debris exhibited medium plasticity, consisting of non-swelling clays and sand. The compressive strength of cement-stabilized samples yielded the best results, reaching 1.90 MPa for the DC5 sample. The thermal conductivity of samples stabilized with cement increased, contrasting with lime and straw, which had the opposite effect.

Introduction

Morocco has had a clay-based material industry since the 1970s. However, the actual development of it began in the 1990s with the implementation of the national construction project known as "200 000 housings"[1,2]. Clay is typically used in artisanal or semi-industrial ceramics, or even in its raw form [3,4]. Clay-based ceramics in Morocco are primarily funded and overseen by small businesses and artisans. Despite their contributions, the national production is still regarded as insufficient and below the desired standard. This situation prompted Morocco to import ceramic materials from neighboring countries like Spain and Italy.[1]. The increases in demand for clayey construction material will be growing due to the devastating earthquake of 8 September 2023 that struck the El Haouz region of Morocco situated at about 70 km south-west of Marrakech, took a heavy toll in terms of both loss of life and damage to infrastructure and people's homes [5]. According to El Bairi et al., (2023), almost 3000 people were killed and more than 5600 injured, and about 50000 homes were totally or partially destroyed [5]. These disastrous consequences mean that this issue is of vital importance to Morocco, and it is therefore vital to supplement existing regulations and update the legislation governing the sector.

Earthen construction has advantages in terms of ecology, society, economy, and culture [6,7]. Each inhabited continent boasts a heritage of earthen architecture, with 20% of the structures listed on the UNESCO World Heritage List constructed using rammed earth materials [4]. Several studies have focused on clays stabilization with cement or lime and natural plant fibres (i.e. wheat straw) to enhance their durability [8–16].

Incorporating lime into the soil has the effect of improving its properties, reducing both swelling and plasticity [8]. Ouedraogo et al., [17], The findings indicate that a combination of 5% cement and lime was effective in improving both the mechanical and thermal properties of the studied soils. Additionally, in the same study, it was determined that approximately 5% of lime was adequate for the short-term reaction, representing the initial lime consumption.

The Ksar Ait Benhaddou is on the UNESCO World Heritage List since 1987. Visitors come from all over the world to see this building. Moreover, it is recognized as a cinematographic location, having been used for the filming of numerous masterpieces. In order to protect the site's natural elements and cultural legacy, the Center for the Restoration and Rehabilitation of the Architectural Heritage of the Atlas and Sub-Atlas areas (CERKAS[1]) has been conducting a number of restoration interventions on the Ksar since the beginning of the 1990s. The United Nations Development Program (UNDP), the Moroccan Ministry of Culture, and UNESCO's technical assistance provided funding for the intervention [18].

The aim of this research is to supply sustainable and suitable earthen material, based on the reuse of clayey ruin material, to restore the Ksar Ait Benhaddou monument. Characterization of the earthen material of the wall of this monument, which has fallen into ruin was previously performed [4]. The additives we have already chosen are cement, lime and wheat straw.

Lime and portland cement, both calcium-based stabilizers, can be incorporated into clayey soils for improvement. This enhancement involves four mechanisms: cation exchange, flocculation and agglomeration, hydration cementitious (geopolymerization of C-S-H gel), and pozzolanic reaction. Portland cement supplies the necessary chemical components for all four processes. Notably, lime is unable to complete the cementitious hydration process [19], Through the carbonation process of portlandite (calcite neoformation), it solidifies the mortar [20].

This analysis will include volume shrinkage, variations in humidity, and alterations in thermal conductivity. The assessment of stabilization-induced changes will involve the examination of scanning electron microscope images and X-ray diffraction data. Combining these tests with those conducted in previous research will enhance our comprehension of the stabilization mechanism, considering physical, chemical, and microstructural perspectives.

Material and methods
Preparation of bricks specimens
A representative sample of debris from the Ksar Ait Benhaddou was collected, and three specimens were created for each aging period as outlined below:

---

[1] Centre de Conservation, de restauration et de réhabilitation des ksours et kasbahs des zones atlassiques et subatlassiques (CERKAS Ouarzazate)

The debris sample was mixed with water, variable amounts of binders: portland cement, lime, and straw (Table 1). Only 5% of mineral binders (lime and portland cement) were added to earthen debris for economic and environmental coherence. The 5% of mineral binders is adequate to stabilize earthen materials effectively [17,21,22].

Due to the limited use of a hydraulic binder (5%), we opted for pre-curing the mixtures before forming the brick specimens. This method ensures that the microstructure of the hardened earth specimens remains intact without causing increased pliability in the clay. Furthermore, this strategy prevents the evaporation of mixing water, crucial for the consolidation of hydraulic binders and clay particles.

Table 1: Type and quantities of binders employed in the production of brick specimens.

	Brick specimens	Debris (%)	Cement (%)	Lime (%)	Straw (%)
Bricks stabilized without pre-curing	$DC_{NC}1$	99	1	-	-
	$DC_{NC}2$	98	2	-	-
	$DC_{NC}3$	97	3	-	-
	$DC_{NC}4$	96	4	-	-
	$DC_{NC}5$	95	5	-	-
Bricks stabilized with pre-curing	DC1	99	1	-	-
	DC2	98	2	-	-
	DC3	97	3	-	-
	DC4	96	4	-	-
	DC5	95	5	-	-
	DL1	99	-	1	-
	DL2	98	-	2	-
	DL3	97	-	3	-
	DL4	96	-	4	-
	DL5	95	-	5	-
	DS1	99	-	-	1
	DS1.5	98.5	-	-	1.5
	DS2	98	-	-	2
	DS2.5	97.5	-	-	2.5
	DS3	97	-	-	3

Experimental methods

The absorbance of water test allows for the measurement of capillary water absorption. By measuring the mass gain of the specimen in a crystallizer with water positioned at 1 cm above the specimen's underside, the water absorption values of stabilized specimens were determined [22,23].

The geometric parameters (length, breadth, and height) of the specimens were measured with 0.02 mm of errors using a sliding caliper to determine their volume.

Loss on ignition at 105°C for 2H is used to determine moisture content of samples. The moisture reduction percentage change of specimens as a function of curing time was calculated as the following formula (1):

$$Mr\ (\%) = (M\ \%\ (4\ D) - M\ \%\ (0\ D))/\ (M\ \%\ (0\ D)) * 100. \tag{1}$$

As:

- Mr: Moisture reduction
- M: Moisture
- D: Days

The thermal conductivity of stabilized samples was determined utilizing the Hot Disk device model TPS 2500 meter (Al Akhawayn University, Ifrane).

Scanning Electron Microscopy was performed on stabilized specimens using a Quanta 200 model at the Technical Support Units for Scientific Research (CNRST, Rabat).

SEM images were taken by the scanning electron microscope for mixtures that gave the best results for compressive strength (DC_5, DL_1, DS_2, DC_{NC5}) to fully understand what is happening microstructurally

Raw material characterisation

The thermal conductivity of the untreated debris sample demonstrates only slight changes over different aging periods. (Fig. 1). Treating with a precuring process has the potential to enhance physical strength [24].

Though the change in thermal conductivity is slight, it could imply differences in bricks formed without additives. Therefore, the chosen duration for the remaining tests is three days.

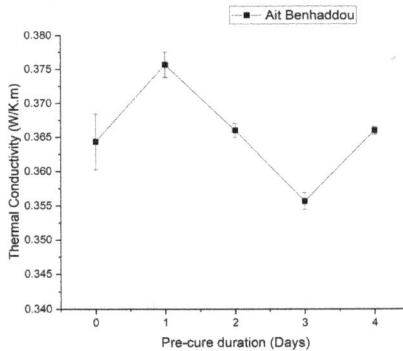

Fig. 1 : Variations in thermal conductivity observed in the untreated sample in relation to pre-curing durations.

Characterisation of stabilized bricks

The introduction of the stabilizer, along with an extended curing period, leads to a reduction in volume shrinkage. (Fig. 2). The type of stabilizer influences the mechanical properties: volume shrinkage decreases by 0.69% per day for lime and 0.88% for straw, and rises to 0.96% per day for cement without curing, and 1.16% for cement with curing.

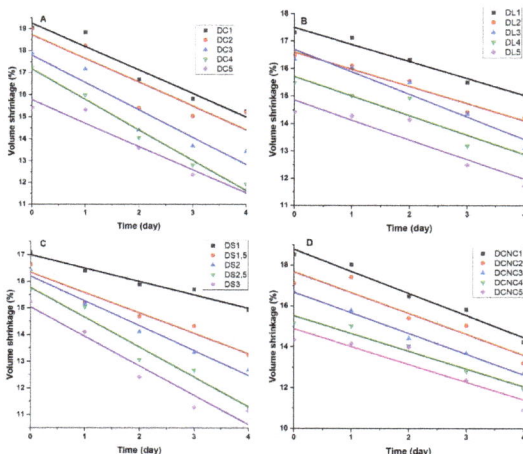

Fig. 2: Changes in volume shrinkage concerning curing duration and the type and amount of additives. A: portland cement (DC); B: lime (DL); C: straw (DS); D: cement without pre-cure.

Table 2: Relationship between changes in volume shrinkage and the curing time of the additives used to manufacture brick specimens.

Additives (%)	Cement		Lime		Straw		Cement (without pre-cure)	
	Slope	R^2	Slope	R^2	Slope	R^2	Slope	R^2
1	-1.06	0.93	-0.61	0.97	-0.50	0.97	-1.07	0.97
1.5	-	-	-	-	-0.76	0.94	-	-
2	-1.07	0.81	-0.62	0.95	-0.93	0.98	-1.01	0.89
2.5	-	-	-	-	-1.11	0.96	-	-
3	-1.23	0.89	-0.81	0.93	-1.10	0.94	-1.00	0.99
4	-1.38	0.98	-0.70	0.89	-	-	-0.86	0.95
5	-1.06	0.95	-0.72	0.85	-	-	-0.86	0.85
Average	-1.16		-0.69		-0.88		-0.96	

The changes in volume shrinkage as a function of the type of stabilizer, calculated from the slopes of the volume shrinkage curves, and their correlation coefficient R^2 are given in **Table 1**. Volumetric shrinkage followed the same trend for samples stabilized with lime (0.69% / day) and straw (0.88% / day). However, it is twice as high for cement (1.16% / day) than for cement without pre-cure (0.96%/day). Cement, unlike the other stabilizers, allows the grains of the material to clump together and improves workability with a long curing time. As a result, the volumetric shrinkage of these specimens is greatest.

The addition of 5% of cement decreases the volume shrinkage less than 15.5% after 4 days of pre-curing, which significantly decreases the occurrence of cracks. The composition of the clays employed affects how shrinkage develops [22,25], indeed, as demonstrated in our previous research, samples obtained from the remains of the wall of the Kasbah of Ait Benhaddou do not exhibit the presence of swelling clay[4]. The amount of shrinkage is significantly reduced by the addition of lime up to 5%, which lessens the likelihood of cracking [26]. Low R^2 values for the shrinkage caused by straw addition can be explained by large and quick water evaporation via capillary action from specimens stabilized by straw as opposed to diffusion through pores and cracks. By integrating natural fibers, earth plasters in straw bale constructions experience significantly less shrinkage [14].

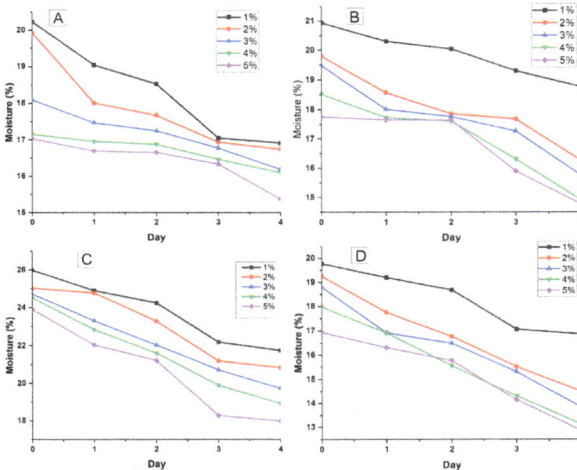

Fig. 3: Variations in moisture development for the specimens relative to curing duration and the type and amount of additives. A: lime (DL); B: cement (DC); C: straw (DS); D: cement without pre-cure (DCNC).

The moisture level diminishes with the duration of pre-curing and the proportion of the additive (Fig. 3). The decrease is more important for straw than for cement without precure and lime, although the evolution of this decrease is almost the same according to the quantity of the added admixture, the lowest decrease was observed for cement-stabilized samples. After 4 days of the pre-cure duration, the moisture decreased (Fig. 4) respectively with a slope of -2.27, -2.10, -2.29 and -1.43, depending on the amount of straw, cement without pre-cure, lime, and cement added.

Zeolite water, adsorbed water, water combined with aluminous silicate hydrates (C-A-H) or calcium silicates (C-S-H) are some of the sources of residual moisture in the paste. This reduction in moisture during the curing process may be caused by pozzolanic processes in addition to evaporation [27]. The low amount of residual moisture compared to straw with lime and cement, which are hydraulic binders, could be explained by the production of stable hydrates at 105°C. For the last reason, evaporative drying during curing may account for the drop in cement paste moisture.

Mediterranean Architectural Heritage - RIPAM10

Materials Research Proceedings 40 (2024) 186-197

Materials Research Forum LLC

https://doi.org/10.21741/9781644903117-20

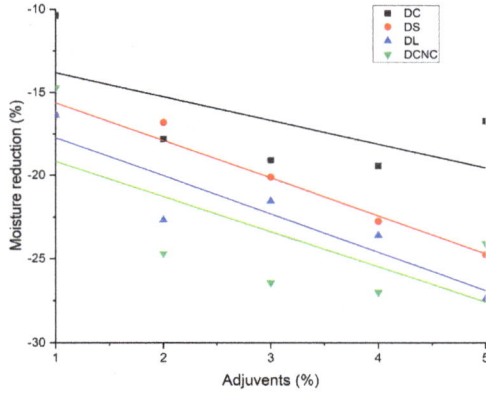

Fig. 4: Moisture reduction as a function of the quantity of additives for the specimens. A: DL : lime; DC : cement ; DS: straw; DC_{NC}: cement without pre-cure.

Fig. 5 illustrates the thermal conductivity results of pastes pre-cured for three days before shaping brick specimens (DL, DC, and DS), as well as those stabilized with cement without pre-curing (DCNC). The figure indicates that the incorporation of cement, regardless of curing, resulted in an elevated thermal conductivity with an increasing percentage of added cement. Conversely, the introduction of lime and wheat straw had an opposing effect.

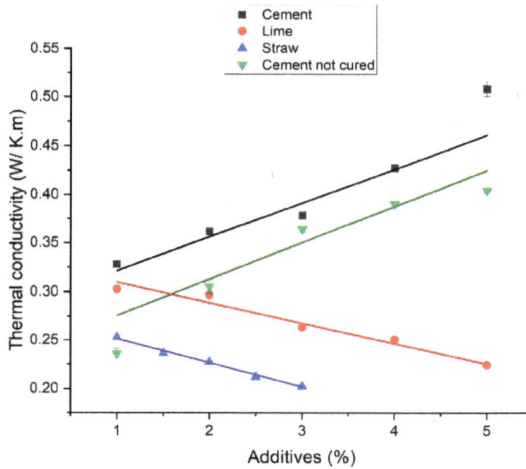

Fig. 5: Fluctuations in thermal conductivity based on the additions made to the specimens.

Mediterranean Architectural Heritage - RIPAM10

Materials Research Forum LLC

Materials Research Proceedings 40 (2024) 186-197

https://doi.org/10.21741/9781644903117-20

Table 3: R^2 and slope values derived from the changes in thermal conductivity curves concerning the amount of aggregate introduced into brick specimens.

	Slope $(W.K^{-1}.m^{-1}.\%^{-1})$	R^2
Lime (DL)	-0.02123	0.98
Straw (DS)	-0.02487	0.99
Cement (DC)	0.03487	0.92
Cement without pre-cure (DC_{NC})	0.03728	0.85

The thermal conductivity variation is influenced by the type of added aggregate. The dimensions of the aggregate and the presence of admixture in the mortar structure contribute to the reduction in thermal conductivity, with larger sizes having a more pronounced impact. Indeed, introducing an aggregate to the debris diminishes thermal conductivity by reducing particle-to-particle contact. [22].

Regarding lime and straw, which had no impact on the compressive strength as already shown in our previous work [4], the conductivity experiences a linear decrease with the added material quantity at a consistent rate. The thermal conductivity of cement surpasses that of the reference sample as the added quantity increases, leading to the consolidation of the matrix. Samples stabilized with cement without pre-curing showed the same fluctuation profile, these samples' thermal conductivity values, however, are lower than those of the cement-stabilized and pre-cured specimens. The drop in the thermal conductivity value is likely caused by pores present in the samples that were not pre-cured.

Considerably reduced thermal conductivity values compared to those documented by Ouedraogo et al., [17] are compatible with the outcomes produced using cement and lime. Due to the insulating properties of the cellulose in the straw and its increased porosity, the addition of straw reduces thermal conductivity [28,29].

Mediterranean Architectural Heritage - RIPAM10
Materials Research Proceedings 40 (2024) 186-197

Materials Research Forum LLC
https://doi.org/10.21741/9781644903117-20

Monitoring of stabilization-induced changes

Fig. 6: SEM images of earthen material debris from the Ksar Ait Benhaddou A: DC_5 B: DL_1 C: DS_2 D: DC_{NC5}

Sample DC5 (Fig 6-A) displays a gel-like substance enveloping the grain surface when cement is added. This substance is believed to be calcium silicate hydrate (C-S-H) gel, formed through the hydration of the primary phases of portland cement, belite, and alite. The grains are closely bound to each other, eliminating visible separation, and the particles exhibit significant cohesion with pore sizes ranging from 0.2 to 0.5 μm.

Sample DC5 (Fig 6-A) displays a gel-like substance enveloping the grain surface when cement is added. This substance is believed to be calcium silicate hydrate (C-S-H) gel, formed through the hydration of the primary phases of portland cement, belite, and alite. The grains are closely bound to each other, eliminating visible separation, and the particles exhibit significant cohesion with pore sizes ranging from 0.2 to 0.5 μm. [11,30].

The straw-stabilised samples have similar mineralogical composition as raw debris sample (Fig.7) of our previous work [4]. The addition of straw reveals a fractured texture. The addition of 2-3% of straw accelerated the formation of cracks in silty clay soil [13].

Comparison of SEM images revealed that cement-stabilized samples DC_5 showed better cohesion and less porosity than DL_1, DS_2, and DC_{NC5}, which explains the enhanced mechanical strength [4]. and thermal results (Fig. 5).

Mediterranean Architectural Heritage - RIPAM10
Materials Research Proceedings 40 (2024) 186-197

Materials Research Forum LLC
https://doi.org/10.21741/9781644903117-20

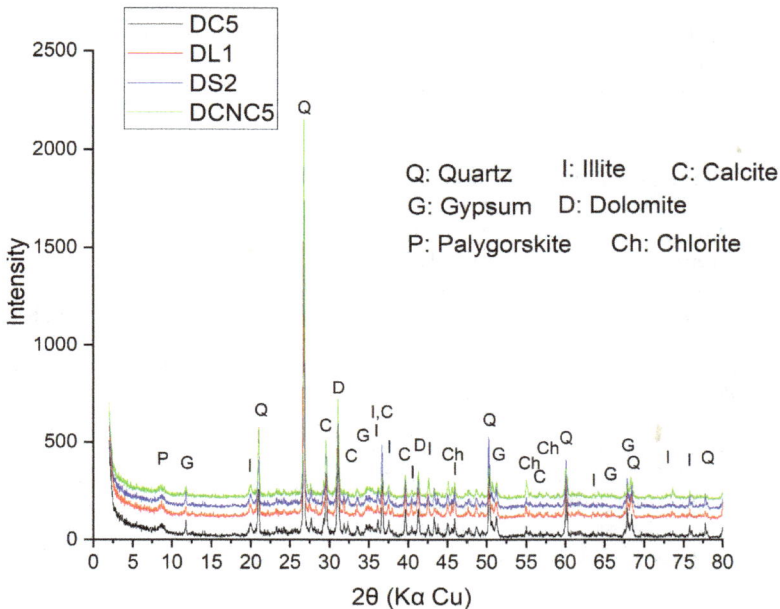

Fig. 7: XRD spectra of stabilized specimens

All stabilized specimens and the raw debris sample show similar mineralogical composition (Fig. 7), suggesting that the stabilisation mechanism is agglomeration of the sample grains around the different grains of hydraulic binders or fibers of straw used to stabilise them, as no new phases were created.

Conclusion

Samples from the Ait Benhaddou Kasbah wall debris were stabilized using portland cement, lime, and straw with the aim of improving their mechanical and potentially thermal properties and efficiency.

Volume shrinkage decreases with the addition of stabilizer and the duration of curing. The shrinkage kinetics is almost high for cement, with or without precure, compared to lime and wheat straw. Cement is a stronger hydraulic binder than lime, while the speed of wheat straw can be explained by significant evaporation at the beginning of the precure as well as the effect of water drainage through the straw channels.

Both the proportion of the additive and the pre-curing duration have an effect on the decrease of moisture content. The increase is more significant for straw than for cement without precure and lime. The least level of decrease was seen in cement-stabilized samples, even if the evolution of this decrease is almost the same for the other admixture.

The size of the binder in the structure of clayey specimens are the causes of the decrease in thermal conductivity; the larger the size, the more pronounced the loss. Indeed, the addition of a binder to the debris decreases the thermal conductivity because it decreases the contact between the particles

Cement, reducing brick porosity, enhances thermal conductivity, as evidenced by SEM images illustrating improved cohesion and, consequently, higher compressive strength. Conversely, lime and straw lead to a decrease in both thermal conductivity and compressive strength.

During setting, no mineralogical composition occurred suggesting that the stabilization mechanism is an agglomeration of the grains of the debris samples around the different stabilizers.

References

[1] M. El Ouahabi, VALORISATION INDUSTRIELLE ET ARTISANALE DES ARGILES DU MAROC, Université de Liège , 2013.

[2] C.M.F. Vieira, R. Sánchez, S.N. Monteiro, Characteristics of clays and properties of building ceramics in the state of Rio de Janeiro, Brazil, Constr Build Mater. 22 (2008) 781–787. https://doi.org/10.1016/J.CONBUILDMAT.2007.01.006

[3] G. El Boukili, M. Lechheb, M. Ouakarrouch, A. Dekayir, F. Kifani-Sahban, A. Khaldoun, Mineralogical, physico-chemical and technological characterization of clay from Bensmim (Morocco): Suitability for building application, Constr Build Mater. 280 (2021) 122300. https://doi.org/10.1016/J.CONBUILDMAT.2021.122300

[4] M. Lechheb, A. Harrou, G. El Boukili, M. Azrour, A. Lahmar, M. El Ouahabi, E.K. Gharibi, Physico-chemical, mineralogical, and technological characterization of stabilized clay bricks for restoration of Kasbah Ait Benhadou- Ouarzazate (south-east of Morocco), Mater Today Proc. 58 (2022). https://doi.org/10.1016/j.matpr.2022.01.459

[5] K. El Bairi, O. Al Jarroudi, S. Afqir, Morocco earthquake: mitigating the impact on patients with cancer, The Lancet. 402 (2023) 1323–1324. https://doi.org/10.1016/S0140-6736(23)02019-6.

[6] R. Anger, Approche granulaire et colloïdale du matériau terre pour la construction, INSA , 2011.

[7] A. Vissac, A. Bourgès, D. Gandreau, R. Anger, L. Fontaine, argiles & biopolymères-les stabilisants naturels pour la construction en terre, CRAterre éditions, 2017.

[8] M. Al-Mukhtar, A. Lasledj, J.F. Alcover, Behaviour and mineralogy changes in lime-treated expansive soil at 20 °C, Appl Clay Sci. 50 (2010) 191–198. https://doi.org/10.1016/J.CLAY.2010.07.023

[9] R. Jauberthie, F. Rendell, D. Rangeard, L. Molez, Stabilisation of estuarine silt with lime and/or cement, (2010). https://doi.org/10.1016/j.clay.2010.09.004

[10] J.E. Oti, J.M. Kinuthia, Stabilised unfired clay bricks for environmental and sustainable use, Appl Clay Sci. 58 (2012) 52–59. https://doi.org/10.1016/J.CLAY.2012.01.011

[11] A. Harrou, E. Gharibi, H. Nasri, N. Fagel, M. El Ouahabi, Physico-mechanical properties of phosphogypsum and black steel slag as aggregate for bentonite-lime based materials, Mater Today Proc. 31 (2020) S51–S55. https://doi.org/10.1016/J.MATPR.2020.05.819

[12] A. Harrou, E.K. Gharibi, Y. Taha, N. Fagel, M. El Ouahabi, Phosphogypsum and Black Steel Slag as Additives for Ecological Bentonite-Based Materials: Microstructure and Characterization, Minerals 2020, Vol. 10, Page 1067. 10 (2020) 1067. https://doi.org/10.3390/MIN10121067

[13] S.F. Mousavi, M. Moazzeni, B. Mostafazadeh-Fard, M.R. Yazdani, Effects of rice straw incorporation on some physical characteristics of paddy soils, J. Agr. Sci. Tech. 14 (2012) 1173–1183.

[14] T. Ashour, An experimental study on shrinkage of earth plaster with natural fibres for straw bale buildings, Taylor & Francis. 3 (2010) 299–304. https://doi.org/10.1080/19397038.2010.504379

[15] S.K. Meena, R. Sahu, R. Ayothiraman, Utilization of Waste Wheat Straw Fibers for Improving the Strength Characteristics of Clay, Journal of Natural Fibers. (2019). https://doi.org/10.1080/15440478.2019.1691116

[16] F. Le Digabel, N. Boquillon, P. Dole, B. Monties, L. Averous, Properties of thermoplastic composites based on wheat-straw lignocellulosic fillers, J Appl Polym Sci. 93 (2004) 428–436. https://doi.org/10.1002/APP.20426

[17] K.A.J. Ouedraogo, J.E. Aubert, C. Tribout, G. Escadeillas, Is stabilization of earth bricks using low cement or lime contents relevant?, Constr Build Mater. 236 (2020) 117578. https://doi.org/10.1016/J.CONBUILDMAT.2019.117578

[18] S. Moriset, M. Boussalh, Ksar Aït Ben Haddou patrimoine mondial : Plan de gestion 2007-2012, CRAterre-ENSAG, 2007.

[19] M.Yıldız, A. S. Soğancı, Effect of freezing and thawing on strength and permeability of lime-stabilized clays, Scientia Iranica. 19 (2012) 1013–1017. https://doi.org/10.1016/J.SCIENT.2012.06.003

[20] M.R. Thompson, Lime Reactivity of Illinois Soils, Journal of the Soil Mechanics and Foundations Division. 92 (1966) 67–92. https://doi.org/10.1061/JSFEAQ.0000911

[21] E. Hamard, B. Cazacliu, A. Razakamanantsoa, J.C. Morel, Cob, a vernacular earth construction process in the context of modern sustainable building, Build Environ. 106 (2016) 103–119. https://doi.org/10.1016/J.BUILDENV.2016.06.009

[22] H. Houben, H.C. Guillaud, Traité de construction en terre, Éd Parenthèses, 2006.

[23] J. Norton, P. Oliver, Building with Earth: a handbook, Intermediate Technology Publications; 2nd edition, 1997.

[24] S. Wang, H. Li, S. Zou, L. Liu, C. Bai, G. Zhang, L. Fang, Experimental study on durability and acoustic absorption performance of biomass geopolymer-based insulation materials, Constr Build Mater. 361 (2022) 129575. https://doi.org/10.1016/J.CONBUILDMAT.2022.129575

[25] Doat. P, Hays. A, Houben. H, Matuk. S, Vitoux. F, Construire En Terre, Ed Alte et Par, 1979.

[26] S.K. Dash, M. Hussain, Influence of lime on shrinkage behavior of soils, Journal of Materials in Civil Engineering. 27 (2015). https://doi.org/10.1061/(ASCE)MT.1943-5533.0001301

[27] J. Wang, C. Yu, W. Pan, Life cycle energy of high-rise office buildings in Hong Kong, Energy Build. 167 (2018) 152–164. https://doi.org/10.1016/J.ENBUILD.2018.02.038

[28] K. El Azhary, Y. Chihab, M. Mansour, N. Laaroussi, M. Garoum, Energy Efficiency and Thermal Properties of the Composite Material Clay-straw, Energy Procedia. 141 (2017) 160–164. https://doi.org/10.1016/J.EGYPRO.2017.11.030

[29] M. Ouedraogo, K. Dao, Y. Millogo, J.E. Aubert, A. Messan, M. Seynou, L. Zerbo, M. Gomina, Physical, thermal and mechanical properties of adobes stabilized with fonio (Digitaria exilis) straw, Journal of Building Engineering. 23 (2019) 250–258. https://doi.org/10.1016/J.JOBE.2019.02.005

[30] S. Oumnih, N. Bekkouch, E.K. Gharibi, N. Fagel, K. Elhamouti, M. El Ouahabi, Phosphogypsum waste as additives to lime stabilization of bentonite, Sustainable Environment Research. 1 (2019) 1–10. https://doi.org/https://doi.org/10.1186/s42834-019-0038-z

Mediterranean Architectural Heritage - RIPAM10

Materials Research Forum LLC

Materials Research Proceedings 40 (2024) 198-206

https://doi.org/10.21741/9781644903117-21

Mechanical and Thermal Properties of Goulmima's Earth Bricks

M. LECHHEB[1]*, M. CHRACHMY[1], M. BEN BAAZIZ[1], G. EL BOUKILI[2],
H. OUALLAL[1], M. AZDOUZ[1], A. LAHMAR[3], A. BENAZZOUK[4], and M. AZROUR[1]

[1]Materials Sciences for Energy and Sustainable Development Team, Department of Chemistry, FST Errachidia, Moulay Ismail University, Errachidia, Morocco

[2] Team of Modeling and Simulation of Mechanical and Energetic, Physical Department, Faculty of Sciences, Mohammed V University, Rabat, Morocco

[3]Laboratory of Condensed Matter Physics, University of Picardie Jules Verne, Amiens, France

[4] Laboratoire des Technologies Innovantes (UR-UPJV, 3899) / University of Picardie Jules Verne – Amiens – France

*m.lechheb@edu.umi.ac.ma

Keywords: Fired Clay Bricks, Date Palm Seed, Compressive Strength, Shrinkage

Abstract. Earthen constructions are prevalent in the Drâa-Tafilalet region of Morocco, yet there is a notable absence of comprehensive studies on the mud bricks used in these structures. This research endeavors to characterize a quarry utilized by Goulmima residents for mud brick production. Various techniques, including physico-chemical analysis (XRD, IR, FX, etc.), microscopic examination (SEM), and geotechnical assessments (Atterberg limits, grain size, etc.), were employed to analyze samples from the site. The bricks, formed at different firing temperatures, were blended with naturally crushed date palm seeds. The chemical composition, firing temperatures, and organic matter percentage directly influence mechanical properties and thermal conductivity. Material compressive strength exhibited a proportional increase with rising firing temperatures, peaking at 850°C. Conversely, elevating the percentage of crushed palm seed had an adverse impact on mechanical strength.

Introduction

The current global growth model faces challenges due to the concerning impacts of global warming. Historically, this model has prioritized profit and gain, often neglecting environmental consequences. In Morocco, building construction stands as a major contributor to energy consumption, representing approximately 33% of the country's total energy expenditures (26% for residential structures and 7% for tertiary buildings) [1].

Improving traditional approaches and utilizing naturally available resources can serve as a viable alternative for achieving sustainable development. Some advantages of constructing with earth include its abundance, nearly unlimited recyclability, cost-effective building methods, hygroscopic and thermal properties, and positive impacts on occupant health. [2–4].

Producing environmentally friendly building materials can involve utilizing waste materials instead of clay in the manufacturing process. [5,6].

By adopting this approach, the adverse environmental impacts associated with such waste can be alleviated, and the excessive exploitation of natural clay deposits can be curbed, preventing further harm to the ecosystem. This can lead to the creation of more environmentally friendly products. [7,8].

Per the European standard TS EN771-1, 2005, it is imperative that a brick maintains a compressive strength of no less than 7 MPa. Therefore, incorporating crushed palm seed (CPS) into clay bricks should not exceed 15%. Any surpassing of this threshold resulted in the creation of bricks lacking the necessary compressive strength, thus failing to meet the standards set for conventional bricks [9].

De Silva and Perera conducted a study on the water absorption characteristics of fired brick specimens that integrated waste rice husk ash [8]. When 10% rice husk ash was added, the water absorption value was 27%, while the reference samples had a water absorption percentage of 21%. When Eliche Quesada [10] Studied how clay bricks absorb water when infused with wood ash at a temperature of 900 °C. Findings indicated that the initial absorption value rose to 22% with a 30% additive content, contrasting with reference samples that had a water absorption rate of 15%. Put differently, the addition of CPS causes the pores and capillary channels of the material to open more, which allows water to penetrate the brick matrix and increases the water absorption values.

In this work, we are interested in earth construction techniques, and more specifically in characterizing samples from the Goulmima region using different analytical methods and determining the stabilization effect of the designed bricks with crushed palm seeds as additive. The stabilization effect was monitored through the mechanical and thermal characteristics of the materials to validate their use as a construction material.

Materials and methods
Raw materials
Crushed palm seeds used during this investigation were sourced from a regional producer in the Errachidia area, whereas the clay was acquired in the southern Moroccan Atlas (Goulmima). Standardized sieves were employed following the AFNOR procedure for the crushing of the raw materials, the length of the chosen particles is less than 0.5 cm, the percentages added were 0, 5, 10,15, 20%.

Characterization of the raw material
Evaluating the quality and suitability of clay materials for load-bearing, (and/or) non-load-bearing walls is essential through compressive strength testing. [11] and using a sedimentometry technique for the 80 μm portion [12].

The XRD patterns were recorded using a Shimadzu 6100 diffractometer, equipped with a diffracted beam monochromator and Ni-filtered $CuK\alpha$ radiation ($\lambda = 1.5406$Å), operating at 40 kV and 30 mA, the 2θ angle was swept between 2° and 70° with a counting time of 1 s in steps of 0.02°.

Using a TESWELL machine with a 20 KN capacity and a 0.5 mm/min speed, the mechanical resistance tests were performed on stabilized specimens.

Results and discussion
Grain Size Analysis
Fig. 1 illustrates the dispersion of particle sizes within the earthen components of Goulmima. The results indicate that our clay consists of 71.01% sand, 20.10% silt, and 8.15% clay. The literature claims that the final produced bricks' qualities are directly impacted by the raw materials particle size distribution [13–15] .

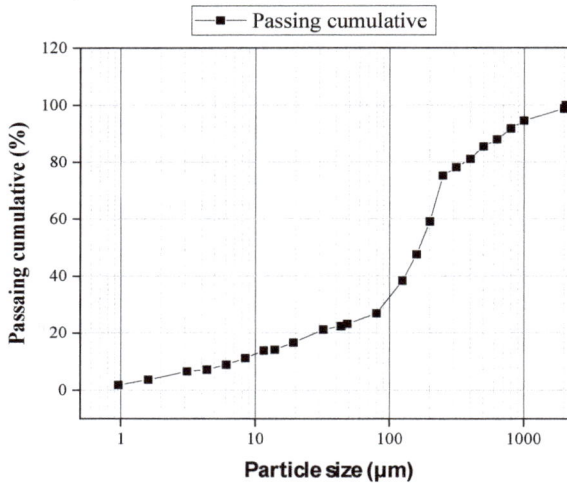

Fig. 1. Particle size analysis curve

Firing Shrinkage

Fig. 2 shows the progression of firing shrinkage with the proportion of CPS applied.

It was discovered that firing shrinkage reduces with increasing CPS concentration up to 20%, which is advantageous for building materials. Bricks containing 20% CPS shrunk by 0.5%, while the reference sample, which did not include CPS, shrank by 2.6%. Therefore, the mixture's plasticity limit is lowered when CPS is added to clay [16], Faria et al. [17] , Khoudja et al. [18] and Tjaronge et al. [19] discovered similar outcomes.

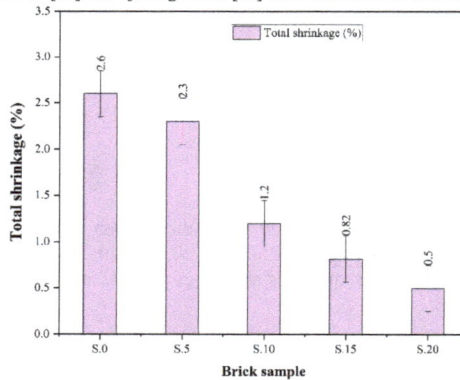

Fig. 2. Changes in overall shrinkage with the incorporation of Crushed Palm Seeds

The obtained shrinkage values for various CPS percentages ranged from 2.6% to 0.5%, demonstrating that all bricks fall within the specified clay brick limit of less than 8% [20].

Mediterranean Architectural Heritage - RIPAM10

Materials Research Forum LLC

Materials Research Proceedings 40 (2024) 198-206

https://doi.org/10.21741/9781644903117-21

Porosity, Water Absorption and Bulk Density

Fig. 3 and **4** show the changes in bulk density and porosity with the addition of CPS percentage, respectively. The quantity of CPS and bulk density were shown to be negatively correlated; that is, the brick samples became lighter as the amount of CPS increased. In actuality, the bulk density of the brick, with a 20% addition, increased from 2015 kg/m^3 in the reference clay brick to 1543 kg/m^3. On the other hand, porosity increases when CPS is added. Porosity specifically increases to 33% when 20% CPS is added, compared to the reference sample's 17.37%. The dehydroxylation reaction of carbonate in CPS and the burning of organic matter are responsible for the appreciable increase in porosity. As a result, brick samples become more porous and have less density. The findings exhibit an ideal relationship with earlier research. Eliche-Quesada et al.[10] , for example, examined the changes in bulk density and porosity of burned brick specimens when biomass ash was added. When 50% of waste additive was added, the density value decreased to 1270 kg/m^3 from 1754 kg/m^3 in the reference samples. The evolution of porosity with the addition of bottom ash was examined by Suctu et al.[21]. According to the results, the apparent porosity of the burnt brick samples increased when residual ash was added. Indeed, the porosity of the reference brick burned at 950 °C was 23.2%, and it increased to 26% with the addition of 30% of clinker. The bulk density of burned brick samples is reduced by the addition of biomass ash, as demonstrated by Casa and Castro [22] and other investigations [10,17,18,22–24].

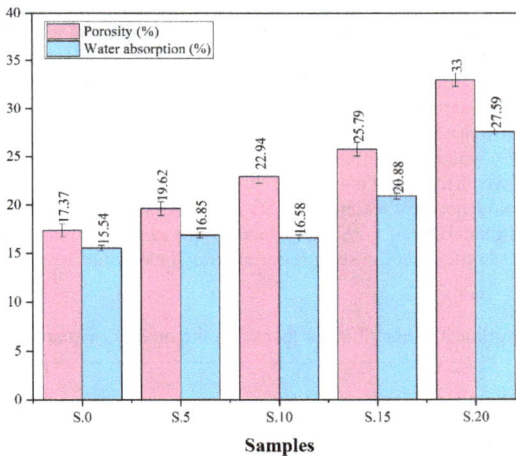

Fig. 3. Variation of open porosity, and water absorption with the addition of CPS

Materials Research Forum LLC

https://doi.org/10.21741/9781644903117-21

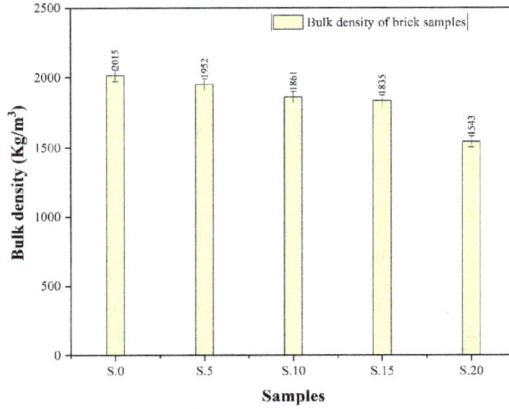

Fig. 2. Variation of bulk density with the addition of CPS

Fig. 3 shows the water absorption values of the brick samples. The addition of CPS appears to improve water absorption, as can be demonstrated. The brick without additives had the lowest value at 15.54%, while the brick with 20% CPS had the highest value at 27.59%. Since material porosity and water absorption are closely correlated, this conclusion is consistent with the bulk density figures. These conclusions are validated by the literature study. For example, the water absorption of burned brick specimens with waste rice husk ash addition was studied by De Silva et Perera [8]. When 10% rice husk ash was added, the water absorption value was 27%, while the reference samples had a water absorption percentage of 21%. The water absorption of bricks that are resistant to severe weather should not be more than 17%, and that of bricks that are resistant to moderate weather should not be greater than 22%, according to the ASTMC67-07a standards: 2003 [25]. When less than 10% of CPS was added to our samples, the water absorption results demonstrated that the samples met the strict requirements for weather-resistant bricks.

Compressive Strength
The compressive strength test is essential for assessing the quality of shaped bricks[10,26].

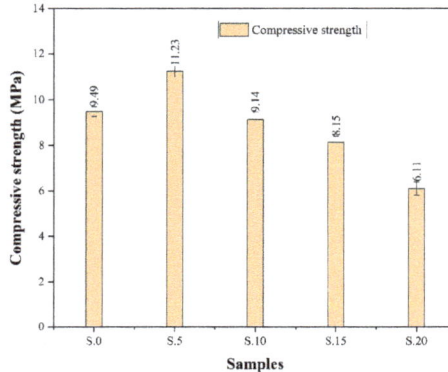

Fig. 3. Variation of compressive strength with the addition of CPS

Mediterranean Architectural Heritage - RIPAM10   Materials Research Forum LLC
Materials Research Proceedings 40 (2024) 198-206   https://doi.org/10.21741/9781644903117-21

The compaction resistance test results are displayed in **Fig. 5**. In this study, only 5% of CPS was added to bricks to improve their compressive strength by around 25The addition of CPS up to 5% results in an increase in compressive strength to a maximum of 11.23 MPa for S.5 sample. As a result, the compressive strength decreases below this rate. A significant amount of silica in the CPS used in this study may have contributed to the geopolymerization process, which forms a compact and strong geopolymer by hardening and forming a viscous cementitious slurry from aluminate and silicate [26–28]. However, it is best to avoid adding a lot of CPS as this can reduce the bricks' compressive strength. This might be the result of CPS breaking down, as it contains some organic material that burns when it is fired [18]. In accordance to the European Standard [29], The compressive strength outcomes for all bricks (excluding S.20) meet the necessary minimum threshold of 7 MPa, consistent with the conclusions drawn from previous literature [26–30].

X-Ray Diffraction
The bulk raw sample's XRD data are shown in Figure 6. It is evident that the primary constituents of Goulmima's sample include quartz, calcite, dolomite, kaolinite, and trace amounts of gibbsite. Because of its limited plasticity, low water absorption, and low shrinkage characteristics, kaolinite is regarded as a non-expandable clay. [31,32] . This is because their tetrahedral and octahedral sheets have strong interlayer bonds (Van Der Waals forces for kaolinite), which stop clay from expanding and absorbing water. This implies that the mechanical properties of this soil are generally good. [32,33].

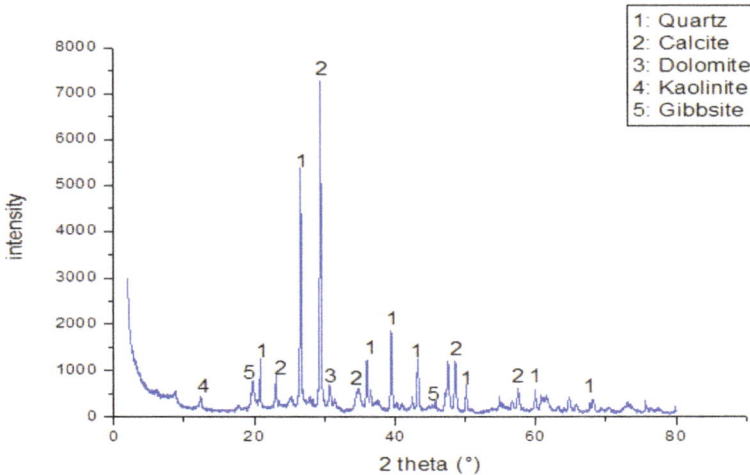

Conclusion
This investigation establishes the viability of utilizing crushed date seeds as a clay additive to enhance the quality of clay bricks. The deliberate selection of date seeds is rooted in the prolific production of dates in the Errachidia region, resulting in a notable surplus of date stones as a byproduct. The study introduces a practical approach to systematically integrate date stone powder into the terracotta brick manufacturing process, effectively closing the recycling loop for this waste material. This integration not only addresses the waste predicament in the clay brick industry but also elevates the environmental sustainability profile of the resultant bricks.

References

[1] M. Boumhaout, L. Boukhattem, H. Hamdi, B. Benhamou, F. Ait Nouh, Thermomechanical characterization of a bio-composite building material: Mortar reinforced with date palm fibers mesh, Constr Build Mater. 135 (2017) 241–250. https://doi.org/10.1016/J.CONBUILDMAT.2016.12.217

[2] S. Muguda, G. Lucas, P.N. Hughes, C.E. Augarde, C. Perlot, A.W. Bruno, D. Gallipoli, Durability and hygroscopic behaviour of biopolymer stabilised earthen construction materials, (2020). https://doi.org/10.1016/j.conbuildmat.2020.119725

[3] The properties of earth as a building material, in: Building with Earth, Birkhäuser Basel, 2006: pp. 19–35. https://doi.org/10.1007/3-7643-7873-5_2

[4] T. Morton, Earth masonry: Design and construction guidelines, IHS BRE Press, 2008.

[5] N. Jannat, A. Hussien, B. Abdullah, A. Cotgrave, Application of agro and non-agro waste materials for unfired earth blocks construction: A review, Constr Build Mater. 254 (2020) 119346. https://doi.org/10.1016/J.CONBUILDMAT.2020.119346

[6] L. Pérez-Lombard, J. Ortiz, C. Pout, A review on buildings energy consumption information, Energy Build. 40 (2008) 394–398. https://doi.org/10.1016/J.ENBUILD.2007.03.007

[7] P. Muñoz Velasco, M.P. Morales Ortíz, M.A. Mendívil Giró, L. Muñoz Velasco, Fired clay bricks manufactured by adding wastes as sustainable construction material – A review, Constr Build Mater. 63 (2014) 97–107. https://doi.org/10.1016/J.CONBUILDMAT.2014.03.045

[8] G.H.M.J.S. De Silva, B.V.A. Perera, Effect of waste rice husk ash (RHA) on structural, thermal and acoustic properties of fired clay bricks, Journal of Building Engineering. 18 (2018) 252–259. https://doi.org/10.1016/J.JOBE.2018.03.019

[9] Standard - Specification for masonry units - Part 1: Clay masonry units SS-EN 771-1:2011 - Swedish Institute for Standards, SIS, (2011).

[10] D. Eliche-Quesada, M.A. Felipe-Sesé, J.A. López-Pérez, A. Infantes-Molina, Characterization and evaluation of rice husk ash and wood ash in sustainable clay matrix bricks, Ceram Int. 43 (2017) 463–475. https://doi.org/10.1016/J.CERAMINT.2016.09.181

[11] Norme NF P94-056: Sols : reconnaissance et essais - Analyse granulométrique - Méthode par tamisage à sec après lavage., (1996).

[12] Norme NF P94-057: Sols : reconnaissance et essais - Analyse granulométrique des sols - Méthode par sédimentation, (1992).

[13] G. El Boukili, M. Lechheb, M. Ouakarrouch, A. Dekayir, F. Kifani-Sahban, A. Khaldoun, Mineralogical, physico-chemical and technological characterization of clay from Bensmim (Morocco): Suitability for building application, Constr Build Mater. 280 (2021) 122300. https://doi.org/10.1016/J.CONBUILDMAT.2021.122300

[14] M. Lechheb, A. Harrou, G. El Boukili, M. Azrour, A. Lahmar, M. El Ouahabi, E.K. Gharibi, Physico-chemical, mineralogical, and technological characterization of stabilized clay bricks for restoration of Kasbah Ait Benhadou- Ouarzazate (south-east of Morocco), Mater Today Proc. 58 (2022). https://doi.org/10.1016/j.matpr.2022.01.459

[15] M. Hajjaji, S. Kacim, M. Boulmane, Mineralogy and firing characteristics of a clay from the valley of Ourika (Morocco), Appl Clay Sci. 21 (2002) 203–212. https://doi.org/10.1016/S0169-1317(01)00101-6

Mediterranean Architectural Heritage - RIPAM10

Materials Research Forum LLC

Materials Research Proceedings 40 (2024) 198-206

https://doi.org/10.21741/9781644903117-21

[16] H. Zouaoui, J. Bouaziz, Performance enhancement of the ceramic products by adding the sand, chamotte and waste brick to a porous clay from Bir Mcherga (Tunisia), Appl Clay Sci. 143 (2017) 430–436. https://doi.org/10.1016/J.CLAY.2017.04.015

[17] K.C.P. Faria, R.F. Gurgel, J.N.F. Holanda, Recycling of sugarcane bagasse ash waste in the production of clay bricks, (2012). https://doi.org/10.1016/j.jenvman.2012.01.032

[18] D. Khoudja, B. Taallah, O. Izemmouren, S. Aggoun, O. Herihiri, A. Guettala, Mechanical and thermophysical properties of raw earth bricks incorporating date palm waste h i g h l i g h t s, (2020). https://doi.org/10.1016/j.conbuildmat.2020.121824

[19] M.W. Tjaronge, M.A. Caronge, Physico-mechanical and thermal performances of eco-friendly fired clay bricks incorporating palm oil fuel ash, Materialia (Oxf). 17 (2021) 101130. https://doi.org/10.1016/j.mtla.2021.101130

[20] A. Srisuwan, N. Phonphuak, Physical property and compressive strength of fired clay bricks incorporated with paper waste, Journal of Metals, Materials and Minerals. 30 (2020) 103–108. https://doi.org/10.55713/JMMM.V30I1.598

[21] M. Sutcu, E. Erdogmus, O. Gencel, A. Gholampour, E. Atan, T. Ozbakkaloglu, Recycling of bottom ash and fly ash wastes in eco-friendly clay brick production, (2019). https://doi.org/10.1016/j.jclepro.2019.06.017

[22] J.A. De La Casa, E. Castro, Recycling of washed olive pomace ash for fired clay brick manufacturing, Constr Build Mater. 61 (2014) 320–326. https://doi.org/10.1016/J.CONBUILDMAT.2014.03.026

[23] J. Munir, M. Lachheb, N. Youssef, Z. Younsi, A Comprehensive Review of the Improvement of the Thermal and Mechanical Properties of Unfired Clay Bricks by Incorporating Waste Materials, Buildings 2023, Vol. 13, Page 2314. 13 (2023) 2314. https://doi.org/10.3390/BUILDINGS13092314

[24] O.J. Oyedepo, L.M. Olanitori, S.P. Akande, Performance of coconut shell ash and palm kernel shell ash as partial replacement for cement in concrete, Journal of Building Materials and Structures. 2 (2015) 18–24. https://doi.org/10.34118/jbms.v2i1.16

[25] M.I. Carretero, M. Dondi, B. Fabbri, M. Raimondo, The influence of shaping and firing technology on ceramic properties of calcareous and non-calcareous illitic-chloritic clays, (n.d.).

[26] G. El Boukili, M. Ouakarrouch, M. Lechheb, F. Kifani-Sahban, A. Khaldoune, Recycling of Olive Pomace Bottom Ash (by-Product of the Clay Brick Industry) for Manufacturing Sustainable Fired Clay Bricks, Silicon. 14 (2022) 4849–4863. https://doi.org/10.1007/s12633-021-01279-x

[27] L. Pérez-Villarejo, D. Eliche-Quesada, J. Martín-Pascual, M. Martín-Morales, M. Zamorano, Comparative study of the use of different biomass from olive grove in the manufacture of sustainable ceramic lightweight bricks, Constr Build Mater. 231 (2020) 117103. https://doi.org/10.1016/J.CONBUILDMAT.2019.117103

[28] A. Bhatt, S. Priyadarshini, A. Acharath Mohanakrishnan, A. Abri, M. Sattler, S. Techapaphawit, Physical, chemical, and geotechnical properties of coal fly ash: A global review, Case Studies in Construction Materials. 11 (2019) e00263. https://doi.org/10.1016/J.CSCM.2019.E00263

[29] EN 771-1:2011+A1:2015 - Specification for masonry units - Part 1: Clay masonry units, (n.d.).

[30] F. Pr, Experimental Investigation of the Effect of Fired Clay Brick on Partial Replacement of Rice Husk Ash (RHA) with Brick Clay, 2 (2017)

[31] Doat. P, Hays. A, Houben. H, Matuk. S, Vitoux. F, Construire En Terre, Ed Alte et Par, 1979.

[32] J. Morel, C. Kouakou, Performances mécaniques de l'adobe, 3èmes Échanges Transdisciplinaires Sur Les Constructions En Terre Crue. Table - Ronde de Toulouse. (2009) 17–26.

[33] M. Olivier, A.M.-Bull.L.Lab.P. et Chaussées, Le matériau terre: Essai de compactage statique pour la fabrication de briques de terre compressées, Bull. Liaison Lab. Ponts et Chaussées. (1986).

Mediterranean Architectural Heritage - RIPAM10
Materials Research Proceedings 40 (2024) 207-217

Materials Research Forum LLC
https://doi.org/10.21741/9781644903117-22

Assessment of the Mechanical and Thermal Properties of Bricks based on Clay and Stabilized with Reed Fibers from the Drâa-Tafilalet Region

Azzeddine El GHOMARI[1,a,*], Amine TILIOUA[2,b,*], Mohammed TOUZANI[1]

[1]Research Team in Mechanics, Energy, Automatic Systems and Sustainable Development (MESADD), Mechanics, Energy Efficiency and Renewable Energies Laboratory (L.M.3.E.R.), Department of Physics, Faculty of Sciences and Techniques Errachidia, Moulay Ismaïl University of Meknès, B.P. 509, Boutalamine, Errachidia, Morocco

[2]Laboratory of Research Team in Thermal and Applied Thermodynamics (2.T.A.), Mechanics Energy Efficiency and Renewable Energies (L.M.3.E.R.), Department of Physics, Faculty of Sciences and Techniques Errachidia, BP 509, Boutalamine, University Moulay Ismail of Meknès, Errachidia 52000, Morocco.

[a] azzeddine.elghomari@gmail.com, [b] a.tilioua@umi.ac.ma

Keywords: Reed Fiber, Compressive Strength, Tensile Splitting Strength, Thermal Conductivity, Clay-Based Bricks

Abstract. In the field of sustainable construction, it is imperative to find environmentally-friendly materials with optimized performance. The present study examines a comprehensive exploration of the untapped potential for using natural resources in the Drâa-Tafilale region, focusing on the incorporation of reed fibers into traditional clay bricks, a naturally abundant and renewable resource. The design of reed fiber-stabilized bricks was carried out using six distinct mixes, mainly differentiated by their reed fiber content, ranging from 0% to 5%. Initial observations highlighted the promising potential of reed fibers. Increasing the reed fiber content in a mix improves the mechanical strength of clay brick compositions (at 1% reed fiber). However, above this threshold, the inclusion of reed fibers led to a decrease in tensile and compressive strength. In addition, the thermal conductivity of the samples decreased with increasing fiber content. Furthermore, brick density decreased with increasing fiber percentage.

Introduction

In light of growing energy demands and environmental issues, the global focus on energy-efficient buildings has intensified. In Morocco, [1] a significant dependency on energy imports has driven the creation of a national energy strategy, which places special emphasis on the expansion of renewable energy sources and the enhancement of energy efficiency in all sectors, particularly in the construction industry. Despite being a major energy consumer, the construction sector presents a noteworthy potential for energy conservation through the efficient oversight of technical equipment and the optimization of building structures.

The use of natural materials in construction has garnered significant attention, driven by a growing commitment to sustainability and energy efficiency. Among these materials, clay stands out as a readily available and cost-effective resource, prompting extensive research, particularly into its remarkable thermal conductivity and acoustic insulating properties[2] [3] [4]. Its wide accessibility makes it an environmentally responsible alternative to more resource-intensive construction materials. render it a highly appealing option for construction, especially in regions characterized by scorching climates like the Draa-Tafilalet area. However, it is undeniable that, when compared to industrialized construction materials, exhibits several limitations in construction. These materials have significantly lower tensile and flexural strengths compared to

Mediterranean Architectural Heritage - RIPAM10   Materials Research Forum LLC
Materials Research Proceedings 40 (2024) 207-217   https://doi.org/10.21741/9781644903117-22

industrialized counterparts, rendering them weak against stretching and bending forces. Furthermore, their vulnerability to erosion when exposed to moisture and rain poses a substantial concern, potentially compromising structural integrity. Additionally, the volumetric instability of earthen materials, with the potential for expansion or shrinkage in response to moisture fluctuations, can lead to ongoing maintenance and structural issues. These limitations underscore the need for meticulous planning and construction techniques to ensure the long-term stability and durability of structures employing earthen materials.

In recent years, there has been a notable focus from researchers on investigating and conducting laboratory assessments of the physical and mechanical attributes of adobe materials [5] [6] [7] [8], showcasing their capacity to enhance various crucial properties, including toughness, tensile strength, permeability, and resistance to drying shrinkage cracks. Research involving the utilization of natural fibers has demonstrated encouraging outcomes.S. Ramakrishnan et All [9] found that fibers like banana, jute, and millet improved compressive and tensile strength, alongside reduced thermal conductivity, enhancing resistance to erosion. P. Muñoz et All [10] demonstrated that a 10% inclusion of paper and pulp industry residues (PPR) significantly bolstered compressive strength and lowered thermal conductivity while meeting erosion resistance standards. Millogo, Y et All [11] noted the effectiveness of Hibiscus cannabinus fibers in reinforcing pressed adobe blocks, with a 0.4% fiber content improving compressive strength. Eslami, A. et All [12] explored palm fibers to enhance traditional adobe bricks, with varying optimal contents for different properties. In the research conducted by Hachem, H et All [13], they explored the use of fly bottom ash (FBA) waste and Arundo donax leaves (ADL) as alternatives to cement in mortar composites, with the goal of creating more sustainable building materials. They found that FBA reduced density and thermal conductivity while increasing porosity and water absorption. A 20% FBA replacement achieved a good balance between emissions reduction and strength. The incorporation of ADL fibers further reduced density and thermal conductivity but weakened mechanical properties. However, a low ADL ratio (0.4%) in 20% FBA mortar improved strength and reduced emissions slightly.

One fundamental query permeating this research field concerns the influence of natural fiber inclusion on adobe's mechanical characteristics. Within the literature, an intriguing duality of findings has emerged, giving rise to divergent conclusions. On one hand, certain studies suggest that augmenting fiber content corresponds to an increase in compressive strength [9] [12] [14] [15], signifying the potential for reinforced adobe structures. Conversely, a separate body of research contends that greater fiber content leads to a reduction in compressive strength [16] [17], which poses a challenge to conventional wisdom. Given the diverse and sometimes contradictory findings within the literature, it is clear that additional research is imperative to establish a comprehensive understanding of the ramifications of natural fiber integration on adobe characteristics. A systematic approach to research and data collection is crucial to reconcile the disparities observed thus far, ultimately advancing the field of adobe construction and fostering more informed design and construction practices.

In pursuit of advancement in this direction, this research paper focuses its attention on exploring the impact of reed fiber and quantity on the physical and mechanical characteristics of adobe materials. Specifically, it conducts a series of experiments to assess how the inclusion of reed fibers, at varying percentages, affects the properties of five distinct adobe mixtures that were formulated in the laboratory. The outcomes are then compared to those of conventional adobe materials for reference. The experimental phase of this study includes mechanical evaluations, such as tests for compressive and flexural strength, and the determination of crucial physical attributes like density and absorption. Furthermore, thermal conductivity measurements are carried out to complete the assessment of the adobe samples under scrutiny. The analysis and comparative

Mediterranean Architectural Heritage - RIPAM10
Materials Research Proceedings 40 (2024) 207-217

Materials Research Forum LLC
https://doi.org/10.21741/9781644903117-22

examination of the experimental findings provide valuable insights into the influence of fiber type and quantity on the characteristics of traditional adobe bricks.

Materials and procedures

Materials used

In this study, the focus was on investigating the feasibility of adobe brick production in the Drâa-Tafilalet region, specifically Errachidia Province in Morocco. The study utilized a variety of materials, including soil, water, and reed fiber. The soil collection process was meticulous, with soil taken from the top 10-40 cm of soil to avoid organic-rich soils. The absence of a musty odor in the soil confirmed its suitability for clay brick production due to its non-organic nature. A sedimentation test confirmed a higher percentage of clay and silt, making it ideal for high-quality clay brick production. Subsequently, soil samples were carefully transported to the laboratory, where X-ray diffractometry (XRD) tests were executed to ascertain their chemical and mineralogical compositions. The outcomes of this examination are graphically represented in *(Fig 2 and Table 1),* respectively. To ensure compliance with quality and consistency standards, clean and dependable water from the public water supply system was employed for mixing the clay bricks.

The study's foundation lay in the acquisition of reed fiber from giant reed plants, a process that involved harvesting and a comprehensive drying period. These dried fibers were then stored and subjected to refinement, ultimately achieving the precise length required for the study's unique application. The study focused on using reed fiber derived from the tall giant reed plants, scientifically known as Arundo donax, which are found in various regions globally, including Europe, Asia, Africa, and especially around the Mediterranean (Fig. 1)[18]. The process of preparing the reed fiber involved harvesting these giant reed plants and allowing them to naturally dry for about a year without air conditioning. This extended drying period was crucial to achieve the desired dryness level for effective use. After drying, the reed fibers were stored in approximately 40 cm segments and then carefully refined to reduce their length to a range of 5 to 30 mm. This tailored length range was essential to make the fibers suitable for the specific application intended in the study.

Fig 1. The potential natural habitat distribution of Giant reed (A. donax) [19]

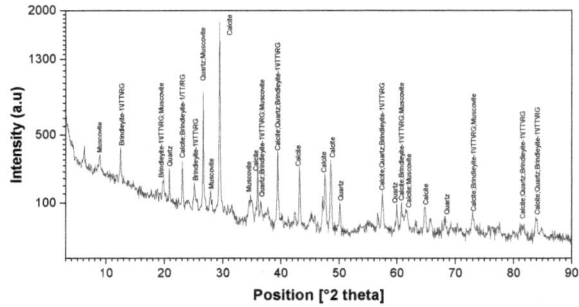

Fig 2. X-ray diffraction pattern

Table 1: Identified Patterns List

Compound Name	Chemical Formula	Scale Factor	Score
Calcite	Ca (CO3)	0,7	79
Quartz	Si O2	0,475	65
Nimesite	(Ni2 Al) (Al Si) O5 (O H)4	0,093	44
Muscovite	H2 K Al3 Si3 O12	0,04	38
Zeolite X	Na17.52 Al24 Si24 O96 H6.48	0,07	20

Samples preparation

Seven distinct types of mixtures were created for thorough thermal and mechanical testing, with each having unique combinations of reed, soil, and water. The aim was to investigate how varying component percentages influenced specific properties. Among these mixtures, one was identified as the reference (C0) with no reed fibers, serving as a consistent benchmark for performance comparisons with the others. *(Table 2 and fig 3)* outline the composition of these mixtures, which include the reference (C0:0%, C1:1%, C2:2%, C3:3%, C4:4%, C5:5%, and C6:5% fiber + 5% cement).

Table 2: Component percentage by mixture

	C0	C1	C2	C3	C4	C5	C6
SOIL(g)	4000	4000	4000	4000	4000	4000	4000
Reed fibre(g)	0	40	80	120	160	200	200
Cement(g)	0	0	0	0	0	0	200

In the experimental setup of this study, custom-designed molds were used to meet the specific requirements of the testing procedures. For thermal assessments, the molds had dimensions of 25 x 25 x 3 cm³ *(fig 3a)*, while for mechanical tests, cylindrical molds with a diameter of 10 cm and a height of 20 cm were employed*(fig 3b)*. Once the molds were filled with the designated components, the specimens underwent a natural drying process in an open-air environment for a period of 28 days.

Fig 3. Samples preparation, a for thermal tests, b for mechanical tests

Experimental procedures
A series of laboratory tests were conducted to comprehensively evaluate both stabilized and non-stabilized clay bricks. These tests were designed to assess various fundamental properties such as compressive strength, indirect (split) tensile strength, and thermal properties.

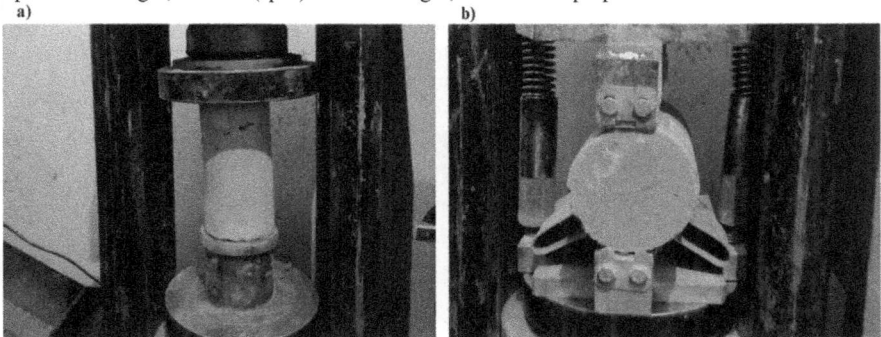

Fig 4. Experimental setup for a. compressive strength, b. tensile splitting

Uniaxial compression tests

In this test, a compression machine was used to assess the compressive strength of clay bricks samples *(Fig.4a)*. The bricks were placed on the machine, and compressive loads were gradually increased until the bricks failed. This process determined the maximum load each brick could withstand. The compressive strength (σ_c) of the adobe bricks was calculated using (Eq1):

$$\sigma_c = F/S.$$ (1)

where F is the maximum load before failure and S is the cross-sectional area of the specimen.

Tensile splitting test

The tensile splitting test is a commonly employed method for assessing the tensile strength of clay-based bricks *(Fig 4b)*. It involves preparing cylindrical specimens with a standard diameter of 10 mm and a length of 20 mm, akin to those used in unconfined compression testing. The testing apparatus used for Uniaxial compression tests (UC) is also utilized for this test, with adjustments made to ensure the samples are loaded horizontally to their axes. A crucial aspect of the test is guaranteeing that the split line, where the tensile forces are applied, aligns along the specimen's diameter. The tensile splitting strength (σ_t) is calculated using the (Eq2)

$$\sigma_t = 2F/\pi Dh.$$ (2)

F represents the maximum load before failure, while D and h stand for the specified specimen diameter and length, respectively.

Thermal test

The thermal conductivity of reed fiber samples was investigated in a specialized insulated house *(fig 5)*, measuring 400 x 400 x 400 mm with a removable lid insulated with 5 cm thick polystyrene. Square openings (210 x 210 mm) in the house's side walls were sealed with the samples, which were held in place by two tensioning screws. The house had foam-insulated orifices in each corner for inserting thermocouples. A 100 W light bulb connected to a thermal regulator was used to maintain the house's internal temperature at around 50°C. Thermocouples were placed inside and outside the sample walls to record temperatures (Tpi internal and Tpe external). The ambient temperature in the laboratory was kept constant at 23°C using an air conditioner.

Mediterranean Architectural Heritage - RIPAM10
Materials Research Proceedings 40 (2024) 207-217

Materials Research Forum LLC
https://doi.org/10.21741/9781644903117-22

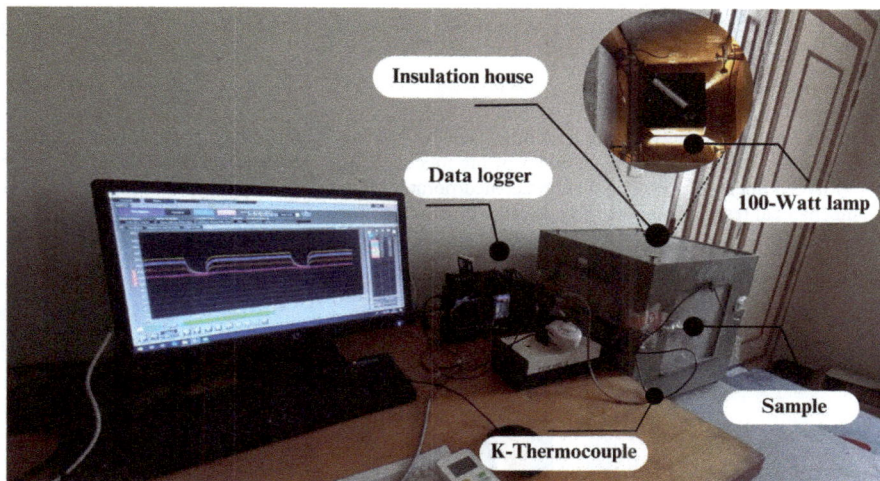

Fig 5. Heat Insulation House

In the realm of thermal analysis and the study of heat transfer, numerous fundamental principles and equations are employed to comprehensively understand and quantitatively describe heat transfer processes. Convection, a vital mechanism, characterizes the exchange of thermal energy between a fluid, such as indoor air, and the surface of a solid, such as an interior wall of samples. This exchange can be effectively calculated using Newton's Law of Cooling or the convective heat transfer equation (Eq 3).

$$\varphi_{cond} = h_i . S(T_{i,air} - T_{i,wall}).$$ (3)

- The convective heat transfer coefficient, $h_i = 8.1$ W/(m²·K) in our study, characterizes heat transfer between interior air and a wall's interior surface.
- S: representing the surface area of the sample where heat flux passes through, in (m²).
- $T_{i,wall}$: representing the interior wall surface temperature, in (K).
- $T_{i,air}$: representing indoor air temperature, in (K).

Therefore, the heat flux through the wall is described by (Eq 4), which outlines the propagation of heat within the sample.

$$\varphi_{conv} = \lambda . S \frac{(T_{i,wall} - T_{e,wall})}{e}.$$ (4)

- λ : which stands for the thermal conductivity of the studied sample, in (W/(m·K)).
- $T_{e,wall}$: wall, representing the exterior wall surface temperature, in (K).
- e, representing the sample thickness,in (m).

To calculate thermal conductivity λ (in W/m.°C), one-dimensional heat transfer has been assumed, Thermal conductivity is calculated using the conservation of heat flow in steady-state thermal conditions:

Mediterranean Architectural Heritage - RIPAM10
Materials Research Proceedings 40 (2024) 207-217

Materials Research Forum LLC
https://doi.org/10.21741/9781644903117-22

$$\lambda = \frac{h_i.e(T_{i,air}-T_{i,wall})}{T_{i,wall}-T_{e,wall}}. \tag{5}$$

Results and discussions

Mechanical Test

In the context of the presented findings, several notable conclusions can be drawn. The results provide a comprehensive assessment of the mechanical properties of clay-based bricks fortified with reed fibers. It is evident that bricks incorporating fibers exhibit superior consistency following mechanical testing compared to pure clay bricks (*Fig 7*).

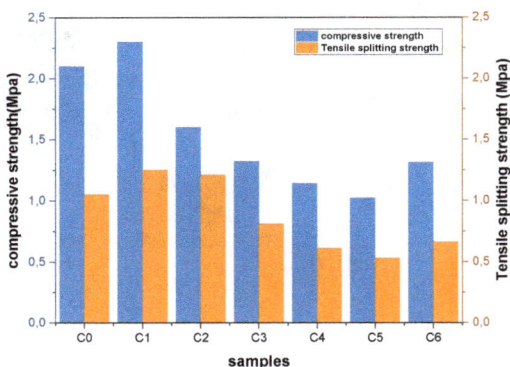

Fig 6. Mechanical strength of samples

Particularly, an initial 1% fiber content (referred to as C1) yields a significant increase of approximately 9.5% in compressive strength and a notable 19% improvement in tensile splitting strength when compared to the reference sample without fibers (designated as C0). This underscores the positive impact of a lower fiber content. However, as the fiber content surpasses 1%, there is a progressive reduction in both compressive and tensile splitting strengths, with the most significant declines occurring at a 5% fiber content. For the sake of clarity and improved presentation, a visual summary is included in Figure 5 for both compression and tensile splitting tests. Notably, the introduction of additional cement at a 5% concentration, combined with a 5% fiber content (referred to as C6), results in an enhanced compressive strength of 4.8% and a remarkable 25.7% increase in tensile splitting strength when compared to the 5% fiber-only sample (C5). This suggests the potential benefits of the composite approach for specific applications.

Fig 7 . Failure of clay brick under tensile force

Thermal test

The results of the assessment of the thermal properties of bricks made from clay and stabilized with varying percentages of reed fibers reveal important insights. As the content of reed fibers increases from C0 (0% fibers) to C6 (5% fibers), the density of the bricks gradually decreases. This reduction in density can be attributed to the lower density of reed fibers compared to clay. In terms of thermal conductivity, an equally intriguing trend emerges. The thermal conductivity of the bricks tends to decrease as the proportion of reed fibers increases (*Fig. 8 and Table 3*). This means that bricks with higher fiber content exhibit improved insulating properties, which is advantageous in applications where thermal insulation is a crucial factor. These findings suggest that the addition of reed fibers not only reduces the density of the bricks but also enhances their thermal insulating capabilities(*Fig 8.b*)., making them potentially suitable for construction projects where thermal efficiency and reduced heat conduction are desirable.

Table 3: Thermophysical properties of samples.

samples	Density	conductivity
C0	2253	0.811
C1	2114	0.753
C2	1986	0.620
C3	1871	0.504
C4	1789	0.472
C5	1687	0.361
C6	1700	0.382

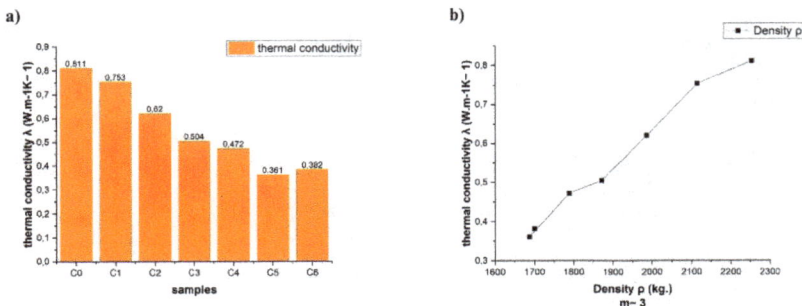

Fig 8 a Thermal conductivity of samples, b Figure 2 shows the variation of thermal conductivity as a function of density.

Conclusion

In conclusion, this study shows an inverse relationship between fiber percentage and thermal conductivity, demonstrating that increasing fiber content leads to a decrease in conductivity. The incorporation of reed fibers in clay mixes led to a reduction in indirect tensile and compressive strength, particularly in mixes containing a high proportion of fibers, i.e. 1%. However, it is important to note that adobes containing fibers maintained better cohesion after mechanical testing compared to clay alone, suggesting that the fibers contributed to enhancing the strength and cohesion of the bricks. This improvement in adobe cohesion due to fibers has important implications for improving the earthquake resistance of adobe structures. In addition, the addition of cement with a 5% fiber content was associated with a significant improvement in compressive and indirect tensile strength. In sum, the incorporation of reed fibers into adobe mixes offers significant potential for enhancing the mechanical properties of these building materials.

References

[1] Moroccan Agency for the Energy Efficiency (AMEE), Regalement Thermique de Construction au Maroc, 2014. www.amee.ma

[2] F. Al-Ajmi, H Abdalla., M. Abdelghaffar, & J. Almatawah (2016). Strength Behavior of Mud Brick in Building Construction. Open Journal of Civil Engineering, 2016, 6, 482-494 http://dx.doi.org/10.4236/ojce.2016.63041

[3] H. Binici, O. Aksogan, D Bakbak., H. Kaplan, &B. Isik (2009). Sound insulation of fibre reinforced mud brick walls. Construction and Building Materials, 23(2), 1035-1041. https://doi.org/10.1016/j.conbuildmat.2008.05.008

[4] H Binici, O. Aksogan, M. N Bodur, E. Akca, & S. Kapur (2007). Thermal isolation and mechanical properties of fibre reinforced mud bricks as wall materials. Construction and Building Materials, 21(4), 901-906. https://doi.org/10.1016/j.conbuildmat.2005.11.004

[5] F Jové-Sandoval., M. M. Barbero-Barrera, & N. Flores Medina (2018). Assessment of the mechanical performance of three varieties of pine needles as natural reinforcement of adobe. Construction and Building Materials, 187, 205-213. https://doi.org/10.1016/j.conbuildmat.2018.07.187

[6] S. Hejazi M., Sheikhzadeh, M., Abtahi, S. M., & Zadhoush, A. (2012). A simple review of soil reinforcement by using natural and synthetic fibers. Construction and Building Materials, 30, 100-116. https://doi.org/10.1016/j.conbuildmat.2011.11.045

[7]H. Danso, , D. B Martinson., , M Ali., &, J. B Williams. (2015). Physical, mechanical and durability properties of soil building blocks reinforced with natural fibres. Construction and Building Materials, 101, 797-809. https://doi.org/10.1016/j.conbuildmat.2015.10.069

[8]A Laborel-Préneron., , J. Aubert, , C Magniont., , C. Tribout &, A Bertron. (2016). Plant aggregates and fibers in earth construction materials: A review. Construction and Building Materials, 111, 719-734. https://doi.org/10.1016/j.conbuildmat.2016.02.119

[9] S. Ramakrishnan, , S. Loganayagan, , G Kowshika., , C Ramprakash., &, M Aruneshwaran. (2020). Adobe blocks reinforced with natural fibres: A review. Materials Today: Proceedings, 45, 6493-6499. https://doi.org/10.1016/j.matpr.2020.11.377

[10] P. Muñoz, , V Letelier., , L. Muñoz, &, M Bustamante. (2020). Adobe bricks reinforced with paper & pulp wastes improving thermal and mechanical properties. Construction and Building Materials, 254, 119314.https://doi.org/10.1016/j.conbuildmat.2020.119314

[11] Y. Millogo, , J. Morel, , J. Aubert, &, K Ghavami. (2014). Experimental analysis of Pressed Adobe Blocks reinforced with Hibiscus cannabinus fibers. Construction and Building Materials, 52, 71-78. https://doi.org/10.1016/j.conbuildmat.2013.10.094

[12] A. Eslami, , H. Mohammadi, &, H. Mirabi Banadaki (2022). Palm fiber as a natural reinforcement for improving the properties of traditional adobe bricks. Construction and Building Materials, 325, 126808. https://doi.org/10.1016/j.conbuildmat.2022.126808

[13] H. Hachem, , I. Mehrez, , I. Boumnijel, , A. Jemni, &, D. Mihoubi (2022). Sustainable Approach of Using Arundo donax Leaves Reinforced Cement Mortar/Fly Bottom Ash Composites. ACS Omega 2023 8 (13), 12039-12051. DOI:10.1021/acsomega.2c07818

[14] A. Vatani Oskouei, Afzali, M., &, M. Madadipour (2017). Experimental investigation on mud bricks reinforced with natural additives under compressive and tensile tests. Construction and Building Materials, 142, 137-147. https://doi.org/10.1016/j.conbuildmat.2017.03.065

[15], A Vatani Oskouei., , M. Afzali, &, M Madadipour. (2017). Experimental investigation on mud bricks reinforced with natural additives under compressive and tensile tests. Construction and Building Materials, 142, 137-147. https://doi.org/10.1016/j.conbuildmat.2017.03.065

[16], S. Serrano, , C. Barreneche, &, L. F Cabeza. (2016). Use of by-products as additives in adobe bricks: Mechanical properties characterisation. Construction and Building Materials, 108, 105-111. https://doi.org/10.1016/j.conbuildmat.2016.01.044

[17] Ş. Yetgin, , Ö. ÇAVDAR, &, A Çavdar. (2008). The effects of the fiber contents on the mechanic properties of the adobes. Construction and Building Materials, 22(3), 222-227. https://doi.org/10.1016/j.conbuildmat.2006.08.022

[18] S. C. Nunes, A. P Gomes., P. Nunes, M. Fernandes, A. Maia, , E. Bacelar, , J. Rocha, , R. Cruz, , A. Boatto, , A. P Ravishankar., , S Casal., , S. Anand, , V. D. Bermudez, &, A. L Crespí. (2022). Leaf surfaces and neolithization - the case of Arundo donax L. Frontiers in Plant Science, 13, 999252. https://doi.org/10.3389/fpls.2022.999252

Mediterranean Architectural Heritage - RIPAM10
Materials Research Proceedings 40 (2024) 218-225

Materials Research Forum LLC
https://doi.org/10.21741/9781644903117-23

Traditional Earth Architecture as a Tool for Sustainability and Adaptation to Climate Change of Heat and Cold Extremes

Khalid EL HARROUNI[1,a] *, Hassane KHARMICH[1,b] and Khadija KARIBI[1,c]

[1]Ecole Nationale d'Architecture de Rabat, BP 6372, Rabat Instituts, Rabat, Morocco

[a]k.elharrouni@enarabat.ac.ma, [b]h.kharmich@enarabat.ac.ma, [c]k.karibi@enarabat.ac.ma

Keywords: Bioclimatic Architecture, Traditional Materials, Earth, Bioclimatic Tools, Thermal Comfort

Abstract. The design of sustainable architectural and urban spaces should be one of the essential pillars of any strategy for sustainable development and adaptation to climate change, particularly for the population living in rural areas who suffer from cold during winter and heat during the summer. This paper focuses on the traditional earth-based materials buildings and tries to see to what extent the building envelope could be improved to achieve and further confirm the objectives: improving thermal comfort and reducing heat loss through the traditional envelope (walls, roof, glazing, low floor). The paper is based on bioclimatic architecture principles and adopts passive energy efficiency in two different climatic contexts, hot and cold. The analysis of the approach method includes three issues: 1) the bioclimatic analysis of the environment/site including the building ambiance; 2) thermal comfort; and 3) thermal performance. The methodological tools are based on the bioclimatic analysis of the site and the ambiance for the first two issues; and the prescriptive approach of Moroccan thermal regulation for the third issue. The built environment constructed with traditional materials, once improved, is able to prove that it is respectful of the environment and without any risk to the user's health. In addition, this traditional architecture confirms the objectives of sustainable development.

Introduction

Climate change is one of our century's most complex societal and environmental challenges, and Morocco is not immune to its effects, particularly cold and heat. As a result, the energy issue is at the heart of discussions between the various stakeholders, given the importance of the social, economic, and environmental stakes for a sustainable architecture capable of withstanding more extreme climatic conditions, very cold in winter, and very hot in summer.

The training of architects and urban planners has a major role to play in mastering these social, environmental, and energy transition challenges. The aim is to provide them with the key knowledge they need to take account of the challenges of the energy transition in their professions and to identify the techniques and methods for integrating these challenges into their professional practices.

Moreover, sustainability requires a global understanding of systems and interdisciplinary collaboration between architecture, urban climate design, economics, and engineering (civil, thermal, and energy engineering, etc.). The design of sustainable architectural and urban spaces for users must be one of the essential pillars of any strategy for sustainable development and the fight against climate change, particularly for people in rural areas who suffer from the cold in winter and the heat in summer.

In Morocco, people are beginning to understand that investment in environmental quality, thermal and energy performance, is an asset and a source of sustainable savings. Energy labeling, which will directly impact the value of Morocco's built environment, will certainly bring about a change in mentality and culture among the population and those involved in the act of building.

Materials Research Forum LLC
https://doi.org/10.21741/9781644903117-23

Heritage built with traditional materials, once improved, can prove that it is environmentally friendly and poses no risk to the health of users. Sustainable development considerations call for energy efficiency on the one hand and a drastic reduction in greenhouse gas emissions on the other.

The United Nations Sustainable Development Goals (SDGs) and the National Sustainable Development Strategy remind us of the goals, specific objectives, and challenges, including SDG7: Access to sustainable energy; SDG11: Make cities and human settlements inclusive, safe, resilient, and sustainable; and SDG13: Take urgent action to combat climate change and its impacts.

Today, buildings are generally connected to the electricity grid and sometimes exploit other energy sources to meet user needs. This means they have to adapt to the local climate and resources while respecting the principles of bioclimatic architecture and thermal regulations during the design phase. The latter must satisfy two essential requirements to achieve the building's energy efficiency objectives: reduced energy consumption and improved thermal comfort, by using the Moroccan building energy efficiency regulations [1].

This paper focuses on the built environment constructed using traditional earth-based materials and looks at how we can improve the building envelope to further confirm the objectives mentioned: improving thermal comfort and reducing heat loss through the traditional envelope (walls, roof, glazing, low floor). To achieve this, we apply the principles of bioclimatic architecture and adopt passive energy efficiency measures in two different climatic contexts, hot and cold.

The analysis of the approach uses three thematic inputs: 1) bioclimatic analysis of the environment/site and building; 2) thermal comfort; and 3) thermal performance. The methodological tools are based on those relating to the bioclimatic analysis of the site and environment for the first two issues (thermal comfort indices from the Thermal Comfort Standard, NF ISO 7730) [2]; psychrometric/bioclimatic charts by Givoni [3], Szokolay [4]; Mahoney Tables [5]; and the prescriptive approach of the Moroccan thermal regulations for the third issue.

Principles of Bioclimatic Architecture

The building envelope (exterior walls, openings, roof, and basement floor) is a major lever for reducing energy needs and improving energy efficiency. A well-designed building envelope is crucial for providing lasting protection to the building's occupants from various environmental factors such as wind, cold, rain, frost, heat, and so on. In winter, it minimizes heat loss to the outside, while maximizing solar energy gain through glazed surfaces. Conversely, in summer, a good building envelope helps to keep the interior cool.

In both cases, the insulating capacity of the materials making up the envelope plays a decisive role. A good building envelope is also an effective way of improving interior comfort for occupants while minimizing the impact of construction on the environment (ecological materials, integration into the landscape, reduction of noise pollution).

Architectural choices have a major impact on a building's thermal performance. This is why, right from the pre-project phase, it is important to apply a few principles that will reduce a building's heat loss in winter and prevent it from overheating in summer. The main recommendation concerns the orientation of the building and the presence of openings to optimize solar gains. There are other generally accepted rules of the art, such as the compactness of the construction, which limits the surface area of walls in contact with the outside for a given volume, and therefore heat loss, as well as the layout of interior spaces. The architecture of the building must also anticipate summer comfort by protecting openings and bay windows.

Construction techniques and materials are constantly evolving to provide ever greater comfort and energy savings. The principles of bioclimatic architecture and thermal regulations are simple and easy to implement.

Bioclimatic analysis and local materials in two relevant climatic sites in Morocco: Midelt (cold climate) and Errachidia (warm climate)

Each site has its characteristics closely linked to the environment and climate. The architectural design must take these considerations into account to benefit from the advantages and protect against the constraints and disadvantages.

Figure 1: Built heritage in rural areas in the Province of Midelt

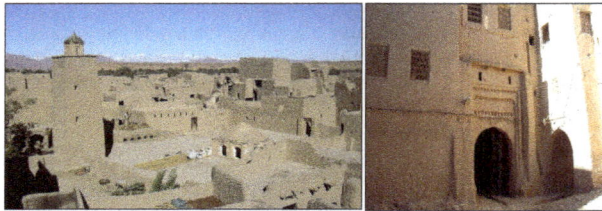

Figure 2: Earth-built heritage in the Errachidia Province

Bioclimatic analysis tools of the sites and interior ambiances:

Concept of thermal comfort:

Thermal comfort is defined as a state of satisfaction with the thermal environment. It is determined by the dynamic balance established by heat exchange between the body and its environment. In other words, it means being in good thermal conditions everywhere and at any time, in terms of air temperature (Ta in °C), relative humidity (RH in %), speed of air movement (Va in m/s), …. If we think in terms of the first two parameters, thermal comfort would be characterized by an average air temperature of 24°C and average relative humidity of around 48%, knowing that $20°C \leq Ta \leq 28°C$ and $28\% \leq RH \leq 68\%$. These intervals may undergo a slight change when the speed of air movement is added.

Comfort indices and strategies for evaluating thermal comfort:

The two indices, PMV- predicted mean vote, and PPD - PPD-predicted percentage dissatisfied, are part of the Thermal Comfort Standard (NF ISO 7730).

The following application allows these indices to be calculated for the two climatic contexts, cold in Midelt during the winter and hot in Errachidia during the summer. The simulation is done with the following parameters:

-In Errachidia, user activity is light with summer clothing; the air temperature is taken equal to 30°C and the average radiation temperature approximately 29°C; the airspeed is 1 m/s and the relative humidity is 45%.

Mediterranean Architectural Heritage - RIPAM10   Materials Research Forum LLC
Materials Research Proceedings 40 (2024) 218-225   https://doi.org/10.21741/9781644903117-23

-In Midelt, user activity is also light with winter clothing; the air temperature taken is equal to 18°C and the average radiation temperature approximately 19°C; the airspeed is 0.2 m/s and the relative humidity is 50%.

PMV takes into account various environmental and personal factors. The formula to calculate PMV is complex and typically requires the use of specialized software or calculators. Here is a simplified form of the equation:

$$PMV = M - W. \tag{1}$$

Where:

M represents the heat balance between the human body and its environment, taking into account factors like metabolic rate, clothing insulation, and air temperature.

W represents the work done by the body to maintain thermal balance.

In practice, detailed calculations for M and W are more involved and require input data on factors such as air temperature, humidity, clothing insulation, metabolic rate, and air velocity.

PPD is typically calculated based on the PMV. The formula for calculating PPD is also complex and involves statistical models. A simplified form of the equation is as follows:

$$PPD = 100 - 95 \times \exp(-0.03353 \times PMV^4 - 0.2179 \times PMV^2). \tag{2}$$

This equation provides an estimate of the percentage of occupants who may be dissatisfied with the thermal conditions in the space based on the PMV value.

The simulation results show that we are a little far from a comfortable situation if nothing is done. We have also noticed that the indices remain in the same order of estimation for the two climates: a PMV of 1 (1.02) and a PPD close to 25% (26.9%) for Errachidia and a PMV of -1 (-0.92) and a PPD also close to 25% (22.9%) for Midelt.

We then focus on the bioclimatic approach using psychrometric charts and other bioclimatic tools. The climatic study through monthly climatic data (variations of the maximum and minimum average temperature, the maximum and minimum average relative humidity, the average precipitation, and the direction of the prevailing winds) of the nearest station characterized by its geographical position (longitude, latitude, and altitude), allows to draw the Bioclimatic Chart or Psychrometric Chart. It is a global decision support tool for the bioclimatic project making it possible to establish the degree of necessity for implementing major options to maintain thermal comfort of the environment.

This bioclimatic approach makes it possible to identify optimal comfort for the different months of the year. A comfort zone indicates the limits, for temperature, air humidity, etc., within which the climate is considered comfortable. The comfort zone is, among other things, used as an instrument to know when to use an active form of heat or cold production.

Givoni and Szokolay bioclimatic charts:

The bioclimatic diagrams of Givoni and Szokolay find their usefulness as soon as the climatic conditions deviate from the polygon or comfort zone: they suggest the constructive and functional solutions that must be adopted to design a suitable building: insulation of the envelope, ventilation, thermal inertia, solar protection, use of passive systems and, if necessary, active ones (heating and cooling).

The architectural and technical arrangements, as well as the interventions planned for the environments to remedy fluctuating climatic conditions, will correspond to the zones of influence, established by the Givoni chart and taken up by Szokolay.

From a perspective of comparison, we instead present the bioclimatic diagram of Szokolay applied to the climates of Midelt and Errachidia, which gives the same recommendations as those

Mediterranean Architectural Heritage - RIPAM10   Materials Research Forum LLC
Materials Research Proceedings 40 (2024) 218-225   https://doi.org/10.21741/9781644903117-23

of Givoni, but with more precision at the level of the comfort zone framed by the main parameters of comfort: the minimum and maximum temperature and relative humidity for each climatic context. For example, the Szokolay diagram specifies the average comfort limits between 19.5°C and 25.5°C with an average of 40% relative humidity in Midelt and 21°C and 27°C with an average of 35 % relative humidity in Errachidia.

The Szokolay diagram recommendations for sites and indoor environments in Midelt are:
-Strong thermal inertia for summer;
-Passive heating through internal supplies for October, April and May;
-Active heating for the winter season and quite cold months like November and March.

The Szokolay diagram recommendations for the sites and interior environments in Errachidia are:
-Natural ventilation with dehumidification especially for September;
-High thermal inertia with night ventilation during summer;
-Passive heating through internal supplies for fairly cold months;
-Active heating during the night for the winter season.

Mahoney Tables:

The bioclimatic approach based on the Szokolay diagram is supplemented by the use of Mahoney Tables intended to summarize and analyze the climatic data of the place, to formulate and prioritize recommendations for ambiance. These tables are considered to be a simple tool to assist in bioclimatic architectural design, which makes it possible to have the first qualitative recommendations in terms of architectural arrangements, in particular simple indications on the mass plan, the shape of the building, the orientation, the inertia of the roof and the external and internal walls, the size and percentage of the openings to the surface area of the facades.

The following is a summary of the Mahoney Tables recommendations for the site of Midelt:
-Compact ground plan with north-south orientation (along the east-west longitudinal axis);
-Spacing between buildings: Compact mass plan;
-Openings: Average openings (20 to 40%);
-Walls: Massive exterior walls, phase shift time of more than 8 hours;
-Roofs: Thick, heavy roofs, phase shift time of more than 8 hours.

The following is a summary of the Mahoney Tables recommendations for the site of Errachidia:
-Compact ground plan with interior courtyard;
-Openings: Average openings (20 to 40%);
-Walls: Massive exterior walls, phase shift time of more than 8 hours;
-Roofs: Thick, heavy roofs, phase shift time of more than 8 hours;
-Space to sleep at night outdoors.

Inertia and thermal phase shift: the case of raw earth for example:

The thermal phase shift of a material or an element of the building envelope (wall or roof) plays a major role in the summer thermal comfort of the building. It represents the time between when the temperature is highest outside and when it is highest indoors. This is the time it takes for the heat to penetrate inside. This often makes it possible to avoid using costly air conditioning.

The phase shift time depends on the thickness as well as the thermal conductivity of the materials. It therefore depends on the speed of heat diffusion through the wall made up of the material or layer materials. Then the diffusivity and the phase shift time are given by the following equations:

$$a = \lambda \ / \ (\rho \ x \ C) \ (m^2/hours). \tag{3}$$

$$\text{the phase shift time} = 1.38 \ x \ e \ x \ \sqrt{a} \ (hours). \tag{4}$$

Where:
-ρ is the density of the material (kg/m^3);
-C is the specific heat capacity of the material (J/kg/K) (Basic relationship: 1 KWh = 3600 KJ);
-λ is the thermal conductivity of the material (W/m/K);
-e is the wall thickness (m).

Raw earth has an important thermal property, inertia because this material has the conditions necessary to obtain it: thermal capacity, thermal diffusivity, and thermal effusivity. For example, 40 cm thick rammed earth has a thermal phase shift of around 12 hours, and the thermal capacity is about 510 W/m^3 °C for rammed earth and 380 W/m^3 °C for adobe. Thanks to its low thermal diffusivity of 2.53 10^{-3} m^2/h (approximately 5.92 10^{-3} m^2/h for fired brick and 5 to 6 10^{-3} m^2/h for stone), the earth offers the advantage of damping and a significant phase shift in variations and external thermal inputs.

Thermal performance of construction elements made from local earth-based materials:

The thermal quality of an element of the built environment envelope can be evaluated through the U value, called thermal conductance or surface transmission coefficient of the wall, roof, or window, expressed in W/m^2 °C; its inverse expresses the overall thermal resistance. For a composite construction element separating two ambiances, the U value is calculated according to the following formula:

$$1/U = 1/h_i + \Sigma(e/\lambda) + \Sigma R + 1/h_e. \tag{5}$$

Where:
-e/λ is the thermal resistance specific to the different layers of materials constituting the composite construction element;
-R is the thermal resistance of other composite element materials;
-$1/h_i$ and $1/h_e$ are the interior and external surface thermal resistances;
For a vertical construction element (wall, glazing): $1/h_i + 1/h_e = 0.17$ m^2°C/W.
For a horizontal construction element (roof, upward flow): $1/h_i + 1/h_e = 0.14$ m^2°C/W.
For a horizontal construction element (floor, downward flow): $11/h_i + 1/h_e = 0.22$ m^2°C/W.

An estimation according to experience can be put forward to assess the value of U and therefore the thermal quality of each element of the envelope, and all issues will depend on the climatic context:
- Poor thermal quality: U value greater than 1.5 W/m^2 °C;
-Average thermal quality: U between 1 and 1.5 W/m^2 °C;
- Fairly good thermal quality: U between 0.8 and 1. W/m^2 °C;
- Good thermal quality: U between 0.5 and 0.8 W/m^2 °C;
- Very good thermal quality: U between 0.3 and 0.5. W/m^2 °C;
- Excellent thermal quality: U value less than 0.3 W/m^2 °C;

From the Thermal Regulation of Constructions in Morocco, two methods of verifying the conformity of buildings about thermal and energy performance: the performance approach and the prescriptive approach. The so-called performance approach defines the annual heating and air conditioning needs (expressed in kWh/m^2.year), according to the climatic zone and the type of building. The reference indoor temperatures for heating and air conditioning are 20°C and 26°C respectively. For humidity, the reference values are 55% and 60% respectively for winter and summer.

The so-called prescriptive approach consists of setting the technical specifications expressed, for each type of building and each climatic zone, in the form of maximum thermal transmission coefficients (U value) of the roof, exterior walls, low floor, windows as well as the equivalent solar

Materials Research Forum LLC

https://doi.org/10.21741/9781644903117-23

factor of windows and the thermal resistance of low floor, depending on the Overall Rate of Bay Windows. The prescriptive approach is only applicable in the case where this rate is less than 45%.

Application to the envelope of the earth-built heritage:

The configurations of the construction elements including the thicknesses of the different layers of materials (Fig. 3a and Fig. 3b) seem to respect the thermal regulations of the normal envelope of a traditional earth-based built framework in the 2 sites.

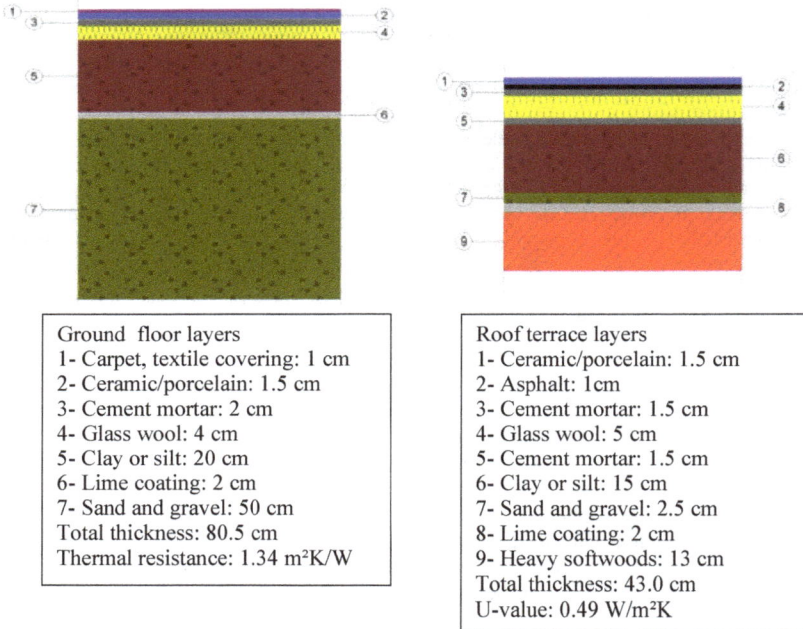

Ground floor layers
1- Carpet, textile covering: 1 cm
2- Ceramic/porcelain: 1.5 cm
3- Cement mortar: 2 cm
4- Glass wool: 4 cm
5- Clay or silt: 20 cm
6- Lime coating: 2 cm
7- Sand and gravel: 50 cm
Total thickness: 80.5 cm
Thermal resistance: 1.34 m²K/W

Roof terrace layers
1- Ceramic/porcelain: 1.5 cm
2- Asphalt: 1cm
3- Cement mortar: 1.5 cm
4- Glass wool: 5 cm
5- Cement mortar: 1.5 cm
6- Clay or silt: 15 cm
7- Sand and gravel: 2.5 cm
8- Lime coating: 2 cm
9- Heavy softwoods: 13 cm
Total thickness: 43.0 cm
U-value: 0.49 W/m²K

Figure 3a: Thermal characteristics: the ground floor thermal resistance and the roof terrace U value according to the prescriptive approach. [1]

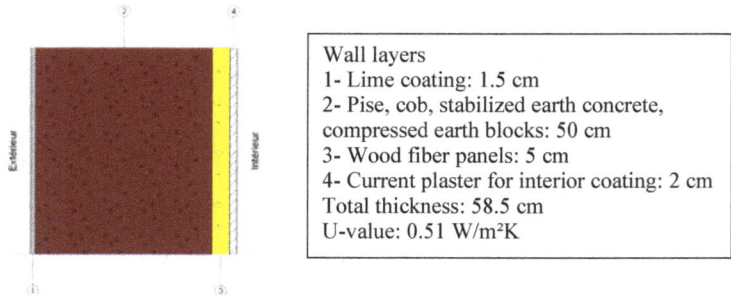

Wall layers
1- Lime coating: 1.5 cm
2- Pise, cob, stabilized earth concrete, compressed earth blocks: 50 cm
3- Wood fiber panels: 5 cm
4- Current plaster for interior coating: 2 cm
Total thickness: 58.5 cm
U-value: 0.51 W/m²K

Figure 3b: Thermal characteristics: the wall U value according to the prescriptive approach. [1]

Mediterranean Architectural Heritage - RIPAM10
Materials Research Proceedings 40 (2024) 218-225

Materials Research Forum LLC
https://doi.org/10.21741/9781644903117-23

The glazing of this built frame must be double-glazing for both climates. Other bioclimatic considerations already mentioned and recommended should also be applied in an integrated manner with the provisions of thermal regulations.

Conclusion

The climatic dimension and the consideration of climate are at the heart of architectural design and push traditional architecture to deal with extreme climates, hot during the summer and cold during the winter. Climate architecture or bioclimatism concretizes this action by trying to combine and apply the principles of bioclimatic architecture, the use of bioclimatic analysis and evaluation tools based on climate data, and finally compliance with regulations, the thermal current of the envelope of the traditional built frame. This, which is considered a third skin, after ours and clothing, plays the role of mediator between the exterior climate and a comfortable and efficient interior environment in terms of thermal and energy.

Traditional earth-based architecture is certainly characterized by a great inertia of the structure making it possible to ensure thermal comfort and the reduction of energy consumption (electricity, firewood, etc.) or even the limitation of the use of active heating systems or air conditioning. The thermophysical properties of the raw earth are significantly improved by the gradual addition of natural products like as olive pomace, date palm fibers, or straw.

References

[1] Royaume du Maroc, Bulletin Officiel, Décret N° 2.13.874 Approuvant le Règlement Général de Construction Fixant les Règles de Performance Energétique des Constructions et Instituant le Comité National de l'Efficacité Energétique dans le Bâtiment. 6306 (2014) 4256-4269.

[2] ISO 7730, Ambiances Thermiques Modérées – Détermination des Indices PMV et PPD et Spécifications des Conditions de Confort Thermique, AFNOR, Paris, 1994.

[3] B. Givoni, Comfort, climate analysis and building design guidelines, in: Energy and Buildings Vol. 18, 1992, pp. 11-23. https://doi.org/10.1016/0378-7788(92)90047-K

[4] S.V. Szokolay, Introduction to Architectural Science. The Basis of Sustainable Design. Architectural Press, 2008. https://doi.org/10.4324/9780080878942

[5] O.H. Koenigsberger, T.G. Ingersoll, A. Mayhew, S.V. Szokolay, Manual of Tropical Housing and Building: Climatic Design. India: Orient Longman, 1973.

Mediterranean Architectural Heritage - RIPAM10
Materials Research Proceedings 40 (2024) 226-232

Materials Research Forum LLC
https://doi.org/10.21741/9781644903117-24

Geopolymers: An Eco-Friendly Approach to Enhancing the Stability of Earthen Constructions

Ilham MASROUR[a] *, Khadija BABA[b] and Khaoula DOUGHMI[c]

Civil Engineering and Environment Laboratory, Civil Engineering, Water, Environment and Geosciences Centre (CICEEG), Mohammadia School of Engineering, Mohammed V University Rabat, Morocco

[a] Ilham_masrour@um5.ac.ma, [b]khadija.baba@est.um5.ac.ma, [c]khaoula_doughmi@um5.ac.ma

Keywords: Earthen Constructions, Geopolymers, Compression Strength, Stabilizing Earthen Constructions

Abstract. Earthen constructions, characterized by their historical longevity and adaptability to various environments, constitute an essential part of the global architectural heritage. These structures offer environmental advantages by utilizing local resources, but they also face challenges such as weather sensitivity, vulnerability to earthquakes, and degradation over time. Preserving these constructions while meeting modern sustainability standards poses a crucial challenge. In this context, geopolymers emerge as innovative solutions for stabilizing earthen constructions. A sustainable alternative is provided by geopolymers, which are composed of fly ash and ground granulated blast furnace slag to enhance soil cohesion and strength. This review article aims to provide an insightful perspective on compression tests specific to various types of geopolymers. The objective is to guide the choice of the method for stabilizing earthen constructions based on available resources.

Introduction

The pressure on the demand for construction materials is significantly intensified by population growth. This can lead to deforestation, contributing to the reduction of natural carbon sinks and increased exploitation of fossil fuels. This phenomenon is one of the major issues causing greenhouse gas emissions, thereby driving climate change. To address these challenges, it is crucial to promote sustainable construction practices, such as earth construction.

In 1982, earthen construction was one of the most widely used building materials for houses worldwide, with approximately one-third of the global population residing in such dwellings [1]. This practice has seen a decline in prevalence, and it is estimated that almost 25% of the world's population lives in houses made of earthen material. Recyclability, durability, high thermal capacity, and cost-effectiveness during construction are all advantages that rammed earth construction offers [2] . This building material is an intriguing choice due to its many benefits. Nonetheless, it is crucial to reinforce and stabilize it to increase its resistance to external forces.

The construction technique of rammed earth relies on the level of soil compaction, whether done manually or mechanically [3]. Generally, it can be utilized either by pouring it directly into a formwork to build walls or by molding bricks. The latter approach involves creating blocks out of soil, water, and a binder, assembling them in molds, and then compressing them with a hydraulic press.

An exploration of various earth construction methods, such as rammed earth, and earth block construction, reveals the crucial role of stabilization in ensuring the durability and strength of these structures. Chemical soil stabilization emerges as a promising technique for enhancing structural safety and reducing repair and rehabilitation costs. Cement and lime, among the most common chemical binding agents for stabilizing expansive soils [4], carry a negative environmental impact

Mediterranean Architectural Heritage - RIPAM10   Materials Research Forum LLC
Materials Research Proceedings 40 (2024) 226-232   https://doi.org/10.21741/9781644903117-24

due to their energy-intensive manufacturing processes and greenhouse gas emissions, contributing to global warming [5].

Scientific research is increasingly turning to new, innovative stabilization pathways. Among these alternatives are:

Stabilization using geopolymers: An innovative approach to consolidating earth structures involves the use of geopolymers. Geopolymers are inorganic materials, typically aluminosilicate materials, synthesized from an activator solution and solid aluminosilicate materials. Geopolymers are characterized by their long-range covalent networks, which impart stabilization properties that improve the mechanical strength of earth constructions while minimizing the impact on the ecosystem

Understanding Geopolymers: The Science of Geopolymerization

The term geopolymer was originally named and developed by Professor Joseph Davidovits. He defined them as new materials for coatings and adhesives, new binders for fibrous composites, waste encapsulation and a new cement for concrete [6].

The semi-crystalline or amorphous form in a three-dimensional network is a key feature of geopolymers, which are green inorganic polymers made of aluminosilicate constituents and are characterized accordingly [7].

The term "geopolymer" was not used until 1978 by French chemist Joseph Davidovits. The term refers both to the inorganic nature of these materials and their polymer-like structure.

At the outset, the exploration into geopolymers concentrated on employing natural source materials like kaolin, calcined clays, silica fume and metakaolin. In more recent times, investigations have expanded to include industrial byproducts, for instance, fly ash [8–10], clay-based slag [11,12],Enhancing its sustainability with regard to environmental aspects, accessibility, and economic influence is a priority.

Geopolymers are characterized by their general chemical formula: $M^+ n\{(SiO_2)_z, AlO_2\}n$, w H_2O with n the degree of polymerization, z the Si/Al molar ratio and M^+ the monovalent cation. The formation of geopolymers is the result of polycondensation and geopolymerization reactions, in which SiO_4 and AlO_4 tetrahedra form a three-dimensional network. The alkali cation M^+ ensures neutrality by compensating for the charge deficit created by the substitution of an Si^{4+} cation by an Al^{3+} cation.

Geopolymer composites: a sustainable solution for soil stabilization:

Alkali-activated materials (AAM), also known as inorganic polymer materials called geopolymers, can be synthesized from natural materials and waste through an alkaline or acidic activation reaction.The advantages of Geopolymers, which are fire-resistant, chemical corrosion resistant, have high mechanical strength, and excellent durability[13,14].Optimal performance and low carbon dioxide emissions have been the main reasons geopolymer materials have been considered as substitutes for Ordinary Portland Cement (OPC) in the early 1980s.

Fly ash geopolymer:

As per the investigation conducted by R.K. Preethi et al. [15], investigates the durability of alkali-activated compressed earth cylindrical samples, which include a 30% clay fraction and 20% kaolin clay, incorporating varying levels of fly ash. The findings indicate that increasing the fly ash content enhances moisture resistance across specimens, independent of the alkali activator's molar concentration. As the fly ash content goes up from 5 to 15%, the compressive strength of natural soil samples increases by roughly 90%. With a clay content of 30%, fly ash percentage of 15%, and 12 M NaOH, the highest recorded strength is around 4 MPa. Additionally, strength improves with increasing molarity, from 8 molars to 12 molars. For samples with 20% kaolin, a nearly twofold increase in wet strength is achieved by increasing the fly ash content from 4 to 15%, which reaches around 3.15 MPa. In addition, as the molarity increases from 8 to 12 M, strength increases by about 70% with 4% fly ash and 60% with 15% fly ash.

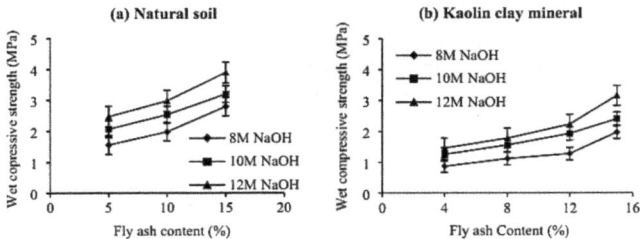

Fig. 2. *Compressed earth specimens made with (a) natural soil or (b) Kaolin clay mineral exhibit differences in strength based on the fly ash content.*

Hany et al.[3] investigated the impact of adding fly ash (FA) and an alkaline solution (AS) on the strength of raw earth bricks. They observed a significant early-age (7-day) strength increase in bricks with FA. Additionally, the long-term (28-day) strength of dried bricks was satisfactory. However, bricks exposed to humid conditions experienced a decrease in long-term strength due to leaching of the alkaline solution. In conclusion, FA addition strengthens raw earth bricks, but its effectiveness depends on environmental conditions.

Ground granulated blast furnace slag (GGBS)-based geopolymer:
Ground Granulated Blast Furnace Slag (GGBS) is a by-product of steel production. It is a granular, glassy material that is rich in silica and aluminates.

Mostafa Zamanian et al. [16] assess stabilization effectiveness primarily through the strength of cementitious bonds, geopolymerization, and suction stress. Curing conditions, especially in hot and dry environments, impact the quality of bonds, with GGBFS-based geopolymer samples showing enhanced compressive strength due to inter-particle geopolymer bonds. Cold curing conditions are discouraging. Suction stress from water evaporation also influences compression strength. GGBFS-based geopolymer samples are more sensitive to curing conditions compared to stabilized Ordinary Portland Cement (OPC) samples. Binder content significantly affects improvement efficiency, varying with curing conditions. In hot and dry conditions, incorporating WTTF fibers enhances UCS, but in moderate and cold conditions, it slightly decreases. To summarize, stabilized samples' strength is affected by binder type, curing conditions, binder content, and fiber incorporation.

For A. Neha VIVEK and al.[17] the compressive strength of wet-tested rammed earth cylinders after 28 days was examined, revealing a more than 100% increase with the addition of cement and Ground Granulated Blast Furnace Slag (GGBS). A significant improvement in wet compressive strength is observed with an increased GGBS percentage, reaching 3.51 MPa from an initial 0.83 MPa. The highest wet compressive strength (3.51 MPa) is achieved with a mix containing 12.4% clay and 20% GGBS. This improvement is attributed to a denser soil matrix due to hydration and slow pozzolanic reactions between soil compounds and GGBS. The initial tangent modulus (ITM) for GGBS specimens ranges from 1770 MPa to 3573 MPa, increasing by 74% with a higher GGBS percentage. Failure modes during compression tests correspond to a brittle type of failure.

Mediterranean Architectural Heritage - RIPAM10

Materials Research Forum LLC

Materials Research Proceedings 40 (2024) 226-232

https://doi.org/10.21741/9781644903117-24

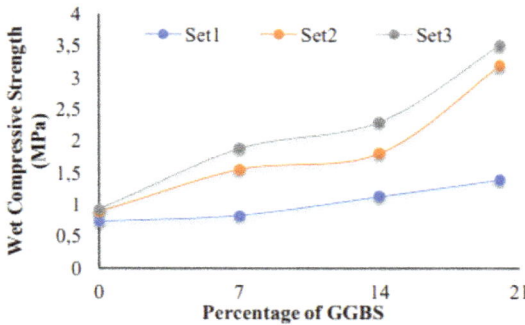

Fig. 3. *Impact of GGBS content on wet compressive strength*

Pozzolana geopolymer:

Pozzolana is a natural volcanic rock that has been utilized in construction for centuries. Abundant in silica and aluminates, it has the ability to react with hydrated lime to create cementitious compounds. These cementitious compounds are accountable for the strength and durability properties of pozzolana.

Rolande Aurélie Tchouateu Kamwa et al. [18] exhibit the outcomes of compression strength tests conducted on Compressed Earth Blocks (CEB) that were stabilized with a geopolymer binder that was activated with phosphoric acid-activated pozzolana in both dry and wet conditions. The mass ratio of phosphoric acid solution to pozzolana was kept constant at 0.8. Overall, the addition of stabilizers is the main factor that influences the strength of geopolymer samples. At 25°C, compressive strength in dry conditions gradually increases with the addition of stabilizer, reaching 20.6 MPa for a 20% by weight stabilizer content. At 70°C, the strength reaches its maximum at 42.8 MPa with a 15% by weight stabilizer content, decreasing to 36.9 MPa with additional stabilizer at 20% by weight. Cohesion between the stabilizer products and unreacted particles influences strength, generating materials with significant mechanical properties. Geopolymeric phosphatic gels derived from pozzolana ensure the stabilization of clayey soils based on CEB. At 70°C, the threshold temperature for the geopolymerization of pozzolanas according to [19], a significant mechanical gain is achieved due to the catalytic action of the polymerization/polycondensation reaction. In an acidic environment, the intra-stabilization of solid cement in CEB is evidenced by its compressive strength at both 25°C (9.7 MPa) and 70°C (10.2 MPa). The weight stabilizer's 20% decrease in strength indicates that a weight stabilization of 15% is sufficient for effective stabilization. The trend for compressive strength in wet conditions is similar to that of dry strength, with a significant decrease of 43.7% at 25°C and 43.8% at 70°C, in accordance with previous studies on phosphatic geopolymer binders. At 70°C, 15% pozzolana by weight seems to be the best option for stabilizing CEB, with a stronger strength than that of CEB stabilized by OPC and other geopolymer binders. These CEB can be used as construction bricks in areas subjected to significant weathering, while others are suitable as load-bearing masonry elements.

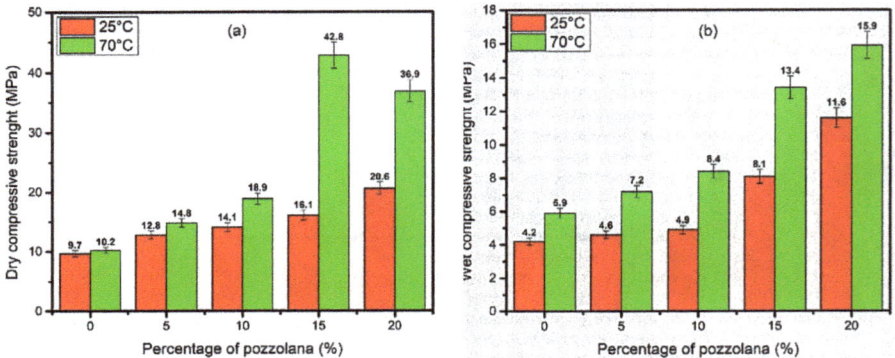

Fig. 4. *Compressive strength of stabilized Compressed Earth Blocks (CEBs) after curing at 25 and 70 °C under dry (a) and wet (b) conditions.*

Conclusion

Pioneered by Professor Joseph Davidovits, geopolymers have emerged as a promising avenue for sustainable construction materials. Defined by their inorganic nature and polymer-like structure, they offer a greener alternative to conventional materials. Initially focused on natural sources, recent research incorporates industrial byproducts like fly ash and ground granulated blast furnace slag, enhancing sustainability, accessibility, and economic viability.

Geopolymer composites, particularly alkali-activated materials, boast impressive qualities like fire and chemical resistance, high mechanical strength, and excellent durability. These properties have positioned them as viable substitutes for Ordinary Portland Cement since the early 1980s, aligning with global efforts to reduce carbon dioxide emissions.

Extensive research on fly ash and ground granulated blast furnace slag-based geopolymers showcases their versatility. Fly ash enhances moisture resistance and compressive strength, while the impact of ground granulated blast furnace slag depends on curing conditions and binder content. Their applications in soil stabilization and raw earth brick construction offer practical solutions, with environmental influences affecting their effectiveness.

Furthermore, utilizing pozzolana, a natural volcanic rock, as a geopolymer stabilizer expands the application scope. Phosphoric acid-activated pozzolana displays significant improvements in compressive strength, highlighting its potential in construction, particularly in challenging weathering conditions.

In essence, geopolymers have established themselves as a sustainable solution for construction, offering diverse applications and the potential to address environmental concerns associated with traditional building materials. The availability of the material used for this stabilization will determine which stabilization method to use, as both methods have a low greenhouse gas footprint.

References

[1] R.A. Silva, D.V. Oliveira, T. Miranda, N. Cristelo, M.C. Escobar, E. Soares, Rammed earth construction with granitic residual soils: The case study of northern Portugal, Constr. Build. Mater. 47 (2013) 181–191. https://doi.org/10.1016/j.conbuildmat.2013.05.047

[2] S. Samadianfard, V. Toufigh, Energy Use and Thermal Performance of Rammed-Earth Materials, J. Mater. Civ. Eng. 32 (2020) 04020276. https://doi.org/10.1061/(ASCE)MT.1943-5533.0003364

[3] E. Hany, N. Fouad, M. Abdel-Wahab, E. Sadek, Investigating the mechanical and thermal properties of compressed earth bricks made by eco-friendly stabilization materials as partial or full replacement of cement, Constr. Build. Mater. 281 (2021) 122535. https://doi.org/10.1016/j.conbuildmat.2021.122535

[4] D. Nigitha, N. Prabhanjan, Efficiency of cement and lime in stabilizing the black cotton soil, Mater. Today Proc. 68 (2022) 1588–1593. https://doi.org/10.1016/j.matpr.2022.07.286.

[5] A.A. Disu, P.K. Kolay, A Critical Appraisal of Soil Stabilization Using Geopolymers: The Past, Present and Future, Int. J. Geosynth. Ground Eng. 7 (2021) 23. https://doi.org/10.1007/s40891-021-00267-w

[6] J. Davidovits, GEOPOLYMER Chemistry & Applications, 2ème édition, Institut géopolymère, 2008.

[7] Z. Zhang, H. Yu, M. Xu, X. Cui, Preparation, characterization and application of geopolymer-based tubular inorganic membrane, Appl. Clay Sci. 203 (2021) 106001. https://doi.org/10.1016/j.clay.2021.106001

[8] H.M. Giasuddin, J.G. Sanjayan, P.G. Ranjith, Strength of geopolymer cured in saline water in ambient conditions, Fuel 107 (2013) 34–39. https://doi.org/10.1016/j.fuel.2013.01.035.

[9] J.E. Oh, P.J.M. Monteiro, S.S. Jun, S. Choi, S.M. Clark, The evolution of strength and crystalline phases for alkali-activated ground blast furnace slag and fly ash-based geopolymers, Cem. Concr. Res. 40 (2010) 189–196. https://doi.org/10.1016/j.cemconres.2009.10.010.

[10] S. Aydın, B. Baradan, Mechanical and microstructural properties of heat cured alkali-activated slag mortars, Mater. Des. 35 (2012) 374–383. https://doi.org/10.1016/j.matdes.2011.10.005

[11] M.O. Yusuf, M.A. Megat Johari, Z.A. Ahmad, M. Maslehuddin, Influence of curing methods and concentration of NaOH on strength of the synthesized alkaline activated ground slag-ultrafine palm oil fuel ash mortar/concrete, Constr. Build. Mater. 66 (2014) 541–548. https://doi.org/10.1016/j.conbuildmat.2014.05.037

[12] E. Hany, N. Fouad, M. Abdel-Wahab, E. Sadek, Compressive strength of mortars incorporating alkali-activated materials as partial or full replacement of cement, Constr. Build. Mater. 261 (2020) 120518. https://doi.org/10.1016/j.conbuildmat.2020.120518.

[13] J. Aliques-Granero, M.T. Tognonvi, A. Tagnit-Hamou, Durability study of AAMs: Sulfate attack resistance, Constr. Build. Mater. 229 (2019) 117100. https://doi.org/10.1016/j.conbuildmat.2019.117100

[14] M. Vafaei, A. Allahverdi, P. Dong, N. Bassim, Acid attack on geopolymer cement mortar based on waste-glass powder and calcium aluminate cement at mild concentration, Constr. Build. Mater. 193 (2018) 363–372. https://doi.org/10.1016/j.conbuildmat.2018.10.203.

[15] R.K. Preethi, B.V. Venkatarama Reddy, Experimental investigations on geopolymer stabilised compressed earth products, Constr. Build. Mater. 257 (2020) 119563. https://doi.org/10.1016/j.conbuildmat.2020.119563

[16] M. Zamanian, M. Salimi, M. Payan, A. Noorzad, M. Hassanvandian, Development of high-strength rammed earth walls with alkali-activated ground granulated blast furnace slag (GGBFS)

and waste tire textile fiber (WTTF) as a step towards low-carbon building materials, Constr. Build. Mater. 394 (2023) 132180. https://doi.org/10.1016/j.conbuildmat.2023.132180.

[17] N. Vivek A, P. Kumar P, H. Reddy M, Mineral Katkılarla Stabilize Edilmiş Sıkıştırılmış Toprağın Uzun Vadeli Dayanımı ve Performansı Üzerine Deneysel Çalışma, El-Cezeri Fen Ve Mühendis. Derg. (2022). https://doi.org/10.31202/ecjse.1094013

[18] R.A. Tchouateu Kamwa, S. Tome, J. Chongouang, I. Eguekeng, A. Spieß, M.N.A. Fetzer, K. Elie, C. Janiak, M.-A. Etoh, Stabilization of compressed earth blocks (CEB) by pozzolana based phosphate geopolymer binder: Physico-mechanical and microstructural investigations, Clean. Mater. 4 (2022) 100062. https://doi.org/10.1016/j.clema.2022.100062

[19] Reactivity of volcanic ash in alkaline medium, microstructural and strength characteristics of resulting geopolymers under different synthesis conditions | Journal of Materials Science, (n.d.). https://link.springer.com/article/10.1007/s10853-016-0257-1 (accessed February 4, 2024).

Mediterranean Architectural Heritage - RIPAM10
Materials Research Proceedings 40 (2024) 233-239

Materials Research Forum LLC
https://doi.org/10.21741/9781644903117-25

Naturally Strengthening Rammed Earth: The Promising Potential of Biopolymers

Ilham MASROUR[a] *, Khadija BABA[b] and Sana SIMOU[c]

Civil Engineering and Environment Laboratory, Civil Engineering, Water, Environment and Geosciences Centre (CICEEG), Mohammadia School of Engineering, Mohammed V University Rabat, Morocco

[a]Ilham_masrour@um5.ac.ma, [b]khadija.baba@est.um5.ac.ma, [c] sana_simou2@um5.ac.ma

Keywords: Earthen Constructions, Biopolymer, Compression Strength, Stabilizing Earthen Constructions

Abstract. Sustainable construction has become a global imperative due to the growing awareness of the harmful environmental impacts of the construction industry. The use of cement and lime in traditional methods of stabilizing earth constructions is a significant problem due to their high carbon footprint. This article examines an ecological alternative to stabilizing earth structures with biopolymers. These Bio-based materials can be used to reduce the environmental impact of the construction industry while also ensuring the structure's stability and durability. The purpose of this article is to examine the mechanical properties of biopolymers in the context of stabilizing earth construction. The objective is to guide the decision on which stabilization method to use for earth construction based on the available resources.

Introduction

Over the past few decades, extensive research has been undertaken on natural materials combined with unconventional technologies due to the growing emphasis on sustainability and the use of recyclable natural resources. The widespread availability, affordability, and ease of implementation have made earth construction more popular in this context. The materials are exceptional in terms of sustainability, with low energy consumption during production and high recyclability. The importance of earth construction can be seen in its applications in traditional architecture, modern low-cost social housing, and heritage restoration [1].

There are different earth construction techniques such as wattle and daub, cob, rammed earth (including earth shotcrete), as well as adobe bricks and compressed earth blocks (CEB).

It has been nearly six millennia since the adoption of the wattle and daub construction method [2], Involves compacting soil against a lattice made of interwoven wooden strips. Wattle and daub is similar to a technique called 'tabique' used in Portugal [3].

On the contrary, cob entails blending earth with straw and water to create layered walls [4].

Rammed earth involves compacting moist earth (stabilized or not) into a wooden formwork, while earth shotcrete involves pre-stabilizing the earth before spraying it onto an internal formwork. A blended earth wall system, incorporating aspects of rammed earth, cob, and wattle and daub, was recently employed in Portugal.

Adobe is an uncomplicated and age-old method of earth construction [5], uses wooden molds filled with wet earth, dried in the sun [6]. As adobe dries, surface shrinkage cracks may emerge, leading some authors [7,8] to propose the incorporation of straw or other vegetable fibers as a preventive measure.

Compressed earth blocks (CEB) represent an evolution of adobe bricks, compressing earth in a mold using a specific device, with manual or mechanical pressure. These blocks exhibit greater weight and strength compared to adobe bricks. Different machines, such as the CINVA-Ram, the Astram, Over the years, various machines such as the CETA-Ram, the Brepak multi-block, and the CTA Triple-Block Press have been developed for the production of Compressed Earth Blocks (CEB) [9].

Earth construction, while promising for its affordability and eco-friendliness, often relies on environmentally harmful chemical stabilizers like cement and lime. These materials create a double whammy for the environment: cement production spews large amounts of CO_2, a major greenhouse gas, while lime extraction and processing are energy-intensive, further contributing to emissions. This dependence on chemical stabilizers also drains natural resources and worsens environmental degradation. To counteract this worrying trend, exploring and implementing sustainable alternatives for stabilizing earth constructions is crucial.[10,11]

Increasingly, scientific research is focusing on new and innovative stabilization pathways. Among these alternatives is:

Natural stabilization using biopolymers: Earthen constructions can be stabilized through the use of biopolymers, natural polymers produced by the cells of living organisms. These biopolymers bind to clay, imparting increased strength. They play an essential role in preserving and maintaining the durability of structures in the face of adverse weather conditions and other degrading factors. These diverse stabilization approaches offer varied solutions to reinforce earthen constructions, enhancing their resistance, durability, and environmental friendliness [12].

Biopolymers

Biopolymers are polymers of natural origin, often derived from renewable raw materials such as plant, animal, algal, fungal or bacterial sources. Their characteristics can be diverse, including plant or animal origin, hydrophilicity, hydrophobicity, or amphiphilicity.

According to Vissac and colleagues [13], biopolymers can be classified according to the nature of the molecule responsible for improving cohesion and the physico-chemical mechanisms involved.

Polysaccharides are complex carbohydrates made up of a large number of simple sugar molecules called monosaccharides, linked together by glycosidic bonds. These bonds are formed by the loss of a water molecule when two monosaccharide molecules are combined. The frequently encountered glycosidic bonds include α-1,4-, β-1,4-, and α-1,6-glycosidic [14]. A straight or branched chain can be connected to the structural unit. In the chain structure that is straight, the structural units are linked together by α-1,4- or β-1,4-glycosidic linkages (such as starch). In the branched structure, the structural units are linked together by α-1,6 glycosidic bonds (such as cellulose) [13,15].

There are two distinct categories of polysaccharides: heteropolysaccharides and homopolysaccharides. A single monosaccharide is what makes up the latter. for example, starch, cellulose, and glycogen. Heteropolysaccharides, on the other hand, are made up of several types of monosaccharide, as is the case with hyaluronic acid (HA), chondroitin and alginate. The latter can be classified according to two main roles: the structural function, observed in polysaccharides such as cellulose and chitin, or the energy storage function, illustrated by examples such as starch [16–18].

Polysaccharides can create microscopic reinforcements among clay particles, enhancing cohesion and making the soil stronger and more resistant. Additionally, they modify the consistency of fresh mortar, forming gels that improve adhesion and ease of application.

Lipids, including fats and other organic compounds, have low solubility in water and can be hydrophobic or amphiphilic based on their physical properties.

A chain of amino acids found in animals or plants is what proteins are made up of. Some play a structural role, such as collagen, which contributes to the formation of skin and bone, while others transport molecules, such as albumin, found in eggs and blood, and casein, found in milk [19].The functionality of a protein, determined by its surface characteristics, depends on the sequence of amino acids and their spatial organization [13].

Their interaction with clays is significant: the hydrophilic parts adsorb onto fine clay particles, while the hydrophobic parts, which are repelled by water, move towards the outside of the clay particle, where they are protected from water. Proteins act as glues for clays and can also have a hydrophobic effect, reducing the water sensitivity of stabilised earth plasters [13,20].

Tannins, natural compounds found in plants, show promise for soil stabilization. Soluble in water, they act like "claws," capturing metal ions in clay soils, weakening ionic strength, and promoting clay dispersion. Additionally, tannins react with iron to release multivalent ions, enhancing attraction between clay platelets and improving water resistance. This dual action—dispersion through their structure and cohesion through released iron ions—positions tannins as a potential eco-friendly alternative to traditional chemical stabilizers for enhancing clay soil strength.

Biopolymer composites: A sustainable method for stabilizing soils

Earthen construction combines traditional methods with modern mechanical testing, relying on factors like soil composition and moisture for stability. The integration of biopolymers is pivotal for soil stabilization, offering eco-friendly solutions that enhance cohesion and strength. This innovative approach aligns with sustainable practices and has the potential to revolutionize traditional methods in earthen construction.

Polysaccharides

The study conducted by C. Galán-Marín et al.[21] aims to stabilize soils using polymers and natural fibers to create a durable, non-toxic, and locally sourced composite construction material. Compression tests reveal significant results, showing a marked improvement in compression strength with the addition of natural fibers such as wool and alginate. The synergistic combination of wool and alginate nearly doubles the soil's strength, providing promising prospects for practical applications in construction.

Another investigation led by Yahor Trambitski and his team [22] explores the compression strength of clay composites treated with biopolymer solutions, including starch, alginate, and chitosan. The results indicate an increase in compression strength with the addition of starch and alginate, while chitosan shows mixed results, suggesting a less effective interaction with clay particles. These studies present intriguing possibilities for the development of more sustainable and environmentally friendly construction materials.

Proteins

C. Kraus and coauthors [23] conducted tests, presenting results on the compressive strength of rammed earth stabilized with blood. The initial compressive strength of the control cylinders was 257 psi (1.77 MPa), exceeding that of the blood-stabilized cylinders, which measured 187 psi (1.29 MPa) after one week. However, by the fourteenth day, the blood-stabilized cylinders gradually narrowed the difference, reaching an average compressive strength of 518 psi (3.57 MPa), demonstrating comparability to the control group's 521 psi (3.59 MPa) with no statistically significant difference. By the twenty-eighth day, the blood-stabilized rammed earth exhibited a remarkable 36% improvement over the control, attaining a pressure of 973 psi (6.71 MPa), in contrast to the control group's 625 psi (4.31 MPa). This highlights the significant enhancement in compressive strength through blood stabilization, especially notable when compared to less reactive soils, where blood-stabilized samples exhibited over a 200% improvement over controls in a prior pilot experiment.

Mediterranean Architectural Heritage - RIPAM10   Materials Research Forum LLC
Materials Research Proceedings 40 (2024) 233-239   https://doi.org/10.21741/9781644903117-25

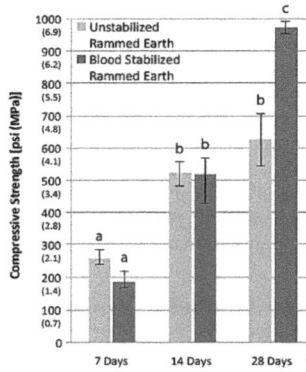

Fig. 1: *Comparing the compressive strength of unstabilized and blood-stabilized rammed earth [23].*

Complex molecules

Younoussa Millogo and colleagues [24] aimed to develop cost-effective, durable blocks with high water resistance and thermal properties conducive to superior indoor comfort. Figure 2 illustrates the variations in compressive strength relative to different levels of cow-dung content in adobes, showing a consistent pattern of increased mechanical parameters with higher cow-dung additions. The improved compressive strength is attributed to reduced porosity and crack prevention within the adobes. Figure 3 establishes a correlation between compressive strength and water absorption, indicating a robust connection. The introduction of cow dung enhances compressive strength by reducing porosity through microbial debris and amine organic compounds. The non-digested fibers enhance adhesion to the clay matrix, creating a rough surface that inhibits crack propagation. Additionally, the silicate amine formed in the fermentation process binds soil particles together, reducing pore size and number, ultimately boosting compressive strength.

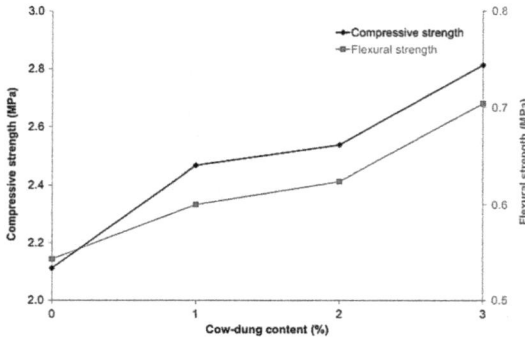

Fig. 2: *The evolution of the flexural and compressive strength of adobes has been influenced by the content of cow dung [24]*

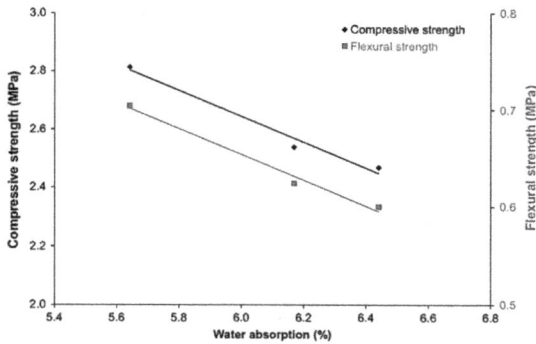

Fig. 3: *The correlation of water absorption and compressive (or flexural) strength in adobes [24]*

Conclusion

In summary, the integration of diverse biopolymers—polysaccharides, lipids, proteins, and tannins—provides a sustainable and multifaceted approach to soil stabilization in earthen construction. These natural compounds enhance soil cohesion, strength, and water resistance while offering eco-friendly alternatives to traditional stabilizers.

Studies, such as those by C. Galán-Marín and Yahor Trambitski, highlight the efficacy of biopolymer composites, incorporating materials like wool and alginate, to significantly improve compression strength. The innovative use of blood-stabilized rammed earth by C. Kraus demonstrates a remarkable 36% improvement in compressive strength over traditional methods.

Younoussa Millogo's work introduces cost-effective, durable blocks with enhanced water resistance through the incorporation of cow dung, showcasing the transformative potential of biopolymers in construction methodologies.

In essence, the integration of biopolymers presents a sustainable and transformative force in soil stabilization, shaping a more resilient and eco-friendly future for earthen construction.

In perspective, a promising avenue for future research would involve exploring the synergistic combination of various biopolymers to further optimize the properties of earthen construction materials. This approach could potentially lead to significant advancements, further enhancing the cohesion, strength, and durability of these materials, thereby opening new possibilities for more sustainable and environmentally friendly construction solutions.

References

[1] F. Jové-Sandoval, M.M. Barbero-Barrera, N. Flores Medina, Assessment of the mechanical performance of three varieties of pine needles as natural reinforcement of adobe, Constr. Build. Mater. 187 (2018) 205–213. https://doi.org/10.1016/j.conbuildmat.2018.07.187

[2] T. Graham, Wattle and Daub: Craft, Conservation and Wiltshire Case Study, (n.d.).

[3] P. Silva, J. Pinto, H. Varum, D. Cruz, D. Sousa, P. Morais, P. Tavares, J. Louzada, P. Silva, J. Vieira, Characterization of Traditional Tabique Constructions in Douro North Valley Region, WSEAS Trans. Environ. Dev. 6 (2010) 105–114.

[4] E. Quagliarini, A. Stazi, E. Pasquali, E. Fratalocchi, Cob Construction in Italy: Some Lessons from the Past, Sustainability 2 (2010) 3291–3308. https://doi.org/10.3390/su2103291.

[5] Mechanical properties of adobe walls in a Roman Republican domus at Suasa - ScienceDirect, (n.d.). https://www.sciencedirect.com/science/article/abs/pii/S1296207409001046 (accessed February 4, 2024).

[6] C.D.F. Rogers, I.J. Smalley, The adobe reaction and the use of loess mud in construction, Eng. Geol. 40 (1995) 137–138. https://doi.org/10.1016/0013-7952(95)00064-X

[7] J. Vargas, J. Bariola, M. Blondet, P.K. Mehta, Seismic strength of adobe masonry, Mater. Struct. 19 (1986) 253–258.

[8] E. Quagliarini, S. Lenci, The influence of natural stabilizers and natural fibres on the mechanical properties of ancient Roman adobe bricks, J. Cult. Herit. 11 (2010) 309–314. https://doi.org/10.1016/j.culher.2009.11.012

[9] F. Pacheco-Torgal, S. Jalali, Earth construction: Lessons from the past for future eco-efficient construction, Constr. Build. Mater. 29 (2012) 512–519. https://doi.org/10.1016/j.conbuildmat.2011.10.054

[10] Efficiency of cement and lime in stabilizing the black cotton soil - ScienceDirect, (n.d.). https://www.sciencedirect.com.eressources.imist.ma/science/article/pii/S2214785322049197 (accessed November 17, 2023).

[11] A.A. Disu, P.K. Kolay, A Critical Appraisal of Soil Stabilization Using Geopolymers: The Past, Present and Future, Int. J. Geosynth. Ground Eng. 7 (2021) 23. https://doi.org/10.1007/s40891-021-00267-w

[12] Stabilisation naturelle pour la construction en terre : argiles et biopolymères, amàco (n.d.). https://amaco.org/publication/publication-stabilisation-naturelle-pour-la-construction-en-terre-argiles-et-biopolymeres/ (accessed November 17, 2023).

[13] A. Vissac, A. Bourgès, D. Gandreau, R. Anger, L. Fontaine, argiles & biopolymères - les stabilisants naturels pour la construction en terre, (n.d.).

[14] Y. Sun, X. Ma, H. Hu, Marine Polysaccharides as a Versatile Biomass for the Construction of Nano Drug Delivery Systems, Mar. Drugs 19 (2021) 345. https://doi.org/10.3390/md19060345

[15] T. Ribeiro, D.V. Oliveira, S. Bracci, The Use of Contact Sponge Method to Measure Water Absorption in Earthen Heritage Treated with Water Repellents, Int. J. Archit. Herit. 16 (2022) 85–96. https://doi.org/10.1080/15583058.2020.1751344

[16] R. Aguilar, J. Nakamatsu, E. Ramírez, M. Elgegren, J. Ayarza, S. Kim, M.A. Pando, L. Ortega-San-Martin, The potential use of chitosan as a biopolymer additive for enhanced mechanical properties and water resistance of earthen construction, Constr. Build. Mater. 114 (2016) 625–637. https://doi.org/10.1016/j.conbuildmat.2016.03.218

[17] J.L. Parracha, A.S. Pereira, R. Velez Da Silva, N. Almeida, P. Faria, Efficacy of iron-based bioproducts as surface biotreatment for earth-based plastering mortars, J. Clean. Prod. 237 (2019) 117803. https://doi.org/10.1016/j.jclepro.2019.117803

[18] C. Cocozza, A. Parente, C. Zaccone, C. Mininni, P. Santamaria, T. Miano, Comparative management of offshore posidonia residues: Composting vs. energy recovery, Waste Manag. 31 (2011) 78–84. https://doi.org/10.1016/j.wasman.2010.08.016

[19] A.E. Losini, A.C. Grillet, M. Bellotto, M. Woloszyn, G. Dotelli, Natural additives and biopolymers for raw earth construction stabilization – a review, Constr. Build. Mater. 304 (2021) 124507. https://doi.org/10.1016/j.conbuildmat.2021.124507

[20] R. Anger, L. Fontaine, Bâtir en terre: du grain de sable à l'architecture, 2009.

[21] C. Galán-Marín, C. Rivera-Gómez, J. Petric, Clay-based composite stabilized with natural polymer and fibre, Constr. Build. Mater. 24 (2010) 1462–1468. https://doi.org/10.1016/j.conbuildmat.2010.01.008

[22] Y. Trambitski, O. Kizinievič, F. Gaspar, V. Kizinievič, J.F.A. Valente, Eco-friendly unfired clay materials modified by natural polysaccharides, Constr. Build. Mater. 400 (2023) 132783. https://doi.org/10.1016/j.conbuildmat.2023.132783

[23] C. Mileto, F. Vegas, L. García Soriano, V. Cristini, eds., Compressive strength of blood stabilized earthen architecture, in: Earthen Archit. Past Present Future, 0 ed., CRC Press, 2014: pp. 233–236. https://doi.org/10.1201/b17392-40

[24] Y. Millogo, J.-E. Aubert, A.D. Séré, A. Fabbri, J.-C. Morel, Earth blocks stabilized by cow-dung, Mater. Struct. 49 (2016) 4583–4594. https://doi.org/10.1617/s11527-016-0808-6

Mediterranean Architectural Heritage - RIPAM10
Materials Research Proceedings 40 (2024) 240-247

Materials Research Forum LLC
https://doi.org/10.21741/9781644903117-26

Evaluating the Influence of Shale Extracted from the Settat Khouribga Region on the Characteristics of Concrete

Ayoub SOUILEH[1*], Latifa OUADIF[1], Driss EL HACHMI[2],
Mohammed CHRACHMY[3]

[1] L3GIE, Mohammadia Engineering School, Mohammed V University in Rabat, Morocco

[2] EMM, Faculty of science, Mohammed V University in Rabat, Morocco

[3] LMEENR, Faculty of Science and Technology, Moulay Ismail University, Errachidia, Morocco

*Ayoub.souileh@research.emi.ac.ma

Keywords: Local Materials, Reinforced Concrete, Clay Shale, Durability, Mechanical Resistance

Abstract. The incorporation of indigenous materials in the production of reinforced concrete offers a twofold prospect: the potential reduction of construction costs and environmental impact, along with the stimulation of economic growth in the source regions. To maximize the utility of shale derived from the Settat-Khouribga area and assess its influence on concrete properties, a series of tests were conducted on the material. Samples underwent meticulous characterization, encompassing mineralogical composition, particle size distribution, and mechanical properties. A comprehensive set of assessments was applied to the concrete samples, including tests for compressive strength, flexural strength, water absorption, porosity, and resistance to reinforcement corrosion over an extended timeframe. Concrete samples containing clay shale demonstrated compressive and flexural strength comparable to or even surpassing that of conventional concrete samples. Additionally, the incorporation of clay shale led to a reduction in porosity and water absorption in concrete, indicating an enhancement in durability.

Introduction

The global construction industry is witnessing a paradigm shift towards sustainable and locally sourced materials to address economic, environmental, and regional development concerns. In this context, the utilization of indigenous materials in concrete production has emerged as a promising avenue, offering potential cost savings, reduced environmental impact, and localized economic benefits. The present study focuses on the evaluation of shale extracted from the Settat-Khouribga region as a key constituent in reinforced concrete.[1]

The Settat-Khouribga area is renowned for its rich shale deposits,[2] presenting a valuable opportunity for exploring the performance of this material in concrete formulations. This research aims to comprehensively assess the impact of Settat-Khouribga shale on the mechanical and durability properties of concrete, thereby contributing to both scientific understanding and practical applications in sustainable construction practices.[3]

To achieve this objective, an extensive series of tests has been conducted on shale samples obtained from the Settat-Khouribga region. The characterization process involves a detailed analysis of mineralogical composition, particle size distribution[4], and mechanical properties. Subsequently, the shale is incorporated into concrete mixtures, and the resulting samples undergo a battery of tests, including assessments of compressive strength, flexural strength, water absorption, porosity, and resistance to reinforcement corrosion[5] over an extended period.

This investigation seeks to provide insights into the feasibility and performance of Settat-Khouribga shale as a supplementary material in concrete production. The findings of this study have the potential to influence construction practices, promote sustainable resource utilization, and contribute to the socio-economic development of the Settat-Khouribga region. Through a rigorous and integrated material characterization approach, this research aims to bridge the gap between theoretical understanding and practical applications, fostering advancements in the utilization of indigenous materials for sustainable construction solutions.[6]

Research methods

The study was conducted in the Settat-Khouribga region, renowned for its substantial shale deposits[7]. Sample collection was meticulously executed across multiple sites within the region to ensure a representative selection of shale specimens. Sampling locations were determined based on geological surveys [8]and accessibility considerations, with an emphasis on capturing variations in composition by collecting samples at different depths.

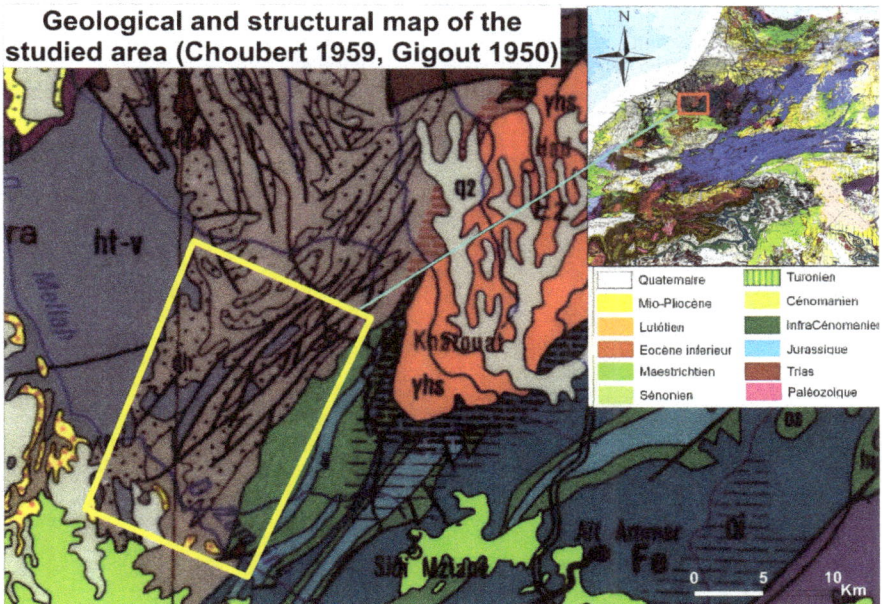

Figure 1: Geological map as a reference of studied area in khouribga Settat region, Morocco

Shale samples underwent a comprehensive characterization process. The mineralogical composition was determined using X-ray diffraction (XRD) analysis, involving the preparation of finely powdered specimens subjected to X-ray radiation for identification and quantification of mineral phases. Particle size distribution analysis, conducted through laser diffraction, provided insights into the granulometry of the shale by dispersing samples in a liquid medium.

Mechanical properties of the shale, including compressive strength and modulus of elasticity, were evaluated using standard testing procedures. Cylindrical samples were prepared and subjected to compressive loading to assess their mechanical behavior.

Concrete mixes were formulated by integrating varying percentages of Settat-Khouribga shale as a substitute for conventional aggregates. The mix design aimed at achieving optimal strength and durability while prioritizing the sustainable use of local resources.

Concrete samples underwent a battery of tests. Compressive strength testing involved casting and curing cubes under standard conditions, with subsequent assessment using a hydraulic press to evaluate load-bearing capacity. [9]

Flexural strength testing utilized prepared beams loaded until failure, with data collected to analyze the impact of shale on flexural performance.

Durability testing included assessments of water absorption, porosity, and resistance to reinforcement corrosion over an extended period[10]. These evaluations were crucial in gauging the durability of concrete when incorporating Settat-Khouribga shale.[11]

Throughout the experimental procedures, strict adherence to international standards and protocols was maintained to ensure the accuracy and reproducibility of results. The chosen methodologies aimed to provide a robust assessment of the impact of Settat-Khouribga shale on concrete properties, combining geological, mineralogical, and mechanical analyses with comprehensive testing of the resulting concrete formulations.[12]

Results and Discussion

1. Mineralogical Composition of khouribga Settat clay Shale:

The X-ray mineralogical study of clay shale samples from the Khouribga-Settat region has yielded significant results related to the focus of our research on the impact of these minerals on concrete properties.[13] Illite, predominant at 23.1%, suggests a substantial clay component in the material, potentially influencing soil plasticity. Minerals such as Kaolinite (15.07%) and Montmorillonite (13.89%) further reinforce this clay component, underscoring the importance of understanding their potential impact on the handling and strength of materials in the construction domain.

The substantial presence of Quartz at 16.13% presents an opportunity to enhance the mechanical properties of concrete, while potassic and plagioclase feldspars, at 8.01% and 6.74% respectively, contribute to the mineralogical diversity that may influence the strength and durability of construction materials. Muscovite (7.61%) and Biotite (2.5%) add a mica dimension to the samples, potentially influencing the stability of embankments and slope design in the realm of civil engineering.

Lastly, the presence of carbonate minerals such as Calcite (3.6%) and Dolomite (1.65%) emphasizes variability in the chemical properties of the samples, which can have implications for the reactivity of soils and aggregates in concrete mixes. These findings guide our understanding toward geotechnical and concrete formulation considerations specific to the Khouribga-Settat region.[14]

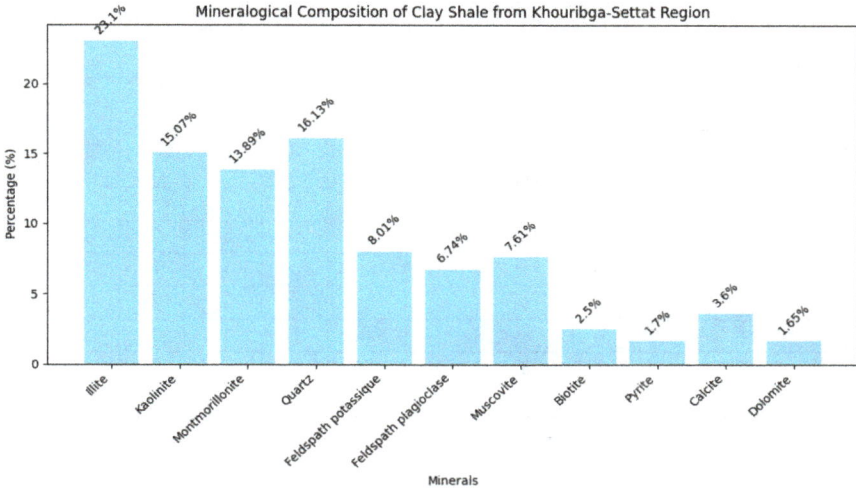

Mineralogical Composition of Settat-Khouribga clay Shale

2. Particle Size Distribution :

The presented table delineates the granulometry, hardness, and plasticity characteristics of distinct samples identified by sample numbers. Granulometry, expressed as the percentage passing through varying sieve sizes from 40 mm to 0.08 mm, signifies the particle size distribution. Mohs' Hardness Scale (MDE) offers insights into the samples' resistance to abrasion, while Liquid Limit (L.A) and Plasticity Index (IP) quantify their plasticity.

Upon scrutiny, the samples exhibit diverse granulometric profiles, depicting the percentage passing through sieves ranging from 40 mm to 0.08 mm. The consistently low MDE values, below 25 for all samples, indicate a relatively modest resistance to abrasion. Plasticity, as gauged by L.A and IP, elucidates the material's capacity for deformation without fracturing.

Remarkably, samples S1, S2, and S5 manifest elevated liquid limits (L.A) and plasticity indices (IP), suggesting an increased potential for deformation. Sample S6 stands out with a higher IP of 20.00, indicating heightened plasticity. These nuanced variations in granulometry, hardness, and plasticity furnish valuable insights for geotechnical considerations, particularly in construction projects where soil behavior assumes pivotal importance. A nuanced understanding of these properties facilitates the formulation of tailored construction materials and the design of foundations that meticulously account for the specific characteristics of the shale samples from the Khouribga-Settat region.[14]

Table 1:Representative table of geotechnical characteristics of shale clay samples

Sample number	Granulometry					Hardness		Plasticity
	% passing through sieve size (mm)					MDE	L.A	IP
	40	20	10	2	0,08	< 25	<30	< 12
	100	60	40	20	2	17	24	18.00
		to 90	to 70	to 48	to 14			
S1	100	70	50	42	35	17	24	18.00
S2	100	66	62	40	34	18	25	19.50
S3	100	71	63	41	36	17	23	18.50
S4	100	68	49	45	32	18	24	18.00
S5	100	65	56	43	35	19	22	19.00
S6	100	67	50	44	36	17	25	20.00

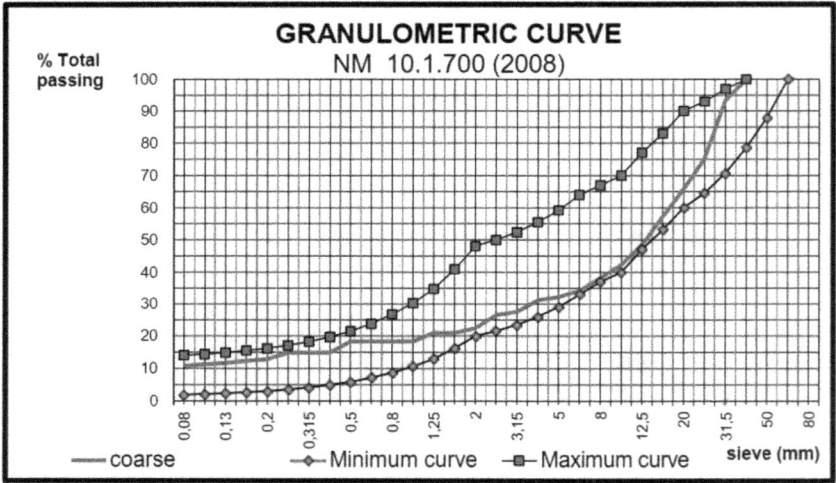

This analysis opens up avenues for further research, suggesting complementary mineralogical and geotechnical analyses for a comprehensive understanding of material properties. Additional tests, such as moisture content analysis and specific gravity measurements, could provide supplementary insights. Overall, this granulometric analysis constitutes a fundamental step for informed decision-making in materials engineering within the construction domain, shedding light on the specificity of clay shale from the Khouribga Settat region.[15]

3. Mechanical Properties of clay Shale:
Concrete samples incorporating Settat-Khouribga shale exhibited promising results in terms of compressive strength, flexural strength, and durability. Table 4 presents a comparison between Conventional concrete and shale-containing concrete, highlighting the approximate performance improvements.

Mediterranean Architectural Heritage - RIPAM10   Materials Research Forum LLC
Materials Research Proceedings 40 (2024) 240-247   https://doi.org/10.21741/9781644903117-26

Table 2: Performance of Concrete Samples

Concrete Property	Concrete Property
Compressive Strength	30 MPa
Elastic Modulus	25 GPa
Flexural Strength	5 MPa
Porosity	8%
Permeability	Low
Tensile Strength	3 MPa
Shrinkage and Swelling	Low

The compressive strength of 30 MPa was determined using the uniaxial compression test method,[16] where concrete samples were subjected to axial loads until failure. The elastic modulus of 25 GPa was assessed through bending tests or ultrasonic tests,[17] providing a measure of the material's stiffness. The flexural strength of 5 MPa was obtained through specific moroccan norme ,(IMANOR), highlighting the concrete's ability to withstand bending forces. The porosity of 8% was quantified using porosimetry techniques or image analysis,[18] offering an indication of the volume of pores in the material. Low permeability was evaluated through water permeability tests, measuring the concrete's ability to resist water penetration. The tensile strength of 3 MPa was measured using direct or indirect tensile tests,[19] providing an assessment of the concrete's ability to withstand tensile forces. Low levels of shrinkage and swelling were evaluated by measuring dimensional changes in samples subjected to controlled drying and humidification conditions. Finally, excellent adhesion between materials was determined through pull-out or shear tests, highlighting the structural integrity of the composite. Overall, this comprehensive geotechnical methodology facilitated a thorough assessment of the mechanical and physical performances of the studied concrete.[20]

The results indicate that concrete samples incorporating Settat-Khouribga shale demonstrated improved mechanical properties compared to conventional concrete[21].

The enhanced performance is attributed to the unique mineralogical composition and particle size distribution of the shale .

Conclusion

The study exploring the integration of Settat-Khouribga shale into reinforced concrete provides a compelling argument for its potential as a sustainable and locally-sourced material. The findings validate the initial hypothesis that Settat-Khouribga shale holds promise as a viable supplementary material in concrete production. The mineralogical composition significantly contributes to improving structural strength, while the particle size distribution enhances packing efficiency within the concrete mix. The heightened performance observed in shale-containing concrete has noteworthy implications for advancing sustainable construction practices, offering a locally-sourced alternative that not only reduces environmental impact but also promotes regional economic development. The study's outcomes emphasize the feasibility and performance of Settat-Khouribga shale as a supplementary material in concrete production. The observed improvements in mechanical properties and durability in concrete with clay shale advocate for its adoption in construction practices. This research advances scientific understanding and holds practical implications for the construction industry, particularly in regions abundant in shale deposits like Settat-Khouribga. Integrating such local materials can stimulate economic growth, reduce environmental impact, and contribute to sustainable development. The results encourage further exploration and application of Settat-Khouribga shale in construction, with the potential to influence construction practices and contribute to the socio-economic development of the region.

Mediterranean Architectural Heritage - RIPAM10   Materials Research Forum LLC
Materials Research Proceedings 40 (2024) 240-247   https://doi.org/10.21741/9781644903117-26

References

[1] , M. VIEIRA, &, A. GONCALVES (2000). Durability of high performance lightweight aggregate concrete. In Second international symposium on structural lightweight aggregate concrete (Kristiansand, 18-22 June 2000) (pp. 767-773).

[2] C Arambourg, J Signeux. & F.M. Bergonioux 1952. Les vertébrés fossiles des phosphates de l'Afrique du Nord. Notes et Mémoires du Service. Géologique du Maroc, 92, 396 p.

[3] N. El Yakoubi 2006. Potentialités d'utilisation des argiles marocaines dans l'industrie céramique : cas des gisements de Jbel Kharrou et de Ben Ahmed (Meseta marocaine occidentale). Thèse de Doctorat ès-sciences, Université Mohamed V, Rabat, 212 p. https://doi.org/10.1016/j.crte.2006.03.017

[4] NM 00.1.004 Tamisage - Analyse granulométrique par tamisage.

[5] NM 10.1.005 Liants hydrauliques - Techniques des essais.

[6] X. Zheng, & B. Zhang (2005). Effect of pre-wetted shale ceramsite on strength and frost-resistance of lightweight aggregate concrete. JOURNAL-CHINESE CERAMIC SOCIETY, 33(6), 758.

[7] E. L El Houssine , A. MRIDEKH, , B. EL MANSOURI, , M. TAMMAL, &, M EL BOUHADDIOUI. (2014). Apport des données géophysiques et géologiques à la mise en évidence de nouveaux éléments structuraux associés à la flexure de Settat (Maroc central) Contribution of geophysical and geological data for the identification of new structural elements related to the Settat flexure (central Morocco). Bulletin de l'Institut Scientifique, Rabat, (36), 109-121.

[8] G Beaudet. 1969. Le plateau central marocain et ses bordures, étude géomorphologique (bordure sud-ouest et sud du plateau central). Thèse de Doctorat Lettres, Université Mohammed V, Rabat, 478 p.

[9] NM 10.1.707 Essais pour déterminer les caractéristiques mécaniques et physiques des granulats - Méthode pour la détermination de la masse volumique en vrac et de la porosité intergranulaire.

[10] NM 10.1.004 Liants hydrauliques – Ciments – Composition, spécifications et critères de conformité.

[11] H Li., B Lai., &, S. Lin (2016). Shale mechanical properties influence factors overview and experimental investigation on water content effects. Journal of Sustainable Energy Engineering, 3(4), 275-298. https://doi.org/10.7569/JSEE.2016.629501

[12] NM 15.2.001 Instruments de pesage à fonctionnement non automatique Partie 1 : Exigences métrologiques et techniques – Essais."

[13] A. Cousture Agniel (2023). Étude de l'activation alcaline des composés calcaires et siliceux pour la formulation de nouveaux matériaux de construction à faible impact environnemental (Doctoral dissertation, CY Cergy Paris Université).

[14] X. G. Zhang, , X. M Kuang, J. H. Yang, &, S. R. Wang (2017). Experimental study on mechanical properties of lightweight concrete with shale aggregate replaced partially by nature sand. Electronic Journal of Structural Engineering, 17, 85-94. https://doi.org/10.56748/ejse.17222

[15] A. A. HAMMAMID AMMAR (2023). Identification des caractéristiques physiques, mécaniques et minéralogiques des matériaux carbonatés destiné aux travaux de BTPH, cas des

carrières d'agrégats de la région d'ibn badais, wilaya de Constantine (Doctoral dissertation, faculté des sciences et de la technologie* univ bba).

[16] NM 10.1.051 Essai pour béton durci - Résistance à la compression des éprouvettes.

[17] NM 10.1.068 Essai pour béton durci - Confection et conservation des éprouvettes pour essais de résistance.

[18] NM 10.1.072 Essai pour béton durci - Masse volumique du béton.

[19] NM 10.1.052 Essai pour béton durci - Résistance en traction par fendage d'éprouvettes.

[20] M TAZIR. (2023). tude comparative des normes régissant la production du béton prêt à l'emploi dans les 3 pays maghrébins.

[21] J. K. Norvell, , J. G. Stewart, , M. C Juenger., & Fowler, D. W. (2007). Influence of clays and clay-sized particles on concrete performance. Journal of materials in civil engineering, 19(12), 1053-1059. https://doi.org/10.1061/(ASCE)0899-1561(2007)19:12(1053)

Mediterranean Architectural Heritage - RIPAM10
Materials Research Proceedings 40 (2024) 248-259

Materials Research Forum LLC
https://doi.org/10.21741/9781644903117-27

Application of Raw Moroccan Clay as a Potential Adsorbent for the Removal of Malachite Green Dye from an Aqueous Solution: Adsorption Parameters Evaluation and Thermodynamic Study

Mohammed CHRACHMY[1,a,*], Rajae GHIBATE[2,b], Najia EL HAMZAOUI[3,c]
Mahdi LECHHEB[1,d], Hassan OUALLAL[1,e], and Mohamed AZROUR[1,f]

[1] Laboratory of Materials Engineering for the Environment and Natural Resources, Faculty of Science and Technology, Moulay Ismail University, Errachidia, 52000, Morocco

[2] Laboratory of Physical Chemistry, Materials and Environment, Faculty of Sciences and Technologies, Moulay Ismail University, Errachidia, 52000, Morocco

[3] Laboratory of Ecology and Biodiversity of Wetlands, Faculty of Sciences, Moulay Ismail University, Meknes, 11201, Morocco

[a] mo.chrachmy@edu.umi.ac.ma, [b] rajae.ghibate@gmail.com
[c] najia.elhamzaoui@gmail.com, [d] m.lechheb@edu.umi.ac.ma
[e] hassanouallalaghbalou@gmail.com, [f] m.azrour@fste.umi.ac.ma

Keywords: Raw Clay, Adsorption, Malachite Green, Parameters Effects, Thermodynamic Study

Abstract. This study uses raw clay from the Es-sifa (REC) territorial commune in the region of Draa-Tafilalet, Morocco, for malachite green (MG) adsorption from an aqueous medium. The characterization of clay was ensured by Fourier Transform Infrared (FTIR) spectroscopy, X-ray diffraction (XRD), and determining the Point of Zero Charge pH. Moreover, five parameters were considered to regulate the adsorption process, including time of contact, initial dye concentration, mass of the adsorbent, stirring speed, and temperature. The finding displayed that equilibrium was reached after only 40 minutes of contact. The maximum adsorption amount achieved was 89.61 mg/g for a dye concentration of 10^{-4} M at 25 °C, using a REC mass of 0.10 g. According to the thermodynamic study, the endothermicity and spontaneity of MG adsorption were disclosed. Malachite green's fixation on the REC surface was verified by comparing the clay FTIR spectra before and after adsorption.

Introduction

Clay is a material that is abundant and cost-effective and is known for its unique properties. Its applications range from producing cement to manufacturing raw or fired building bricks, cosmetics, and therapeutics. Clays are fine-grained minerals that can harden when dried or fired, making them ideal for building bricks. Clays typically contain phyllosilicates and other substances like quartz and calcite, which impart plasticity or hardening properties when dried or fired [1]. Most natural clays are a blend of different mineralogical phases, and their distinct properties arise from differences in structure and charge. Fine-grained silicate minerals found in clays have a negative charge on their structure, which enables them to adsorb positively charged cations like cationic dyes on their surfaces. Generally, clays are characterized by a notable adsorption capacity, given their specific surface area, which may reach up to 800 m^2/g [2].

This work investigates the potential use of Raw Es-sifa Clay (REC) as an adsorbent for treating malachite green polluted solutions. Es-sifa is a Moroccan commune in the Errachidia province, in the Drâa-Tafilalet region.

It is important to note that malachite green (MG) is a harmful pollutant that can cause cancer and mutations in organisms [3, 4]. Therefore, it is essential to remove it from the environment. In

Mediterranean Architectural Heritage - RIPAM10   Materials Research Forum LLC
Materials Research Proceedings 40 (2024) 248-259   https://doi.org/10.21741/9781644903117-27

this respect, the impact of various reactional parameters on MG uptake onto REC was studied. That includes contact time, quantity of clay, MG initial concentration, stirring speed, and temperature. The clay underwent XRD, FTIR, and pH_{pzc} characterization to evaluate its properties. Additionally, the FTIR analysis was conducted before and after adsorption to study any changes that occurred.

Experimental study
Adsorbate
The malachite green oxalate, which has the molecular formula $C_{52}H_{54}N_4O_{12}$, was provided by VWR Chemicals and used without any additional purification. Fig. 1 depicts its chemical structure. The dye powder was weighed accurately and dissolved in demineralized water to create a dye stock solution. From this solution, a series of well-defined concentrations were prepared.

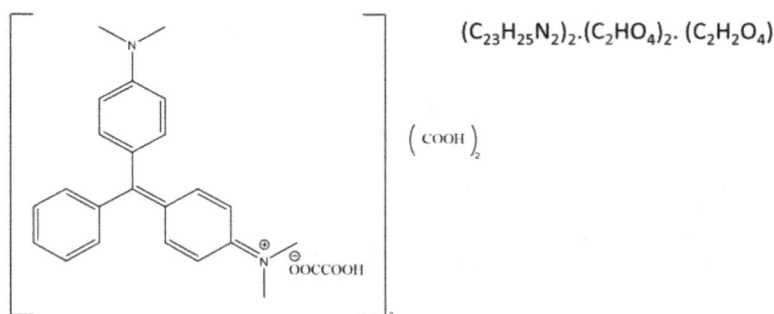

$$(C_{23}H_{25}N_2)_2.(C_2HO_4)_2. (C_2H_2O_4)$$

Figure 1: Malachite green oxalate chemical structure

Clay preparation
The clay used in this work comes from Es-sifa, Morocco. The clay is ground and sieved using ISO standard mesh sieves. Only a portion of particles smaller than 315 μm is kept for adsorption experiments.

Clay particles characterization
The REC was analyzed using X-ray diffraction and IR spectroscopy, as well as by determining pH_{pzc}.

The clay mineralogy was described using the X-ray powder diffraction method. The X-ray diffraction pattern was captured by an X'Pert PRO MPD wide-angle powder diffractometer equipped with a diffracted beam monochromator and Ni-filtered CuKα radiation (λ=1.5406 Å). To take the measurement, the 2θ angle was scanned in increments of 0.02° for 2 seconds, ranging from 10° to 30°.

The clay's vibrational spectra were analyzed with a VERTEX70 model instrument using the Attenuated Total Reflectance method for FTIR, in the 4000-400 cm^{-1} region.

The charge on the surface of clay changes based on the solution pH. Thus, when pH equals pH_{pzc}, it is zero. In order to measure the pH_{pzc}, a series of 50 mL of NaCl solution at the concentration of 0.1 M was prepared, and their pH was adjusted with NaOH or HCl to values between 2 and 12. The sample was then added to each solution at an amount of 0.1 g. The suspensions were agitated at 25°C for 72 hours. The mixtures were then filtered, and the filtrate's equilibrium pH was measured using Accumet AB15 Basic pH meter. The pH_{pzc} is determined by

Mediterranean Architectural Heritage - RIPAM10

Materials Research Proceedings 40 (2024) 248-259

Materials Research Forum LLC

https://doi.org/10.21741/9781644903117-27

finding the intersection of bisector (pHf = pHi) with the curve of the final pH against the initial pH [5].

Batch adsorption experiments

This work aims to study the main adsorption parameters that may affect MG uptake on the REC: time of contact, temperature, stirring speed, initial concentration of adsorbent, and mass of adsorbate.

To examine the impact of the adsorption time on MG removal, a series of experiments was conducted. The experiment involved stirring 0.10 g of REC in 100 mL of a $5{,}10^{-5}$ M MG solution at 250 rpm and 25 °C. The experiments were conducted within a time interval of 5 to 60 minutes. to determine the impact of adsorbent mass, the experiments were performed at 25 °C with a stirring speed of 250 rpm by varying the adsorbent mass from 5 to 25 mg at a concentration of $5{,}10^{-5}$ M.

A number of tests were carried out at 25°C to look into the impact of the initial concentration of malachite green solution. 100 mL of MG solution was mixed with 0.10 g of clay was added to for each experiment, and the mixture was agitated at 250 rpm. The initial concentration was varied between 5, 10^{-6}, and 10^{-4} M.

The effect of stirring speed was evaluated a three speeds 250, 500, and 750 rpm with an MG concentration of 5, 10^{-5} M, using a dose of 1 g/L at 25 °C.

Furthermore, experiments were conducted at 25, 35, and 45°C using 0.10 g adsorbent per 100 mL adsorbate of 5, 10^{-5} M MG concentration with 250 rpm stirring to study the temperature effect.

After each experiment, the mixture is filtrated, and the residual dye concentration is determined with Shimadzu, UV-160 UV/Visible spectrometer at 620 nm.

For calculated the adsorbed amount and adsorption efficiency, the following equations were utilized:

$$q_t = \frac{C_0 - C_t}{m} \times V_{MG} \tag{1}$$

$$R = \frac{C_0 - C_t}{C_0} \times 100 \tag{2}$$

Where R represents the adsorption efficiency (%), q_t, corresponds to the adsorbed amount of dye at t time (mg/g), C_e and C_0 reveal the t time residual and initial concentrations (M), respectively, m refers to the clay mass (g), V_{MG} is the malachite green solution volume (L).

Results and Discussion

Clay particles characterization

Fig. 2 depicts the REC diffractogram, which indicates the presence of several mineral phases. These include silica in the form of quartz (Q), kaolinite (K), dolomite (D), and calcite (C) [6–10].

Mediterranean Architectural Heritage - RIPAM10

Materials Research Proceedings 40 (2024) 248-259

Materials Research Forum LLC

https://doi.org/10.21741/9781644903117-27

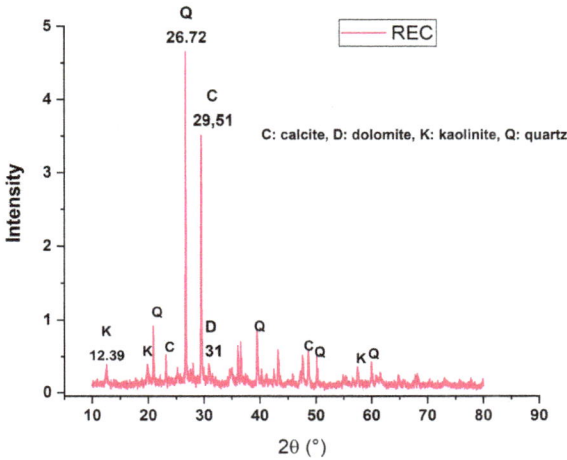

Figure 2: XRD pattern of REC

Fig. 3 presents the infrared spectrum of REC, which displays several absorption bands. The stretching vibration of the hydroxyl group in various settings (Al, AlOH), (Al, MgOH), or (Al, FeOH) is responsible for a band at 3619 cm^{-1} [11,12]. An additional band at 3420 cm^{-1} could be associated with the stretching and bending vibrations of H$_2$O that are adsorbed on the clay's surface and between its layers [13,14]. The deformation and elongation vibration of the calcite (CaCO$_3$) is responsible for the bands at 1799, 1434, 874, and 713 cm^{-1} [15]. The deformation and elongation vibrations of the quartz are allocated for the bands at 778 and 692 cm^{-1} [16]. The bonds that characterize the deformation of the Si-O bond in quartz are respected in environments of 517 and 471 cm^{-1} [17,18].

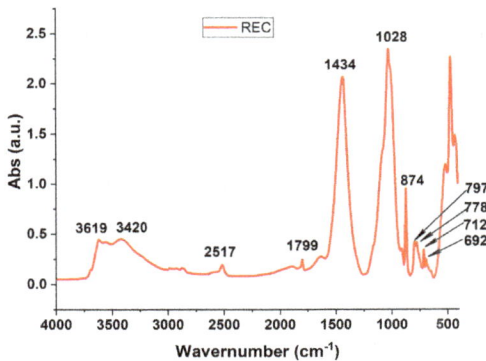

Figure 3: FTIR spectrum of REC

Fig. 4 shows that the pH at which the REC reaches zero charge, also known as pH$_{pzc}$, is 8.41. Below this pH, the REC surface is charged positively, while above it, the surface becomes

Mediterranean Architectural Heritage - RIPAM10   Materials Research Forum LLC
Materials Research Proceedings 40 (2024) 248-259   https://doi.org/10.21741/9781644903117-27

negatively charged. When malachite green is dissolved in water, it releases positively charged malachite green ions and negatively charged oxalate ions. The elimination of these ions can occur at pH both lower and higher than pH_{pzc}, but with different affinities. At a pH higher than pH_{pzc}, retention occurs due to the electrostatic interactions between the cations of malachite green and the negatively charged surface of the adsorbent. Additionally, hydrogen bonds form between the oxalate ions and the OH groups of the adsorbent. However, at a pH lower than pH_{pzc}, elimination takes place due to the hydrogen bond formation between the malachite green ion and the OH groups of the clay. The fixation of the oxalate ion can be done through electrostatic interactions with the surface and/or the formation of hydrogen bonds with the OH groups.

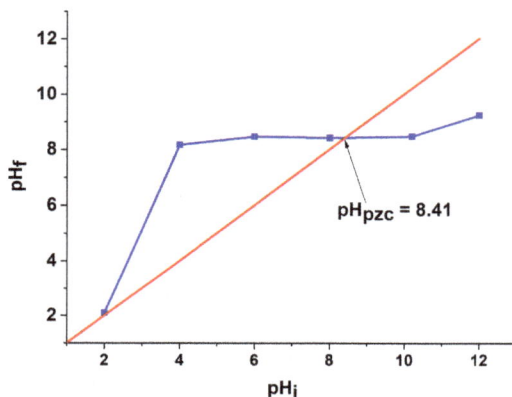

Figure 4: pH_{pzc} determination curve of the REC

Adsorption study
Adsorption is a process where a solute attaches to the surface of an adsorbent. This process is in a dynamic state when it reaches equilibrium. The nature of the adsorbent and the solute, as well as operational factors such as the time of contact, the adsorbate initial concentration, the mass of the adsorbent, the stirring rate, and the temperature of the medium, generally govern adsorption equilibrium. Therefore, it is crucial to study how these operational parameters affect the fixation of malachite green on REC.

Effect of contact time
Fig. 5 illustrate the impact of contact time on the adsorption efficiency of MG onto REC. The adsorbed amount rose rapidly in the first 5 minutes, reaching over 91 %. However, beyond 5 minutes, the adsorption efficiency of MG increased slowly, reaching equilibrium at about 40 minutes. This can be explained by the initial abundance of active sites which gradually became occupied over time, resulting in no significant difference in the adsorption efficiency [19].

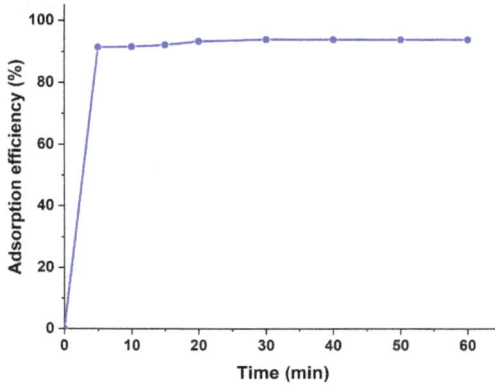

Figure 5: Contact time's impact on MG adsorption onto REC

Effect of adsorbent mass

In order to demonstrate how the adsorbent mass affects the MG retention, three different masses were tested: 0.05, 0.10, and 0.25 g. These masses were added to 100 mL of adsorbate with a concentration of $5,10^{-5}$ M at varying contact times. After 40 minutes, equilibrium was reached for all three masses with different responses, as shown in Fig. 6. The results indicate that as the mass of adsorbent increases, the percentage of MG removal also increases. This can be explained by the increase in retention sites [14,20]. Based on these findings, a mass of 0.10 g REC/100 mL MG was determined to be the optimal mass for further parameter studies.

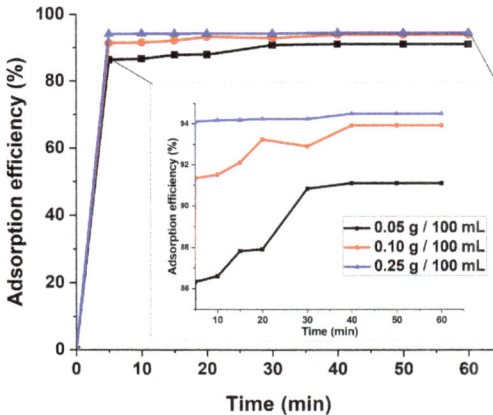

Figure 6: Adsorbent mass's effect on MG adsorption onto REC

Effect of MG initial concentration

The influence of the initial dye concentration on the adsorption of MG onto REC was investigated by altering the concentration within the range of 5.10^{-6} to 10^{-4} M at various contact times, as illustrated in Fig. 7 The results indicated that an elevation in the concentration of the adsorbate led to an increased quantity of adsorption. That can be related to the enhancement of driving force due to the concentration gradient [21]. The 10^{-4} M MG initial concentration yielded the highest adsorption capacity, 89.61 mg/g.

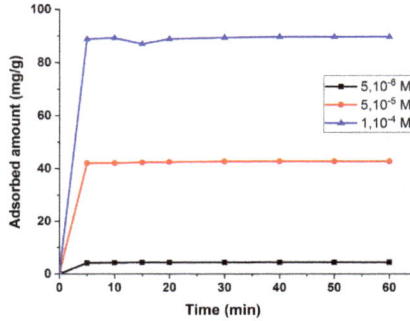

Figure 7: Initial dye concentration's impact on MG adsorption onto REC

Effect of agitation speed

Adsorption experiments were conducted at three different agitation speeds (250, 500, and 750 rpm) to investigate how agitation affects the adsorption of MG by REC. For each experiment, 0.10 g of the clay was added to 100 mL of a $5,10^{-5}$ M adsorbate solution at 25°C. The graphs depicting the variation of the adsorbed amount over time looked similar across all three speeds (Fig. 8). The adsorbed amount slightly increased with higher agitation speed, rising from 42.68 mg/g at 250 rpm to 43.17 mg/g at 750 rpm. This finding indicates that agitation speed does not significantly impact MG adsorption. Therefore, the optimal stirring speed for further parameter studies was selected as 250 rpm since it is sufficient to enhance contact between MG ions and REC particles.

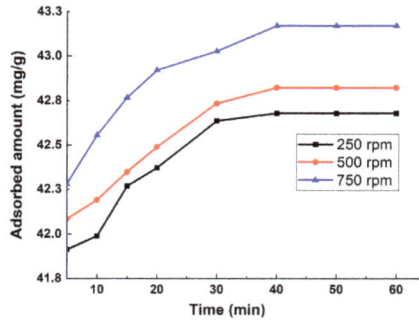

Figure 8: The impact of agitation speed on the MG adsorption onto REC

Effect of temperature

Fig. 9 illustrates how temperature changes affect MG adsorption onto REC. It was found that clay's adsorption capacity marginally increases with temperature. In fact, by raising the temperature from 25 to 45 °C, the amount adsorbed enhanced from 42.68 to 44.43 mg/g. That suggests that the MG adsorption process on REC is endothermic [22].

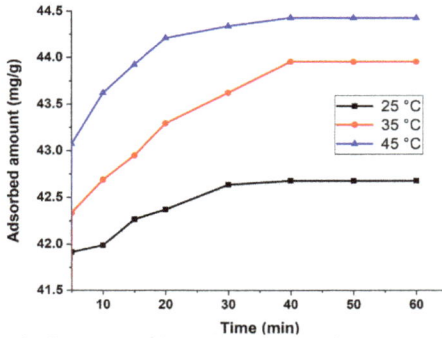

Figure 9: Temperature's impact on MG adsorption onto REC

Thermodynamic study

The thermodynamics of adsorption can be described by three parameters: entropy ($\Delta S°$), standard enthalpy ($\Delta H°$) and standard free energy ($\Delta G°$). The values of $\Delta S°$ and $\Delta H°$ can be determined through the following equation:

$$lnK_d = \frac{-\Delta H°}{RT} + \frac{\Delta S°}{R} \tag{3}$$

Where K_d represents the distribution coefficient ($K_d = q_e/C_e$), R (8314 J/mol K) and T (K) correspond to the gas constant and absolute temperature, respectively.

Plotting ln K_d against the inverse of temperature yields, from the slope and intercept, the values of enthalpy and entropy standards respectively (Fig. 10).

The following formula was used to get the standard free energy, $\Delta G°$:

$$\Delta G° = \Delta H° - T\Delta S° \tag{4}$$

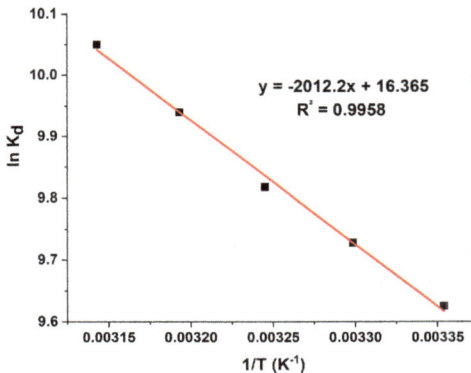

Figure 10: Van't Hoff plot for determining thermodynamic parameters

Mediterranean Architectural Heritage - RIPAM10   Materials Research Forum LLC
Materials Research Proceedings 40 (2024) 248-259   https://doi.org/10.21741/9781644903117-27

Table 1: shows the thermodynamic values obtained.

T (K)	$\Delta G°$ (kJ/mol)	$\Delta H°$ (kJ/mol)	$\Delta S°$ (J/ mol . K)
298	-23,859	16,729	136,067
308	-25,153		
318	-26,583		

According to Table 1, The MG adsorption process on the examined clay is spontaneous, as indicated by the negative values of $\Delta G°$, and this spontaneity increases with temperature [23]. the positive sign's $\Delta S°$ value suggests that adsorption process increases disorder at the interface between the adsorbent and adsorbate Additionally Values of $\Delta H°$ that are positive verify the endothermicity of the adsorption process[21].

FTIR characterization
Fig. 11 shows REC's FTIR spectrum following MG adsorption. Upon analyzing the spectra of the studied clay before and after adsorption (Figs 3 and 11), it was observed that the REC spectra were quite similar. However, after adsorption, the REC spectrum displayed characteristic low-intensity malachite green bands at 1615, 1586, and 1163 cm^{-1}. The first two peaks represent aromatic C-C stretching, while the third is associated with C-N stretching.

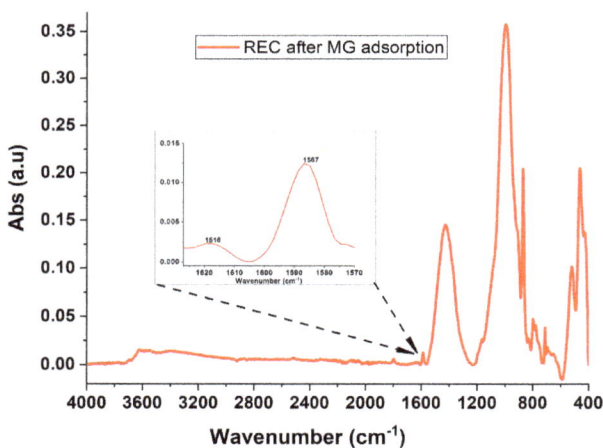

Figure 11: REC's FTIR spectrum following MG adsorption

Comparative study
It would be wise to compare REC's adsorption capacity with other adsorbents studied by other researchers to remove MG. The clay under investigation has a comparatively better adsorption capacity than the others, as indicated in Table 2, which implies that it may be applied as an inexpensive adsorbent for the additional cationic dyes' removal.

Table 2: Analysis of MG adsorption on various adsorbent materials

Adsorbent materials	q_{ads} (mg/g)	Reference
Fly ash	40.65	[24]
Rice husk activated carbon	49.62	[25]
Persian Kaolin	52.91	[26]
Centaurea solstitialis plants	91.00	[27]
Hen feathers	26.10	[28]
Iron humate	19.20	[29]
Bentonite	7.72	[30]
Activated slag	74.20	[31]
Sugar cane dust	4.88	[32]
Raw Es-sifa Clay	89.61	This work

Conclusion

The study shows that raw Es-sifa clay (REC) is effective in adsorbing MG. The adsorption capacity of REC is influenced by various reaction parameters. Among others, time of contact, adsorbent mass, initial concentration of dye, agitation speed, and the temperature of the medium. The MG adsorption process is fast, achieving equilibrium in about 40 minutes at an initial dye concentration of 10^{-4} M and reaching an adsorption capacity of 89.60 mg/g. Furthermore, endothermic and spontaneous adsorption processes are indicated by $\Delta G°$ and $\Delta H°$ negative values, respectively. A positive result for $\Delta S°$ depicted that, during the process, there is more disorder at the interface between the adsorbent and adsorbate. This study highlights that, in comparison to other adsorbents examined by other researchers, REC has a comparatively substantial adsorption capability. That shows that REC has the potential for MG adsorption. Based on these results, REC is a cost-effective option for removing MG dye and other cationic dyes.

References

[1] A. Kausar, M. Iqbal, A. Javed, K. Aftab, Z.-H. Nazli, H.N. Bhatti, S. Nouren, Dyes adsorption using clay and modified clay: A review, J. Mol. Liq. 256 (2018) 395–407. https://doi.org/10.1016/j.molliq.2018.02.034

[2] D.L. Carter, M.M. Mortland, W.D. Kemper, Specific Surface, in: Arnold Klute (Eds), Methods of Soil Analysis: Part 1 Physical and Mineralogical Methods, John Wiley & Sons, Ltd, 1986, pp. 413–423. https://doi.org/10.2136/sssabookser5.1.2ed.c16

[3] X. Jia, J. Li, E. Wang, Lighting-up of the dye malachite green with Mercury(II)–DNA and its application for fluorescence turn-off detection of cysteine and glutathione, Chem. Eur. J. 18 (2012) 13494–13500. https://doi.org/10.1002/chem.201103768

[4] A. Khodabakhshi, M.M. Amin, Determination of malachite green in trout tissue and effluent water from fish farms, Int. J. Environ. Health Eng. 1 (2012) 51–56.

[5] Z. Hicham, Z. Bencheqroun, I.E. Mrabet, M. Kachabi, M. Nawdali, I. Neves, Removal of basic dyes from aqueous solutions by adsorption onto Moroccan clay (Fez city), Mediterr. J. Chem. 8 (2019) 158–167. https://doi.org/10.13171/mjc8319050803hz

[6] H. Ouallal, M. Azrour, M. Messaoudi, H. Moussout, L. Messaoudi, N. Tijani, Incorporation effect of olive pomace on the properties of tubular membranes, J. Environ. Chem. Eng. 8 (2020) 103668. https://doi.org/10.1016/j.jece.2020.103668

[7] S. Mahmoudi, A. Bennour, A. Meguebli, E. Srasra, F. Zargouni, Characterization and traditional ceramic application of clays from the Douiret region in South Tunisia, Appl. Clay Sci. 127–128 (2016) 78–87. https://doi.org/10.1016/j.clay.2016.04.010

[8] I. Akhrif, L. Mesrar, M.E. Jai, M. Benhamou, R. Jabrane, Elaboration and X-Ray Diffraction Techniques Characterization of clay-PEG 6000 Nanocomposites with clay Matrix, Int. J. Multidiscip. Current Res. 3 (2015) 564–571.

[9] P.E. Tsakiridis, M. Samouhos, M. Perraki, Valorization of dried olive pomace as an alternative fuel resource in cement clinkerization, Constr Build Mater. 153 (2017) 202–210. https://doi.org/10.1016/j.conbuildmat.2017.07.102

[10] N. Kaya, M. Atagur, O. Akyuz, Y. Seki, M. Sarikanat, M. Sutcu, M.O. Seydibeyoglu, K. Sever, Fabrication and characterization of olive pomace filled PP composites, Compos. B: Eng. 150 (2018) 277–283. https://doi.org/10.1016/j.compositesb.2017.08.017

[11] T. Sheela, Y.A. Nayaka, Kinetics and thermodynamics of cadmium and lead ions adsorption on NiO nanoparticles, Chem. Eng. J. 191 (2012) 123–131. https://doi.org/10.1016/j.cej.2012.02.080

[12] D. Hank, Z. Azi, S. Ait Hocine, O. Chaalal, A. Hellal, Optimization of phenol adsorption onto bentonite by factorial design methodology, J. Ind. Eng. Chem. 20 (2014) 2256–2263. https://doi.org/10.1016/j.jiec.2013.09.058

[13] O. Abdelwahab, N.K. Amin, Adsorption of phenol from aqueous solutions by Luffa cylindrica fibers: Kinetics, isotherm and thermodynamic studies, Egypt. J. Aquat. Res. 39 (2013) 215–223. https://doi.org/10.1016/j.ejar.2013.12.011

[14] V. Zaspalis, A. Pagana, S. Sklari, Arsenic removal from contaminated water by iron oxide sorbents and porous ceramic membranes, Desalination. 217 (2007) 167–180. https://doi.org/10.1016/j.desal.2007.02.011

[15] J.R.O. Kikouama, K.L. Konan, A. Katty, J.P. Bonnet, L. Baldé, N. Yagoubi, Physicochemical characterization of edible clays and release of trace elements, Appl. Clay Sci. 43 (2009) 135–141. https://doi.org/10.1016/j.clay.2008.07.031

[16] A. Qlihaa, S. Dhimni, F. Melrhaka, N. Hajjaji, A. Srhiri, Caractérisation physico-chimique d'une argile Marocaine [Physico-chemical characterization of a morrocan clay], J. Mater. Environ. Sci. 7 (2016) 1741-1750.

[17] P. Sabbatini, F. Rossi, G. Thern, A. Marajofsky, M.M.F. de Cortalezzi, Iron oxide adsorbers for arsenic removal: A low cost treatment for rural areas and mobile applications, Desalination. 248 (2009) 184–192. https://doi.org/10.1016/j.desal.2008.05.104

[18] K.P. Raven, A. Jain, R.H. Loeppert, Arsenite and Arsenate Adsorption on Ferrihydrite: Kinetics, equilibrium, and adsorption envelopes, Environ. Sci. Technol. 32 (1998) 344–349. https://doi.org/10.1021/es970421p

[19] R. Ghibate, O. Senhaji, R. Taouil, Valuation of Pomegranate Peel for Cationic Dye Removal, Int. J. Eng. Res. Appl. 10 (2020) 19–22.

[20] R. Ghibate, F. Sabry, O. Senhaji, R. Taouil, M. Touzani, State of the art of technologies for Zn^{2+} ions removal from industrial effluents with adsorption: Examination of process parameters (Part I), Int. J. Innov. Res. Technol. Sci. Eng. 2 (2015) 39–48.

[21] R. Ghibate, O. Senhaji, R. Taouil, Kinetic and thermodynamic approaches on Rhodamine B adsorption onto pomegranate peel, Case Stud. Chem. Environ. Eng. 3 (2021) 100078. https://doi.org/10.1016/j.cscee.2020.100078.

[22] M. Messaoudi, M. Douma, N. Tijani, Y. Dehmani, L. Messaoudi, Adsorption process of the malachite green onto clay: Kinetic and thermodynamic studies, Desalination Water Treat. 240 (2021) 191–202. https://doi.org/10.5004/dwt.2021.27688

[23] E. Bazrafshan, P. Amirian, A.H. Mahvi, A. Ansari-Moghaddam, Application of adsorption process for phenolic compounds removal from aqueous environments: a systematic review, Glob. Nest J. 18 (2016) 146–163. https://doi.org/10.30955/gnj.001709

[24] A. Witek-Krowiak, R.G. Szafran, S. Modelski, A. Dawiec, Removal of cationic dyes from aqueous solutions using microspherical particles of fly ach, Water Environ. Res. 84 (2012) 162–170. https://doi.org/10.2175/106143011X13233670703657

[25] Y.C. Sharma, Adsorption characteristics of a low-cost activated carbon for the reclamation of colored effluents containing malachite green, J. Chem. Eng. Data. 56 (2011) 478–484. https://doi.org/10.1021/je1008922

[26] A.R. Tehrani-Bagha, H. Nikkar, N.M. Mahmoodi, M. Markazi, F.M. Menger, The sorption of cationic dyes onto kaolin: Kinetic, isotherm and thermodynamic studies, Desalination. 266 (2011) 274–280. https://doi.org/10.1016/j.desal.2010.08.036

[27] M. Saleh, M. Yalvaç, H. Arslan, M. Gün, Malachite green dye removal from aqueous solutions using Invader Centaurea Solstitialis plant and optimization by response surface method: Kinetic, isotherm, and thermodynamic study, Eur. J. Sci. Tech. (2019) 755–768. https://doi.org/10.31590/ejosat.643238

[28] A. Mittal, Adsorption kinetics of removal of a toxic dye, Malachite Green, from wastewater by using hen feathers, J. Hazard. Mater. 133 (2006) 196–202. https://doi.org/10.1016/j.jhazmat.2005.10.017

[29] P. Janoš, Sorption of basic dyes onto iron humate, Environ. Sci. Technol. 37 (2003) 5792–5798. https://doi.org/10.1021/es020142o.

[30] S.S. Tahir, N. Rauf, Removal of a cationic dye from aqueous solutions by adsorption onto bentonite clay, Chemosphere. 63 (2006) 1842–1848. https://doi.org/10.1016/j.chemosphere.2005.10.033

[31] V.K. Gupta, S.K. Srivastava, D. Mohan, Equilibrium uptake, sorption dynamics, process optimization, and column operations for the removal and recovery of malachite green from wastewater using activated carbon and activated slag, Ind. Eng. Chem. Res. 36 (1997) 5545–5545. https://doi.org/10.1021/ie970824y

[32] S.D. Khattri, M.K. Singh, Colour Removal from dye wastewater using sugar cane dust as an adsorbent, Adsorpt. Sci. Technol. 17 (1999) 269–282. https://doi.org/10.1177/026361749901700404

Mediterranean Architectural Heritage - RIPAM10
Materials Research Proceedings 40 (2024) 260-272

Materials Research Forum LLC
https://doi.org/10.21741/9781644903117-28

Improving Thermal Insulation and Mechanical Properties of Building Bricks made from Moroccan Clay

Mohammed CHRACHMY[1,a] *, Mahdi LECHHEB[1, b], Hassan OUALLAL[1,c],
Najia EL HAMZAOUI[2,3,d], Ayoub SOUILEH[4,e], M'barek AZDOUZ[1,f]
and Mohamed AZROUR[1,g]

[1] Laboratory of Materials Engineering for the Environment and Natural Resources, Faculty of Science and Technology, Moulay Ismail University, Errachidia, Morocco

[2] Laboratory of medical analyses of Meknes Hospital, Health Regional Direction of Fez, Meknes, Morocco

[3] Laboratory of Ecology and biodiversity of wetlands, Faculty of Sciences, Department of Biology, Moulay Ismail University, Meknes, Morocco

[4] L3GIE, Mohammadia Engineering School, Mohammed V University in Rabat, Morocco

[a] mo.chrachmy@edu.umi.ac.ma, [b] m.lechheb@edu.umi.ac.ma
[c] hassanouallalaghbalou@gmail.com, [d] najia.elhamzaoui@gmail.com,
[e] Ayoub.souileh@research.emi.ac.ma, [f] m_azdouz@hotmail.com, [g] m.azrour@fste.umi.ac.ma

Keywords: Bricks, Clay, Compressive Strength, Porosity, Ceramic, Coffee Waste

Abstract. This study aims to improve the thermal of insulation properties of fired bricks based on clay from the Drâa-Tafilalet region (Es-sifa), using coffee waste without losing mechanical properties. Samples were produced on a laboratory scale by shaping, similar to a feasible industrial production technique. The raw material was first characterized using various analytical techniques such as XRD and TGA/DTA. Secondly, we optimized the main parameters affecting the properties of the bricks, such as particle size, aging time, drying time, final sintering temperature and the percentage of agent porosity. These parameters are assessed by mechanical strength and porosity tests. Examination of results the optimization following results: the particle size is less than 180 μm, the aging time and the drying time are three and four days respectively, while the sintering temperature is 1050 °C and the percentage of coffee waste used as a porosity agent is 4%. Introducing coffee waste to the brick reduces weight and improves thermal and acoustic properties by creating pores during firing. These results are very promising for exploiting these materials as base materials in the manufacture of building bricks with important properties.

Introduction

The construction sector is a vital component of the worldwide economy. The demand for eco-friendly building materials has prompted researchers to investigate alternative solutions that can diminish environmental impact and construction expenses, all while delivering mechanical strength, thermal insulation, and acoustic properties. Consequently, the significance of local and traditional materials has increased, driven by their economic advantages, ease of transportation, and environmentally friendly characteristics. This trend persists even with the presence of contemporary materials such as concrete, fiberglass/resin composites, steel, and plastics [1] clay brick remains a popular choice for wall blocks. This is because it is easy to produce, uses readily available raw materials, is affordable, resists fire and rotting, has a long lifespan, and is recyclable [2].

Numerous investigations have been conducted to mold composite bricks and assess their mechanical, physical, and thermal attributes. Thermal insulation is essential in building bricks. It helps to reduce energy consumption, improve performance and prevent heat exchange between the external and internal environments. An interesting approach to thermal insulation is to improve bricks porosity. Incorporating specific waste materials into the raw clay results in the formation of the pore structure, thereby enhancing the thermal insulation properties of clay bricks. Several studies have been conducted in the literature on the topic of waste materials that can be used for various purposes. Some examples of such materials include mahogany sawdust [3,4], limestone and sawdust[5], kaolinitic sand waste[6,7], waste pomace from the wine industry[8], rubber[9], rice husk ash[10], recycled paper processing residues[11], cigarette butts[12], residues from biodiesel production[13], olive mill waste[14,15], Kraft pulp[16], and tea waste[17].

In this work, during the elaboration of bricks from a paste of clay and different percentages of organic additive (coffee waste), we discussed the effectiveness of optimizing certain parameters such as particle size, water volume, mixing time, aging time, drying time and sintering temperature of the properties of the bricks such as porosity and mechanical strength.

Experimental study

Raw Materials
In this work the raw material used is clay (sampled in the Es-sifa district (Morocco)). During firing, porosity decreases due to sintering. To maintain this important property, we used coffee waste as the organic material. The organic additive used is not characterized because it decomposes entirely by combustion during heat treatment.

Characterization of the raw clay
The analytical techniques employed for characterizing the raw clay under investigation were as follows: X-ray diffraction, TDA/GTA thermal analysis.

The XRD pattern was acquired using a X'PERT MPD-PRO wide-angle X-ray powder diffractometer, employing a diffracted beam monochromator and Ni-filtered CuKα radiation (λ=1.5406 Å). The scanning of the 2θ angle ranged from 4° to 30° with a step size of 0.01 /s;

With a simultaneous LABSYS/evo TGA/DTA thermal analyzer, TDA/GTA analyses were realized in an air atmosphere Subjected to linear heating at a rate of 10°/min, within the temperature range from room temperature to 1000°C.

Brick Shaping
Before the bricks can be shaped, a plastic paste must first be prepared. Fig.1, shows the different stages involved in obtaining fired bricks. A good-quality formulation is obtained after several trials, taking into account a number of parameters such as the granulometry of the raw clay, which affects pore size in the ceramic part, during sintering, the aging time (t_v), the drying time (t_s) and the sintering temperature (T_f)[18,19].

The clay used was sieved through different sieves to obtain powders of different granulometries such as:

250 μm < Φ_1< 315 μm
180 μm < Φ_2< 250 μm
160 μm < Φ_3< 180 μm
Φ_4< 315 μm
Φ_5< 250 μm
Φ_6< 180 μm

The clay paste ought to be homogeneous, cohesive, hard, and plastic. These characteristics will be obtained, for an adjusted composition, by the mixing time, the volume of water, and the aging time[20] Each prepared dough is preserved in a refrigerator for aging. The porosity agent used in this work is coffee grounds in varying percentages from 2 to 20%.

Mediterranean Architectural Heritage - RIPAM10   Materials Research Forum LLC
Materials Research Proceedings 40 (2024) 260-272   https://doi.org/10.21741/9781644903117-28

The pastes are then moulded in laboratory moulds. The size of the bricks depends on the dimensions of the mould used: in our case, we used a mould of length L= 70 mm, width l= 30 mm and thickness e=10mm. The shaped bricks are air-dried and then sintered at a sintering temperature (T_f= 800, 850, 900, 950, 1000, and 1050°C).

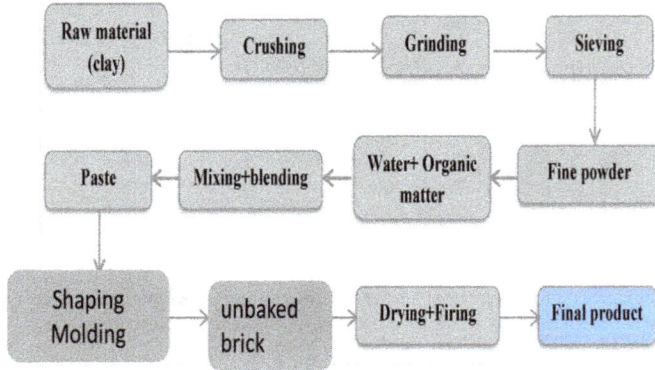

Figure 1: Brick manufacturing process

Thermal treatment
The thermal program we drew up in Fig.2, takes account of the results of differential thermal and gravimetric analysis (DTA - GTA). The temperature was raised slowly to avoid cracking the bricks. The bricks were fired in a programmable Nabertherm electric kiln.

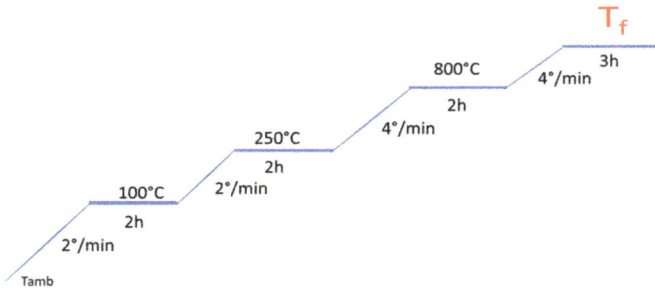

Figure 2: Thermal program drawn up

Bricks characterization
In order to shape bricks with high mechanical strength and porosity, it is necessary to meet the requirements for shaping them from a clay paste. This formulation is obtained after a many tests, taking into consideration a number of parameters for example the granulometry of the powder, the percentage of porosity agent, the aging time (t_v), the drying time (t_s), and the sintering temperature (T_f)[19,20] In this work we studied the variation of porosity and mechanical strength as a function of these parameters.

Materials Research Forum LLC

https://doi.org/10.21741/9781644903117-28

Test of porosity

Porosity is determined using the water impregnation method as per ISO 10545-3. The calculation of porosity is performed through equation 1 [21].

$$P = \frac{M_{sat} - M_{sec}}{M_{sat} - M_h} * 100 \ (\%) \qquad (1)$$

With

M_{sec}, M_h and M_{sa} are respectively the dry mass, the wet mass, and the saturated mass.

Compressive strength

The mechanical strength was determined using the three-point flexural method. The equipment used in the laboratory allows a force F to be applied to the brick placed on two supports (two points) until it breaks, then the rupture force is recorded in Newton. The stress of the breaking is calculated from the equation(2) [22].

$$\sigma = \frac{3 * F * d}{2 * l * e^2} (Mpa) \qquad (2)$$

With:

F: intensity of load applied at fracture (N)
d: distance between the two supports (mm)
e: sample thickness (mm)
l: sample width (mm)

Results and Discussion

Characterization the raw clay

XRD

Upon conducting a mineralogical study of the raw clay (shown in Fig. 4) using X-ray diffraction, several crystalline phases were observed. Quartz, which is considered to be a vital phase, was identified through the 2θ diffraction peaks at 20.84, 26.72, 39.38, and 50.1° [23]. Additionally, the diffraction diagram showed other peaks at 12.39, 29.51, and 31° which characterize kaolinite, calcite, and dolomite, respectively[24–27].

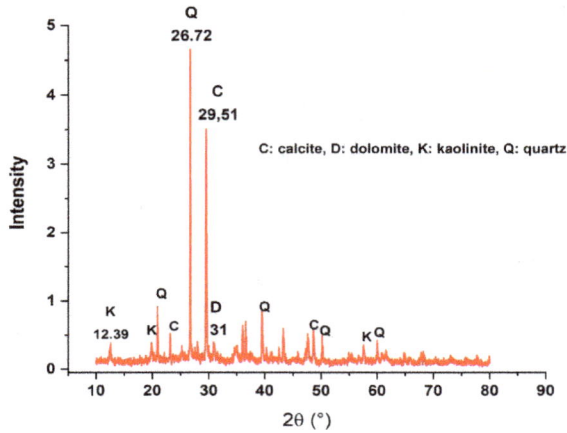

Figure 3: XRD diffractogram of Clay

DTG/DTA

The results of the TGA DTA analysis are shown in Figure 4. The peak identified at 104°C corresponds to the evaporation of moisture and is concomitant with a mass reduction. The peak detected at 332°C is associated with the elimination of organic substances, coinciding with a decrease in mass. The peak low intensity detected at 485°C shows that the clay studied is poor in kaolinite. The peak at 577°C is linked to the conversion of quartz from α to β. The endothermic peak detected at 803°C corresponds to carbonate decomposition. The peak identified at 892 °C attributes the transformation of the meta kaolinite into a spinel aluminium-silicon and the amorphous silica[28,29].

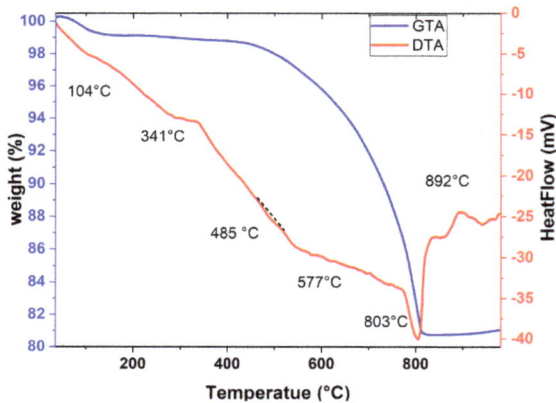

Figure 4: TGA/DTA of raw Clay

Mediterranean Architectural Heritage - RIPAM10

Materials Research Proceedings 40 (2024) 260-272

Materials Research Forum LLC

https://doi.org/10.21741/9781644903117-28

Characterization of bricks

Particle size optimization

The distribution of particle sizes in the clay raw materials utilized in brick manufacturing significantly influences the behavior of bricks throughout the drying and firing processes. Additionally, it impacts various properties of construction materials, such as mechanical characteristics [30]. Figure 5 depicts the fluctuations in porosity and mechanical strength of the manufactured bricks concerning particle size. The results obtained show two aspects.

The first aspect is that the mechanical strength for the grain size. Φ_1, Φ_2 and Φ_3 is low compared to that obtained for Φ_4, Φ_5 and Φ_6, unlike the porosity, which is due to the particle size distribution of the fractions Φ_1, Φ_2 and Φ_3, which is narrow compared to that of the fractions Φ_4, Φ_5 and Φ_6 fractions, on the other hand, is large, in the latter case the apparent density increases because the interstices between the larger particles are filled with In the context of ceramics and metallurgy, it is observed that smaller particles exhibit accelerated sintering kinetics at a given temperature. Consequently, they can be sintered effectively at lower temperatures compared to their larger counterparts[31].

The second aspect is that mechanical strength increases with decreasing grain size, whereas porosity decreases. This can be explained by the contact surface, i.e. the smaller the grain size, the greater the contact surface, which favours the various physicochemical transformations during firing and, consequently, the formation of ceramic bonds, resulting in a denser ceramic part with greater mechanical strength[32]. It can be seen from these results that the optimum particle size for maximum mechanical strength (11.05 MPa) and acceptable porosity (36.5%) is less than 180 µm, which is the size chosen for the rest of our work.

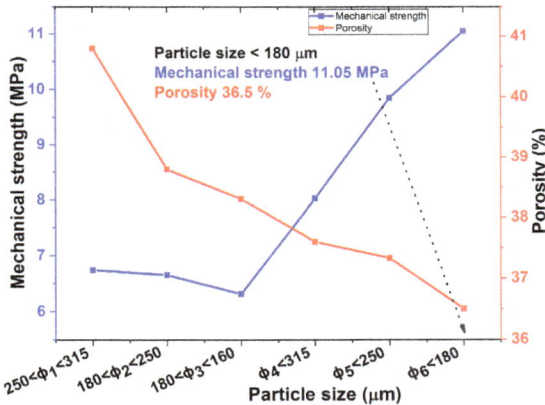

Figure 5: Optimization of the particle size of the bricks are treated at 1000°C/3h.

Optimizing aging time

Aging is an operation that consists of homogenizing the paste, and the migration of organic additives and water leads to an improvement in the quality of the final ceramic part[19], which is the objective of this section. In order to optimize the aging time, we set the drying time at three days and the particle size at less than 180µm in the production of the bricks. Fig.6 illustrates the variation in porosity and mechanical strength as a function of aging time. It can be seen that porosity decreases with aging time, while mechanical strength increases with aging time. This is quite normal, since the homogeneity and hardness of the paste increase with aging time[20]. Examination of the results also shows that after three days the increase in mechanical strength and

Mediterranean Architectural Heritage - RIPAM10   Materials Research Forum LLC

Materials Research Proceedings 40 (2024) 260-272   https://doi.org/10.21741/9781644903117-28

the decrease in porosity are not significant, so three days is considered the most adequate aging time.

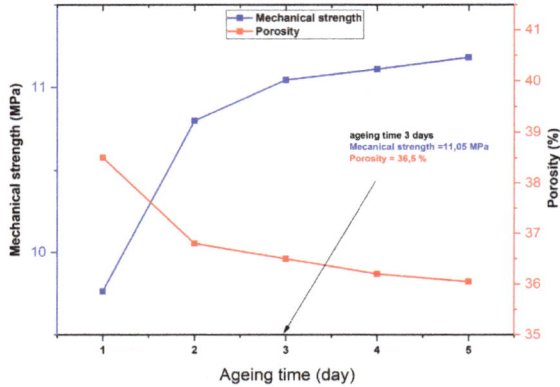

Figure6: optimizing aging time

Optimization of drying time

The purpose of drying shaped products is to eliminate moisture water. This is a delicate and important phase in the manufacturing process, which must be carefully controlled to avoid cracking, significant differential shrinkage and distortion of the ceramic products. Drying is carried out progressively by controlling two parameters, temperature and humidity [33,34] and this stage requires significant energy input. The need to reduce energy consumption and CO_2 emissions has led to the development of new technologies for more efficient drying for this reason, in this work we carried out open-air drying at ambient temperature and the drying time varied from one to five days. The aging time was set at three days and the particle size was less than 180μm. To optimize this parameter, we studied the mechanical strength and porosity as a function of the drying time in Fig.7. During the drying process, we noticed that the bricks contracted and strong volume shrinkage took place. The evacuation of the water caused a contraction of the solid structure, directly linked to the mechanical properties of the material. Figure 9 explains the drying process, with a large quantity of interstitial water evaporating. This leads to the hardening of cured bricks[35].

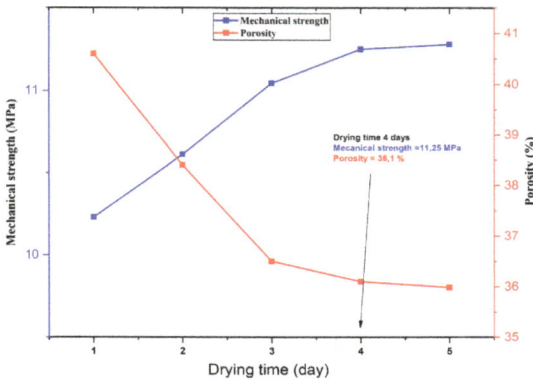

Figure 7: optimizing of drying time.

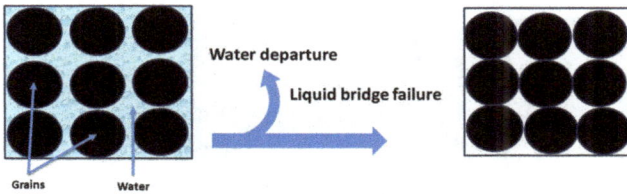

Figure 8: diagram explaining the drying process.

Optimization of sintering temperature

Figure 9 displays how porosity and mechanical strength change with temperature. The findings indicate that the mechanical strength rises proportionally with the increase in sintering temperature, reaching its maximum (11.74 MPa) at 1050°C. On the other hand, the porosity decreases from 39 to 35.27 %, which can be explained, on the one hand, by The porous structure becomes more compact, and the grains aggregate more as the sintering temperature increases [19]. Secondly, the appearance of a glassy phase, as the clay contains large quantities of (sodium oxide, potassium oxide and magnesium oxide) [36], which is confirmed by the appearance of the yellow color after heat treatment in Fig.10 As can be seen from the figure, the bricks are homogeneous and without any kind of macro-defects or breaks. They retain the same parallelepiped shape and the color of the bricks changes from a brick-red hue to a deep yellow as the sintering temperature increases. is increased from 800 to 1050 °C. The color change in bricks after thermic treatment is caused by the oxidation of iron oxide.[37,38]. Given the obtained results, selecting 1050°C as the optimal sintering temperature for the remaining part of our study is a judicious choice.

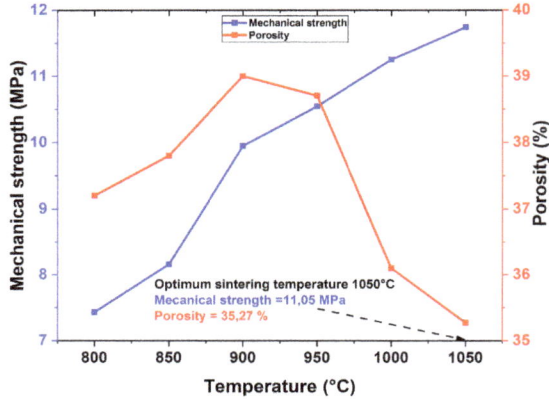

Figure 9: Optimization of sintering temperature

Figure 10: sintered bricks at different temperatures

Optimization of pore-forming agent percentage

The porosity of the bricks is a property that must be taken into account, as it improves the thermal and acoustic insulation of the bricks and increases their lightness, which affects transport costs. In order to increase porosity, coffee waste was added to the mixture in percentages ranging from 0 to 20%. Fig.11, Shows how compressive strength and porosity changes with varying proportions of coffee waste.

Analysis of the results shows that porosity and compressive strength are inversely proportional; in our case, porosity increases and mechanical strength decreases as a function of the quantity of coffee waste. This phenomenon is elucidated by the formation of pores resulting from the combustion of coffee waste during the sintering procedure [39]. The quantity of pores corresponds directly to the percentage of the added material. Combustion of the coffee waste is primordial in energy saving, as combustion is exothermic. The presence of pores leads, on the one hand, to a reduction in the thermal conductivity, on the other hand, to a reduction in density, so the bricks become lighter, which reduces transport costs. The third role affects sound insulation...

The composition that provides favorable porosity without compromising the flexural strength of the produced bricks is 4%, sintered at 1050°C.

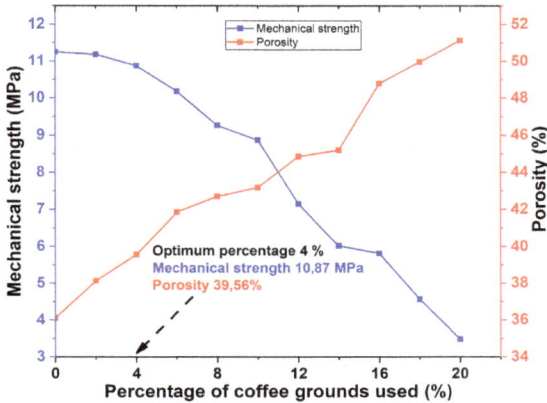

Figure 11: optimization of the organic additive

Conclusion

The use of used coffee grounds as an organic additive in the elaboration process of fired bricks is very important due to the following reasons: it leads to improved mechanical, thermal and acoustic properties and reduced brick weight, as well as additional environmental and economic benefits. This work provides the characterization and development of fired building bricks based on low-cost local raw materials. Optimization of the main parameters in the manufacturing process revealed the following results: the particle size is less than 180 μm, the aging time is three days, the drying time is four days, the sintering temperature is 1050°C and the percentage of coffee waste used as a porosity agent is 4%. The exothermic nature of coffee grounds combustion saves energy during the sintering process.

the incorporation of coffee grounds creates pores in the final product, which increases the porosity and therefore improves the thermal and acoustic insulation and lightness of the bricks. Up to 4% of coffee grounds does not significantly reduce the bending strength of the bricks. Coffee grounds can therefore be used as a good pore-forming additive in clay bodies without any adverse effect on the other properties of the bricks.

References

[1] H. Chemani, B. Chemani, Valorization of wood sawdust in making porous clay brick, Sci. Res. Essays. (n.d.).

[2] M. Saberian, J. Li, A. Donnoli, E. Bonderenko, P. Oliva, B. Gill, S. Lockrey, R. Siddique, Recycling of spent coffee grounds in construction materials: A review, Journal of Cleaner Production. 289 (2021) 125837. https://doi.org/10.1016/j.jclepro.2021.125837

[3] F.O. Aramide, Production and Characterization of Porous Insulating Fired Bricks from Ifon Clay with Varied Sawdust Admixture, JMMCE. 11 (2012) 970–975. https://doi.org/10.4236/jmmce.2012.1110097

[4] E. Bwayo, S.K. Obwoya, Coefficient of Thermal Diffusivity of Insulation Brick Developed from Sawdust and Clays, Journal of Ceramics. 2014 (2014) 1–6. https://doi.org/10.1155/2014/861726

[5] P. Turgut, H. Murat Algin, Limestone dust and wood sawdust as brick material, Building and Environment. 42 (2007) 3399–3403. https://doi.org/10.1016/j.buildenv.2006.08.012

[6] M.S. Elmaghraby, A.I.M. Ismail, Utilization of Some Egyptian Waste Kaolinitic Sand as Grog for Bricks and Concrete, Silicon. 8 (2016) 299–307. https://doi.org/10.1007/s12633-015-9333-4

[7] P. Muñoz, V. Letelier, M.A. Bustamante, J. Marcos-Ortega, J.G. Sepúlveda, Assessment of mechanical, thermal, mineral and physical properties of fired clay brick made by mixing kaolinitic red clay and paper pulp residues, Applied Clay Science. 198 (2020) 105847. https://doi.org/10.1016/j.clay.2020.105847

[8] P. Muñoz, M.P. Morales, M.A. Mendívil, M.C. Juárez, L. Muñoz, Using of waste pomace from winery industry to improve thermal insulation of fired clay bricks. Eco-friendly way of building construction, Construction and Building Materials. 71 (2014) 181–187. https://doi.org/10.1016/j.conbuildmat.2014.08.027

[9] P. Turgut, B. Yesilata, Physico-mechanical and thermal performances of newly developed rubber-added bricks, Energy and Buildings. 40 (2008) 679–688. https://doi.org/10.1016/j.enbuild.2007.05.002

[10] D. Tonnayopas, P. Tekasakul, S. Jaritgnam, Effects of Rice Husk Ash on Characteristics of Lightweight Clay Brick, (n.d.).

[11] M. Sutcu, S. Akkurt, The use of recycled paper processing residues in making porous brick with reduced thermal conductivity, Ceramics International. 35 (2009) 2625–2631. https://doi.org/10.1016/j.ceramint.2009.02.027

[12] A.A. Kadir, A. Mohajerani, Possible Utilization of Cigarette Butts in Light- Weight Fired Clay Bricks, International Journal of Civil and Environmental Engineering. (2010).

[13] D. Eliche-Quesada, S. Martínez-Martínez, L. Pérez-Villarejo, F.J. Iglesias-Godino, C. Martínez-García, F.A. Corpas-Iglesias, Valorization of biodiesel production residues in making porous clay brick, Fuel Processing Technology. 103 (2012) 166–173. https://doi.org/10.1016/j.fuproc.2011.11.013

[14] M. Sutcu, S. Ozturk, E. Yalamac, O. Gencel, Effect of olive mill waste addition on the properties of porous fired clay bricks using Taguchi method, Journal of Environmental Management. 181 (2016) 185–192. https://doi.org/10.1016/j.jenvman.2016.06.023

[15] M.D. La Rubia-García, Á. Yebra-Rodríguez, D. Eliche-Quesada, F.A. Corpas-Iglesias, A. López-Galindo, Assessment of olive mill solid residue (pomace) as an additive in lightweight brick production, Construction and Building Materials. 36 (2012) 495–500. https://doi.org/10.1016/j.conbuildmat.2012.06.009

[16] I. Demir, M. Serhat Baspınar, M. Orhan, Utilization of kraft pulp production residues in clay brick production, Building and Environment. 40 (2005) 1533–1537. https://doi.org/10.1016/j.buildenv.2004.11.021

[17] I. Demir, An investigation on the production of construction brick with processed waste tea, Building and Environment. 41 (2006) 1274–1278. https://doi.org/10.1016/j.buildenv.2005.05.004

[18] H. Ouallal, M. Azrour, M. Messaoudi, H. Moussout, L. Messaoudi, N. Tijani, Incorporation effect of olive pomace on the properties of tubular membranes, Journal of Environmental Chemical Engineering. 8 (2020) 103668. https://doi.org/10.1016/j.jece.2020.103668

[19] H. Ouallal, Valorisation de deux Argiles marocaines : Application à l'élimination du phénol en milieu aqueux par adsorption et par filtration sur membrane, Moulay Ismail University, 2021.

Materials Research Forum LLC
https://doi.org/10.21741/9781644903117-28

[20] S. Khemakhem, ELABORATION DE MEMBRANES DE MICROFILTRATION ET D'ULTRAFILTRATION EN CERAMIQUE A BASE D'ARGILE TUNISIENNE, (n.d.).

[21] I. Demir, Effect of organic residues addition on the technological properties of clay bricks, Waste Management. 28 (2008) 622–627. https://doi.org/10.1016/j.wasman.2007.03.019

[22] M. Sawadogo, L. Zerbo, M. Seynou, S. Brahima, R. Ouedraogo, Technological properties of raw clay based ceramic tiles: Influence of talc, Scientific Study & Research. 15 (2014) 231–238.

[23] J. Baliti, A. Asnaoui, S. Abouarnadasse, l'élimination du bleu de méthylène par une argile naturelle de Taza en milieu aqueux, International Journal of Innovative Resear ch in Advanced Engineering. 1 (2014).

[24] H. Ouallal, M. Azrour, M. Messaoudi, H. Moussout, L. Messaoudi, N. Tijani, Incorporation effect of olive pomace on the properties of tubular membranes, Journal of Environmental Chemical Engineering. 8 (2020) 103668. https://doi.org/10.1016/j.jece.2020.103668

[25] M. Lechheb, A. Harrou, G. El Boukili, M. Azrour, A. Lahmar, M. El Ouahabi, E.K. Gharibi, Physico-chemical, mineralogical, and technological characterization of stabilized clay bricks for restoration of Kasbah Ait Benhadou- Ouarzazate (south-east of Morocco), Materials Today: Proceedings. 58 (2022) 1229–1234. https://doi.org/10.1016/j.matpr.2022.01.459

[26] S. Mahmoudi, A. Bennour, A. Meguebli, E. Srasra, F. Zargouni, Characterization and traditional ceramic application of clays from the Douiret region in South Tunisia, Applied Clay Science. 127–128 (2016) 78–87. https://doi.org/10.1016/j.clay.2016.04.010

[27] P.E. Tsakiridis, M. Samouhos, M. Perraki, Valorization of Dried Olive Pomace as an alternative fuel resource in cement clinkerization, Construction and Building Materials. 153 (2017) 202–210. https://doi.org/10.1016/j.conbuildmat.2017.07.102

[28] H. Ouallal, Y. Dehmani, H. Moussout, L. Messaoudi, M. Azrour, Kinetic, isotherm and mechanism investigations of the removal of phenols from water by raw and calcined clays, Heliyon. 5 (2019) e01616. https://doi.org/10.1016/j.heliyon.2019.e01616

[29] S. Iaich, Y. Miyah, F. Elazhar, S. Lagdali, M. El-Habacha, Low-cost ceramic microfiltration membranes made from Moroccan clay for domestic wastewater and Congo Red dye treatment, DWT. 235 (2021) 251–271. https://doi.org/10.5004/dwt.2021.27618

[30] M. Dondi, B. Fabbri, G. Guarini, Grain-size distribution of Italian raw materials for building clay products: a reappraisal of the Winkler diagram, Clay Miner. 33 (1998) 435–442. https://doi.org/10.1180/000985598545732

[31] R.L. Coble, Effects of Particle-Size Distribution in Initial-Stage Sintering, J American Ceramic Society. 56 (1973) 461–466. https://doi.org/10.1111/j.1151-2916.1973.tb12524.x

[32] J. Huang, H. Chen, J. Yang, T. Zhou, H. Zhang, Effects of particle size on microstructure and mechanical strength of a fly ash based ceramic membrane, Ceramics International. 49 (2023) 15655–15664. https://doi.org/10.1016/j.ceramint.2023.01.157

[33] E. Portuguez, Gouttes millimétriques d'eau en milieu confiné : comportement au cours du séchage, (n.d.).

[34] E. Keita, Physique du séchage des sols et des matériaux de construction, (n.d.).

[35] Z. Ge, Y. Feng, H. Zhang, J. Xiao, R. Sun, X. Liu, Use of recycled fine clay brick aggregate as internal curing agent for low water to cement ratio mortar, Construction and Building Materials. 264 (2020) 120280. https://doi.org/10.1016/j.conbuildmat.2020.120280

Mediterranean Architectural Heritage - RIPAM10   Materials Research Forum LLC
Materials Research Proceedings 40 (2024) 260-272   https://doi.org/10.21741/9781644903117-28

[36] M. El Ouahabi, L. Daoudi, F. De Vleeschouwer, R. Bindler, N. Fagel, Potentiality of Clay Raw Materials from Northern Morocco in Ceramic Industry: Tetouan and Meknes Areas, JMMCE. 02 (2014) 145–159. https://doi.org/10.4236/jmmce.2014.23019

[37] R. Mouratib, B. Achiou, M.E. Krati, S.A. Younssi, S. Tahiri, Low-cost ceramic membrane made from alumina- and silica-rich water treatment sludge and its application to wastewater filtration, Journal of the European Ceramic Society. 40 (2020) 5942–5950. https://doi.org/10.1016/j.jeurceramsoc.2020.07.050

[38] S. Lagdali, Y. Miyah, M. El-Habacha, G. Mahmoudy, M. Benjelloun, S. Iaich, M. Zerbet, M. Chiban, F. Sinan, Performance assessment of a phengite clay-based flat membrane for microfiltration of real-wastewater from clothes washing: Characterization, cost estimation, and regeneration, Case Studies in Chemical and Environmental Engineering. 8 (2023) 100388. https://doi.org/10.1016/j.cscee.2023.100388

[39] H. Elomari, B. Achiou, A. Karim, M. Ouammou, A. Albizane, J. Bennazha, S. Alami.Younssi, I. Elamrani, Influence of starch content on the properties of low cost microfiltration membranes, Journal of Asian Ceramic Societies. 5 (2017) 313–319. https://doi.org/10.1016/j.jascer.2017.06.004

Mediterranean Architectural Heritage - RIPAM10
Materials Research Proceedings 40 (2024) 273-283

Materials Research Forum LLC
https://doi.org/10.21741/9781644903117-29

Insight on the Natural Moroccan Clay Valorization for Malachite Green Adsorption: Kinetic and Isotherm Studies

Hassan OUALLAL[1,a,*], Mohammed CHRACHMY[1,b], Najia EL HAMZAOUI[2,c], Mahdi LECHHEB[1,d], Rajae GHIBATE[3,e], Houssam EL-MARJAOUI[4,f], and Mohamed AZROUR[1,g]

[1]Laboratory of Materials Engineering for the Environment and Natural Resources, Faculty of Science and Technology, Moulay Ismail University, Errachidia, 52000, Morocco

[2]Laboratory of Ecology and Biodiversity of Wetlands, Faculty of Sciences, Moulay Ismail University, Meknes, 11201, Morocco

[3]Laboratory of Physical Chemistry, Materials, and Environment, Faculty of Sciences and Technologies, Moulay Ismail University, Errachidia, 52000, Morocco

[4]Laboratoire de Spectrométrie des Matériaux et Archéomatériaux (LASMAR), Unité de Recherche Labéllisée par CNRST (URL-CNRST N°7), Université Moulay Ismail, Faculté des Sciences, Zitoune BP 11201, 50000 Meknès, Morocco

[a]hassanouallalaghbalou@gmail.com, [b]mo.chrachmy@edu.umi.ac.ma, [c]najia.elhamzaoui@gmail.com, [d]m.lechheb@edu.umi.ac.ma, [e]rajae.ghibate@gmail.com, [f]elmarjaoui1993@gmail.com, [g]m.azrour@fste.umi.ac.ma

Keywords: Natural Clay, Adsorption, Malachite Green, Kinetic Modeling, Isotherm Study

Abstract. This work investigates the potential of Natural Moroccan Clay (NMC) sourced from the Draa-Tafilalet region for removing malachite green from aqueous media through adsorption. X-ray diffraction (XRD) and scanning electron microscopy combined with energy-dispersive X-ray microanalysis (SEM/EDX) were used to characterize the clay. Malachite green (MG) was subjected to batch adsorption experiments, and a kinetic study was carried out at three concentrations (5, 50, and 100 mg/L). The results showed that adsorption is typically fast for all three concentrations and that the adsorbed amount rises with time and dye concentration. Equilibrium is reached within just 40 minutes. The kinetics adsorption at varying MG concentrations were modeled using non-linear and linear forms of pseudo-first order and pseudo-second order and with intraparticle diffusion models. The non-linear form of the pseudo-second-order model was found to be best suited to describing this adsorption process. The isotherm was studied using two models: Freundlich and Langmuir. According to the error functions analysis, the Langmuir model is well suited for equilibrium data fitting. Investigated clay attained an adsorption capacity of 214 mg/g. SEM/EDX characterization of NMC before and after adsorption confirms the malachite green adsorption on the NMC surface. These results illustrate the effectiveness of the investigated clay as a cost-effective adsorbent.

Introduction

Nowadays, water vulnerability is heightened by climate change and the expansion of the industrial world, increasing volumes of effluent discharged into the natural medium. That is detrimental to the environment and, by extension, to public health [1,2]. Among the pollutants present in these effluents are dyes. The present study focuses on the remediation of water contaminated with malachite green (MG), one of the dyes recognized for its mutagenic and carcinogenic effects on living beings, once released into the environment [1,3]. Several treatment technologies have been proposed for MG removal by other researchers, including advanced oxidation processes [4], membrane separation [5], and adsorption [6]. In this context, clays have found a way to be used as

Mediterranean Architectural Heritage - RIPAM10   Materials Research Forum LLC
Materials Research Proceedings 40 (2024) 273-283   https://doi.org/10.21741/9781644903117-29

an adsorbent, given their particular physico-chemical characteristics, abundance, and low cost. Indeed, some clay materials have demonstrated excellent adsorption capacity, high surface reactivity, and notable cation exchange potential [7,8].

This study aims to valorize natural Moroccan clay (NMC) from the Es-sifa commune in the Draa-Tafilalet region as an adsorbent for treating water containing MG dye. The clay studied was characterized by SEM/EDX and XRD. Furthermore, kinetic and isotherm studies were conducted to understand better the reaction mechanism of MG fixation on the NMC surface. The adequacy of the investigated models was ensured and evaluated by error functions.

Materials and Methods

Adsorbate

Malachite green oxalate ($C_{52}H_{54}N_4O_{12}$) was supplied by VWR Chemicals. Fig. 1 illustrates its chemical structure. A precisely weighed mass of dye powder is dissolved in demineralized water to prepare a dye stock solution. Different concentrations were prepared from this solution.

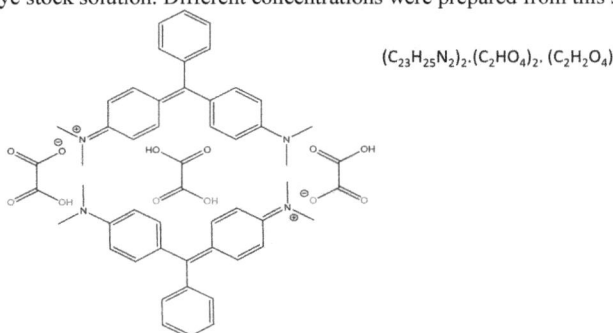

$(C_{23}H_{25}N_2)_2.(C_2HO_4)_2. (C_2H_2O_4)$

Figure 1: Malachite green oxalate chemical structure

Clay preparation

The NMC used in the adsorption experiments was collected from Es-sifa, located in the province of Errachidia, Morocco. The natural clay was crushed and passed through ISO standard mesh sieves, and only particles smaller than 315 µm were utilized for the experiments.

Clay Characterization

The NMC was characterized using XRD and SEM/EDX. The NMC mineralogy was analyzed by X-ray powder diffraction. The XRD pattern was obtained using an X'Pert PRO MPD diffractometer instrument with a wavelength of 1.5406 Å of CuKα radiation. The scanning of the 2θ angle was conducted within the range of 10 to 30 with an increment of 0.02 for 2 seconds.
The morphology of NMC was examined using SEM/EDX analysis (JSM-IT500HR) at 10 kV acceleration voltage under room temperature. NMC sample after MG adsorption was first metalized and fixed to carbon support before being analyzed.

Adsorption experiments

The experiments of adsorption were executed in batch mode to determine equilibrium time and observe the influence of time on MG. In each experiment, 100 mg of NMC was introduced into 100 mL of solutions with varying MG concentrations (5, 50, and 100 mg/L). These mixtures were maintained at different contact times (5 to 60 min) and underwent agitation at 250 revolutions per minute at a temperature of 25 C. For the isotherm study, experiments were conducted at 25 °C with varying concentrations from 5 to 300 mg/L.

For every test, 100 mL of the MG solution was mixed with 100 mg of clay. These mixtures were immediately shaken at 250 rpm for 60 minutes. Following each experiment, the mixture is filtered, and the residual dye concentration is measured with a Shimadzu UV-160 UV/Visible spectrometer at 620 nm. According to the equation below, the adsorbed quantity was calculated:

$$q_t = \frac{C_0 - C_t}{m_{clay}} \times V_{MG} \tag{1}$$

Where
m_{clay}: mass of clay (g)
V_{MG}: volume of malachite green solution (L)
C_0: initial concentrations of MG
C_t: concentrations of MG at t time (M)
q_t: adsorption quantity at t time (mg/g)

Kinetic modeling
The kinetic data was fitted with non-linear and linear equations of the pseudo-first-order model (PFOM) and pseudo-second-order model (PSOM), along with the model of intraparticle diffusion (PDM). This was done to find out how malachite green was adsorbed onto the studied clay. The table below outlines these models.

Table 1: kinetic models of adsorption

Model	Non-linear equation	Linear equation	Parameters	Reference
Pseudo-first order	$q_t = q_e(1 - e^{-K_1 t})$	$ln(q_e - q_t) = ln(q_e) - K_1 t$	K_1	[9]
Pseudo-second order	$q_t = \frac{q_e^2 K_2 t}{q_e K_2 t + 1}$	$\frac{t}{q_t} = \frac{1}{q_e^2 K_2} + \frac{1}{q_e} t$	K_2	[9]
Intraparticle diffusion	–	$q_t = K_{id} t^{1/2} + C_i$	K_{id}, C_i	[10]

q_e (mg/g) represents the equilibrium adsorption quantity. C is the thickness of boundary layer. K_1 (min^{-1}), K_2 (min.g/mg), and K_{id} (mg/g.min$^{1/2}$) correspond to the adsorption rate constants of the PFOM, PSOM, and IPDM, respectively.

Equilibrium Isotherm Modeling
For the isotherm study, solutions of MG with varying concentrations (5 - 300 mg/L). The dye adsorption was carried out under optimized conditions of 0.10 g of adsorbent mass, agitation speed of 250 rpm, and pH of the medium. The finding was plotted with Langmuir and Freundlich models. The relevant parameters were calculated from the non-linear plots using equations below:

$$q_e = \frac{q_{max} K_L C_e}{1 + K_L C_e} \tag{2}$$

$$q_e = K_F C_e^{1/n} \tag{3}$$

C_e (mg/L) stands for the dye concentration that is still present at equilibrium. q_{max} (mg/g) represents the maximum adsorption capacity. The Langmuir constant is designated by K_L (L/mg). The K_F ((mg/g)(mg/L)$^{-1/n}$) and n stand for the constants of Freundlich.

Mediterranean Architectural Heritage - RIPAM10
Materials Research Proceedings 40 (2024) 273-283

Materials Research Forum LLC
https://doi.org/10.21741/9781644903117-29

Error analysis

The adaptation of the models was deemed through error function analysis, which included the standard deviation (Δq (%)), Chi-square (χ^2), and the coefficient of determination (R^2) [11,12]

$$\Delta q(\%) = 100 \times \left\{ \frac{\sum_{i=1}^{n}\left[(q_{exp}-q_{cal})\big/q_{exp}\right]^2}{n-1} \right\}^{1/2} \tag{5}$$

$$\chi^2 = \sum_{i=1}^{n} \frac{(q_{exp}-q_{cal})^2}{q_{cal}} \tag{6}$$

$$R^2 = \frac{\sum_{i=1}^{n}(q_{cal}-\overline{q_{exp}})^2}{\sum_{i=1}^{n}(q_{cal}-\overline{q_{exp}})^2+\sum_{i=1}^{n}(q_{cal}-q_{exp})^2} \tag{4}$$

Results and Discussion

Clay characterization

Fig. 2 displays the diffractogram of NMC, revealing the existence of various mineral phases. These encompass silica in the form of quartz (Q), kaolinite (K), dolomite (D), and calcite (C) [13–16].

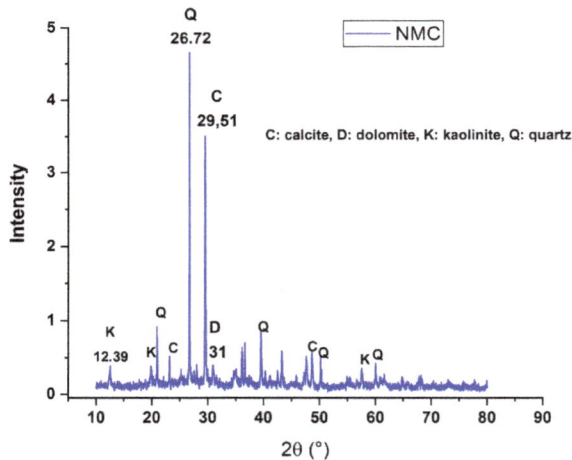

Figure 2: XRD pattern of NMC

Kinetic study

The malachite green adsorption onto NMC was studied to determine if equilibrium had been reached. The adsorption was conducted at different concentrations (5, 50, and 100 mg/L) in a time-dependent study (Fig.3). The findings showed that the adsorption of MG on NMC occurred rapidly within the first few minutes of contact at all three concentrations. However, as time progressed, adsorption became slower. Equilibrium was reached at 40 minutes. A possible explanation for this is the abundance of active sites on the surface of the NMC, which gradually demineralize until they become saturated. The quantity adsorbed rises with the adsorbate concentration, according to

Mediterranean Architectural Heritage - RIPAM10   Materials Research Forum LLC
Materials Research Proceedings 40 (2024) 273-283   https://doi.org/10.21741/9781644903117-29

an analysis of kinetic data. That makes sense since the concentration gradient is getting steeper, which encourages the dye to diffuse toward the adsorbent.[17]

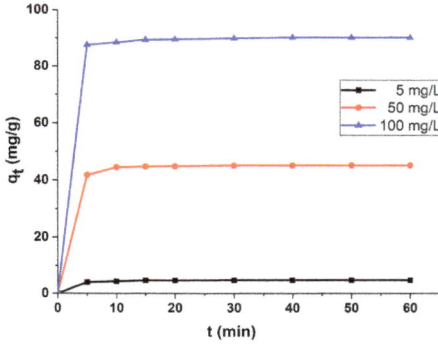

Figure 3: Kinetic of MG adsorption onto NMC at different dye concentrations

The kinetics of MG adsorption by raw clay were modeled using both linear and nonlinear forms of PFOM and PSOM as illustrated in Figs. 4 and 5. Table 2 displays the parameters for these models. The findings indicate that there is a weak correlation between the linear and nonlinear forms of the PFOM. Therefore, MG adsorption onto NMC does not follow PFOM kinetics. However, two forms of the PSOM showed correlation coefficient values (R^2) close to unity, revealing the best fit of this model to the MG adsorption kinetic. Furthermore, the predicted equilibrium amounts of MG adsorbed with the nonlinear form of the PSOM are very close to those measured experimentally, which is well supported by the low values of Δq and $\chi 2$. Accordingly, this form provides a better description of MG adsorption on NMC.

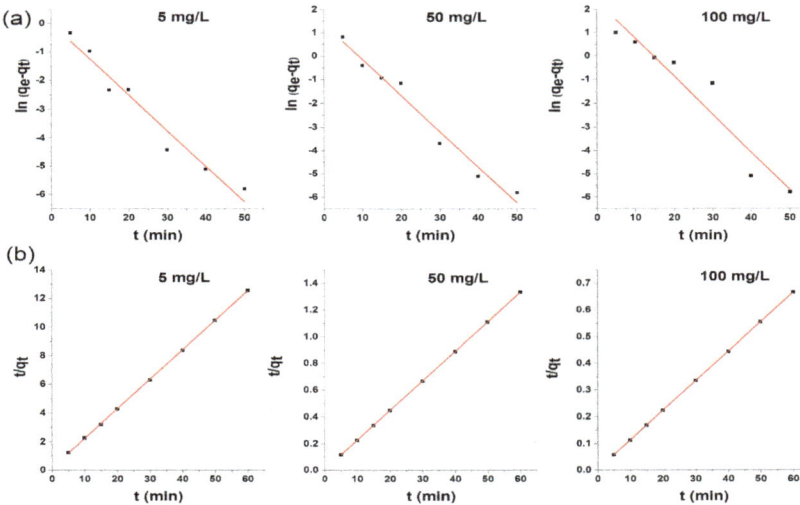

Figure 4: The plots of the linear form of PFOM (a) and PSOM (b) at different MG concentrations

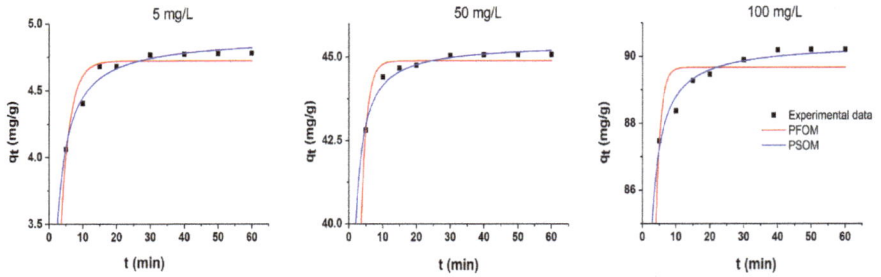

Figure 5: The plots of the nonlinear form of PFOM and PSOM at different MG concentrations

Table 2: PFOM and PSOM kinetic parameters at different MG concentrations using linear and non-linear forms.

MG concentration	Parameter & statistical error	PFOM		PSOM	
		LF	NLF	LF	NLF
5 mg/L	$q_{e,\,cal}$(mg/g)	0.99	4.72	4.86	4.90
	K_1 (min^{-1}) / K_2 (min.g.mg^{-1})	0.1250	0.3740	0.2579	0.2006
	R^2	0.9599	0.9969	0.9999	0.9993
	χ^2	124.26	0.0130	0.0073	0.0031
	Δq (%)	88.54	2.08	1.59	0.97
50 mg/L	$q_{e,\,cal}$(mg/g)	3.96	44.88	45.27	45.41
	K_1 (min^{-1}) / K_2 (min.g.mg^{-1})	0.1526	0.6111	0.1023	0.0772
	R^2	0.9753	0.9998	1.0000	0.9999
	χ^2	3 537.2	0.0075	0.0084	0.0028
	Δq (%)	99.60	0.49	0.54	0.30
100 mg/L	$q_{e,\,cal}$(mg/g)	10.7423	89.6762	90.5797	90.4141
	K_1 (min^{-1}) / K_2 (min.g.mg^{-1})	0.1614	0.7360	0.0512	0.0610
	R^2	0.9257	0.9996	1.0000	1.0000
	χ^2	4 875.4	0.0293	0.0061	0.0035
	Δq (%)	96.51	0.69	0.32	0.24

Fig. 6 shows multi-linear intraparticulaire diffusion plots, which do not cross the origin, suggesting that other rate-step processes control the MG adsorption for the three concentrations than intraparticle diffusion. The steeply sloping initial linear section reflects external mass transfer, the second half shows intraparticle diffusion, and the plateau region represents equilibrium [10]

Therefore, the boundary layer thickness (C) and the rate constant of the intraparticle diffusion (K_{id}) were generated from the intercept and the slope of the middle linear section (Table 3).[18]

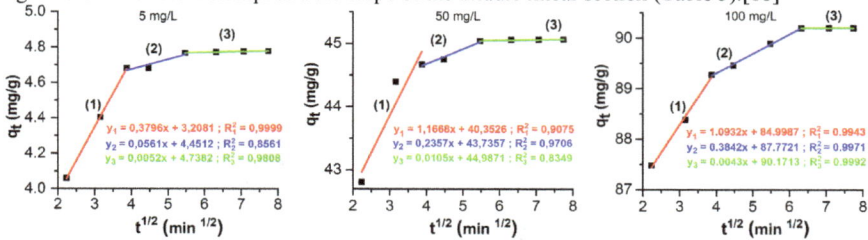

5 mg/L
$y_1 = 0{,}3796x + 3{,}2081 ; R_1^2 = 0{,}9999$
$y_2 = 0{,}0561x + 4{,}4512 ; R_2^2 = 0{,}8561$
$y_3 = 0{,}0052x + 4{,}7382 ; R_3^2 = 0{,}9808$

50 mg/L
$y_1 = 1{,}1668x + 40{,}3526 ; R_1^2 = 0{,}9075$
$y_2 = 0{,}2357x + 43{,}7357 ; R_2^2 = 0{,}9706$
$y_3 = 0{,}0105x + 44{,}9871 ; R_3^2 = 0{,}8349$

100 mg/L
$y_1 = 1.0932x + 84.9987 ; R_1^2 = 0.9943$
$y_2 = 0.3842x + 87.7721 ; R_2^2 = 0.9971$
$y_3 = 0.0043x + 90.1713 ; R_3^2 = 0.9992$

Figure 6: Intraparticulaire diffusion plots of MG adsorption onto NMC at different concentrations

Table 3: Kinetic parameters from the intraparticulaire diffusion at different MG concentrations

Step	Parameter	Concentration (mg/L)		
		5	50	100
First step	$K_{id,1}$ (mg/g.min$^{1/2}$)	0.3796	1.1668	1.0932
	C_1	3.2081	40.3526	84.9987
	R^2	0.9999	0.9075	0.9943
Second step	$K_{id,2}$ (mg/g.min$^{1/2}$)	0.0561	0.2357	0.3842
	C_2	4.4512	43.7357	87.7721
	R^2	0.8561	0.9706	0.9971
Third step	$K_{id,3}$ (mg/g.min$^{1/2}$)	0.0052	0.0105	0.0043
	C_3	4.7382	44.9871	90.1713
	R^2	0.9808	0.8349	0.9992

Isotherm study

Fig. 7 shows the experimental results of the isotherm study, with their predictions by the nonlinear forms of the Freundlich and Langmuir models. Table 4 summarizes the results of this study. Analysis of the error functions R^2, Δq, and χ^2 clearly shows that Langmuir model most appropriate for describing the MG adsorption isotherm on the clay studied. That suggests that MG is adsorbed as a monolayer on energetically comparable sites without any interaction between the adsorbed MG molecules[19]. The maximum adsorption quantity predicted by this model is 245.67 mg/g.

Table 4: Freundlich and Langmuir isotherm constants and their error analysis

$q_{max, exp}$ (mg/g)	Langmuir					Freundlich					
	$q_{max, cal}$ (mg/g)	K_L (L/mg)	R^2	χ^2	Δq (%)	$q_{max, cal}$ (mg/g)	K_F (mg/g)	$1/n$	R^2	χ^2	Δq (%)
214.51	245.67	0.05	0.9911	0.92	21.94	216.51	31.87	0.43	0.9901	11.45	86.91

Mediterranean Architectural Heritage - RIPAM10   Materials Research Forum LLC

Materials Research Proceedings 40 (2024) 273-283   https://doi.org/10.21741/9781644903117-29

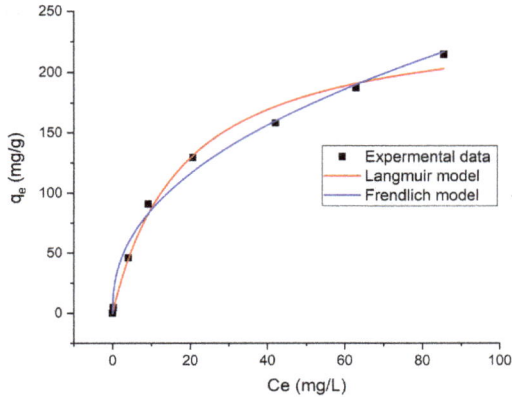

Figure 7: Adsorption isotherm of MG at 25°C and their fitting through Freundlich and Langmuir models

The dimensionless constant R_L is an essential characteristic of the Langmuir isotherm calculated using K_L and C_0 as per the following equation:

$$R_L = \frac{1}{1 + K_L C_0} \tag{7}$$

The adsorption procedure can be categorized depending on the R_L value. It's favorable if the R_L value is between 0 and 1, linear if R_L is around 1, irreversible if R_L is zero, and unfavorable if R_L is greater than 1[20].

Fig. 8 illustrates how R_L changes with the initial concentration. As the figure depicts, the R_L values range from 0 to 1, indicating a favorable MG adsorption process onto NMC.

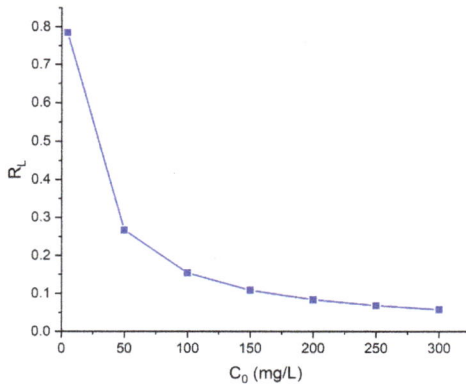

Figure 8: The plot of the R_L values as a function of concentration for MG adsorption onto NMC

Mediterranean Architectural Heritage - RIPAM10

Materials Research Forum LLC

Materials Research Proceedings 40 (2024) 273-283

https://doi.org/10.21741/9781644903117-29

ESM/EDX characterization

In Fig. 9, the scanning electron microscope images show a clear difference in the surface appearance of the adsorbent before and after adsorption. The surface of the clay is now covered with malachite green molecules, as confirmed by the comparison of the EDX spectra of NMC before and after adsorption. The spectra analysis revealed an increase in carbon intensity, with the percentage of carbon rising from 7.6% to 23%. This increase could be attributed to the organic nature of the malachite green dye, which has bonded to the clay surface.

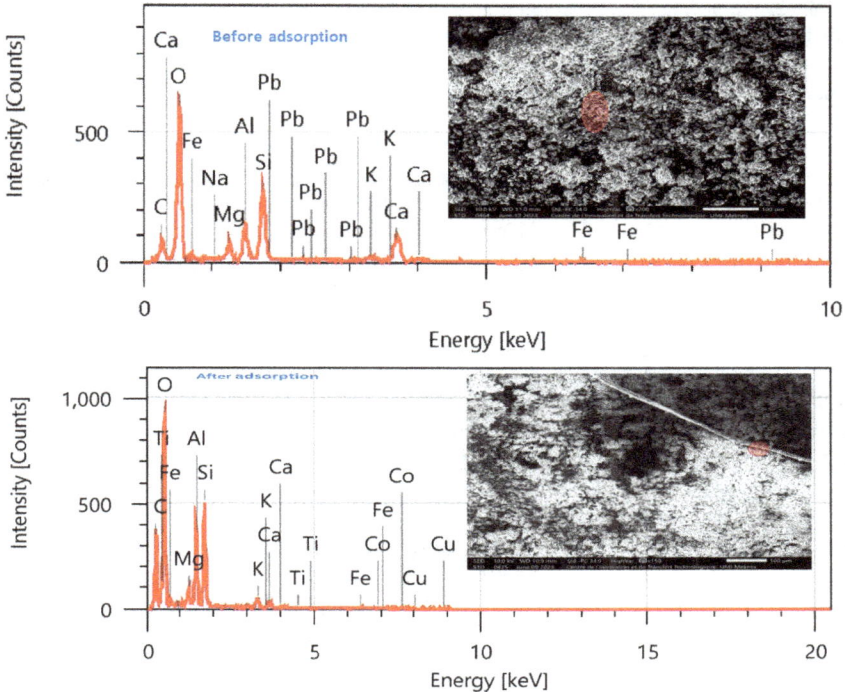

Figure 9: SEM images and EDX spectra of NMC before and after MG adsorption

Conclusion

This study investigates the effectiveness of natural Moroccan clay from the Draa-Tafilalet region in removing malachite green from aqueous solutions. XDR analysis revealed that the clay predominantly consists of kaolinite, calcite, dolomite, and quartz. The findings suggest a favorable adsorption potential for malachite green dye. As the concentration increases from 5 to 100 mg/L, the adsorption quantity escalates from 4.78 to 90.20 mg/g. The kinetics of adsorption align precisely with the pseudo-second-order model. Additionally, it was observed that intraparticle diffusion does not singularly govern the process. The Langmuir model proves most apt for fitting equilibrium data, revealing a maximum adsorption capacity of 214 mg/g for the clay. SEM/EDX analysis conducted before and after adsorption corroborated the malachite green adsorption onto the clay surface. EDX spectra analysis indicated a rise in carbon intensity, underscoring the organic nature of the malachite green dye. The examined clay emerges as a cost-effective adsorbent for removing malachite green from aqueous solutions.

References

[1] X. Jia, J. Li, E. Wang, Lighting-Up of the Dye Malachite Green with Mercury(II)–DNA and Its Application for Fluorescence Turn-Off Detection of Cysteine and Glutathione, Chem. Eur. J. 18 (2012) 13494–13500. https://doi.org/10.1002/chem.201103768

[2] H. Worku, Rethinking urban water management in Addis Ababa in the face of climate change: An urgent need to transform from traditional to sustainable system, Environmental Quality Mgmt. 27 (2017) 103–119. https://doi.org/10.1002/tqem.21512

[3] M. Amin, A. Khodabakhshi, Determination of malachite green in trout tissue and effluent water from fish farms, Int J Env Health Eng. 1 (2012) 10. https://doi.org/10.4103/2277-9183.94394

[4] J.A. Bañuelos, O. García-Rodríguez, A. El-Ghenymy, F.J. Rodríguez-Valadez, L.A. Godínez, E. Brillas, Advanced oxidation treatment of malachite green dye using a low cost carbon-felt air-diffusion cathode, Journal of Environmental Chemical Engineering. 4 (2016) 2066–2075. https://doi.org/10.1016/j.jece.2016.03.012

[5] P.O. Oladoye, T.O. Ajiboye, W.C. Wanyonyi, E.O. Omotola, M.E. Oladipo, Insights into remediation technology for malachite green wastewater treatment, Water Science and Engineering. 16 (2023) 261–270. https://doi.org/10.1016/j.wse.2023.03.002

[6] M.T.M. Hussien Hamad, Optimization study of the adsorption of malachite green removal by MgO nano-composite, nano-bentonite and fungal immobilization on active carbon using response surface methodology and kinetic study, Environ Sci Eur. 35 (2023) 26. https://doi.org/10.1186/s12302-023-00728-1

[7] T. Zhao, S. Xu, F. Hao, Differential adsorption of clay minerals: Implications for organic matter enrichment, Earth Sci Rev . 246 (2023) 104598. https://doi.org/10.1016/j.earscirev.2023.104598

[8] H. Ouallal, Y. Dehmani, H. Moussout, L. Messaoudi, M. Azrour, Kinetic, isotherm and mechanism investigations of the removal of phenols from water by raw and calcined clays, Heliyon. 5 (2019) e01616. https://doi.org/10.1016/j.heliyon.2019.e01616

[9] H. Moussout, H. Ahlafi, M. Aazza, H. Maghat, Critical of linear and nonlinear equations of pseudo-first order and pseudo-second order kinetic models, Karbala International Journal of Modern Science. 4 (2018) 244–254. https://doi.org/10.1016/j.kijoms.2018.04.001

[10] R. Ghibate, O. Senhaji, R. Taouil, Kinetic and thermodynamic approaches on Rhodamine B adsorption onto pomegranate peel, CSC EE. 3 (2021) 100078. https://doi.org/10.1016/j.cscee.2020.100078

[11] M. Aazza, H. Ahlafi, H. Moussout, H. Maghat, Adsorption of metha-nitrophenol onto alumina and HDTMA modified alumina: Kinetic, isotherm and mechanism investigations, J. Mol. Liq. 268 (2018) 587–597. https://doi.org/10.1016/j.molliq.2018.07.095

[12] H.N. Tran, Y.-F. Wang, S.-J. You, H.-P. Chao, Insights into the mechanism of cationic dye adsorption on activated charcoal: The importance of π–π interactions, *PSEP*   107 (2017) 168–180. https://doi.org/10.1016/j.psep.2017.02.010

[13] S. Iaich, Y. Miyah, F. Elazhar, S. Lagdali, M. El-Habacha, Low-cost ceramic microfiltration membranes made from Moroccan clay for domestic wastewater and Congo Red dye treatment, DWT. 235 (2021) 251–271. https://doi.org/10.5004/dwt.2021.27618

[14] M.E. Ouardi, L. Saadi, M. Waqif, H. Chehouani, I. Mrani, M. Anoua, A. Noubhani, Characterization of the bouchane phosphate (Morocco) and study of the elements of the main control elements of its calcination, Phys. Chem. News (2010).

[15] H. Ouallal, M. Azrour, M. Messaoudi, H. Moussout, L. Messaoudi, N. Tijani, Incorporation effect of olive pomace on the properties of tubular membranes, JECE . 8 (2020) 103668. https://doi.org/10.1016/j.jece.2020.103668

[16] P.S. Nayak, B.K. Singh, Instrumental characterization of clay by XRF, XRD and FTIR, Bull Mater Sci. 30 (2007) 235–238. https://doi.org/10.1007/s12034-007-0042-5

[17] M. Abewaa, A. Mengistu, T. Takele, J. Fito, T. Nkambule, Adsorptive removal of malachite green dye from aqueous solution using Rumex abyssinicus derived activated carbon, Sci Rep. 13 (2023) 14701. https://doi.org/10.1038/s41598-023-41957-x

[18] N.S. Randhawa, N.N. Das, R.K. Jana, Adsorptive remediation of Cu(II) and Cd(II) contaminated water using manganese nodule leaching residue, Desalination and Water Treatment. 52 (2014) 4197–4211. https://doi.org/10.1080/19443994.2013.801324

[19] P. Saha, S. Chowdhury, S. Gupta, I. Kumar, Insight into adsorption equilibrium, kinetics and thermodynamics of Malachite Green onto clayey soil of Indian origin, Chem. Eng. J, 165 (2010) 874–882. https://doi.org/10.1016/j.cej.2010.10.048

[20] K. Kalpana, S. Arivoli, K. Veeravelan, Activated Carbon Merremia emarginata Adsorption Capacities of Low-cost Adsorbent for Removal of Methylene Blue, Eco. Env. & Cons,. 28 (2022) S425–S430. https://doi.org/10.53550/EEC.2022.v28i08s.064

Mediterranean Architectural Heritage - RIPAM10
Materials Research Proceedings 40 (2024) 284-293

Materials Research Forum LLC
https://doi.org/10.21741/9781644903117-30

Valorization of Clays in the Removal of Organic Pollutants by Adsorption and Membrane Filtration: A Comparative Study

Hassan OUALLAL[1,a,*], Mohammed CHRACHMY[1,b], Najia EL HAMZAOUI[2,c], Mahdi LECHHEB[1,d], M'barek AZDOUZ[1,e], Abdekrim BATAN[1,f] and Mohamed AZROUR[1,g]

[1]Laboratory of Materials Engineering for the Environment and Natural Resources, Faculty of Science and Technology, Moulay Ismail University, Errachidia, 52000, Morocco

[2]Laboratory of Ecology and biodiversity of wetlands, Faculty of Sciences, Department of Biology, Moulay Ismail University, Meknes, Morocco

[a]h.ouallal@umi.ac.ma, [b]mo.chrachmy@edu.umi.ac.ma, [c]najia.elhamzaoui@gmail.com, [d]m.lechheb@edu.umi.ac.ma, [e]m_azdouz@hotmail.com, [f]a.batan@umi.ac.ma, and [g]m.azrour@fste.umi.ac.ma

Keywords: Clay, Ceramic Membrane, Membrane Filtration, Adsorption, Phenol

Abstract. Within the framework of the valorization of a clay from the Draa-Tafilalet region and following the study of the retention of phenol in aqueous solution which was carried out previously by adsorption on a clay, we have proceeded in this present work to the elimination of phenol by filtration on a mineral membrane already elaborated on the basis of the same clay and characterized by several tests in a previous work. A phenolic solution of known concentration was treated by tangential filtration using a laboratory-scale micro-pilot, with a filtering surface area of approximately 0.0072 m². To make a comparative study of the retention of phenol by the two techniques mentioned above, the membrane was first characterized by the water permeation test, and then the tests of filtration which were carried out at a circulation pressure of between 0.5 and 1 bar at room temperature. Secondly, the results obtained were compared with those obtained by adsorption. The results obtained showed that the filtration process is very effective at a pressure of around 0.5bar. A comparison of the results obtained for the yield of phenol elimination by adsorption and by filtration revealed that the adsorption technique can reduce the concentration of phenol by up to 97%, while the filtration technique also recorded a very high percentage of around 90%.

1. Introduction

Today, the world's water problem is not only about the quantity available but also the quality of water [1]. The increase in human and industrial activity is leading to a significant amount of wastewater being discharged, polluting and damaging aquatic environments and ecosystems[2,3]. For a long time, considerable endeavors have been dedicated to devising and adopting processes that are both more effective and economically efficient in treating liquid effluents [4–6]. These processes include adsorption and membrane technology.

Membrane technology is widely used in many separation processes, for example, the food industry, the chemical industry, seawater desalination, industrial effluent treatment, etc. Because of the mechanical, thermal, chemical, and long-life performance that mineral membranes can offer, they are more popular with professionals than organic membranes [1]. Membranes are available in a variety of shapes, including flat and tubular, etc...., with varying diameters. Membrane filtration is a process that works by applying a pressure difference that allows the solvent to pass through a membrane with pores of a specific size to retain the solute. Tangential filtration stands out as the frequently employed method due to its cost-effectiveness in terms of energy. It possesses the added benefit of postponing and restricting the accumulation of deposits on the membrane, through the fouling phenomenon generated by the tangential flow of fluid across the membrane surface. The formation of deposits on the membrane leads to fouling of the pores, and this phenomenon causes variations in permeability and selectivity. To combat fouling, various cleaning techniques have been used, such as periodic backwashing that improves permeability, when the filtration rate is high the backwashing is even more obvious [7,8]. Photocatalytic materials are also used because of their effectiveness in combating fouling, for example, TiO_2 They have a photocatalytic capacity to decompose organic matter, microorganisms, and pollutants, reducing their adsorption on the surface of the membrane. [9,10]. Chemical cleaning of the membrane is also used and is still manner necessary regularly, their conditions depend on both the nature of the fouling and the membrane.

Adsorption is one of the easily implementable technologies; it is commonly used for water treatment. Activated carbon stands out as a commonly utilized adsorbent due to its elevated capacity for adsorbing organic contaminants[11,12]. However, its usage for treating substantial quantities of contaminated water is restricted by its elevated cost. Various researchers have explored the quest for economical and efficient alternative adsorbents, preferably sourced from natural substances and local waste [13,14]. Clays, owing to their abundance and cost-effectiveness, rank among the frequently employed inorganic materials.

This bibliographic review aims to present the studies undertaken in the field of organic pollutant elimination. The adsorption of phenol by clay was the subject of our most recent work research [15].

The membrane cost is contingent on both the raw material prices and the energy consumption during the sintering process [16]. One significant hurdle in membrane processes involves utilizing or developing membranes crafted from materials that impart novel properties to the membrane processes. [17]. Due to its low cost, natural abundance, and the benefit of densification at comparatively lower sintering temperatures in comparison to the mentioned oxides [18]. The development of membranes based on natural clay has been the subject of our previous work [17].

The primary aim of this study is to use clay in the treatment of a phenolic solution using the adsorption technique and filtration on an M_8 membrane that we developed previously in a work that has already been published [17]. In this work, a study was carried out on reducing the concentration of phenol in an aqueous solution by tangential filtration using a laboratory-scale micro-pilot. The filtering surface area is approximately 0.0072 m^2, and the results were compared with those obtained using the adsorption technique in the previous work.

2. Materials and Methods

2.1 Raw materials
The clay used for the development of the membranes was collected in the Draa-Tafilalet region, it is characterized by several analysis techniques, namely X-ray fluorescence, Fourier transform infrared spectroscopy (FTIR), X-ray diffraction (XRD), thermogravimetric and differential thermal analysis (TGA/DTA). The choice of this clay is due to its abundance in nature [17].

2.2 Manufactured membranes
The M_8 membranes were shaped by extrusion, and were characterized by several tests: porosity test, chemical resistance test, mechanical strength test, and scanning electron microscopy test[17].

2.3 Test water permeation
The permeation test was carried out on drinking tap water using a laboratory-scale filtration pilot Fig.1. The pilot has been fitted with an adjustable-flow electric pump (P) with a maximum pressure of 3 bar, a pressure gauge (m), a safety valve (V), a membrane module (M), a pressure regulator (R), a filtered water (permeate) recovery tank (B) and a drinking water feed tank (FW).

Mediterranean Architectural Heritage - RIPAM10

Materials Research Proceedings 40 (2024) 284-293

Materials Research Forum LLC

https://doi.org/10.21741/9781644903117-30

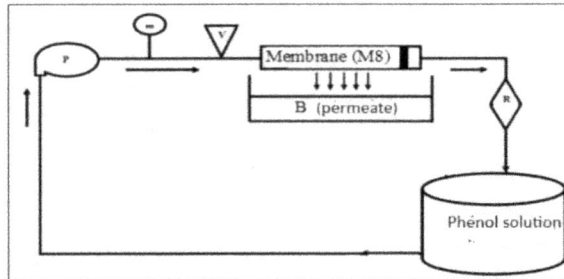

Fig 1: Micro-pilot of tangential filtration

The permeate flow rate is calculated using the following equation:

$$D = \frac{V}{t*S} \ (L/h.m^2) \tag{1}$$

D (L/h.m²) is the Permeate volume flow rate, t (hour) is the time, V (Liter) is the Volume, and S (m²) is the Filter surface area.

The calculation for the filtering surface of each membrane is as follows:

$$S = 2.\pi.r.l = \pi.D_{int}.l \ (m^2) \tag{2}$$

With, l (m) and Dint (m) are the length and inside diameter of the membrane respectively.

2.4 UV-visible spectrophotometric properties of phenol

Before beginning the study of the elimination of phenol by membrane filtration, the first approach consists of determining the UV-visible spectrophotometric properties of phenol, i.e. determining the λ_{max} for which the absorbance is maximum and verifying the validity of the Beer-Lambert law. To determine λ_{max} of phenol, measurements were made on a series of daughter solutions prepared by dilution of the mother solution. Measurements of absorbance as a function of wavelength (200-700nm) enabled us to deduce λ_{max}. The curve shown in Fig.2 represents the variation in absorbance as a function of wavelength.

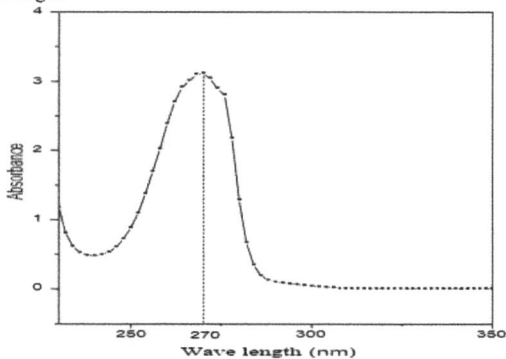

Fig 2: Determination the λ_{max} of phenol

Mediterranean Architectural Heritage - RIPAM10   Materials Research Forum LLC
Materials Research Proceedings 40 (2024) 284-293   https://doi.org/10.21741/9781644903117-30

2.4.1 Calibration curve

Taking previous results into account, we drew the calibration curve for phenol in order to determine the range of concentrations for which Beer-Lambert's law is respected (obtaining a straight line). The calibration was carried out using solutions of different concentrations, prepared from a stock solution. Table 1 gives the absorbance values as a function of the solutions at different concentrations prepared by dilution. Fig. 3 shows the calibration curve in the form of a straight line and verifies Beer-Lambert's law.

Table 1: absorbances after dilution

C (mg/l)	10^{-4}	4.10^{-4}	8.10^{-4}	10^{-4}	2.10^{-3}	6.10^{-3}
Absorbance	0.233	0.675	1.166	1.378	2.72	3.518

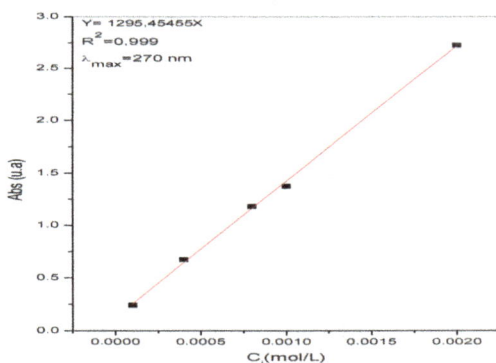

Figure 3: Phenol calibration curve

2.5 Treatment of the phenolic solution

The first stage consists of preparing a phenolic solution of known concentration. The second stage involves filtering the previous solution Fig 4. The concentration of the filtrates obtained is determined by visible UV. The efficiency of this membrane was evaluated by the flow rate of the permeate and the elimination yield as a function of time.

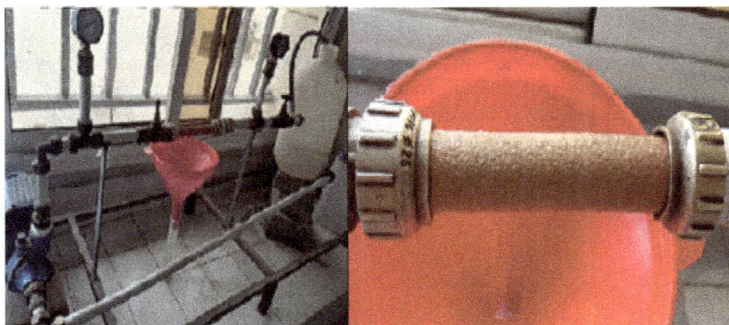

Fig 4: Treatment of the phenolic solution

3. Results and Discussion

3.1 Water permeation test

3.1.1 Variation in flow rate as a function of time D=f (t)

Fig. 5 shows the variation in flow rate as a function of time for M_8 treated at 1000°C [16], with a circulation pressure of 1 to 2 bar. Examination of the results obtained shows that the water permeation flow rate values are higher at the start of filtration for different pressure values from 1 to 2 bar. The M_8/1000°C/1bar flow rate = 37.9 L/h.m^2, M_8/1000°C/1.5bar flow rate = 13.15 L/h.m^2 and M_8/1000°C/2bar flow rate = 15.36 L/h.m^2. This is due to the pores being partially blocked by suspended solids or colloids present in the drinking water used. After a filtration time equal to 68, 97 and 90 min respectively for a pressure equal to 1 , 1.5 and 2 bar, we observe a stabilisation of the flow rate, which is explained by the concentration polarisation phenomenon. In other words, during the circulation of water at a given pressure, all the species present create an accumulation of matter at the water-membrane interface. Concentration polarisation refers to the formation of a very thin layer of water, whose concentration is greater than that of the feed solution (tap water). This layer remains immobile and thus reduces the apparent size of the pores, a phenomenon that has been well explained by several researchers [19].

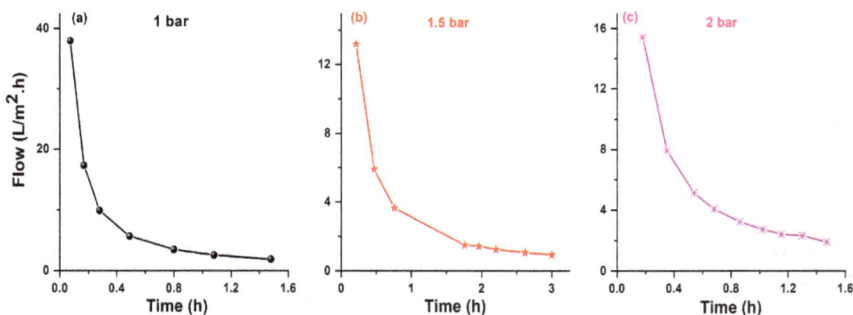

Fig 5: *Evolution of membrane water permeability flow rate as a function of Time*

3.1.2 Variation in flow rate as a function of pressure D=f(P)

The variation in flow rate as a function of pressure is shown Fig. 6. The decrease in flow rate at P=1.5 bar can be explained by the low driving force of the feed flow created by this pressure to remove the material accumulated at the water-membrane interface during filtration at P= 1bar. This material forms what is known as a deposit (rearrangement of suspended particles and colloids present in the water), so during filtration there will be an increase in deposit resistance. In the case of P = 2bar, there is an increase in deposition, unlike the case of 1.5 bar, which is due to the partial shearing of the layer formed during the previous filtrations.

Mediterranean Architectural Heritage - RIPAM10

Materials Research Proceedings 40 (2024) 284-293

Materials Research Forum LLC

https://doi.org/10.21741/9781644903117-30

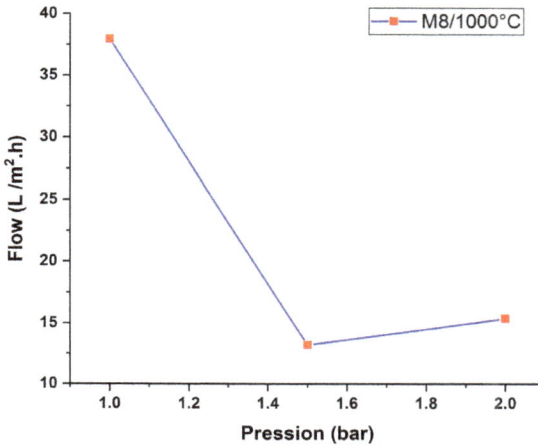

Fig 6: Flow rate evolution of membrane water permeability as a function of pressure

3.2 Polluted solution treatment tests

3.2.1 Changes in permeate flow rate as a function of time

For all the 0.5 and 1 bar pressure values, the permeate flow rate for this membrane decreased as a function of time and stabilized after 31 and 75 min of filtration respectively at 2.21 L/m^2.h and 5.16 $L/h.m^2$. The decrease in permeate flow rate is due to the progressive blocking of the pores by the retained phenol molecules on the external surface of the pores and on the internal surface of the membrane, leading to the phenomenon of concentration polarization.

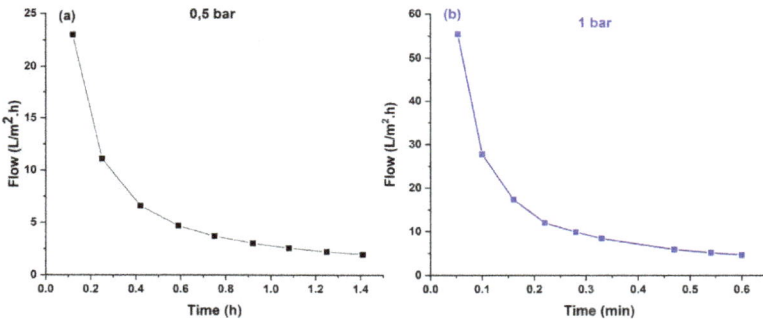

Fig 7: Evolution of membrane permeability flow rate as a function of time

3.2.2 Phenol retention efficiency

The visible UV spectrum of the phenol stock solution (510-4mol/L) shows a characteristic band at λ= 270 nm (Fig.2). The absorbance of permeates obtained after filtration at circulation pressures of 0.5 and 1bar, corresponding to the same value of λ, were determined by UV visible analysis.

To determine the concentrations of permeates we plotted the calibration curve (Fig.3) using standard phenol solutions of different concentrations. The retention efficiency of phenol as a function of time was calculated by equation 3:

$$R(\%) = \frac{C_i - C_{res}}{C_i} * 100 \tag{3}$$

C_i: initial concentration (mol/L)

C_{res}: concentration of phenol in the permeate (mol/L)

Analysis of the results obtained after filtration at P= 0.5bar shows that the yield of the reduction in the concentration of phenol in aqueous solution can reach 90% (Figure 8). The low absorbances of the permeates at λmax compared with the absorbance of the stock solution confirm these results (Fig. 9 and 10).

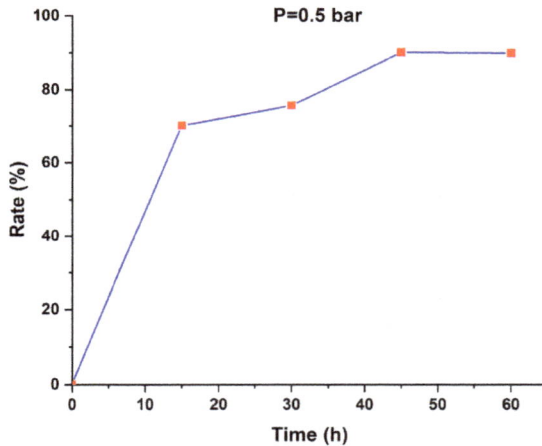

Fig 8: Evolution of phenol elimination yield in aqueous medium as a function of Ttime

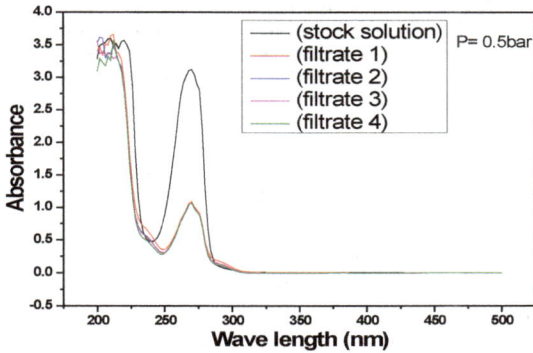

Fig 9: Illustration of UV analysis of the stock solution and permeates at P=0.5 bar

However, at P=1bar no elimination was recorded. Indeed the curves obtained by the UV analysis, illustrating the variation in the absorbances of the permeates are superimposed with that of the stock solution (Fig.2) which shows that there is no elimination at P=1bar.

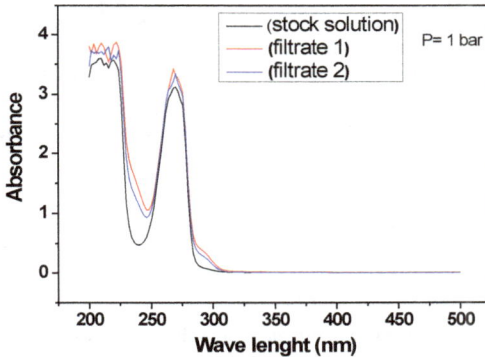

Fig 10: Illustration of UV analysis of the stock solution and of the permeates at P=1bar

3.3 Comparative study of phenol retention by adsorption and filtration

A comparison of the results obtained for the yield of phenol elimination by adsorption and filtration revealed that the yield of phenol retention by filtration on the M_8 membrane recorded a very high percentage of around 90%, while that obtained by adsorption on ART clay was only 51.8%[15].

Conclusion

In this study, we showcased the potential of developing cost-effective mineral membranes utilizing Moroccan clay. Additionally, this research has provided insights into various phenomena associated with the treatment of liquid effluents overall, with a specific focus on phenol elimination through filtration.

Mediterranean Architectural Heritage - RIPAM10   Materials Research Forum LLC
Materials Research Proceedings 40 (2024) 284-293   https://doi.org/10.21741/9781644903117-30

Examining the findings reveals that M_8 membranes, sintered at 1000°C, demonstrate favorable permeability and exceptional performance in phenol removal. This underscores the versatility of clays, suggesting their utility not only as raw materials for constructing building materials but also as effective adsorbents for retaining pollutants in aqueous environments. Consequently, we can deduce that the filtration membranes devised in the laboratory hold promise for widespread application in the treatment of liquid effluents.

References

[1] I. Tilla, D. Blumberga, Qualitative indicator analysis of a sustainable remediation, Energy Procedia. 147 (2018) 588–593. https://doi.org/10.1016/j.egypro.2018.07.075

[2] S. Kansal, M. Singh, D. Sud, Studies on photodegradation of two commercial dyes in aqueous phase using different photocatalysts, Journal of Hazardous Materials. 141 (2007) 581–590. https://doi.org/10.1016/j.jhazmat.2006.07.035

[3] K. Buruga, H. Song, J. Shang, N. Bolan, T.K. Jagannathan, K.-H. Kim, A review on functional polymer-clay based nanocomposite membranes for treatment of water, Journal of Hazardous Materials. 379 (2019) 120584. https://doi.org/10.1016/j.jhazmat.2019.04.067

[4] S. Evangelista, G. Viccione, O. Siani, A new cost effective, long life and low resistance filter cartridge for water treatment, Journal of Water Process Engineering. 27 (2019) 1–14. https://doi.org/10.1016/j.jwpe.2018.11.004

[5] N.H. Shalaby, E.M.M. Ewais, R.M. Elsaadany, A. Ahmed, Rice husk templated water treatment sludge as low cost dye and metal adsorbent, Egyptian Journal of Petroleum. 26 (2017) 661–668. https://doi.org/10.1016/j.ejpe.2016.10.006

[6] D. Zou, M. Qiu, X. Chen, E. Drioli, Y. Fan, One step co-sintering process for low-cost fly ash based ceramic microfiltration membrane in oil-in-water emulsion treatment, Separation and Purification Technology. 210 (2019) 511–520. https://doi.org/10.1016/j.seppur.2018.08.040

[7] Y. Gao, Y. Zhang, M. Dudek, J. Qin, G. Øye, S.W. Østerhus, A multivariate study of backpulsing for membrane fouling mitigation in produced water treatment, Journal of Environmental Chemical Engineering. 9 (2021) 104839. https://doi.org/10.1016/j.jece.2020.104839

[8] E. Mohammad-Pajooh, D. Weichgrebe, G. Cuff, B.M. Tosarkani, K.-H. Rosenwinkel, On-site treatment of flowback and produced water from shale gas hydraulic fracturing: A review and economic evaluation, Chemosphere. 212 (2018) 898–914. https://doi.org/10.1016/j.chemosphere.2018.08.145

[9] A. Lee, J.W. Elam, S.B. Darling, Membrane materials for water purification: design, development, and application, Environ. Sci.: Water Res. Technol. 2 (2016) 17–42. https://doi.org/10.1039/C5EW00159E

[10] H. Zhang, A.U. Mane, X. Yang, Z. Xia, E.F. Barry, J. Luo, Y. Wan, J.W. Elam, S.B. Darling, Visible-Light-Activated Photocatalytic Films toward Self-Cleaning Membranes, Adv Funct Materials. 30 (2020) 2002847. https://doi.org/10.1002/adfm.202002847

[11] Y. Jeong, S. Lee, S. Hong, C. Park, Preparation, characterization and application of low-cost pyrophyllite-alumina composite ceramic membranes for treating low-strength domestic wastewater, Journal of Membrane Science. 536 (2017) 108–115. https://doi.org/10.1016/j.memsci.2017.04.068

[12] H. Zouaoui, J. Bouaziz, Physical and mechanical properties improvement of a porous clay ceramic, Applied Clay Science. 150 (2017) 131–137. https://doi.org/10.1016/j.clay.2017.09.002

[13] M. Ghulam, K. Wyns, S. Janssens, A. Buekenhoudt, V. Meynen, Evaluation of the fouling resistance of methyl grafted ceramic membranes for inorganic foulants and co-effects of organic foulants, Separation and Purification Technology. 193 (2018) 29–37. https://doi.org/10.1016/j.seppur.2017.11.015

[14] C.-Y. Huang, C.-C. Ko, L.-H. Chen, C.-T. Huang, K.-L. Tung, Y.-C. Liao, A simple coating method to prepare superhydrophobic layers on ceramic alumina for vacuum membrane distillation, Separation and Purification Technology. 198 (2018) 79–86. https://doi.org/10.1016/j.seppur.2016.12.037

[15] H. Ouallal, M. Chrachmy, M. Azrour, M. Lechheb, A. EL-kordy, Y. Dehmani, H. Mousout, M. Azdouz, Study of acid treatmenteffect of a natural red clay onto physicochemical and adsorption propertie, Desalination and Water Treatment Ref.: TDWT-2023-0531.R2. (n.d.).

[16] B. Das, B. Chakrabarty, P. Barkakati, Separation of oil from oily wastewater using low cost ceramic membrane, Korean J. Chem. Eng. 34 (2017) 2559–2569. https://doi.org/10.1007/s11814-017-0185-z

[17] H. Ouallal, M. Azrour, M. Messaoudi, H. Moussout, L. Messaoudi, N. Tijani, Incorporation effect of olive pomace on the properties of tubular membranes, Journal of Environmental Chemical Engineering. 8 (2020) 103668. https://doi.org/10.1016/j.jece.2020.103668

[18] A.R.S. Ross, Identification of Histidine Phosphorylations in Proteins Using Mass Spectrometry and Affinity-Based Techniques, in: Methods in Enzymology, Elsevier, 2007: pp. 549–572. https://doi.org/10.1016/S0076-6879(07)23027-7

[19] A. Bes-Piá, J.A. Mendoza-Roca, L. Roig-Alcover, A. Iborra-Clar, M.I. Iborra-Clar, M.I. Alcaina-Miranda, Comparison between nanofiltration and ozonation of biologically treated textile wastewater for its reuse in the industry, Desalination. 157 (2003) 81–86. https://doi.org/10.1016/S0011-9164(03)00386-2

Mediterranean Architectural Heritage - RIPAM10
Materials Research Proceedings 40 (2024) 294-302

Materials Research Forum LLC
https://doi.org/10.21741/9781644903117-31

Structural Characterization of Stone Flooring "Calçada Marble", in Lisbon, Portugal

Mohamed EL OMARI[1,a], Mohamed EL AMRAOUI[1,b] *, Imane FIKRI[1,c],
Lahcen BEJJIT[1,d], Mustapha HADDAD[1,e], Saadia AIT LYAZDI[1,f],
Aumeur EL AMRANI[1,g]

[1]Laboratory of Spectrometry, Materials and Archaeomaterials (LASMAR), URL CNRST N° 7
Faculty of Sciences, Moulay Ismail University of Meknes, P.O Box 11201 Zitoune, Meknes,
Morocco

[a]moh4med.elomari@gmail.com, [b]m.elamraoui@umi.ac.ma, [c]imane.fikri16@gmail.com,
[d]l.bejjit@umi.ac.ma, [e]mhaddad22@yahoo.fr, [f]aitlyazidisaadia@yahoo.fr,
[g]a.elamrani@fste.umi.ac.ma

Keywords: Raman Spectroscopy, ATR-FTIR, XRD, Characterization, Mineralogical Phase, Flooring, Calçada

Abstract. The Lisbon "Calçada" sidewalks have a traditional style made of small flat and irregular cobblestones, representing magnificent mosaics. This work presents the characterization and identification of the mineralogical phases of sidewalk stones, originating from Lisbon, Portugal. Samples referenced MEKPBL (white) and MEKPNL (grey-black) were studied by means of Raman spectroscopy, ATR-FTIR and XRD. All Raman spectra show a strong band around 1089 cm^{-1} which can be attributed to the v_1-symmetric stretching mode of $(CO3)^{2-}$, indicating that calcite is the main crystal phase in both samples. All Raman spectra of MEKPNL show also graphite fingerprints localized around 1329 cm^{-1} and 1607 cm^{-1}, which are characteristic respectively of the D and G peaks of the carbonaceous material. These carbonaceous materials are the substance responsible of the grey-black color to these stones. Furthermore, the Raman spectra of the MEKPNL sample show also a weak Raman band located at 460 cm^{-1} which can be attributed to the symmetrical stretching of Si-O-Si, characteristic of the α-quartz. Furthermore, ATR-FTIR and XRD analyses supported the Raman results.

Introduction

Since antiquity, stone's mechanical and thermal properties have allowed it to be used for many applications [1, 2], particularly in funerary art as structural elements in buildings and architectural monuments, as well as in the paving and ornamentation of sidewalks and roads [2, 3]. Therefore, in the 19th century, a traditional flooring style that is part of Portugal's cultural heritage, called "calçada portuguesa", was introduced to most of the pavements in Portugal regions. It consists of small, flat stones arranged irregularly to form patterns or images, creating a mosaic-like appearance [2-5]. These stones come from various quarries, with some found in Portuguese regions such as Estremoz, Borba, and Vila Viçosa, while others are found in different countries, such as Italy, Greece and Spain [2]. Furthermore, this style of pavement can also be found in cities worldwide, from Olivenza (under Spanish colonial era), to former Portuguese colonies like Goa in India, and even the countries influenced by Portuguese culture, such as Brazil and Mozambique [3].

The investigation of the mineralogical composition of decorative stones, even those used in paving, requires the integration of multiple analysis techniques. Their correlation can provide crucial knowledge about these materials and manufacturing techniques, as well as solve problems related to their conservation/restoration of historical monuments, sculptures and architectural structures, including floor pavements. In this context, we find that Raman spectroscopy, ATR-

FTIR (Attenuated total reflectance-Fourier transform infrared) spectroscopy and XRD (Powder X-ray diffraction) analysis, are widely used techniques to identify the crystalline phases of ornamental stones [6, 7]. Additionally, Raman spectroscopy has been applied as a tool for examining the phases of different types of carbonaceous matter present in rocks and stones, such as kerogen, amorphous carbon, carbon, solid bitumens as well as graphite [8].

Hence, the work here aims to characterize and identify the mineralogical composition of two decorative stones coming from Lisbon; they are used to ornament the pavements of Calçada. The analysis was carried out using a combination of structural techniques including Raman spectroscopy, ATR-FTIR spectroscopy and XRD.

Materials and Methods

Materials

Two natural stones of almost identical size, originating from Lisbon, Portugal; they are the subject of a study using a multi-technique approach. These samples stones are referenced in LASMAR laboratory as MEKPNL and MEKPBL. The color of MEKPBL is white, while MEKPNL varies from grey to black. The stone samples were manually crushed using an agate mortar and pestle. The obtained powders were used for Raman and XRD analysis. For ATR-FTIR analysis, the powders were sieved to a grain size of ≤ 50 μm. Photos and a brief description of the stones studied are given in Table 1.

Table 1: Samples codes, photos, color and their origin.

Sample	Photo	Color	Origin
MEKPBL		White	Lisbon, Portugal
MEKPNL		Gray-black	Lisbon, Portugal

This type of stones has many heritage uses, particularly for paving the Portuguese Calçada. Fig.1 shows a particular section of the Lisbon pavement, made from small flat stones (white and black) arranged in traditional mosaic patterns.

Mediterranean Architectural Heritage - RIPAM10

Materials Research Proceedings 40 (2024) 294-302

Materials Research Forum LLC

https://doi.org/10.21741/9781644903117-31

Fig. 1: Center of Lisbon and particular sections of Calçada pavement in Lisbon.

Characterization methods

Micro-Raman spectroscopy

Raman spectra were carried out using the Renishaw RM 1000 confocal spectrometer. It's equipped with an (He-Ne) laser excitation source operating at 632.8 nm, an electrically cooled CCD detector and a DMLM confocal microscope with different objectives (10x, 20x, 50x and 100x). The spectra were recorded in the wavenumber range of 100 - 2000 cm^{-1} with a resolution less than of 2 cm^{-1}. Calibration was performed using the 520 cm^{-1} Raman line of a silicon wafer. Repeated acquisition using the highest magnification was accumulated to improve the signal to noise ratio. The Laser power was set between 10 and 50 mW in order to avoid degradation and decomposition of sample. Data were collected and plotted using the WIRE 3.4 software.

ATR-FTIR spectroscopy

Infrared spectra were obtained from powdered samples using an Alpha-Bruker spectrometer equipped with an ATR accessory equipped with a diamond crystal. The spectra were recorded in the wavenumber range from 400 to 4000 cm^{-1} with a spectral resolution of 4 cm^{-1}.

X-ray Diffraction (XRD)

The stone samples were characterized in powder form through X-ray diffraction at room temperature. This analysis was carried out using an ADVANCE D8 X-ray diffractometer, available at the Innovation and Technology Transfer Centre (CITT) of Moulay Ismaïl University, Morocco. It's equipped with Cu-Kβ radiation (wavelength of 1.5418 Å with 40 mA and 45 kV). The diffraction patterns were performed in the 2θ range of 5° - 80° with a step of 2° per minute. In addition, the various Bragg reflections are indexed using the RUFF database and the ICDD PDF-4 files [9, 10].

Results and discussion

Micro-Raman analyses

Fig. 2 presents the Raman spectra of the MEKPBL and MEKPNL samples, recorded in 150-2000 cm^{-1} range. It's evident from the spectra that both stones exhibit characteristic bands around 160, 286, 717 and 1089 cm^{-1}. The stronger band at 1089 cm^{-1} is attributed to the v_1-symmetric stretching mode of CO$_3^{2-}$ ionic group. The band at 717 cm^{-1} is assigned to the v_4- symmetric bending mode of CO_3^{2-}. The two bands at 160 and 286 cm-1 are related to the external vibrations, the relative translation between the CO_3^{2-} group and the Ca^{2+} cation (Ca^{2+}, CO_3^{2-})). These bands are the characteristics of the calcite ($CaCO_3$), which means that both samples can be classified as limestones composed mainly of calcite ($CaCO_3$) [1, 11-13].

The typical Raman bands of calcite ($CaCO_3$) observed in the studied limestones are consistent with those reported by S. Gunasekaran et al. (2006) of a natural limestone sourced from India, which is characterized at room temperature through FT-Raman spectroscopy, FTIR spectroscopy and XRD [12]. According to their Raman results, the external vibration modes (T (Ca^{2+}, CO_3^{2-})) were observed at 162 and 288 cm^{-1}. The symmetric bending mode (v_2 (CO_3^{2-})) at 716 cm^{-1}, while the symmetric stretching mode v_1 (CO_3^{2-}) at 1092 cm^{-1}.

An additional wide band is observed in both stones at 460 cm^{-1}, which can be assigned to the symmetrical stretching mode of Si-O-Si; it's the characteristic of α-quartz (SiO_2) [14- 16].

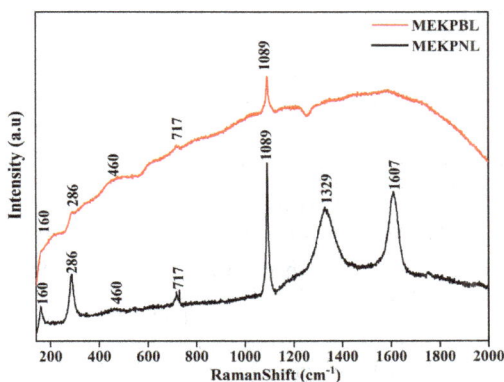

Fig. 2: Raman spectra of MEKPNL and MEKPBL samples.

The Raman spectrum of MEKPNL stone exhibits also two supplementary bands located around 1329 cm^{-1} and 1607 cm^{-1}; which are the fingerprints of the graphite (C) [1, 8, 17]. The first one at 1329 cm^{-1} is attributed to the D band, linked to disordered carbon or defective graphitic structures. Whereas those at 1607 cm^{-1} can be assigned to G band, the first-order Raman band of graphitic matter [8, 17, 21]. We can notice that graphite is the responsible element for the gray-black color of MEKPNL [1, 8, 17].

The Raman signals corresponding to graphite(C), detected in the MEKPNL spectrum are similar to those reported in the works of Raneri S. et al. (2020) [17]. The mentioned study focused on the characterization of three decorative black limestones from Tolfa region in Italy using Raman spectrometry. Consequently, the Raman band positions of these three black limestones were observed as follows: 1345, 1339 and 1334 cm^{-1} for the D bands and 1607, 1607 and 1609 cm^{-1} for the G bands, respectively.

ATR-FTIR analyses
ATR-FTIR spectra recorded in the 400-4000 cm^{-1} range of the MEKPBL and MEKPNL stones are illustrated in Fig. 3; they showed absorption bands at 711, 872, 1401 and 1794 cm^{-1}. The two strongest bands at 1401 cm^{-1} and 872 cm^{-1} are attributed to the asymmetric stretching (v_3) and bending (v_2) modes of the CO_3^{2-} ionic group, respectively. The band at 711 cm^{-1} is attributed to

Mediterranean Architectural Heritage - RIPAM10   Materials Research Forum LLC
Materials Research Proceedings 40 (2024) 294-302   https://doi.org/10.21741/9781644903117-31

the ν_4-asymmetric bending of the CO_3^{2-} group. The weak line at 1794 cm^{-1} is assigned to the combined modes ($\nu_1 + \nu_4$) [12, 13]. These bands are characteristic of calcite ($CaCO_3$).

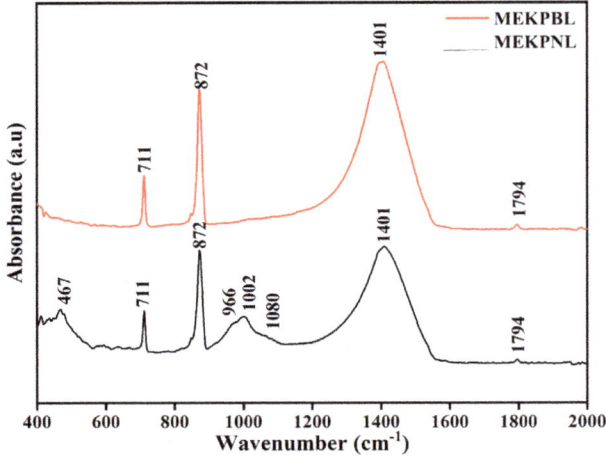

Fig. 3: ATR-FTIR spectra of MEKPNL and MEKPBL samples.

In addition, the spectrum of MEKPNL displays bands located at 467, 966, 1002 cm^{-1} and 1080 cm^{-1}, characteristic of the α-quartz (SiO_2) [14, 18, 19]. The band at 467 cm^{-1} is attributed to the asymmetric bending vibrations mode of Si-O-Si groups [18-20]. The both bands that are appear at 1002 and 1080 cm^{-1} are assigned to the asymmetric stretching mode of Si-O groups [18, 20], while the band at 966 cm^{-1} is attributed to the bending vibration of Si-O-(H-H_2O) [20].

The assignment of Raman and infrared bands for the two stones studied are given in Table 2, besides the identification of the mineralogical phases achieved through Raman and ATR-FTIR spectroscopy.

Table 2: Bands assignment and mineralogical compositions of Raman and ATR-FTIR analyses of MEKPBL and MEKPNL stones.

Sample	Band positions (cm⁻¹)		Assignment	Mineralogical composition
	Raman	Infrared		
MEKPBL	160-286	—	T (Ca²⁺, CO₃²⁻)	Calcite (CaCO₃)
	717	711	v_4-(CO₃²⁻)	
	—	872	v_2-(CO₃²⁻)	
	1089	—	v_1 (CO₃²⁻)	
	—	1401	v_3 (CO₃²⁻)	
	—	1794	v_{1+} v_4	
	460	—	Symmetric stretching of Si-O-Si	α- quartz (SiO₂)
MEKPNL	160-286	—	T (Ca²⁺, CO₃²⁻)	Calcite (CaCO₃)
	717	711	v_4-(CO₃²⁻)	
	—	872	v_2-(CO₃²⁻)	
	1089	—	v_1 (CO₃²⁻)	
	—	1401	v_3 (CO₃²⁻)	
	—	1794	v_{1+} v_4	
	460	—	Symmetric stretching of Si-O-Si	α- quartz (SiO₂)
	—	467	Asymmetric bending of Si-O-Si	
	—	966	bending of Si-O-(H-H₂O)	
	—	1002-1080	Asymmetric stretching of Si-O-Si	
	1329	—	D band	Graphite (C)
	1607	—	G band	

X-ray diffraction analyses

Fig. 4 presents the X-ray Diffraction patterns of the studied stones. As shown, both MEKPNL and MEKPBL samples exhibit a predominant line located at 2θ = 29.4°, attributed to calcite (CaCO₃), along with other peaks clearly observed and related to same component. MEKPBL and MEKPNL are calcitic stones. In addition, a weak line is observed in the MEKPBL sample at 2θ = 26.65°, that may correspond to α-quartz (SiO₂). The diffraction lines attributed to calcite and quartz are in good agreement with those reported in other works [1, 12, 21].

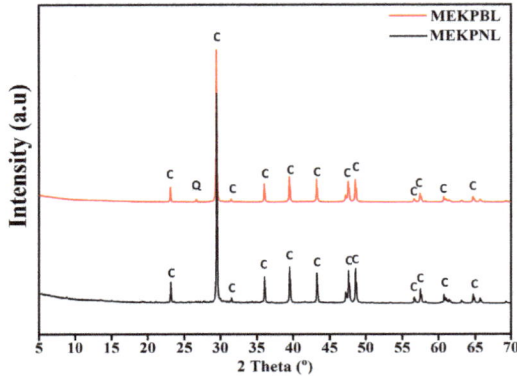

Fig. 4: XRD patterns of MEKPBL and MEKPNL samples. [C-calcite, Q- α-quartz]

The identification of the mineralogical compositions of the two studied stones MEKPBL and MEKPNL, through Raman spectroscopy, ATR-FTIR spectroscopy and XRD analysis are combined and summarized in Table 3.

Table 3: Mineralogical compositions of MEKPBL and MEKPNL stones.

Sample code	Origin	Main Phase	Other Phases
MEKPNL	Lisbon, Portugal	Calcite ($CaCO_3$)	α-quartz (SiO_2), Graphite (C)
MEKPBL	Lisbon, Portugal	Calcite ($CaCO_3$)	α-quartz (SiO_2)

The predominance of calcite ($CaCO_3$) indicates that the stones are limestone-based, which can have important implications durability and strength of the pavement [22, 23]. The presence of α-quartz (SiO_2) increases their impact resistance and strength [22, 23]. Additionally, the detection of graphite (C) through Raman spectroscopy suggests potential contributions to the mechanical properties [24, 25].

Conclusion

This study is part of an extensive study of two ornamental paving stones, originating from Lisbon, Portugal. The main objective is to determine the physical and chemical properties of these materials, where we characterize them through of multi-analytical techniques such as XRF, XRD, FORS, Raman spectroscopy, ATR-FTIR spectroscopy, EPR, and others methods can be applied.

The primary study, our focus was to characterize them in order to identify their mineralogical compositions. We used a combination of three structural spectroscopic techniques, including Raman spectroscopy, ATR-FTIR and XRD. This correlation shows that the calcite ($CaCO_3$) is the main mineralogical phase in the two identified limestones, with the presence of the α-quartz (SiO_2) in low content. Furthermore, micro-Raman spectroscopy show their ability to identify the graphitic nature of the MEKPNL stone, which is the responsible for its gray-black color.

References

[1] S. Khrissi, M. Haddad, L. Bejjit, S. A. Lyazidi, M. El Amraoui, C. Falguères, Raman and XRD characterization of Moroccan Marbles, IOP Conf. Ser: Mater. Sci. Eng. 186, (2017) 012028. https://doi.org/10.1088/1757-899X/186/1/012028

[2] J. Menningen, S. Siegesmund, L. Lopes et al. The Estremoz marbles: an updated summary on the geological, mineralogical and rock physical characteristics. Environ Earth Sci 77, (2018) 191. https://doi.org/10.1007/s12665-018-7328-3

[3] C.M. da Silva, S. Pereira. Walking on Geodiversity: the Artistic Stone-Paved Sidewalks of Lisbon (Portugal) and Their Heritage Value. Geoheritage 14, (2022) 98. https://doi.org/10.1007/s12371-022-00733-5

[4] D. Pacheco, J. Cotas, A. Domingues, S. Ressurreição, K. Bahcevandziev, L. Pereira. Chondracanthus teedei var. lusitanicus: the nutraceutical potential of an unexploited marine resource. Marine Drugs, 19(10), (2021) 570. https://doi.org/10.3390/md19100570

[5] Y. Chen, J. Fan. An applied study of Calçada Portuguesa in a former Portuguese colonial cityTake Goa, India as an example. India as an example. SSRN. (June 1, 2022). http://dx.doi.org/10.2139/ssrn.4145685

[6] A. Rousaki, P. Vandenabeele. In situ Raman spectroscopy for cultural heritage studies. Journal of Raman Spectroscopy, 52(12), (2021) 2178-2189. https://doi.org/10.1002/jrs.6166

[7] A. Boukir, S. Fellak, P. Doumenq. Structural characterization of Argania spinosa Moroccan wooden artifacts during natural degradation progress using infrared spectroscopy (ATR-FTIR) and X-Ray diffraction (XRD). Heliyon, 5(9). (2019).

https://doi.org/10.1016/j.heliyon.2019.e02477

[8] J. Jehlička, A. Šťastná, R. Přikryl. Raman spectral characterization of dispersed carbonaceous matter in decorative crystalline limestones. Spectrochimica Acta Part A: Molecular and Biomolecular Spectroscopy, 73(3), (2009) 404-409. https://doi.org/10.1016/j.saa.2008.09.006

[9] https://rruff.info/ (accessed August 20, 2023)

[10] www.icdd.com/pdfsearch/ (accessed September 25, 2023).

[11] N. Buzgar, A. I. Apopei. The Raman study of certain carbonates. Geologie Tomul L, 2(2), (2009) 97-112. https://www.researchgate.net/publication/210110408

[12] S. Gunasekaran, G. Anbalagan, S. Pandi. Raman and infrared spectra of carbonates of calcite structure. Journal of Raman Spectroscopy: An International Journal for Original Work in all Aspects of Raman Spectroscopy, Including Higher Order Processes, and also Brillouin and Rayleigh Scattering, 37(9), (2006) 892-899. https://doi.org/10.1002/jrs.1518

[13] Y. Kim, M. C. Caumon, O. Barres, A. Sall, J. Cauzid. Identification and composition of carbonate minerals of the calcite structure by Raman and infrared spectroscopies using portable devices. Spectrochimica Acta Part A: Molecular and Biomolecular Spectroscopy, 261, (2021) 119980.. https://doi.org/10.1016/j.saa.2021.119980

[14] M. Ostroumov, E. Faulques, E. Lounejeva. Raman spectroscopy of natural silica in Chicxulub impactite, Mexico. Comptes Rendus Geoscience, 334(1), (2002) 21-26.. https://doi.org/10.1016/S1631-0713(02)01700-5

[15] R. Palmeri, M. L. Frezzotti, G. Godard, R. J. Davies. Pressure-induced incipient amorphization of α-quartz and transition to coesite in an eclogite from Antarctica: a first record and some consequences. Journal of Metamorphic Geology, 27(9), (2009) 685-705.. https://doi.org/10.1111/j.1525-1314.2009.00843.x

[16] S. K. Sharma, A. K. Misra, S. Ismail, U. N. Singh. Remote Raman spectroscopy of various MIXED and composite mineral phases at 7.2 m distance. 37th Lunar and Planetary Science Conference, (2006). https://ntrs.nasa.gov/20060010179

[17] S. Raneri, F. Košek, L. Lazzarini, D. Wielgosz-Rondolino, J. Jehlicka, F. Antonelli. Raman spectroscopy as a tool for provenancing black limestones (bigi morati) used in antiquity. Journal of Raman Spectroscopy, 52(1), (2021) 241-250. https://doi.org/10.1002/jrs.5948

[18] P. Makreski, G. Jovanovski, T. Stafilov, B. Boev. Minerals from Macedonia XII. The dependence of quartz and oral color on trace element composition-AAS, FT IR and MICRO-RAMAN spectroscopy study. Bulletin of the Chemists and Technologists of Macedonia, 23(2), (2004) 171-184. https://eprints.ugd.edu.mk/id/eprint/2449

[19] F. Bosch-Reig, J. V. Gimeno-Adelantado, F. Bosch-Mossi, A. Doménech-Carbó. Quantification of minerals from ATR-FTIR spectra with spectral interferences using the MRC method. Spectrochimica Acta Part A: Molecular and Biomolecular Spectroscopy, 181, (2017) 7-12 https://doi.org/10.1016/j.saa.2017.02.012

[20] I. Ramalla, R. K. Gupta, K. Bansal. Effect on superhydrophobic surfaces on electrical porcelain insulator, improved technique at polluted areas for longer life and reliability. Int. J. Eng. Technol, 4(4), (2015) 509. https://doi.org/10.14419/ijet.v4i4.5405

[21] O. Beyssac, B. Goffé, J. P. Petitet, E. Froigneux, M. Moreau, J. N. Rouzaud. On the characterization of disordered and heterogeneous carbonaceous materials by Raman spectroscopy. Spectrochimica Acta Part A: Molecular and Biomolecular Spectroscopy, 59(10), (2003) 2267-2276. https://doi.org/10.1016/S1386-1425(03)00070-2

[22] T. M. B. Senarathna, SHMPK. Janith, A. Dassanayake, S. P. Chaminda, C L. Jayawardena, Correlations between durability, mineralogy and strength properties of limestone (2021). https://www.researchgate.net/publication/361683014

[23] P. Q. Zhao, L. C. Lu, S. D. Wang. Influence of high-silicon limestone on mineral structure and performance of belite-barium calcium sulphoaluminate clinker. Advanced Materials Research, 168, (2011) 460-465. https://doi.org/10.4028/www.scientific.net/AMR.168-170.460

[24] V. Černý, G. Yakovlev, R. Drochytka, Š. Baránek, L. Mészárosová, J. Melichar, R. Hermann. Impact of Carbon Particle Character on the Cement-Based Composite Electrical Resistivity. Materials, 14(24), (2021) 7505. https://doi.org/10.3390/ma14247505

[25] Z. Ren, J. Sun, W. Tang, X. Zeng, H. Zeng, Y. Wang, X. Wang. Mechanical and electrical properties investigation for electrically conductive cementitious composite containing nano-graphite activated magnetite. Journal of Building Engineering, 57, (2022) 104847. https://doi.org/10.1016/j.carbon.2004.12.033

Mediterranean Architectural Heritage - RIPAM10
Materials Research Proceedings 40 (2024) 303-310

Materials Research Forum LLC
https://doi.org/10.21741/9781644903117-32

Fire Induced Microstructural Changes in Local Building Materials: Cases of White Marble and Limestone

Laila AKRAM[1,a *], Imane FIKRI[1,b], Abdelkhalek KAMMOUNI[1,c], Salam KHRISSI[2,d], Mustapha HADDAD[1,e] and Saadia AIT LYAZIDI[1,f]

[1]Laboratory of Materials and Archaeomaterials Spectrometry (LASMAR), URL-CNRST, N° 7, Faculty of Sciences-Moulay Ismail University, Meknes, Morocco

[2]National School of Applied Sciences-Cadi Ayyad University, Process Engineering & Advanced Materials, Safi, Morocco

[a]l.akram@edu.umi.ac.ma, [b]imane.fikri16@gmail.com, [c]a.kammouni@umi.ac.ma, [d]khrissi_salam@hotmail.com, [e]m.haddad@umi.ac.ma, [f]s.aitlyazidi@umi.ac.ma

Keywords: Ancient Buildings, Limestone and Marble, Fire Exposure, Characterization Techniques, Thermal Stability

Abstract. The aim of this work is to evaluate the degradation state of natural stones after their exposure to fire. These building and decorative materials, widely used in the architectural heritage, suffer irreversible damage when exposed to high temperatures. Therefore, knowledge of their residual durability is crucial in order to determine whether the post-fire building structure should be restored, reinforced or demolished. For this purpose, limestones (calcarenites) and white marbles collected from local quarries were subjected to heating-cooling cycles in a muffle furnace at various temperatures up to 1100°C. After each exposure, the selected samples were characterized at room temperature using X-ray diffraction (XRD), micro-Raman and ATR-FTIR infrared techniques. The results obtained showed that the mineralogical nature of both calcareous and marble natural stones is a key factor in their thermal stability when exposed to high temperatures. Above 570°C, natural stones undergo calcite decarbonation at different temperature ranges. Marble, which is mineralogically monophasic, underwent decomposition at 800°C, similar to calcite in its pure state. Calcarenite was decomposed at a much lower temperature of about 700 °C. This study classifies marble as more thermally stable than calcarenite.

Introduction

Since antiquity, natural stone has been widely used as a building and decorative material in ancient edifices. It continues to play an important role in the restoration and replacement of damaged parts. However, the fire of the Great Mosque of Taroudant (Morocco), the main mosque of the Saadian dynasty, in 2013, and the fire of Notre-Dame Cathedral in Paris in 2019 have brought back the problem of preserving built heritage in the face of serious risks. It's therefore extremely important to collect experimental data on the behavior of natural stone at high temperatures in order to diagnose structures that have suffered fire damage.

When exposed to fire, natural stone undergoes a significant increase in temperature, resulting in irreversible changes in its mineralogical composition and structural properties. Several studies have examined the effects of high temperatures on the residual physical and mechanical properties of various types of stone [1-5]. The works of Zhang et al [1] and Vigroux et al focused on calcareous and siliceous stones [2]; they analyzed the effect of high temperatures on the physico-mechanical properties of seven stones with different mineralogical and physical properties. The evolution of these properties up to 1000 °C in four limestones and four marbles with different textural and structural properties was also studied by Ozguven et al. [5]. The primary colour changes that stones are likely to undergo when exposed to high temperatures have also been

Mediterranean Architectural Heritage - RIPAM10
Materials Research Proceedings 40 (2024) 303-310

Materials Research Forum LLC
https://doi.org/10.21741/9781644903117-32

identified in some studies. A non-destructive technique using colorimetric measurements to assess the extent of stone degradation, and more specifically the highest temperature reached after exposure to fire, was investigated by Beck et al. [6] and Ozguven et al. [7]. However, there are few studies on the thermal behavior of natural stones using Raman and ATR-FTIR vibrational techniques, as well as XRD. The aim of this work is to provide additional experimental data on the physico-chemical behavior of natural stones when exposed to temperatures ranging from room temperature (RT) to 1100°C. Hence, two natural stones commonly used in ancient buildings were selected: white marble and limestone/calcarenite. The investigation combined techniques that were effective and complementary in assessing the high-temperature behavior of the building stones considered: ATR-FTIR, micro-Raman and XRD.

Materials and methods

Samples

Two series of natural stones were investigated: white marble and limestone collected from different regions in Morocco (Fig. 1). The white marbles were sampled from Sidi Lamine quarries in the Khénifra region. The rock presents a white color with locally heterogeneous levels of yellow shades. Previous studies have shown that it consists mainly of calcite with traces of dolomite [8]. The limestones are sedimentary rocks extracted from a quarry near the city of Meknes, Morocco. The rock displays a yellowish-ochre hue and exhibits visible heterogeneity. Two distinct facies are discernible: one is compact and fine, while the other is porous and grainy. Preliminary analyses indicate that it is primarily composed of calcite, with an average quartz content and traces of kaolinite clay.

Only the results of a single sample, which is highly representative of the others, will be presented for each series.

(a)   (b)

Fig. 1. The studied samples: white marble (a) and limestone (b).

Heat treatment protocol

Heating-cooling cycles were carried out using a programmable muffle furnace (Nabertherm). The thermal protocol involved a gradual increase in temperature at a constant rate of 1°C/min until the target temperature was reached, followed by a 4-hour isothermal stage to ensure uniform sample temperature. To prevent thermal shock, the cooling process was dependent on the furnace's inertia. Various temperatures were considered on the basis of TGA/DTA sample analysis: 80, 100, 200, 350, 450, 570, 800, and 1100°C. To deepen structural analyses only 570°C, 800°C, and 1100°C were chosen as they correspond to the physico-chemical transformations present in the studied natural stones. This option allows for the understanding of actual temperatures achieved during a fire [2].

Instrumentation

An Alpha II Bruker spectrometer equipped with an ATR accessory with a diamond crystal was used to collect infrared spectra in the 4000-400 cm^{-1} frequency range with a spectral resolution of 4 cm^{-1}. Powder micro-samples were pressed onto the surface of the crystal.

Mediterranean Architectural Heritage - RIPAM10
Materials Research Proceedings 40 (2024) 303-310

Materials Research Forum LLC
https://doi.org/10.21741/9781644903117-32

Raman spectra were obtained using a portable Sequentially Shifted Excitation Raman Spectrometer (SSERS) Bravo from Brucker. The spectra were collected, on compacted pellets, with excitation ranging from 700 to 1100 nm.

A Bruker D8 Advance X-ray diffractometer with a Cu Kα anode ($\lambda = 1.5418$ Å) operating at 40 kV and 30 mA was used to obtain XRD diffractograms of the powders. The measurements were carried out at room temperature over an angular range of 10-70 degrees at 2θ with a step size of 0.02 degrees.

Results and discussion

Infrared spectroscopy ATR-ATR

The ATR-FTIR analysis of limestone and white marble before heat treatment reveals the presence of calcite, identified by its typical intense bands at 1417, 874, and 711 cm^{-1} (Fig. 2). They are respectively attributable to the asymmetric $v_3(C - O)$ stretch, the $v_2(C - O)$ out-of-plane bend and the $v_4(C - O)$ in-plane bend of CO_3^{2-} carbonate group [9–11]. The limestone spectrum also exhibits a band at 1028 cm^{-1}, probably associated with the Si-O plane stretching vibration mode of clays in the form of kaolinite [12,13]. Doublet at 797-777 cm^{-1} and bands at 1165, 1082 and 693 cm^{-1} can be attributed to quartz in small quantities [10,11,14].

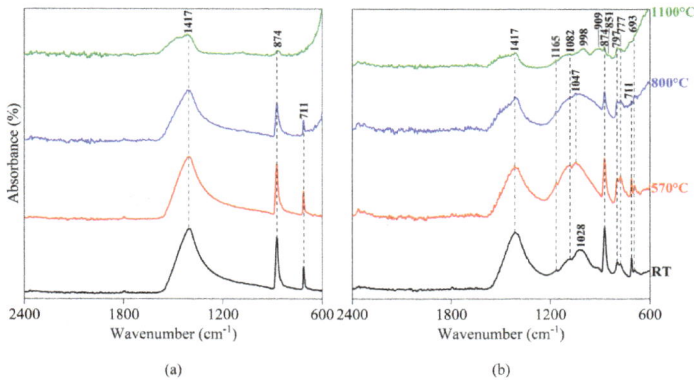

Fig. 2. ATR-FTIR spectra at room temperature (RT) of white marble (a) and limestone (b) after heat treatment at different temperatures.

The spectra collected at room temperature after heat treatment are compared with those obtained at the initial state (RT) before heating. After heat treatment of limestone at 570°C, the band associated with kaolinite (1028 cm^{-1}) shifts to 1047 cm^{-1} (Fig. 2b), indicating kaolinite ($Al_2Si_2O_5(OH)_4$) dehydroxylation to metakaolin ($Al_2Si_2O_7$) [13]. At 800°C, a decrease in the intensity of the calcite characteristic bands is observed. At this temperature calcite decomposes into calcium oxide, releasing carbon dioxide and forming portlandite as a result of (CaO) interaction with air moisture [2,5]. The characteristic bands of lime and portlandite, located respectively at 3460 and 3644 cm^{-1}, are outside the explored spectral range [11,15]. At 1100°C, typical calcite modes become extremely weak (v_2, v_3) or disappear (v_4) ; calcite decarbonation process is complete or almost complete. This may indicate a phase transition [16]. The emerging bands at 998, 909 and 851 cm^{-1} indicate the apparition of a new phase, called gehlenite [17,18], resulting from the reaction between metakaolin and lime according to the following reaction [11,19]:

$$Al_2Si_2O_7 + 2CaO \rightarrow Ca_2Al(AlSiO_7) + SiO_2. \tag{1}$$

Micro-Raman spectroscopy

The micro-Raman spectra of untreated stones corroborate the results obtained by IR (Fig. 3), showing typical bands of calcite at 1086 and 712 cm^{-1}, attributed respectively to the stretching (v_1) and bending (v_4) symmetric modes of the CO_3^{2-} carbonate group [8,20]. Low intensity signals associated with quartz are also present in limestone. In particular, we distinguish the peak at 464 cm^{-1} corresponding to the symmetric stretching of $Si - O - Si$, as well as a weak line at 392 cm^{-1} associated with the lattice or network mode [8,21,22]. The kaolinite could not be identified because it occurs as a trace element in limestone.

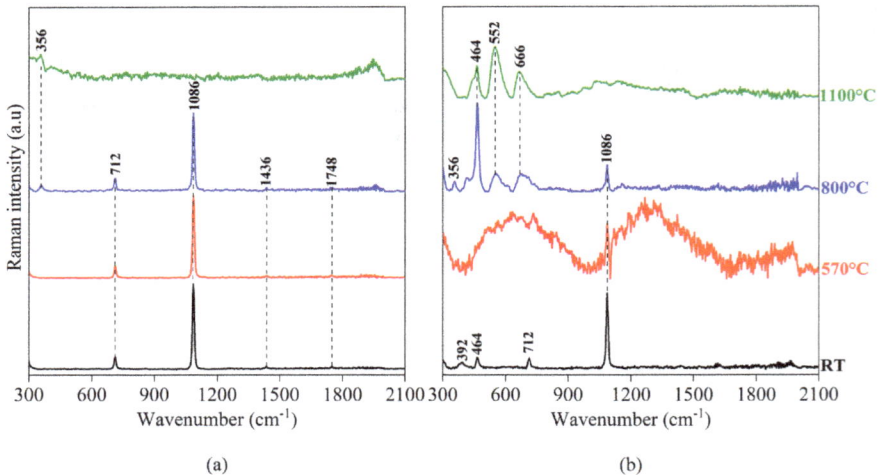

Fig. 3. Micro-Raman spectra recorded at room temperature of white marble (a) and limestone (b) after heat treatment at different temperatures.

Heat treatment at 570°C caused significant changes in the limestone sample. Only the peak attributed to calcite remained distinct at this temperature, while signals attributed to quartz were disturbed by fluorescence phenomena. With heat treatment at 800°C, the calcite peak at 712 cm^{-1} disappeared and the intensity of its main band located at 1086 cm^{-1} decreased (Fig. 3b) while new bands emerged at 666 and 356 cm^{-1}, probably related to the formation of gehlenite and portlandite [21, 23, 24]. As previously stated in the infrared section, gehlenite is formed by the reaction of metakaolin and lime. The hydration of lime is also facilitated by the humidity in the surrounding air during cooling, resulting in the formation of portlandite [2,25]. Another band is observed at 552 cm^{-1}. It can be attributed to the presence of aluminum or silicon compounds [26]. As a result, the intensity of the main silica band around 464 cm^{-1} increased, followed by a decrease as the heat treatment temperature increased to 1100°C. This phenomenon may be due to the partial decomposition of calcite and the α-β phase transition of quartz [27]. At 1100°C, the calcite and portlandite bands disappear, while the band attributed to gehlenite becomes slightly more distinct. This suggests that the lime produced is involved in the formation of gehlenite. These results support the hypothesis of calcite decomposition already corroborated in the infrared section.

For white marble (Fig. 3a), heat treatment at 570°C and 800°C revealed a good thermal stability of this stone. After reaching a temperature of 1100°C, decarbonation of calcite occurs, resulting in the formation of portlandite identified by the band at 356 cm^{-1}.

X-ray diffraction analysis
Figure 4 shows the diffractograms of the studied stones recorded at room temperature after heat treatment at different temperatures. After treatment at 570°C, limestone diffractogram shows significant increase of the quartz characteristic line and disappearance of the low intensity peak at 12.42°2θ related to kaolinite [28]. Additionally, there is a decrease in the intensity of the peak at 29.48°2θ related to calcite (Fig. 4b) [8, 10, 11]. This is as a consequence of kaolinite dehydroxylation and phase transition of quartz [27, 29].

No significant changes were observed in the case of white marble treated at 570°C. However, after treating at 800°C, XRD analysis showed an important reduction in calcite lines and the appearance of new ones at 37.4 and 18.09° 2θ. These lines illustrate the decomposition of calcite and the formation of lime and portlandite [1,15]. The formation of portlandite occurs as soon as the lime reacts with atmospheric water vapor. After treatment at 1100°C, calcite is completely transformed into lime. In addition to peaks assigned to lime and portlandite, the limestone diffractogram also displays peaks attributable to gehlenite, due to the reaction of lime with metakaolin [19,30].

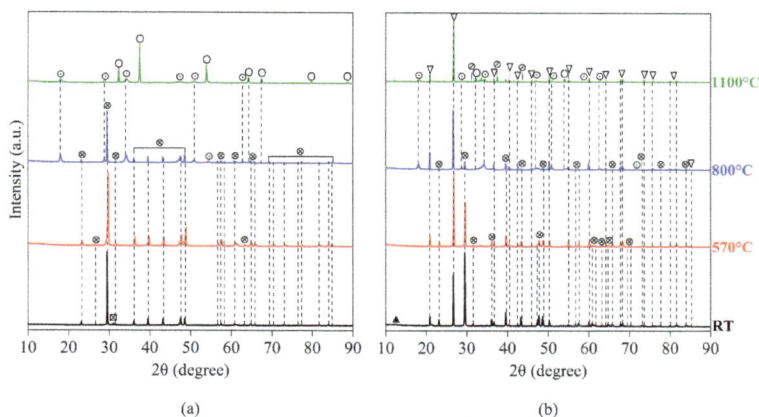

Fig. 4. Diffractograms performed at room temperature of white marble (a) and limestone (b) after heat treatment at different temperatures.

Conclusion
The investigation focused on the structural diagnosis of the effects of high-temperature exposure on building and decorative materials in architectural heritage. Specifically, the study examined the behavior of limestone and white marble when exposed to fire. The stones are sampled from quarries in Morocco.

The ATR-FTIR, Raman, and XRD structural analyses were found in perfect agreement and showed that exposure up to 570°C has very little influence on the microstructure of these materials. However, at higher temperatures, calcite, which is the main mineral in these stones, begins to decompose. In the case of marble, identified as monophasic, calcite showed a decomposition temperature around 800°C, similarly to pure calcite. Nonetheless, in the case of calcarenite, the presence of secondary minerals proved to be a stimulating factor for the relatively early decomposition of calcite. This calcite breaks down makes the limestone less resistant to temperature compared to marble, making it more vulnerable to fire.

Acknowledgments

Authors acknowledge the financial support from the Moroccan CNRST (PPR/2015/75 project and statute fund of the URL 7).

References

[1] Y. Zhang, X. Ta, S. Qin, Effect of heat treatment on physico-mechanical behaviour of a natural building stone: Laizhou dolomite marble, J. Build. Eng. 47 (2022) 103885. https://doi.org/10.1016/j.jobe.2021.103885

[2] M. Vigroux, J. Eslami, A.L. Beaucour, A. Bourgès, A. Noumowé, High temperature behaviour of various natural building stones, Constr. Build. Mater. 272 (2021) 121629. https://doi.org/10.1016/j.conbuildmat.2020.121629

[3] Q. Guo, H. Su, J. Liu, Q. Yin, H. Jing, L. Yu, An experimental study on the fracture behaviors of marble specimens subjected to high temperature treatment, Eng. Fract. Mech. 225 (2020) 106862. https://doi.org/10.1016/j.engfracmech.2019.106862

[4] W. Zhang, Q. Sun, S. Zhu, B. Wang, Experimental study on mechanical and porous characteristics of limestone affected by high temperature, Appl. Therm. Eng, 110 (2017) 356–362. https://doi.org/10.1016/j.applthermaleng.2016.08.194

[5] A. Ozguven, Y. Ozcelik, Effects of high temperature on physico-mechanical properties of Turkish natural building stones, Eng. Geol. 183 (2014) 127-136. https://doi.org/10.1016/j.enggeo.2014.10.006

[6] K. Beck, S. Janvier-Badosa, X. Brunetaud, Á. Török, M. Al-Mukhtar, Non-destructive diagnosis by colorimetry of building stone subjected to high temperatures, Eur. J. Environ. Civ. 20 (2016) 643-655. https://doi.org/10.1080/19648189.2015.1035804

[7] A. Ozguven, Y. Ozcelik, Investigation of some property changes of natural building stones exposed to fire and high heat, Constr. Build. Mater. 38 (2013) 813-821. https://doi.org/10.1016/j.conbuildmat.2012.09.072

[8] S. Khrissi, M. Haddad, L. Bejjit, S. Ait Lyazidi, M. El Amraoui, C. Falguères, Étude spectrométrique de marbres du Maroc central, L'Anthropologie, 121 (2017) 55-62. https://doi.org/10.1016/j.anthro.2017.03.004

[9] T. Lamhasni, H. El-Marjaoui, A. El Bakkali, S. Ait Lyazidi, M. Haddad, A. Ben-Ncer, F. Benyaich, A. Bonazza, M. Tahri, Air pollution impact on architectural heritage of Morocco: Combination of synchronous fluorescence and ATR-FTIR spectroscopies for the analyses of black crusts deposits, Chemosphere, 225 (2019) 517-523. https://doi.org/10.1016/j.chemosphere.2019.03.109

[10] A. Di Salvo Barsi, M.A. Trezza, E.F. Irassar, Comparison of dolostone and limestone as filler in blended cements, Bull. Eng. Geol. Environ. 79 (2020) 243-253. https://doi.org/10.1007/s10064-019-01549-4

[11] S. Petlitckaia, A. Gharzouni, E. Hyvernaud, N. Texier-Mandoki, X. Bourbon, S. Rossignol, Influence of the nature and amount of carbonate additions on the thermal behaviour of geopolymers: A model for prediction of shrinkage, Constr. Build. Mater. 296 (2021) 123752. https://doi.org/10.1016/j.conbuildmat.2021.123752

[12] E. Gasparini, S.C. Tarantino, P. Ghigna, M.P. Riccardi, E.I. Cedillo-González, C. Siligardi, M. Zema, Thermal dehydroxylation of kaolinite under isothermal conditions, Appl. Clay. Sci. 80 (2013) 417-425. https://doi.org/10.1016/j.clay.2013.07.017

[13] R. Deju, C. Mazilu, I. Stanculescu, C. Tuca, Fourier transform infrared spectroscopic characterization of thermal treated kaolin, Rom. Rep. Phys. 72 (2020) 806.

[14] Z. Xing, R. Hébert, A.L. Beaucour, B. Ledésert, A. Noumowé, Influence of chemical and mineralogical composition of concrete aggregates on their behaviour at elevated temperature, Mater. Struct. 47 (2014) 1921-1940. https://doi.org/10.1617/s11527-013-0161-y

[15] Y. Tang, J. Xu, J. Zhang, Y. Lu, Biodiesel production from vegetable oil by using modified CaO as solid basic catalysts, J. Clean. Prod. 42 (2013) 198-203. https://doi.org/10.1016/j.jclepro.2012.11.001

[16] S. Gunasekaran, G. Anbalagan, Spectroscopic study of phase transitions in natural calcite mineral, Spectrochim. Acta A: Mol. Biomol. Spectrosc. 69 (2008) 1246-1251. https://doi.org/10.1016/j.saa.2007.06.036

[17] N.V. Chukanov, A.D. Chervonnyi, Infrared Spectroscopy of Minerals and Related Compounds, Springer Mineralogy. Cham: Springer International Publishing, 2016. https://doi.org/10.1007/978-3-319-25349-7

[18] I. Perná, M. Šupová, T. Hanzlíček, Gehlenite and anorthite formation from fluid fly ash, J. Mol. Struct. 1157 (2018) 476-481. https://doi.org/10.1016/j.molstruc.2017.12.084

[19] P. Ptáček, T. Opravil, F. Šoukal, J. Havlica, R. Holešinský, Kinetics and mechanism of formation of gehlenite, Al–Si spinel and anorthite from the mixture of kaolinite and calcite, Solid. State. Sci. 26 (2013) 53-58. https://doi.org/10.1016/j.solidstatesciences.2013.09.014

[20] I. Fikri, M. El Amraoui, M. Haddad, A.S Ettahiri, C. Falguères, L. Bellot-Gurlet, T. Lamhasni, S. Ait Lyazidi, L. Bejjit, Raman and ATR-FTIR analyses of medieval wall paintings from al-Qarawiyyin in Fez (Morocco), Spectrochim. Acta. A: Mol. Biomol. Spectrosc. 280 (2022) 121557. https://doi.org/10.1016/j.saa.2022.121557

[21] Š. Pešková, V. Machovič, P. Procházka, Raman spectroscopy structural study of fired concrete, Ceram. - Silik. 55 (2011) 410-417.

[22] N. Böhme, K. Hauke, M. Neuroth, T. Geisler, In situ Raman imaging of high-temperature solid-state reactions in the CaSO4–SiO2 system, Int. J. Coal. Sci. Technol. 6 (2019) 247-259. https://doi.org/10.1007/s40789-019-0252-7

[23] M. Chollet, M. Horgnies, Analyses of the surfaces of concrete by Raman and FT-IR spectroscopies: Comparative study of hardened samples after demoulding and after organic post-treatment, Surf. Interface. Anal. 43 (2011) 714-725. https://doi.org/10.1002/sia.3548

[24] T. Schmida, P. Dariz, Shedding light onto the spectra of lime: Raman and luminescence bands of CaO, Ca(OH)2 and CaCO3, J. Raman. Spectrosc. 46 (2014) 141-146. https://doi.org/10.1002/jrs.4622

[25] F. Sciarretta, J. Eslami, A.L. Beaucour, A. Noumowé, State-of-the-art of construction stones for masonry exposed to high temperatures, Constr. Build. Mater. 304 (2021) 124536. https://doi.org/10.1016/j.conbuildmat.2021.124536

[26] M. Pinet, D.C. Smith, B. Lasnier, Utilité de la microsonde Raman pour l'identification non-destructive des gemmes, La Microsonde Raman en Geologie, ed. N. Hors-Serie, Revue de Gemmologie, Paris, p. 11, 1992.

Mediterranean Architectural Heritage - RIPAM10   Materials Research Forum LLC
Materials Research Proceedings 40 (2024) 303-310   https://doi.org/10.21741/9781644903117-32

[27] M. Tufail, K. Shahzada, B. Gencturkan, J. Wei, Effect of Elevated Temperature on Mechanical Properties of Limestone, Quartzite and Granite Concrete, Int. J. Concr. Struct. Mater. 11 (2017) 17-28. https://doi.org/10.1007/s40069-016-0175-2

[28] A. Aboulayt, M. Riahi, M. Ouazzani Touhami, H. Hannache, M. Gomina, R. Moussa, Properties of metakaolin based geopolymer incorporating calcium carbonate, Adv. Powder. Technol. 28 (2017) 2393-2401. https://doi.org/10.1016/j.apt.2017.06.022

[29] W. Li, Q. Li, Y. Qian, F. Ling, R. Liu, Structural properties and failure characteristics of granite after thermal treatment and water cooling, Geomech. Geophys. Geo-energ. Geo-resour. 9 171 (2023) 1-18. https://doi.org/10.1007/s40948-023-00716-y

[30] M. El Ouahabi, L. Daoudi, F. Hatert, N. Fagel, Modified mineral phases during clay ceramic firing, Clays Clay Miner. 63 (2015) 404-413. https://doi.org/10.1346/CCMN.2015.0630506

Mediterranean Architectural Heritage - RIPAM10   Materials Research Forum LLC
Materials Research Proceedings 40 (2024) 311-318   https://doi.org/10.21741/9781644903117-33

Investigating Cultural Dimensions in Sustainable Tourism: Carrying Capacity Assessment and GIS-based Management Strategies for Historic Buildings

Sana SIMOU[1*], Khadija BABA[1], Yassine RAZZOUK[1], Abderrahman NOUNAH[1]

[1] Civil Engineering and Environment Laboratory, Civil Engineering, Water, Environment and Geosciences Centre (CICEEG), Mohammadia School of Engineering, Mohammed V University Rabat, Morocco

sanae.simou@gmail.com

Keywords: Carrying Capacity, GIS Management, Sustainable Tourism, Historic Buildings

Abstract. This research delves into the sustainability of tourism at Tour Hassan, a historic building in Morocco, by examining its carrying capacity and implementing GIS-based management strategies. As a significant cultural heritage site, Tour Hassan draws a considerable number of visitors. The study involves a comprehensive assessment of the site's carrying capacity, encompassing visitor flow analysis, infrastructure evaluation, and spatial limitations. Informed management strategies are then devised based on these findings. Leveraging Geographic Information Systems (GIS), data-driven decision-making processes are employed to optimize visitor experiences while safeguarding the historical building's integrity. This case study provides valuable insights into the effective management of tourism in historic sites, ensuring their long-term conservation and sustainable development.

Introduction

Historic buildings and sites are invaluable cultural assets that not only showcase a region's rich heritage but also attract a significant number of tourists [11]. These visitors contribute to the local economy and foster cultural exchange. However, managing tourism in historic buildings presents unique challenges. Striking a balance between accommodating visitors, maintaining the integrity of the historic fabric, and ensuring sustainable practices requires careful planning and effective management strategies [8,17].

Historic buildings serve as cultural touchstones, embodying the identity and history of a place. Their architectural beauty and historical significance attract a steady stream of visitors who seek to experience the past firsthand. However, the preservation of these structures while accommodating tourism poses several challenges. Excessive visitor numbers can lead to overcrowding, wear and tear, and damage to the fragile historic fabric [4,7]. Furthermore, inadequate infrastructure and poor management can undermine the visitor experience and detract from the authenticity of the site. Therefore, it becomes essential to evaluate the carrying capacity of these sites to strike a balance between tourism and conservation [2].

In recent years, Geographic Information Systems (GIS) have emerged as a valuable tool for assessing the carrying capacity of tourist destinations and facilitating informed decision-making processes [1,21]. GIS has revolutionized the field of tourism management by providing a spatial framework to analyze, visualize, and manage various aspects of a destination. By incorporating geospatial data, GIS enables the assessment of visitor flow patterns, infrastructure requirements, and the impacts of tourism on the historic building and its surroundings [13]. Through spatial analysis and modeling, GIS-based management strategies can be developed to optimize visitor experiences while preserving the integrity of the site [6]. This research examines the application of GIS-based management strategies in evaluating the carrying capacity of historic buildings for

Mediterranean Architectural Heritage - RIPAM10   Materials Research Forum LLC
Materials Research Proceedings 40 (2024) 311-318   https://doi.org/10.21741/9781644903117-33

sustainable tourism, with a focus on the case study of Tour Hassan, a significant cultural landmark in Morocco.

Tour Hassan, located in Rabat, Morocco, represents an ideal case study for examining the application of GIS-based management strategies in the context of historic buildings. Tour Hassan is a historic tower and minaret that forms part of an incomplete mosque dating back to the 12th century. It holds great cultural and historical significance, attracting a substantial number of visitors each year [24]. However, the management of tourism in Tour Hassan poses several challenges, including maintaining visitor safety, ensuring sustainable access, and preserving the fragile historic fabric.

By applying GIS techniques, the carrying capacity and the density of tourist in Tour Hassan can be assessed. GIS allows for the analysis of visitor flow patterns, identifying areas of congestion, and potential bottlenecks. This information can guide the development of strategies to manage visitor numbers, such as implementing timed entry systems, controlling access to sensitive areas [19,20], and optimizing visitor routes [25]. GIS can also aid in infrastructure planning, ensuring that facilities such as visitor centers, parking areas, and amenities are strategically located to enhance the visitor experience while minimizing the impact on the historic building [23].

Furthermore, GIS-based management strategies can help in monitoring and preserving the historic fabric of Tour Hassan. By mapping the site's condition and vulnerability, GIS can provide insights into areas that require immediate attention and conservation efforts. It can assist in identifying potential risks and developing mitigation measures to protect the site from natural disasters or human-induced threats. Additionally, GIS can aid in the documentation and visualization of the site's historical evolution, enhancing interpretive experiences for visitors.

Materials and methods

GIS-data preparation

Geographic Information Systems (GIS) provide us with powerful tools for the storage, manipulation, analysis, and production of geographic information. These systems offer two key capabilities. Firstly, GIS allows us to model reality by organizing information into multiple layers, enabling autonomous analysis or the exploration of various dimensions and aspects of a given territory. This layered approach allows for a comprehensive understanding of spatial relationships and patterns. Secondly, GIS enables the incorporation of geographic information and its associated attributes, allowing us to link diverse data sets to specific locations on the map [10]. This integration of geographical information and its attributes enhances the depth and richness of analysis, facilitating a more comprehensive assessment of the characteristics, dynamics, and complexities of a particular area or region. Through these capabilities, GIS serves as a valuable tool in spatial analysis, planning, and decision-making processes across various fields and industries.

The present study follows the concept introduced by Glasson [9], which suggests that the impact of visitors on a site is site-specific rather than uniform throughout the entire area. The carrying capacity of the historic building, Hassan Tour, is currently a concern due to the significant number of tourists it attracts each year, leading to irreversible and ongoing damage to its architectural features. Evaluating the carrying capacity requires considering spatial and temporal variations and adjusting it mathematically based on the distribution of activities on the site.

Figure 1: Realistic photos (a) and virtual view (b) of the Hassan Tour site located in Rabat city (Morocco)

The number of annual visitors to Hassan Tour is sourced from the Directorate of Heritage in Rabat, Morocco. Hassan Tour is one of the most historically significant monuments. To estimate the number of potential visitors, formulas 1 and 2 [15] are employed, utilizing known sample averages (x and y) and expected averages ($\bar{x}$ and $\bar{y}$).

$$(1) \quad a = \bar{y} - b\bar{x} \quad \text{and} \quad (2) \quad \frac{\sum(x-\bar{x})(y-\bar{y})}{\sum(x-\bar{x})}$$

The capacity of different areas and the rotation coefficient for daily visits are calculated using modified formulas from Boullon [5]. Formula (3) establishes an adjusted average individual standard (AVIS) by considering the area of each monument and its relation to the largest area (2177.49), with a psychological comfort standard of 8m². The visitors' area is calculated in ArcGIS 10.4.1 by mapping each monument as a polygon. The psychological comfort of visitors is taken into account, considering factors such as restroom availability, ticket office location, and the presence of only one entrance. This study adopts an average individual standard of 8m², aligning with psychological comfort limitations.

$$(3) \quad \textbf{Adjusted average individual standard (AVIS)} = \left. \left(\frac{\text{area of a monument}}{2177,49} \times 100 \right) \middle/ 8m^2 \right.$$

The next step involves calculating the potential carrying capacity (PCC) by multiplying the Basic Carrying Capacity (BCC) by the adjusted Rotation Coefficient (ARC). The PCC represents the maximum number of tourists that can be accommodated in a specific tourist area within a day. BCC is determined by dividing the monument's area by AVIS, while ARC is calculated as the area of monuments divided by the total area of monuments, multiplied by 5, representing the number of daily hours the site is open divided by the duration of a tour (2 hours) [21,22].

(4) Potential Carrying Capacity (PCC) = Basic Carrying Capacity (BCC) X Adjusted Rotation Coefficient (ARC)

Which ARC is:

$$(4.1) \quad \text{Adjusted rotation coefficient (ARC)} = \frac{\text{Area of monuments}}{\text{Total area of monuments}} * 5$$

Where 5 is the number of daily hours the site is open to visitors (10 h) divided by the time of the tour (2 h).

And BCC is:

$$(4.2) \quad \text{Basic carrying capacity (BCC)} = \frac{\text{Area of monuments}}{\text{AVIS}}$$

The density of visitors on the site is estimated using Kernel Density, a statistical method that generates a density surface based on the spatial coordinates of visitors, highlighting areas of concentration [15]. The density at a given point x is determined using formula (5), where k represents the nuclear function, h is the bandwidth, and (x _ xi) refers to the distance between the evaluation point and the event.

$$(5) \quad fn(x) = \frac{1}{nh}\sum_{i=1}^{n}\left[k\left(\frac{x-xi}{h}\right)\right]$$

Where k is the nuclear function; h > 0, for the bandwidth; and (x - x_i) refers to the distance between the evaluation point x and the event x_i.

In study, we utilize GIS and mathematical modeling techniques to assess the carrying capacity of the historic building, Hassan Tour. It considers site-specific impacts, estimates potential visitor numbers, calculates carrying capacities, and evaluates visitor density using Kernel Density estimation.

Management strategy development
Based on the findings from data analysis, spatial modeling, and risk assessment, informed management strategies were developed. These strategies aimed to optimize visitor experiences, protect the historic fabric, and ensure the sustainable management of tourism in Tour Hassan. Strategies included timed entry systems, visitor route optimization, targeted conservation efforts, and infrastructure improvements.

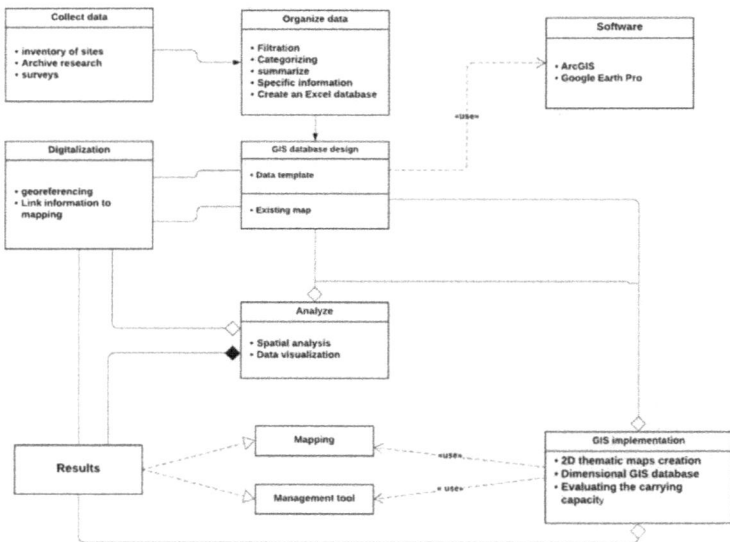

Figure 2: Diagram illustrating the construction of a historical heritage management database

Results and discussions
The need to address tourist activities as a management concern arises due to the potential cultural and environmental pressures exerted on the Hassan Tour. However, the application of methods and tools for assessing carrying capacity in such contexts is limited. This poses a significant challenge for planning and decision-making in sustainable tourism. Destinations that attract mass tourism, particularly historic and heritage sites like Hassan Tour, require policies and practices to

Mediterranean Architectural Heritage - RIPAM10   Materials Research Forum LLC
Materials Research Proceedings 40 (2024) 311-318   https://doi.org/10.21741/9781644903117-33

monitor carrying capacity. Establishing a scientific foundation and fostering collaborative efforts is essential. In the case of historic and heritage sites, mass tourism often leads to high visitor numbers concentrated around monuments, especially during short visits. The Hassan Tour faces similar concerns. Morocco, in general, has limited experience in implementing carrying capacity management in tourism destinations, likely due to the ambiguities associated with the concept and operational difficulties. The three components of carrying capacity are given varying importance based on the destination's characteristics, the type of tourism, and the interface between tourism and the environment [12,18].

To provide an example, between Mai and September 2019, the Hassan Tour site received a total of 60,550 visitors. The percentage of and density of tourists is relatively consistent across the various areas and historical assets of the Hassan Tour site. **Figure 3** provides a visual representation of the percentage of tourists at the site. From the map, it is evident that visitors tend to congregate around the minaret, mosque, and mausoleum, which are situated within the site. Notably, the courtyard of the site does not appear to be overly crowded with tourists.

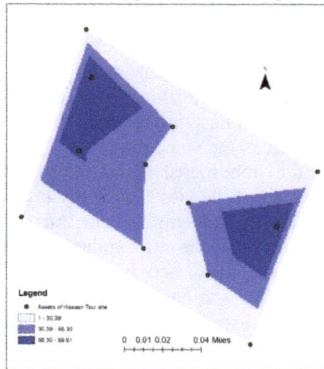

Figure 3: Map of percentage of visitors in Hassan Tour site

Figure 4 provides a detailed illustration maps of the density and carrying capacity of tourists visiting the Hassan Tour site during a specific period. The figure highlights that the density of tourists is directly associated with the notable attractions present within the site. The mausoleum and mosque exhibit the highest density of visitors, followed by the minaret then the courtyard. Regarding the carrying capacity, based on the information provided in **Figure 4**, we can infer that the carrying capacity of the Hassan Tour site appears to be influenced by the density of tourists. The figure suggests that the mausoleum and mosque experience a higher number of visitors, indicating that these areas may approach or reach their carrying capacity during the specified period. Meanwhile, the minaret exhibits a relatively lower density, suggesting that it may have a lower impact on the carrying capacity of the site.

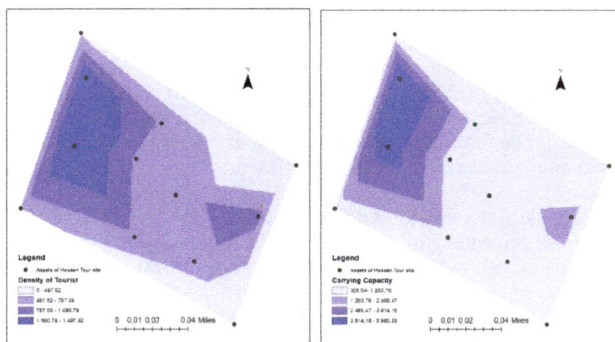

Figure 4: Maps of the density and the carrying capacity of visitors in Hassan Tour site

Conclusion

This study provided valuable insights into the carrying capacity of the Hassan Tour site and its implications for visitor management. The evaluation of carrying capacity shed light on the distribution of visitors across the site, emphasizing the need for spatial and temporal considerations [14]. By incorporating concepts such as the standard area of visitors, as highlighted by previous research, the study highlighted the importance of adapting carrying capacity calculations to account for the unique characteristics of the site. The study underscored the negative impact of excessive visitor numbers on the historic site, emphasizing the urgency of implementing effective management strategies. The preservation of the site's authenticity and outstanding universal value requires careful planning and control of visitor activities. The findings serve as a wake-up call for authorities and stakeholders to prioritize sustainable tourism practices and strike a balance between visitor satisfaction and the long-term preservation of the site's heritage.

The development of a comprehensive database for managing historic buildings and sites represents a significant step forward [16]. It offers new avenues for future research and facilitates informed decision-making in heritage management. By integrating the results obtained through GIS technology, the study opens up possibilities for creating user-friendly applications that enhance the viewing, modification, and overall management of the historic site. This technological integration will empower authorities to implement adaptive and proactive measures to protect and transmit the heritage to future generations.

Overall, this study highlights the importance of adopting a multidisciplinary approach that combines GIS analysis, carrying capacity assessment, and heritage management principles. The insights gained from this research provide a solid foundation for developing effective policies and practices that ensure the sustainable preservation and enjoyment of historic sites like Chellah. The findings will serve as a valuable resource for authorities, planners, and researchers working towards the preservation and responsible tourism development of similar heritage sites worldwide.

Acknowledgements

We would like to thank the Heritage Directorate of Rabat (Morocco), and the Higher School of Technology of Sale, for their collaboration and support in providing data and materials for this study. Their assistance has been invaluable in conducting our research. We greatly appreciate their contributions, which have enhanced the quality and depth of our work.

References

[1] Marwa Abdel-Fadeel, Samar Saad, and Souad Omran. 2013. *Opportunities and challenges of using GIS in sustainable tourism development: the case of Egypt.*

[2] Reem Almasri and Abdelkader Ababneh. 2021. Heritage Management: Analytical Study of Tourism Impacts on the Archaeological Site of Umm Qais—Jordan. *Heritage* 4, (September 2021), 2449–2469. https://doi.org/10.3390/heritage4030138

[3] Qadar Bakhsh Baloch, Syed Naseeb Shah, Nadeem Iqbal, Muhammad Sheeraz, Muhammad Asadullah, Sourath Mahar, and Asia Umar Khan. 2023. Impact of tourism development upon environmental sustainability: a suggested framework for sustainable ecotourism. *Environ Sci Pollut Res* 30, 3 (January 2023), 5917–5930. https://doi.org/10.1007/s11356-022-22496-w

[4] Bouzid Boudiaf. 2019. TOURISM AND THE HERITAGE CONSERVATION OF THE HISTORICAL AREAS. TOURISM AND THE HERITAGE CONSERVATION OF THE HISTORICAL AREAS. (October 2019).

[5] Roberto C. Boullón. 1985. *Planificación del espacio turístico*. Editorial Trillas.

[6] Lu Chen, Hongying Wang, and Jing Meng. 2023. Image Analysis of Spatial Differentiation Characteristics of Rural Areas Based on GIS Statistical Analysis. *Electronics* 12, 6 (January 2023), 1414. https://doi.org/10.3390/electronics12061414

[7] Yuri Yurievich Filippov. 2021. Significance of historical and cultural architectural heritage of Kaliningrad region. *E3S Web Conf.* 291, (2021), 05016. https://doi.org/10.1051/e3sconf/202129105016

[8] María García-Hernández, Manuel De la Calle-Vaquero, and Claudia Yubero. 2017. Cultural Heritage and Urban Tourism: Historic City Centres under Pressure. *Sustainability* 9, 8 (August 2017), 1346. https://doi.org/10.3390/su9081346

[9] John Glasson. 1995. Towards visitor impact management: Visitor impacts, carrying capacity and management responses in Europe's historic towns and cities. *(No Title)* (1995).

[10] Michael F. Goodchild. 2005. Geographic Information Systems. In *Encyclopedia of Social Measurement*, Kimberly Kempf-Leonard (ed.). Elsevier, New York, 107–113. https://doi.org/10.1016/B0-12-369398-5/00335-2

[11] Gulnara Ismagilova, L.N. Safiullin, and Ilshat Gafurov. 2015. Using Historical Heritage as a Factor in Tourism Development. *Procedia - Social and Behavioral Sciences* 188, (May 2015). https://doi.org/10.1016/j.sbspro.2015.03.355

[12] Zelenka Josef and Kacetl Jaroslav. 2014. The Concept of Carrying Capacity in Tourism. *Amfiteatru Economic* 16, (May 2014), 641–654.

[13] Verka Jovanović and Angelina Njeguš. 2008. The Application of GIS and its component in Tourism. *Yugoslav journal of operations research* 18, (January 2008), 261–272. https://doi.org/10.2298/YJOR0802261J

[14] Enrique Jurado, Macarena Tejada, Fernando Almeida García, J. González, Rafael Macias, Jesus Delgado, F. Gutiérrez, Gonzalo Gutierrez Fernandez, Mariano Luque, Gonzalo Malvarez, Oscar Gutiérrez, F. Navas, Francisco Ruiz, Jose Ruiz-Sinoga, and F. Becerra. 2012. Carrying capacity assessment for tourist destinations. Methodology for the creation of synthetic indicators applied in a coastal area. *Tourism Management* 33, (December 2012). https://doi.org/10.1016/j.tourman.2011.12.017

[15] Ahmad Makhadmeh, Mohammad Al-Badarneh, Akram Rawashdeh, and Abdulla Al-Shorman. 2020. Evaluating the carrying capacity at the archaeological site of Jerash (Gerasa) using mathematical GIS modeling. *The Egyptian Journal of Remote Sensing and Space Science* 23, 2 (August 2020), 159–165. https://doi.org/10.1016/j.ejrs.2018.09.002

[16] Rocío Mora, Luis Javier Sánchez-Aparicio, Miguel Ángel Maté-González, Joaquín García-Álvarez, María Sánchez-Aparicio, and Diego González-Aguilera. 2021. An historical building information modelling approach for the preventive conservation of historical constructions: Application to the Historical Library of Salamanca. *Automation in Construction* 121, (January 2021), 103449. https://doi.org/10.1016/j.autcon.2020.103449

[17] Monika Murzyn-Kupisz. Cultural, economic and social sustainability of heritage tourism: Issues and challenges.

[18] Martina Pásková, Geoffrey Wall, David Zejda, and Josef Zelenka. 2021. Tourism carrying capacity reconceptualization: Modelling and management of destinations. *Journal of Destination Marketing & Management* 21, (September 2021), 100638. https://doi.org/10.1016/j.jdmm.2021.100638

[19] Y. Razzouk, M. Ahatri, K. Baba, and A. El Majid. 2023. The Impact of Bracing Type on Seismic Response of the Structure on Soft Soil. *Civil Engineering and Architecture* 11, 5 (2023), 2706–2718. https://doi.org/10.13189/cea.2023.110534

[20] Y. Razzouk, M. Ahatri, K. Baba, and A.E. Majid. 2023. Optimal Bracing Type of Reinforced Concrete Buildings with Soil-Structure Interaction Taken into Consideration. *Civil Engineering Journal (Iran)* 9, 6 (2023), 1371–1388. https://doi.org/10.28991/CEJ-2023-09-06-06

[21] Sana Simou, Khadija Baba, and Abderrahman Nounah. 2022. A GIS-based Methodology to Explore and Manage the Historical Heritage of Rabat City (Morocco). *J. Comput. Cult. Herit.* 15, 4 (December 2022), 74:1-74:14. https://doi.org/10.1145/3517142

[22] Rosigleyse Sousa, Luci Pereira, Rauquirio Marinho da Costa, and José Jiménez. 2014. Tourism carrying capacity on estuarine beaches in the Brazilian Amazon region. *Journal of Coastal Research* 70, (April 2014), 545–550. https://doi.org/10.2112/SI70-092.1

[23] Daminda Sumanapala and Isabelle D. Wolf. 2022. Introducing Geotourism to Diversify the Visitor Experience in Protected Areas and Reduce Impacts on Overused Attractions. *Land* 11, 12 (December 2022), 2118. https://doi.org/10.3390/land11122118

[24] UNESCO. 2011. Rabat, Capitale moderne et ville historique: un patrimoine en partage. Retrieved from https://whc.unesco.org/fr/list/1401/

[25] Isabelle Wolf, Gregory Brown, and Teresa Wohlfart. 2017. Applying Public Participation GIS (PPGIS) to inform and manage visitor conflict along multi-use trails. *Journal of Sustainable Tourism* (August 2017). https://doi.org/10.1080/09669582.2017.1360315

Mediterranean Architectural Heritage - RIPAM10
Materials Research Proceedings 40 (2024) 319-322

Materials Research Forum LLC
https://doi.org/10.21741/9781644903117-34

Automation of Historical Buildings: Historical Building Information Modeling (HBIM) based Virtual Reality (VR)

Keltoum OUMOUMEN[1,a] *, Fatima ABOUBANE[1,b], Younes ECH-CHARQY[1,c]

[1]Energy, Engineering, Materials and Systems Laboratory (LGEMS), Department of Civil Engineering, National School of Applied Sciences, Ibn Zohr University, Morocco

[a]keltoum.oumoumen@edu.uiz.ac.ma, [b]fatima.aboubane@edu.uiz.ac.ma, [c]y.echcharqy@uiz.ac.ma

Keywords: Building Information Modeling (BIM), Historical Building Information Modeling (HBIM), Virtual Reality (VR), Built Heritage

Abstract. Every civilization's constructed legacy, identity, and history require the cooperation of several specialists and a historical database to be preserved and passed on to future generations. The difficulties of maintaining and preserving these old buildings have been successfully handled by the integration of contemporary technology. Historical Building Information Modeling (HBIM), a branch of BIM that deals with the complexity of built heritage, acquires, manages, models, and documents specific data. Virtual reality (VR) provides the benefits of intuition, realism, and teamwork. Combining HBIM and VR improves historical building visualization, analysis, comprehension, and communication, resulting in a more effective conservation process. This paper, reviews the capabilities of VR and HBIM technologies, focusing on their impact when combined in heritage conservation practices. The aim is to determine the current state of development of VR technology in the field of existing historical buildings and to identify the challenges and limitations of its application.

Introduction

The built heritage, the identity of every civilization, witnessing every country's rich cultural and architectural history [1], needs great efforts of preservation to protect this richness from extinction and help pass it on to future generations. Given the uniqueness of each historic building, the requirement for the collaboration of several specialists, and a historical database and records, the heritage conservation process becomes complex [2]. Calling for the digitization of the maintenance phase using advanced technologies such as building information modeling (BIM) and virtual reality (VR).

While BIM, an innovative method of information and data management of a building lifecycle, fulfills usual, new, and Smart construction needs. Historic buildings, characterized by complex shapes and irregular materials carrying symbolic meanings and originating from historic changes [2] require a sub-field of BIM designated the "Historic Building Information Modeling" to provide the missing historical aspect in the standard BIM [3]. The HBIM intervenes to treat existing buildings and to preserve the architecture and structure of heritage buildings without affecting their identity. The VR technology complements HBIM by offering intuitive, realistic, and collaborative simulations improving the heritage building's visualization, analysis, understanding, and communication.

To study the usefulness of Virtual Reality technology and its development in the field of existing buildings, this paper was structured as follows, the research generalities including definitions of the HBIM and VR technologies are presented in Section 2, and the data collection methodology and publication distribution are detailed in Section 3. Section 04 contains a discussion of the challenges of VR implementation in the field of existing building conservation and its future developments. The conclusion is finally mentioned in Section 5.

Mediterranean Architectural Heritage - RIPAM10   Materials Research Forum LLC
Materials Research Proceedings 40 (2024) 319-322   https://doi.org/10.21741/9781644903117-34

Generalities

VR

Virtual Reality (VR) including various applications and experiences giving to users virtual immersive environments (VIE), is derived from the "Man-Machine Graphical Communication System", and it is a part of the continuum RV [4] divided into four levels: Reality Environment (RE), Augmented Reality (AR), Augmented virtuality (AV) and Virtual Environment (VE). Achieving the VR level depends on the VE quality, its visuality and immersion, and the level of the user's sensation and feedback realism. The VR use in the AEC industry was dedicated to design, safety, project management, site layout optimization, collaboration, and understanding of complex designs [7]. This field's research has been developing since 1995, determining future directions including adaptive design, information systems, integrated human factors in construction training, facility management, and industry adoption, along with types of VR devices such as stationary, head-based, and hand-based displays[8]. Performance evaluation of hardware components and their effect on the VR experience is an area of continuous study [9].

HBIM

The HBIM is the extension of the BIM technology for existing historical buildings, focusing on the accurate collection and management of the various types of information and data related to built heritage [10], including structural, architectural, technical, and historical which are characterized by their complexity, uniqueness, and inter-project differences. First identified in 2009 by Murphy et al [10] to establish a vast database serving to heritage management, preservation, and communication. It includes essential data for heritage building preservation, such as geometric data, architectural syntax, materials characterizing, degradation representation and typologies, damage study, and environmental parameters [11]. The HBIM technology enables this data to be documented and integrated into a digital computer model accessible to all stockholders of the project, offering benefits in terms of sustainable preservation, lifecycle assessment, 3D visualization, and structural analysis. The HBIM process includes data acquisition, data processing, and the creation of a parametric database, producing a complete HBIM model.

Research methodology and publications distribution

This study was achieved by analyzing 57 research papers related to the intersection of VR with BIM and 13 papers on its intersection with HBIM. These papers included journal articles, conference papers, and academic documents from the Science Direct database. The paper selection process involved different steps, starting with keyword searches for "VR", "Virtual Reality", "BIM" and "HBIM", subsequently refined based on an examination of keywords, titles, and abstracts as shown in Fig 1.

The analysis revealed an increasing publishing trend relating to VR in the context of BIM picking in 91 articles in 2022. In contrast, VR publications in HBIM were comparatively limited, starting in 2018 and increasing gradually to seven articles in 2022, as illustrated in Fig 2.

Mediterranean Architectural Heritage - RIPAM10

Materials Research Forum LLC

Materials Research Proceedings 40 (2024) 319-322

https://doi.org/10.21741/9781644903117-34

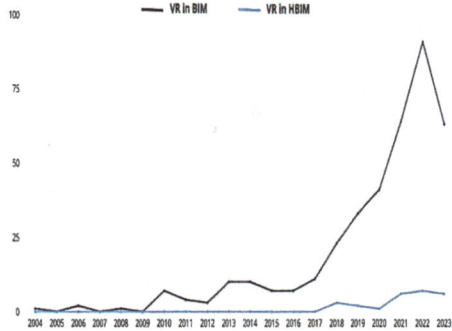

Figure 1: The VR literature progress in the BIM and HBIM context

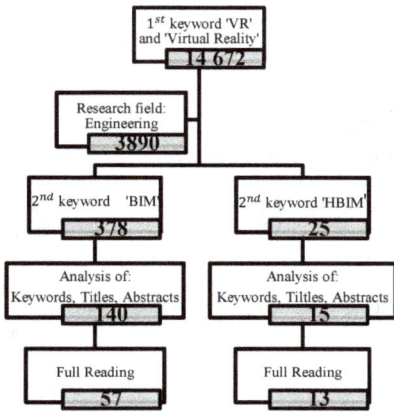

Figure 2: Paper extraction methodology

Discussion

The application of VR in the Architecture, Engineering, and Construction (AEC) industry has varied purposes in the context of BIM compared to the context of HBIM. In the context of BIM, VR is largely applied in areas such as design, construction safety, training, building management, stakeholder coordination, and energy optimization. While the VR in the context of HBIM is currently used specifically for virtual representations, risk management, operation, and maintenance of historical buildings. The historical buildings, being a unique construction with special architectural features, materials, construction methods, and history, conserving the historic integrity of these buildings has more of stakeholder's interest and priority over other applications. As well as this uniqueness generates difficulty in applying modern techniques, lack of detailed information required, and limits to structural modification, in addition to the need for a substantial renovation budget, specialized tools and expertise, and detailed regulations.

HBIM technology creating detailed models of historical buildings, serving two main purposes: operation and maintenance. In the operational phase, it focuses on the use of story-telling for educational and tourism purposes, while in the maintenance phase, it aims to provide predictive maintenance of structural and architectural aspects, although this area requires more research and development. VR technology, while valuable for visualizing building elements, is more widespread as well in the operation phase than in the maintenance phase. Its integration into HBIM makes it possible to move from short-term objectives to a long-term approach aimed at understanding the building's behavior over time, with detailed models in the structural and MEP fields. To achieve this, a comprehensive digitization process is suggested, using a centralized model and collaborative platforms such as BIM 360, with continuous updates by architectural, structural, and engineering professionals for efficient maintenance. VR plays a crucial role in simplifying the interpretation of technical details for those unfamiliar with the HBIM method, but several challenges need to be overcome, divided into three main phases: the HBIM process itself, the transfer of data to VR platforms, and the lack of compatible VR platforms with civil engineering needs.

VR applications in BIM and HBIM face similar challenges, such as high device costs, software limitations, data complexity, interoperability issues, data management and security issues, and standardization issues. This underlines the importance of further research into virtual reality simulation and the effective preservation of historic buildings using VR.

Conclusion

The application of VR technology in the context of historic building information modeling (HBIM) presents both opportunities and challenges for the use and preservation of historic buildings. It enables visualization and interaction with historic buildings, offering benefits in education, tourism, and maintenance through VR presentations. However, it faces challenges related to the complex structural and architectural aspects, unique materials, and historical significance of these buildings, as well as issues of data collection, interoperability, standardization, and expertise. These challenges also have an impact on the development of VR application areas such as structural maintenance.

To fully exploit the potential of VR in HBIM and extend its applications, further research, and development are essential. This includes automating historic documentation processes, developing standard data and formats specific to historic buildings, and improving the accessibility of the technology and associated tools.

References

[1]   I. Younus, W. Al-Hinkawi, et S. Lafta, « The role of historic building information modeling in the cultural resistance of liberated city », *Ain Shams Eng. J.*, p. 102191, févr. 2023. https://doi.org/10.1016/j.asej.2023.102191.

[2]   A. Osello, G. Lucibello, et F. Morgagni, « HBIM and Virtual Tools: A New Chance to Preserve Architectural Heritage », *Buildings*, vol. 8, n° 1, p. 12, janv. 2018. https://doi.org/10.3390/buildings8010012.

[3]   G. Nagy et F. Ashraf, « HBIM platform & smart sensing as a tool for monitoring and visualizing energy performance of heritage buildings », *Dev. Built Environ.*, vol. 8, p. 100056, sept. 2021. https://doi.org/10.1016/j.dibe.2021.100056.

[4]   S. Benford, C. Greenhalgh, G. Reynard, C. Brown, et B. Koleva, « Understanding and constructing shared spaces with mixed-reality boundaries », *ACM Trans. Comput.-Hum. Interact.*, vol. 5, n° 3, p. 185-223, sept. 1998. https://doi.org/10.1145/292834.292836.

[5]   M. J. Skibniewski, « INFORMATION TECHNOLOGY APPLICATIONS IN CONSTRUCTION SAFETY ASSURANCE », *J. Civ. Eng. Manag.*, vol. 20, n° 6, p. 778-794, déc. 2014. https://doi.org/10.3846/13923730.2014.987693.

[6]   X. Li, W. Yi, H.-L. Chi, X. Wang, et A. P. C. Chan, « A critical review of virtual and augmented reality (VR/AR) applications in construction safety », *Autom. Constr.*, vol. 86, p. 150-162, févr. 2018. https://doi.org/10.1016/j.autcon.2017.11.003.

[7]   S. Alizadehsalehi, A. Hadavi, et J. C. Huang, « From BIM to extended reality in AEC industry », *Autom. Constr.*, vol. 116, p. 103254, août 2020. https://doi.org/10.1016/j.autcon.2020.103254.

[8]   Y. Zhang, H. Liu, S.-C. Kang, et M. Al-Hussein, « Virtual reality applications for the built environment: Research trends and opportunities », *Autom. Constr.*, vol. 118, p. 103311, oct. 2020. https://doi.org/10.1016/j.autcon.2020.103311.

[9]   W. R. Sherman et A. B. Craig, « Virtual Reality », in *Understanding Virtual Reality*, Elsevier, 2018, p. 780-821. doi: 10.1016/B978-0-12-800965-9.00010-6.

[10]   M. Murphy, E. McGovern, et S. Pavia, « Historic building information modelling (HBIM) », *Struct. Surv.*, vol. 27, n° 4, p. 311-327, janv. 2009. https://doi.org/10.1108/02630800910985108.

[11]   D. P. Pocobelli, J. Boehm, P. Bryan, J. Still, et J. Grau-Bové, « BIM for heritage science: a review », *Herit. Sci.*, vol. 6, n° 1, p. 30, déc. 2018. https://doi.org/10.1186/s40494-018-0191-4.

Mediterranean Architectural Heritage - RIPAM10
Materials Research Proceedings 40 (2024) 323-349

Materials Research Forum LLC
https://doi.org/10.21741/9781644903117-35

Innovation and Resilience in the Redevelopment, Restoration and Digitalisation Strategies of Architectural Heritage

Consiglia MOCERINO[1,a]*, Abderrahim LAHMAR[2], Mohamed AZROUR[2], Abdelilah LAHMAR[3]

[1] Former adjunct professor, PhD Urban Recovery and Regeneration, PDTA-Faculty of Architecture, Sapienza University of Rome, Italy

[2] Materials Sciences for Energy and Sustainable Development Team, Department of Chemistry, Faculty of Sciences and Techniques, Moulay Ismail University, Errachidia, Morocco

[3] Laboratory of Condensed Matter Physics, University of Picardie Jules Verne, Amiens, 33 rue Saint Leu, Amiens 80039, France

[a]l.mocerino12@gmail.com

Keywords: Valorization, Architecture, Restoration, Resilience, Sustainability

Abstract. The architectural heritage is the expression of the vast cultural heritage, as a set of cultural and landscape assets, of the different national or international territorialities, which indicates in the connective of the heterogeneous urban, rural and mountain landscape areas, an inestimable value of the monumental historical buildings. The various transformative factors of administrative management and socio-economic cultural type, monitored by quality and compatibility indicators, distinguish different realities over time, in which technological innovations prevail. The objectives are the valorisation and resilience of the architectural heritage at different scales, for the reduction of seismic vulnerability and prevention against natural disasters (floods, erosions, seismic risk, exposure to ultraviolet rays, etc.) and climate crisis, through a model operational that focuses on various innovative strategies that also guarantee an adaptive reuse with a view to sustainability. The intervention criteria on the historical monumental building are according to a project that distinguishes the historical/cultural values with the use of innovative technologies and energy efficiency with the use of innovative materials, for safety needs, fire resistance, integrability, appearance, etc. according to indicators of environmental sustainability and chemical/physical, dimensional, energetic compatibility, etc. with zero environmental impact. Both in line and in cooperation with ICCROM, UNESCO, UNFCCC, Paris Agreement, ICOMOS, New Technology, COP27 Cleantech, etc. The methodological approach is based on phases of documentary collection, mapping, cognitive analysis, cognitive analysis of the architectural heritage privileging the diagnostic aspect, on a meta-project with verification, interdisciplinarity, monitoring and control of the definitive project. Use of innovative smart non-invasive technologies with properties of durability, flexibility, mechanical resistance, thermal conductivity, etc. with nanobiotechnology, innovative materials such as fiberglass mesh and reinforcing bars, on the internal surfaces, socks of artificial micro steel strands for the consolidation and linking between the pre-existing and new walls, of pillars, columns, etc. Use of technologies, local construction techniques and indigenous materials for raw earth buildings, FRCM (Fibre/Fabric Reinforced Cementitious Matrix/Mortar), FRP (Fiber Reinforced Polymer) composites in epoxy or polyester resins with carbon fibers, glass, aramid and boron polymers, for reinforced concrete products, etc.360° video technology, for immersive use both for document management in the archive database and for interactively viewing reconstructed monuments etc., acquisition of 3D models with non-contact 3D scanning technology to preserve and transmit object data in the future. Redevelopment strategies with energy retrofit as part of urban regeneration,

Mediterranean Architectural Heritage - RIPAM10   Materials Research Forum LLC
Materials Research Proceedings 40 (2024) 319-322   https://doi.org/10.21741/9781644903117-34

sustainable architectural restoration, recovery and consolidation also with a view to new digital reality (DR) and virtual reality (VR) IT strategies. Evaluation systems, analysis of degradation and instability with design, bioclimatic choices, etc. The challenge is to pass on to future generations a sustainable architectural heritage as a document of historical, architectural, artistic, archaeological, etc. value. of cultural heritage.

Introduction

The architectural values of the cultural heritage are highlighted in a set of other historical and landscape values through a language praying for authenticity and originality.

Well these represent a set of works of cultural heritage in which man designates his own artisticity and peculiarity with techniques and technologies, to valuable artefacts that indicate the different artistic and historical cultures, in a connective of popular cultures of intrinsic characteristics that settle in a certain territory.

In particular, the architectural value of monuments can suffer over time forms of dispersion of both the image and the material due to various factors which highlight, first of all, environmental pollutants and land degradation due to certain natural causes which slowly influence on the built.

Especially for seismic and flood disasters as well as for natural ventilation in certain territories and marine exhilaration that act over time or suddenly on the architectural heritage in which the image and consistency need to be passed down for its valorisation and historical document.

Therefore numerous organizations aim at this objective in which architectural value and resilience predominates, in the sphere of conservation of the built environment, such as the ICCROM [1] (International Center for the Study of the Preservation and Restoration of Cultural Property), acronym

of the merger of International Center for Conservation and Center of Rome, which is an intergovernmental organization that stimulates scientific research, promotes the restoration and conservation of monuments and all cultural heritage at an international level for the purpose of their valorization, protection and also for sustainable adaptive reuse.

UNESCO [2] (United Nations Educational, Scientific and Cultural Organization), the United Nations organization, also contributes in an international cooperation of peace and security to the protection, protection and identification of the world's cultural heritage, for the valorization and transmission to future generations .

This also happens through new guides and methodological tools for impact assessment on projects and infrastructures in collaboration with the other consultative bodies of the World Heritage Committee, such as ICOMOS [3,4] (International Council on Monuments and Sites) and IUCN (International Union for Conservation of Nature). In fact, among the many works there is the initiative to reconstruct the medieval Medina of Marrakech (Fig.1) in Morocco among the world heritage assets, heavily damaged, like all the other UNESCO sites in Morocco currently at risk, from the natural seismic event of 8 September 2023 which caused a huge disaster of monuments and archaeological sites, etc. but above all of human lives. Therefore, with the same objectives is the ICOMOS (International Council on Monuments and Sites) as a non-governmental organization, which is characterized above all by the promotion of technologies and methodologies for diagnostics and conservation aimed at the redevelopment with valorisation and revitalization of the architectural heritage and sites of historical, artistic and cultural interest.

Fig.1. Koubba Ba'adiyn, Marrakech-Moroccan Architecture. [5]

The criteria are those of intervention on the built environment, of identification of historical, architectural and cultural values with the use of innovative and efficient technologies and materials, also energetically, for safety, fire resistance, integration, appearance, sustainable and with dimensional compatibility, and above all to zero environmental impact.

Therefore innovation and resilience in architectural heritage, through a methodology of graphic and photographic documentary collection, on-site analysis, etc. cognitive and cognitive analysis favoring diagnostics, to delve into the needs of the restoration of the monument, and in which to identify the architectural and distributional characteristics, analysis of the crack pattern, of an operational model, in which digitalisation becomes an important approach with zero sustainable interventions impact and energy saving and efficiency, assessed on a case-by-case basis, for the purpose of its "recognition" of cultural value as supported by Cesare Brandi in "Theories of restoration" [6] and against climate change as also supported by the 2030 Agenda. Even the same international environmental treaty of Rio de Janeiro in 1992, called the United Nations Framework Convention on Climate Change UNFCCC (United Nations Framework Convention on Climate Change), ratified by the COP (Conference of Parties), like the Paris Agreement of 2015 , to keep the increase in global average temperature below 2°C, while for COP 27 it aims for 1.5°C, etc. for the reduction of greenhouse gases, as it is considered the predominant cause of global warming. Innovative technologies and materials, for the restoration and in particular cases of redevelopment of both historical and modern architectural assets, with their refunctionalization (museums, libraries, legal/educational/administrative units, etc.) which are distinguished case by case, with efficiency energy, represent the intervention criteria on the built environment, for authenticity requirements in the integration of the image, consistency and state of conservation [7] of the architectural work to be handed down, according to a meta-project methodology with potential impact assessment and verification of alternative or mitigation systems for the final project. Use of smart technologies for the restoration and structural consolidation of pre-existing reinforced concrete, 3D models with 3D scanning technology, 360° video technology for immersive use, with DR (digital realities) and VR (virtual realities). New Technology and Cleantech, respectively with artificial intelligence, robotics, etc. with the adoption of durable materials, with chemical-physical compatibility, flexible, reversible, in epoxy or polyester resins with carbon fibres, aramid and boron polymers, FRCM composites, glass, titanium, etc. with innovative products for the structural reinforcement of pre-existing masonry and reinforced concrete buildings. Particular attention is paid in almost all Mediterranean cities to the creation of architecture with raw earth from Italy, Jordan, Morocco, etc. with different construction techniques and mainly the Bisè and the Adobe in which the new strategies aim above all at the adoption of innovative technologies, with the improvement of local materials of the *genius loci,* with new compositions of material mixes and innovative technological intervention devices. For this purpose, the use of clay is increasing, a material that is well suited both for the restoration of the historical architectural heritage and for recovery such as the cladding of some walls, for roofs, etc., and in new buildings. Through a practice, first of reconnaissance of the places and then of the object of restoration, or recovery and in some cases of urban regeneration, through cognitive and cognitive investigations with diagnostic instruments, the state of progress of the degradation is known, aimed at a architectural and conservative restoration project, recovery of the pre-existing with revitalization and urban regeneration, through the adoption of a procedural operational model, in which to highlight the technological innovations in the restoration intervention with resilient and sustainable technologies, both environmentally, economically and socially .

Restoration for sustainable practice

The unitary value of the work evident in a monumental asset, as the result of a restoration intervention, represents the integration of an architectural image of the contextualized asset in which the aesthetic and historical instances are intrinsic, indicating the historicity, the artisticity and the philological identity that brings the degraded work back to its original form.

In Italy, according to the Code of Cultural Heritage and Landscape of the Legislative Decree, 22 January 2004, (updated to 01/08/2023) n. 42, art.29, paragraph 4, " *By restoration we mean direct intervention on the asset through a complex of operations aimed at the material integrity and recovery of the asset itself, at the protection and transmission of its cultural values. In the case of real estate located in areas declared to be at seismic risk according to current legislation, the restoration includes structural improvement interventions*" [8].

Therefore within a unitary project of sustainable restoration the aim is pure conservation, even with structural reinforcements, the protection and museum values of liberation, consolidation and differentiated integration, of the historicity and artisticity of the pre-existing work, up to functional adaptation with objectives of valorisation of the assets and also projected various refunctionalisation purposes including scientific/cultural educational, legal/administrative, etc.

To this end, scientific research and new intervention trends with local and international initiatives including UNESCO (United Nations Educational, Scientific and Cultural Organization), etc. they also aim at reconstruction projects, such as for the medieval Medina of Marrakech, and with the advanced WHEAP program (World Heritage Earthen Architecture Programme), for other interventions for raw earth buildings, etc. such as ICCROM, ICOMOS (International Council on Monuments and Sites), IUCN (International Union for Conservation of Nature), etc. which push for the valorization of cultural heritage with sustainable and resilient redevelopment, etc. This is achieved through new technologies and construction techniques with innovative and eco-sustainable materials, at an international level for the restoration of architectural heritage and as in the Mediterranean areas which include those from Europe, North Africa to Western Asia with the Near East, etc. UNESCO ensures, through the World Heritage Center and the Intergovernmental World Heritage Committee (WHC), the implementation of the World Heritage Convention, ratified in 1972 in Paris, which represents the first instrument for safeguarding the World Heritage for transmission to future generations of cultural assets which are important for maintaining solidarity, development of the planet and peace. In particular, cultural assets are recognized by the Committee (representatives of 21 member countries elected by the General Assembly) with Outstanding Universal Value (OUV) (Fig.2) together with the conditions of authenticity and integrity, which are the basis of the World Heritage Convention and all associated activities, including impact assessment. They are included in the WHL World Heritage List (World Heritage List), established by the Convention, corresponding to at least one of the criteria set out in the guidelines, as the assets that constitute it belong to the world population, regardless of the territories in which they are located and well over 1,100 sites worldwide are recognized as World

Mediterranean Architectural Heritage - RIPAM10
Materials Research Proceedings 40 (2024) 319-322

Materials Research Forum LLC
https://doi.org/10.21741/9781644903117-34

Heritage Sites. In 2023, "Guidance and Toolkit for Impact Assessment in a World Heritage Context" were published by UNESCO, ICCROM, IUCN.

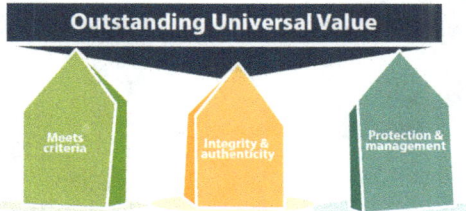

Fig.2. The 'three pillars' of Outstanding Universal Value. [9]

Guide and tools, for project impact assessments, become the reference for achieving damage prevention objectives, and for identifying sustainable operations of places subjected to increasing pressure from development projects both within and around the same sites. The UNESCO guidelines "Operational Guidelines for the Implementation of the World Heritage Convention" Operational Guidelines WHC.23/01 24 September 2023 also represent an important tool mainly for the protection and conservation of global cultural heritage which are periodically updated regarding decisions of the World Heritage Committee [10]. To this end, the restoration needs relating to the architectural heritage are also environmental and contextual and mainly due to climate change, which affect the pre-existing cultural heritage, causing degradation with lesions, microfractures, etc. chemical-environmental, erosion and sometimes irreversible or partially reconstructable damage in historical artefacts, also degradation from natural seismic disruptions, hydrogeological disruptions, anthropic activities and more as among multiple restorations, including that of the church of Santa Maria del Suffragio from 175 in Aquila in Italy, with a neoclassical dome added in the first half of the 19th century, and attributed to Giuseppe Valadier, following the 2009. L'Aquila earthquake, with restoration work completed (Fig.3) (Fig.4), (Fig.5). In this restoration intervention, innovative technologies were applied with anti-seismic reinforcement products of FRMC systems in R.A. fiberglass, alkali resistant with two-component premixed cement mortar.

Certainly each territory is characterized by pre-existing cultural heritage in which the architectural heritage is distinguished by local construction systems and techniques, different technologies and materials for the building typologies, with different intended uses such as residential, tertiary/office, commercial, tourism-receptive, productive, agricultural, etc.

In Italy also under the patronage of the CNR (National Council of Research-Works of Art of Rome and Milan) and of the ICR (Central Institute for Restoration) with the NORMAL Commission (Normalization of Stone Materials) and continuous ongoing research, we have identified unified and specific study methodologies of the alterations of stone materials, with control of the effectiveness of the conservation interventions of the same stone materials of Cultural Heritage in technological innovation and sustainability [13].

Mediterranean Architectural Heritage - RIPAM10   Materials Research Forum LLC
Materials Research Proceedings 40 (2024) 319-322   https://doi.org/10.21741/9781644903117-34

Fig.3. Church of S. Maria del Suffragio, Aquila,
Italy. Seismic failure of the dome. [11]

Fig.4. Church of S. Maria del Suffragio, Aquila,   Fig.5.Church of S.Maria del Suffragio, Aquila
Italy- Dome restoration.[12]   Italy, Façade restoration with work completed. [12]

Globalization has induced major social and economic changes which, with new technological processes, have influenced the typological evolution in the construction sector, with the use of innovative products. These are adopted both for restoration work in the different typologies of conservative, architectural, scientific, functional, philological, critical, and reconstruction, redevelopment, recovery, etc. with a view to sustainability and resilience of the works subject to intervention, a programming that requires a unitary project and in which the various specialist skills are coordinated by interfacing on a scientific/operational level. Therefore, we aim for an operational building model in the design process of architectural restoration of the architectural heritage, based on criteria of minimum intervention, which requires an in-depth study of the monument to "know so as not to intervene" so that good diagnostics can avoid invasive interventions, such as in the field of medicine. Even the criterion of reversibility required in the architectural restoration of cultural heritage and in particular of monuments, must potentially give the possibility of removing, exceptionally, any additions, additions or re-integrations, which have occurred over time during conservation interventions and without causing damage to the original monument. These interventions generally refer to gaps in the wall structure integrated with new, durable, eco-compatible and eco-sustainable green building materials, in a dialogue of artistic continuity between old and new, in various techniques that indicate the intervention of a conservative restoration project in the monument , potentially predisposed to future interventions for re-integration.

For this purpose, the reference is also to multiple solutions of integrations or additions to the pre-existing one, such as removable systems placed dry and without wiring, concealed in the walls of buildings, or to covers or roofs with removable frames in a scheduled maintenance project, etc.

The distinguishability criterion is adopted to highlight the pre-existing historical artistic structure from the restoration interventions [14] that have taken place over the years and the adoption of steel and glass indicates the distinction between load-bearing and load-bearing

Mediterranean Architectural Heritage - RIPAM10   Materials Research Forum LLC
Materials Research Proceedings 40 (2024) 319-322   https://doi.org/10.21741/9781644903117-34

structures, etc. Finally, it is necessary to adopt the compatibility of new construction materials with reference to the chemical-physical one which integrates without damage into the monument. To this end, research advances new sustainable, resilient materials, nanotechnologies and biotechnologies in historic structures such as titanium in the textures of historic walls, the injections of compatible low-carbon and low-cost lime mortars, [15] formulated with high durability, not particularly expensive, appropriately formulated based on the relative functions of the restoration project, replacing traditional nailing systems with reinforced concrete additions, etc.

The partial reconstructions sometimes respect the continuity and assonance of colours, materials, configurations and sometimes continuity with pre-existing construction techniques in the walls, highlighting the innovative quality of the restoration process and in which for structural consolidation, compatible micro pilings and plinth systems are often adopted from which to begin the partial reconstruction. For this purpose, in the thirteenth-century castle of Saliceto in Cuneo, Italy, (Fig.6), the[16] eastern tower known as the technological tower (Fig.7) was rebuilt on its original form, totally destroyed and according to criteria above all of distinguishability in the conservative restoration with adaptive reuse as a new public building, with a steel structure and ventilated walls covered in cedar wood panels, compared to the remaining three towers and the entire monument in load-bearing brick masonry. It contains the new thermal power plant.

Fig.6. Saliceto Castle, Cuneo, Italy. North-East view, restoration with reconstruction of the tower. [16]

Fig.7. Saliceto Castle, Cuneo, Italy: a. East elevation, state of the places, b. Detail of technological tower reconstruction. [16]

Mediterranean Architectural Heritage - RIPAM10
Materials Research Proceedings 40 (2024) 319-322

Materials Research Forum LLC
https://doi.org/10.21741/9781644903117-34

Among the conservative architectural restorations, the reconstruction work on the roof of the Gothic cathedral of Notre-Dame in Paris (Fig.8) and the spire, destroyed by fire, are noteworthy. The cathedral was restored in 1844 by the French architect Viollet le Duc, presenting the plan with a central nave divided into three floors, and two side naves, the central one is approximately 37 m wide, and 125 m long, 32 m high. with columns supporting pointed arches on which the hexapartite vaults originate. Today's conservative restoration underway is based on stylistic choices that recall the original project and materials in wood from centuries-old oaks, originating from French forests, with the reuse of the pre-existing iron reinforcements in the cathedral, in types of iron staples on the top walls of the south nave, in a, instead in b, iron braces of the monolithic columns of the nave, in c iron braces in the choir gallery. (Fig.9. a,b,c).

Fig.8. Notre Dame Cathedral, Paris. Restoration work. [17]

Fig.9.Notre Dame Cathedral, Paris Detail of iron staples: a. south nave, b. monolithic columns nave, c. choir stand.[18]

Fig.10. Notre Dame Cathedral, Paris. Micrograph tests, (a, b, c, d, e). [18]

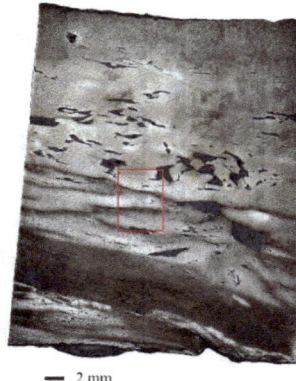

Fig.11. Notre Dame Cathedral, Paris. Oberhoffer on sample NN9-T with mesoscopic microscopic changes in P content, with highrate in lighter areas. [18]

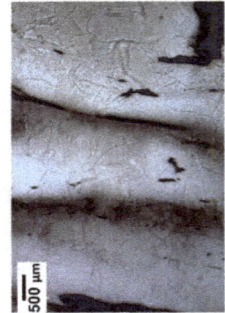

Mediterranean Architectural Heritage - RIPAM10   Materials Research Forum LLC
Materials Research Proceedings 40 (2024) 319-322   https://doi.org/10.21741/9781644903117-34

There are micrograph tests, (Fig.10) a. Micrograph of NS8-L solder with different grades of ferrous alloys (from bottom to top, small SI ferrite, large SI ferrite, carburized ferrite-pearlite). b. S

lag inclusion and porosity (NS106-T1). c. Ghost structure and different grain size of ferrite (NS8-L), d. Lightly carburized area with Widmanstätten ferrite and small SI (NS106T1), e. Area cemented with pearlite and ferrite from Widmanstätten (TRIB01S). Instead mesoscopic and microscopic variations result with Oberhoffer on the NN9-T sample (Fig.11).

The reconstruction is based on innovative digitalisation criteria with 3D model by the French CNR in collaboration with the French Ministry of Culture, based mainly on the Aïoli platform which has the monument, like all heritage assets, at the center of the design process for its conservation [19]. This platform is based on the photogrammetric technique, useful for the 3D model for image purposes, and on the cloud for the processing and sharing of collected data, therefore also based on documentary and photographic collections and digitization with 3D laser scanners, creating the interface between the artefact, object of restoration, and the documentation produced, in a sort of multi-temporal database analysis which also provides follow-up of the degradation and state of conservation.

For conservation techniques, reinforced concrete castings are also adopted in the internal parts of the degraded walls leaving the visibility of the envelope, on the outside with the previous texture, simple injections of lime mortar or "sew-unseed" operations or the use of titanium, which is more compatible with historic walls than many other materials, etc. with refunctionalization and conservation of monuments and low impact, sustainability and resilience objectives in a development of different phases of a unitary integrated project. The criteria are above all safety needs, visual well-being, usability, adaptive reuse, etc. with requirements for fire resistance, demolition, replaceability, etc. for technological quality in restorative practice.

The methodologies consist of preliminary cognitive investigations of the work subject to intervention, on the places and the territorial and urban context with mapping, documentary collection, analysis of constraints, diagnostics, state of conservation, the state of fact and previous interventions. This occurs through an archival documentary collection, with a phase of reading and graphic and photographic survey of the monument, according to its contextual complexity, its intrinsic values as well as material, artistic, cultural and to pass on its conservation. Furthermore, a subsequent descriptive and cognitive analysis is followed, at a methodological level, by planning for a feasibility study with technical requirements, objectives, performance checks, comparison, figures involved, shipbuilding, etc. and then the design one, preceded by a meta-project, as an organization that identifies technological and spatial criteria and requirements, in order to have a variety of design choices including technological, typological, organisational, spatial, dimensional ones, etc.

The subsequent design phase with intervention strategies is divided into different phases with the adoption of innovative technologies and new materials, as established by the Restoration Charters from that of Athens in 1931 to that of Krakow in 2000, including the three Italian Charters, that of Amsterdam and Washington, etc. and the typological choices of the intervention are defined with a network of operational figures involved, a brief in the preliminary, definitive and executive design phases, execution, testing with checks and control, retrocommissioning, certification systems, etc.

Finally, management with monitoring of intelligent systems for energy and electrical management and control with intelligent technologies, safety, security, maintenance, control of internal pollution levels, checks with blower door tests, etc. [20].

Furthermore, according to its typological and techno-constructive characteristics, its chromatic, spatial, decorative, plastic, urban and landscape qualities and what it has affected the monument over time, anthropic activity with architecture, painting, with the photogrammetric for a BIM

Mediterranean Architectural Heritage - RIPAM10   Materials Research Forum LLC
Materials Research Proceedings 40 (2024) 319-322   https://doi.org/10.21741/9781644903117-34

model we move on to the non-destructive diagnostic phase of the damage and then make design choices for the restoration intervention and in which the critical sense of the specialist is highlighted, also with a view to adaptive reuse and valorisation of the monument as well as historical and artistic, for the purposes of transmitting its cultural values, for the good of the community and future generations.   The damage that leads to consequences in the architectural heritage is distinguished by its many and different causes, including the load of the building itself, the action of the wind, thermal oscillations, etc. for which the global warming of the earth is at global level, as referred to by the international associations and mentioned above, by the Paris Agreement of 2015, to keep the increase in the average global temperature below 2°C, for COP 27 aiming for 1.5°C, etc. Furthermore, they affect the degradation of cultural heritage, natural seismic action, biological agents, weeds, alluvial rains, fires, floods, the vulnerability of the territory to natural phenomena, volcanic eruptions, [21] diffused humidity (from environmental causes, from capillary rising, from infiltrations, from run-off, from crumbling fixtures, etc.), with manifestations on the plaster of efflorescence, swelling, crusts, mould, different types of detachment, gaps in the surface layers, resulting in internal structural alterations and with external lesions, and on the same materials, in sight, with continuous superficial or internal lesions, lesions of architraves in brick walls, etc. In this sector, the scientific research by universities, institutions, etc. is fundamental in an interdisciplinarity of skills and interventions for the heritage of cultural heritage in order to pass on the historical, architectural, environmental and landscape heritage to future generations.

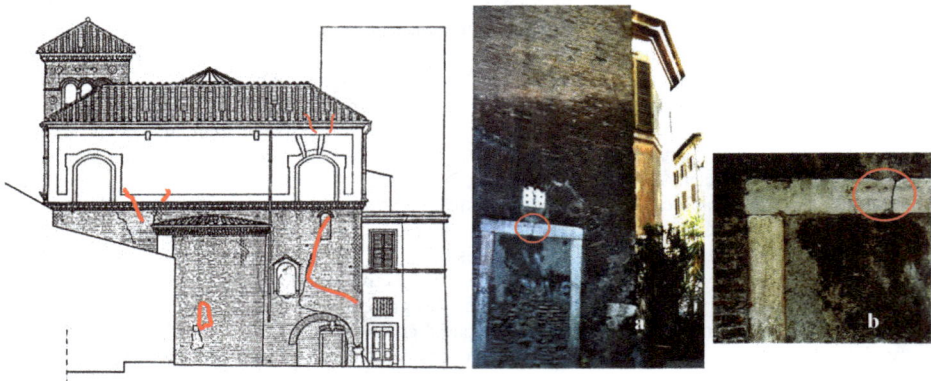

Church of S. Maria della Luce, Trastevere, Rome, Italy: Fig.12. Degradation phenomena on load-bearing brick masonry. [22]

Fig.13, a, b. Details of damage on the façade. [23]

For this purpose, the image in figure12, of a graphic highlights the micro-cracks on the load-bearing masonry in *opus testaceum* with cracks on the arches and architrave in piperine marble on the western apse of the Catholic church of S. Maria della Luce in Trastevere, Rome, built by the roman architect Gabriele Valvassori in 1730 with late Baroque transformations, on a pre-existing medieval early christian plan. This work carried out as part of scientific research and study for a conservative architectural restoration project of the entire church, was the subject of a specialist thesis at the Faculty of Architecture, Sapienza University of Rome, and subsequently reported in a publication.

Mediterranean Architectural Heritage - RIPAM10   Materials Research Forum LLC
Materials Research Proceedings 40 (2024) 319-322   https://doi.org/10.21741/9781644903117-34

Finally, among the further alterations in the architectural artefact, subject to restoration, the effects of condensation due to access of water, of structural voids with porosity and fractures are listed, compromising the mechanical properties and chemical composition of the construction materials of which it is made, and more. Therefore causes that can alter the chemical-physical and mechanical characteristics of the construction components and causing in the monuments, deterioration of the surfaces, load-bearing structures with total or partial structural failures, erosions with removal of materials from the surfaces due to further environmental, mechanical and chemical-physical phenomena, internal alveolizations with micro cavities, decohesion between the materials and components of the structure increasing their porosity [23].

In fact, an advancement of the decohesion causes the digregation of the materials of which the monument to be restored is composed with detachment of plaster even on the facade. Therefore, we aim for a practice of structural consolidation in the unitary architectural restoration project with a master plan and interdisciplinary network of figures involved, with a decisive methodological choice that adopts non-invasive materials and techniques that bring the historical-artistic artefact back to its identity unity. In fact, since the problems resulting from the degradation of monuments [24] affect building typologies with different intended uses, for each of them, in a state of degradation, action is possibly taken with maintenance, prevention and protection interventions, with integration or reconstruction of the gaps within a framework of cracks in the facades, with internal and external paintings that reflect the pre-existing colors with innovative green building materials, with consolidation in the load-bearing structures and innovative technological systems and plant restoration of the entire envelope.

a   b

Church of S. Maria della Luce, Trastevere, Rome, Italy Details (a, b):
Fig.14. a. Structural lesions. [23]   Fig.15. b. Structural lesions with gaps in concrete. [23]

For example, lighting such as heating, security systems, access control, video surveillance, alarm and fire prevention systems, etc. of monuments and buildings of worship, the latter entrusted, in Italy, to the FEC (Fondo Edifici di Culto) and managed by the Ministry of the Interior, with objectives mainly for the restoration, valorisation, protection, seismic safety and promotion , represent a particular complexity for their implementation and wiring, supported in part by projects

Mediterranean Architectural Heritage - RIPAM10   Materials Research Forum LLC
Materials Research Proceedings 40 (2024) 319-322   https://doi.org/10.21741/9781644903117-34

integrated with artificial intelligence and smart building management systems with BAS and BEMS compared to the pre-existing architectural structures, having to reconcile the needs of adaptation to technical regulations, those of distribution in the product.

Therefore, a careful study is required, preceded by a stylistic and historical documentary analysis, on the state of structural degradation and architectural characteristics, aimed at integrating the architectural image with new materials, construction techniques and innovative technologies, such as reading and testimony of the pre-existences of the document, prepared for future restoration interventions with a new design and innovative technologies, with further museum purposes and enhancement of the architectural consistency of the monument [25], frescoes, sculptures, paintings, etc. For this purpose, digitalisation also represents an innovative technology in the practice of the restoration project on which the possible system integrations are based and with the museumisation and other uses of the restoration work which is well prepared for future work.

Innovative strategies in the resilient restoration and redevelopment process

Resilience in the architectural heritage, mainly made up of inorganic materials and characterized by their own chemical constitution, porosity, hardness, the PH represents that ability to materially resist any anthropic, traumatic and environmental event, absorbing energy depending on the different reactions in the material , of an elastic/plastic type and to return to its initial state after the trauma.

In the restoration, recovery and redevelopment of cultural heritage, this ability of the material to resist traumatic, environmental and temporal/anthropic actions and to adapt to changes is assisted above all by new sustainable biotechnologies, which through the use of microorganisms or vital bacteria counteract the deterioration of matter. It is highlighted by the formation of microbial patinas on the surfaces of monuments with losses both on the surfaces, in the casing and in the structural part with sometimes irreversible damage and not yet having reference limit values of polluting effect to which the entire historical and artistic heritage is exposed, both outdoors and indoors.

The diagnostic phase, a principle that arose in Italy from the 1972 Restoration Charter, [26] represents the preliminary activity of restoration investigations of the monument with archaeometric tests of different types, instrumental analyzes of low invasive typologies proceeding from multi/interdisciplinary activities with acquisition of archival, contextual, historical, techno/constructive data and materials on the built heritage, with survey of multiple typologies digitized from HBIM up to 3D laser scans, photogrammetry, and more, restitution and graphic representation, from in situ investigations, from foundations to vertical and horizontal structures. In fact, site analysis involves checking the stability of any slopes with inclinometers, piezometers, etc. with geotechnical, penetrometric, scissometric analyses, seismic and sonic methods, and more, and with laboratory tests for the classification of soils and identification of their mechanical characteristics. For vertical structures including walls, columns and pillars, different types of investigations follow, such as laboratory ones with possible tests directly on samples of stone material (crystalline in nature), for the state of absorption by capillarity BS EN 1925:1999, for condensation, run-off, rainwater, freeze-thaw, with resistance to degradation. Non-invasive microscopic investigations, via optical microscope, scanning electron microscopy (SEM-EDS) (Fig.16) with technologies compliant with ISO measurement methods, which instead of light use electrons to create morphological images of the surface and record physical-microstructural alterations which, integrated with the EDS probe, also performs high-resolution elementary chemical analysis for the composition of stone materials, with the analysis of the fluorescence X-rays emitted from the areas irradiated by the beam. SEM EDS technology allows you to analyze the degradation with possible corrosive phenomena even of metallic materials typically chlorines and sulfur from inorganic acids, or from anthropic phenomena, marine exposures, industrial sites, and more, indicating the type of intergranular corrosion, pit corrosion, with EDS microprobe.

The example of EDS analysis- Electrical contact surface coated in gold and with oxidation - presence of contamination from manipulation (Fig. 16. a, b).

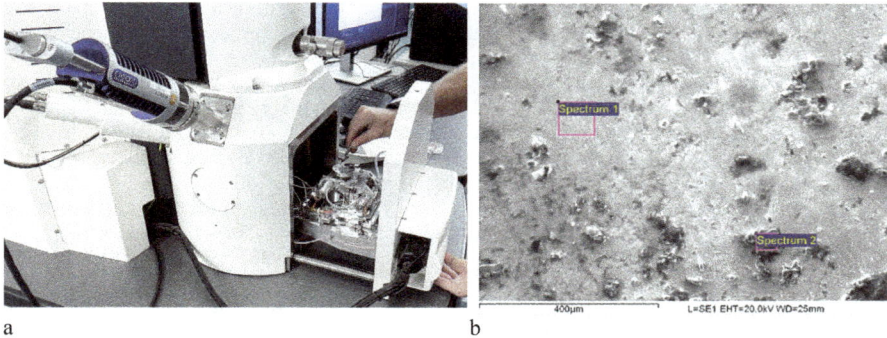

a   b

Fig. 16. Scanning electron microscopy. Morphological and dimensional analysis of the structure of phosphating crystals: a. SEM EDS unit details sample, b. Chemical composition of the coating. The image indicates the correct phosphating application required, such as the drawing.
[27]

Furthermore, X-ray diffraction identifies the composition of stone materials with a state of chemical alteration with efflorescence and surface deposits, calcium oxalate patinas, black crusts, etc. They are followed by other qualitative analyzes with thermography, ultrasonic, skelometric and dynamic tests with injectability tests, magnetometry, and endoscopic tests.

Among these is an endoscopy which, through the endoscope, a flexible tube probe, made of metal called baroscope, made of optical fibers called fiberscope, equipped at both ends with an image camera and at the other with a light source with lenses and mirror, is capable of inspecting the state of conservation of wooden structures or other materials, false ceilings, inaccessible structures, wired systems, cavities, etc.

This is followed by investigations into plant or animal biodetriment which causes chemical, physical and mechanical deterioration, the sclerometric investigation with a steel impact mass adopted for testing the resistance of concrete and quality deficiencies in the masonry.

The thermography technique which instead operates in the band of infrared electromagnetic radiation, is developed through an infrared video camera which records the emissivity of the materials, returning the heat produced through a mapping reworked by a PC in different colours, of thermal dispersions, depending on the surface temperatures of the materials, of walls with different architectural consistencies, different materials adopted, cavities, infills, lines of thermal bridges, demarcation of rising capillary humidity, etc.

Particular interest is also indicated by the micropulse, high-energy LIDAR (Light Detection and Ranging or Laser Imaging Detection and Ranging) fluorescence technique for laser surveying and restitution of 3D/BIM models, sustainable and innovative for the survey of biodeterioration and as a combination of biotechnological technique and physical technique. It also works for quick screening of monuments. Furthermore, the Laser scanner survey to carry out precise and high-performance three-dimensional architectural surveys.

The ultrasonic survey, on the other hand, based on the identification of the speed of wave propagation, is equipped with a detection unit connected to emitting and receiving piezoelectric

Mediterranean Architectural Heritage - RIPAM10   Materials Research Forum LLC
Materials Research Proceedings 40 (2024) 319-322   https://doi.org/10.21741/9781644903117-34

probes, which analyze the mechanical characteristics of the components in masonry, steel, wood, stone, concrete, etc., identifying the stiffness and mechanical resistance of the materials.

Moisture content analyzes are carried out using weight, resistive and capacitive electrical measurements, with calcium carbide analysis. Among the periodic analyses, deformometric readings and structural monitoring are integrated with intelligent wireless sensor technologies, sensor technologies such as SMOOTH, MEMS [28] for intelligent data return, and more. For horizontal structures with vaults and arches, theoretical analyzes with finite element modeling are adopted, with stability checks of load tests on the horizontals and verification of the chains.

For the control and management of safety, fire prevention and air conditioning systems, for safety, scheduled maintenance, structural and environmental monitoring, archiving, non-invasive, which excludes the wiring of electrical wires, and becomes necessary, above all for maintenance, digitalisation with ICT systems, is favoured, detection technologies, drones, satellite, aerial and terrestrial data are adopted in which the technologies are highlighted digital.

Also of importance is the 3D scanning technology for conservation and restoration, for the purposes of monitoring any structural or natural disruptions, erosion, for scheduled maintenance, which is carried out with three-dimensional acquisition of the object of investigation using a structured light or triangulation scanner.

This sustainable, non-invasive scanning operation, which returns a three-dimensional digital model reworked with software, occurs safely for the operators and without contact with the object of investigation, monument or otherwise, preserving it for future generations.

To this end, through HBIM (Heritage Building Information Modeling) for the management, structural monitoring and safety of cultural assets at seismic and hydrogeological risk, etc. or already damaged, for scheduled maintenance, BIM models can be obtained with faithful three-dimensional restitution of the pre-existing one, including the construction systems adopted, the materials, the geometric data of buildings damaged by the earthquake, with a database that can be integrated, updated and replaced depending on the variables received.

Satellite interferometry also allows you to monitor the structural deformations and instability of buildings at risk of seismic, hydrogeological, erosion, etc. and has been used for the 400 European UNESCO sites, such as the Colosseum in Rome (Fig. 17, Fig. 18)) with PROTHEGO (Protection of European Cultural Heritage from GeO-hazards) in addition to the European JPICH PROTHEGO project, for Pompeii, in the center of Rome, etc.

Furthermore, we resort to the use of wireless sensor networks, various intelligent technologies such as BOX-IO etc., on an IoT platform and data analysis, and a web platform and cloud space, which allow us to monitor humidity and temperature levels in a interoperability of eco-systems. For non-invasive structural monitoring, crack meters are adopted in multiple interventions worldwide and as in the Flavian Amphitheater in the archaeological park of the Colosseum in Rome where digitalized surveys with 3D laser scanners have been started (Fig.19,a,b,c) , in which the monitoring system is made up of the Sypeah web platform (System for the Protection and Education of Archaeological Heritage) created in collaboration with the ASI (Italian Space Agency) initiatives of the Italian national body MIBAC of the Ministry of Culture, for a informative webGIS for both monitoring and management and scheduled maintenance, in a collaboration project with research institutions and universities [29]. Also the IBC (Institute for Artistic, Cultural and Natural Heritage) and CNR-ISAC, adheres to these initiatives for the monitoring and management of cultural heritage. In monitoring systems with the crack meter instrument, the possible evolution of crack displacements and instability, etc. is detected. indicating over time, the variation in distance between two initial points of the disconnection, lesion, and more.

Mediterranean Architectural Heritage - RIPAM10

Materials Research Forum LLC

Materials Research Proceedings 40 (2024) 319-322

https://doi.org/10.21741/9781644903117-34

A multi-probe triaxial tiltometer and accelerometer of the monitoring system are connected to analogue communication nodes, consisting of wireless sensors that can be managed remotely, through a Cloud platform with information directly on the online 3D model, about the structures, environmental findings, and more.

Flavian Amphitheater-Colosseum in Rome, Italy :

Fig.17. Perspective view. [30]   *Fig.18. Cavea with new restorations. [30]*

a   b   c

Flavian Amphitheater-Colosseum in Rome, Italy :

Fig.19 (a, b, c) - a. Crack meter on masonry cracks, b. Crack meter detail, c. Geogrà digital system. [30]

These are efficient systems both for monumental, archaeological, museum complexes, and more, in which advantages of process efficiency with modularity of services are distinguished, with a view to improving and optimizing resources, accessibility, security, interoperability of services and between stakeholders and processes.

Therefore eco-sustainable wireless sensors in an integrated management for maintenance, monitoring, access control, smoke and fire detection, etc., also with BMS automation systems for plant monitoring, with a guarantee of using resources more responsibly, with energy efficiency and low consumption, for cultural heritage and in which the architectural heritage demonstrates the quality of the interventions, for the smart and anti-seismic refunctionalization of historic buildings into museums, within an urban regeneration with redevelopment, reuse and valorization of the contextualized asset. Also for user visualization, in a multidisciplinary approach for the restoration, conservation and redevelopment of historic buildings into museum units, with access control, like the museum complexes themselves, the use of digital technologies, in a digital reality (DR), supported by technical means, create a VR virtual reality in which to have an interactive global vision, with the immersive effect of the user's presence (Fig .20, a,b), using 360° video technology. In fact, it is possible to have a vision of closed historical archives, or the reconstruction

Mediterranean Architectural Heritage - RIPAM10　　　　　　　　　　　　Materials Research Forum LLC
Materials Research Proceedings 40 (2024) 319-322　　　　　　　　https://doi.org/10.21741/9781644903117-34

of historical monuments, finds, through a mouse or keyboard of a personal computer using and connecting to a tablet, smartphone, or to a PC or independently without cable, and adopting, in combination, both video/game and metaverse VR viewers, or alternatively, using a smartphone application, for viewing the 360° video. AR viewers, on the other hand, use digital technologies, through an AR app, in which reality is "augmented" overlaid by digital information, such as images, texts, etc., using smartphone or tablet screens and is a technology used in many museums for an immersive experience of augmented reality through digital screens, during exhibitions, events, and more, as well as the recontextualization of historical, archaeological, etc. objects. So by inserting, for example, a Smartguide on your smartphone, for visits, tours, events, you have an interactive experience with an augmented reality visualization.

Fig.20 (a,b)　:　　　　　　　　　　a　　　　　　　　　　　　　　　　b

a. Nxt Museum, Amsterdam.Immersive Museum, b. Batllò house, Barcelona. Augmented Reality.

[31]　　　　　　　　　　　　　　　　　　　　[32]

In a multidisciplinary and innovative approach also with the contribution of digital technologies for cultural heritage, with innovation, museumisation and archaeometry [33], restoration of the architectural heritage, whose project, in the different phases of the building process, aims at interventions resilient with sustainable biotechnologies, innovative strategies with eco-compatible composite materials, etc. in all types of interventions both on wall surfaces, in structures and in some cases in foundations with sustainable nanometric consolidants and protectors, self-cleaning products based on titanium dioxide for concrete of historic buildings, etc. From non-invasive investigations with innovative diagnostic tools and from the activity of interdisciplinary skills that contribute to the identification of damage, degradation phenomena are detected, generated by various main causes (humidity, hydrogeological seismic instability, erosion), both on surfaces and in structures related to contextual and environmental factors and detected above all by the presence of biodeteriogens which initially indicate an effect of aesthetic deterioration on the surfaces. Although it may be found that biopatines are sometimes linked to stone substrates and particular environmental conditions, such as its wavelength, electromagnetic waves of light, which interact with the material, with absorption, reflection, diffraction, and more.

Basically, degradation is indicated as biodeterioration which is distinguished both by the physical/chemical properties of the material of which the restoration work is made and by the environment in which it is contextualized and is caused by microscopic living organisms, such as algae, fungi, bacteria, and more, and those visible such as animals, plants, fungi and higher algae which exert a physical and chemical action on monuments, historical artefacts and in general in building structures. Some examples of physical deterioration are indicated by the phenomenon of

Mediterranean Architectural Heritage - RIPAM10   Materials Research Forum LLC
Materials Research Proceedings 40 (2024) 319-322   https://doi.org/10.21741/9781644903117-34

efflorescence and subefflorescence with salts depositing relatively on the external and internal surfaces and depending on the porosity of the stone, while the freezing/thawing or crystallization of salts is frequent in climates with a high rate of air pollution and cold/humid/rainy ones, while in the warm, less polluted ones the monuments show the presence of biodeteriogens. They produce damage and also waste in the environment, depositing themselves on the stone artefact, and manifesting themselves, albeit slowly over time, in the form of patina, crusts or patches and also producing CO2 in the environment and the set of all the communities that colonize the substrate stone manifest themselves in the form of biofilm, that is an biological structure of microorganisms with a mucilaginous matrix, of 14 polysaccharides (extracellular polymeric substance EPS).

Therefore, in stone artefacts of an organic nature, various phenomena occur such as deformation and swelling which alter the original shape, or phenomena of erosion, exfoliation, gaps, flaking, with loss of material, also surface phenomena with deposit of patina, and other things that overlap with the material, deforming its image. The degradation distinguished into physical, chemical, biological and anthropic has various causes that determine it. In particular, biodeterioration manifests itself both with the physical appearance, through mechanical disintegration (in connections, on the surface, in porosities, an exerted and induced by the pressure of living things (roots, weeds, fungi, microorganisms, and more), which with the chemical aspect due to the exchange of molecules between stone materials and living beings, a less detectable but documented aspect. The two aspects trigger a degradation process with waste, chelating substances, acids which break down the stone material, releasing oxides, salts, etc. nourishing microorganisms and plants that increase the colonization of living things if the artefacts have been covered with polymers, organic materials, guano. For this purpose, lactic acid, deriving from the fermentation of corn, a natural source, was used to obtain a polymer, considered a biodegradable product by the FDA-USA (Federal Drug Administration), easily enamellable, non-toxic, and as a compound of a metal to obtain a plastic material to be used for the protection and conservation of the stone surfaces of historic buildings and all other cultural heritage. It is necessary to prevent biocolonization, so among the innovative products for the conservation of surfaces, which interact less with the environment, we focus on resilient nanocomposites based on TiO2 titanium dioxide nanoparticles which have the ability to give, to stone surfaces , hydrophobic, self-cleaning properties, and inhibit biomass and biocides, photocatalic with respect to pollutants. They are anti-graffiti and water-repellent, while still remaining permeable to water vapour, de-polluting and anti-fouling, durable and removable without removing stone material and do not alter the colour. For this purpose, nanoparticles (SiO_2, TiO_2, Al_2O_3, SnO_2) in polymeric matrices are adopted. Furthermore, adoption of inorganic composite materials of calcium hydroxide for the structural consolidation of carbonate stones, to inorganic silicon-based composites used both for the consolidation and for the protection of siliceous and carbonate stones. Hybrid products based on tetraethylorthosilicate (TEOS), polymers used in the pre-treatment of cleaning as a grafting agent for better adhesion, avoiding infiltration into the pores of the substrate, and nanotechnologies are also effective for both protection and conservation. NA_TiO_2 and $SiO_2-NA\_TiO_2$.

To eliminate microorganisms or slow down their formation, indirect methodologies are adopted for phototrophic microorganisms (sources of light, humidity, nutritional organisms, dirt, dust, and more) which are then followed by direct such as mechanical methodologies. In fact, mechanical, chemical or mixed and physical methods are adopted to remove biofilms, with high and low pressure washing, low pressure water sprays, and with recourse for the former, to the use of scalpels and spatulas, sponges and brushes, pressure guns, scrapers, sleeve/pump/nozzles, jos and rotec systems, microsandblasting, microaspirators, low and high pressure washing. While for the latter there are compress swabs (compresses with ammonium carbonate and adsorbent clays, etc.), physical/mechanical (cellulose pulp/adsorbent clays, gauze, water and more), enzymes, finally the physical ones with laser and ultrasound. Among the many examples we report the removal of microbial biofilms from the stucco ed Roman tuff masonry from the Domus Aurea on the Palatine Hill in Rome, with biocides based on a mix of essential oils (6.1% - v/v) phytoderivative compounds- lavender, white thyme, wild oregano, liquorice extract, for example in 5 liters of water, etc. with the presence of phototrophic biofilms present in the octagonal room and on the intrados of the surface of the stucco-frescoed vault (Research project with ENEA Casaccia, Superintendence of the Archaeological Park of the Colosseum and restorers)[34]. While for the restoration, conservation, restoration and redevelopment of raw earth buildings, research advances with technological innovations that aim to improve materials together with traditional techniques projected towards engineering and innovation with new systems. Therefore innovation for raw earth constructions in which about a third of the world population lives and present in many mediterranean civilizations. In fact they are also widespread in much of Eastern Europe, both in rural and urban contexts, such as the rural towns of central Spain, in Irish cottages, fortified buildings in the Iberian peninsula, the half-timbered buildings of France, of Germany with sustainable buildings of fachwerk construction systems, with wooden load-bearing structures, visible in the facades and filled with clay mixed with straw, terracotta bricks, types of stone, wooden boards. Raw earth buildings are also widespread in Italy (Lombardy, Sardinia, Calabria, Macerata, Ferrara, Chieti, Basilicata, Abruzzo, etc.), as in Morocco, Jordan, Iran, Afghanistan, Yemen, and are adopted for the restoration of buildings multi-storey with decorated facades, vault and dome systems, retaining walls, and more, with sustainable raw earth materials and local construction techniques.

In fact, for the construction of buildings with monolithic, load-bearing walls and with great thermal inertia, different construction techniques are distinguished in raw earth, including *façonage*, for hand-modelled walls, *Adobe* with clay and sand with shredded straw, *Mud , Brick and Toub*, for masonry with block typologies modeled by hand or in wooden molds which can be plastered and stuccoed, the *Cob*, suitable for the integration of architectural details and tested with 3D printing, *Wellerbau, Bauge*, for stacked and compacted masonry. Among the latter, Superadobe or Earthbags is also used, a construction technique that uses polypropylene or jute bags, stacked and compact and held by barbed wire between the various layers in elevation, making it suitable for walls in areas with high seismic risk. The walls created with *Pisé* and *Rammed Earth* techniques, with innovative design and engineering of the formwork, mechanical compaction of the installation, etc. *Tapia, Taipa, Taipa De Pilão, Taipa* are characterized by rammed earth, composed of earth from the soil, below 50 cm in depth, whereby the organic component is eliminated, mixed with water and fibre, inside wooden formwork. These raw earth typologies (Fig.21. a, b), which have performed in the most disparate geographical and climatic contexts and whose construction durability is made possible by the cohesive capacity, resistance and plasticity of the clay, contained in the interior of the earth grain size, they are eco-sustainable, efficient, improving the internal microclimate, reversible and resilient, with acoustic insulating power. They are eco-sustainable, energy saving with approximately 90% using raw earth blocks compared to fired ones, but they may encounter some construction defects such as staggering of the joints between the blocks, roofing beams that unload directly onto the underlying masonry, without the sleeper which allows for a distribution of loads, the lack of a keystone in the arches which become thrusting causing instability in the masonry itself, etc. Or the recess that is formed by the different thickness of the adobe wall, which rest on a foundation or base in stone available locally and laid dry and are walled with earth mortars, the wider it generates in the first row of the wall ashlars, humidity with corrosive phenomena and caused by the stagnation of rainwater. Furthermore, unsuitable connections between perimeter walls with partitions and with corners tend to behave like single, unconnected walls with precarious static suitability. Usually for each row of the *Pisè* a connection is created by staggering the formwork with respect to the underlying level. Furthermore, other alterations may be caused by the precarious construction of the eaves and the

Mediterranean Architectural Heritage - RIPAM10   Materials Research Forum LLC
Materials Research Proceedings 40 (2024) 319-322   https://doi.org/10.21741/9781644903117-34

descendant on the wall, which creates water infiltrations with washout and consequent loss of material, or foundations unsuitable for limiting the capillary rise of water, defects in the construction technique such as for example the poor performance of palm wood with related failures and settlements in the masonry, as well as structural alterations caused by environmental factors, etc. and disruptions associated with natural seismic causes.

From a bottom layer of local stone of no less than 60 cm, which varies depending on the building typology, in an excavation usually about 50 cm deep compared to the walking surface, Pisè walls are raised with an inter-storey height of 2.5 m to 5 m, aligning compatibly with the rows of Pisè blocks, with a thickness of approximately 30-40 cm for the houses called *dâr*, up to three floors. However, for taller buildings, such as the *Kasbahs*, the thickness of the Pisè increases up to 60-100 cm, especially in the cantonal ones, those with greater thickness are evident which sometimes reach 1m, and for the better static stability, before casting in formwork, layers of stone are accommodated. Finally, we also include the heights of the buildings which do not always have dimensional compatibility with the bases, as they are sometimes very high compared to them. In the Draa valley (Fig.22) there are more than 300 ksurs and kasbahs from the 15th century. with the most famous fortifications including Amezrou, Tamnougalt, El Caid Ouslim, Ait Hammou Ousaid and Touririt, built in defense of the Amazigh settlements from attacks by nomadic Amazigh tribes, and which stand out among the most important of the historical architectural heritage of earthen buildings raw. Adobe, which is made with blocks and bedding mortar of a mixture of earth, water and straw, laid only in the horizontal joints of 2-4 cm, for transversal ventilation, is usually adopted for the upper floors, with thicknesses that they vary from 40, 50 or 60 cm and for perimeter walls and to support architraves, wooden floors, etc. or to create columns and walls in the patio and on which decorations and architectural details are created.

Fig.21. Raw earth constructions:
a. Adobe techniques, b. Pisè [35]

Fig.22. Morocco, Draa valley [36]

The new strategies aim to improve interventions in these buildings with a strong state of deterioration of facades with compensation of the gaps, and partially or totally destroyed with the integration or reconstruction of bio-sustainable architectures, through cognitive investigations, up to innovative techniques with the use of energy renewable, suitable reversible materials, and mixtures reinforced with the use of clay, plastic and resilient material, making use of new industrialized technologies for their production, improving the stability and cohesion of the raw earth for the load-bearing structures also susceptible to erosion due to rising humidity, winds, rain, etc.

The methodology of restoration intervention on these construction types is mainly based on scheduled maintenance with continuous renovations and restorations, which, with the help of more innovative interventions, could improve the quality of the earthen artefact with durability and resilience. Therefore the research aims to improve construction techniques on site with the innovation and regeneration of construction materials improved and reinforced by chemical

Mediterranean Architectural Heritage - RIPAM10   Materials Research Forum LLC
Materials Research Proceedings 40 (2024) 319-322   https://doi.org/10.21741/9781644903117-34

stabilizers such as metakaolin, lime, additives with straw, succulents, etc. for flocculation, a phase immediately following coagulation and as a chemical-physical process that favors the sedimentation of the solid parts in the colloids. In fact, in colloidal solutions, the dispersed solid particles present, on their surface, charges of the same sign which prevent their agglomeration, then there is the greater workability of the material, also in order to be able to be used as a consolidant for gaps in historical buildings architectural, integrating the color, verifying the influence of environmental and landscape parameters, such as in marine and desert areas, on the innovative products of reinforced mixtures, as per common research with bilateral agreements CNR (National Research Council) /HCST-NCRD (Jordan, HCST/NCRD - The Higher Council for Science and Technology /National Center for Research and Developmentcon) joint research project "Raw earth constructions: study, recovery and innovation in materials and historical construction techniques" 2019. [37]

Furthermore, for the purposes of protection and conservation, the WHEAP program (World Heritage Earthen Architecture Programme) was launched by UNESCO with an inventory of raw earth heritage assets and recovery projects, identifying 150 earth sites that belong to the World Heritage Sites. and due to the seismic disaster that recently occurred in Morocco, it started a reconstruction program of the medieval Medina of Marrakech, largely destroyed by the earthquake.

Therefore, the use of resilient, eco-compatible biotechnologies, based on non-invasive multifunctional nanocomposites and biocomposites with durability requirements and privileged microbiological techniques, is fundamental, compared to the widespread use of biocides in which the integrability of the monuments is protected, with valorisation and also reuse, the health of operators, in a multi/interdisciplinary and cooperative relationship, furthermore geopolymeric materials are adopted for new raw earth constructions. According to the US international guidelines, 2019 *ACI 434-Acceptance Criteria for Masonry and Concrete Strengthening Using Fiber-Reinforced Cementitious Matrix (FRCM) Composite Systems - ICC Evaluation Service, and RILEM guideline TC 250-CSM &ACI 549 - Guide to Design and Construction of Externally Bonded Fabric-Reinforced Cementitious Matrix (FRCM) and Steel Reinforced Grout (SRG) Systems for Repair and Strengthening Masonry Structures* si adottano nanotecnologie a base di carbonio e PBO per the new fibre-reinforced composites with an inorganic matrix with the function of adhesive and which surpass the FRP (Fibre Reinforced Polymers) composites with an organic nature matrix based on epoxy resins in terms of resilience performance. [38]

In particular, the FRCM (Fiber Reinforced Cementitious Matrix) are composed of nets, made up of meshes with a maximum size of 30x30 mm, with long fiber with a high elastic modulus greater than 15%, tensile strength greater than 20% which represent the reinforcement made up of different materials and of which basalt, aramid, carbon, PBO (Polyparaphenylenebenzobisoxazole), basalt fibers and high-strength steel fibers which, in the form of strands, create corrugated surfaces for the best adhesion between the reinforcement and matrix, alkali-resistant glass, integrated into the inorganic cement or lime mortar matrix. The latter are added with percentages not exceeding 10% to guarantee the durability, permeability and fire prevention properties, and in different formulations, depending on the interventions, thus ensuring effective adhesion to the materials of the substrate to be reinforced and to the structural fibers of the networks. The thicknesses of these reinforcements vary from 5 to 15 mm for single nets, and do not exceed 30 mm for overlapping nets.

Mediterranean Architectural Heritage - RIPAM10
Materials Research Proceedings 40 (2024) 319-322

Materials Research Forum LLC
https://doi.org/10.21741/9781644903117-34

a   b

Batllò house in Barcelona, Spain (a,b) :

Fig.23. a. Façade after the restoration [39]   *Fig. 24. b. Detail of reinforcement of arches, walls and vaults with reinforcement systems including FRCM [40]*

These FRCM materials, among the many restoration and building recovery interventions, were also used in the 2020 restoration of the Batllò house in Barcelona, Spain, an eight-storey building pre-existing in 1877 created by the architect Emilio Sala Cortés, but it was restored and rebuilt, in part, by the catalan architect Antonio Gaudì in the period of modernism, in 1904 with a sandstone façade, in sinuous projecting shapes representing balconies and inclined and forged chimney roofs, in a varied composition of colored mosaics (Fig .23, Fig.24). The restoration work by the architect Joan Olona and others mainly concerns the facade, the main floor with the recovery of almost all the plaster and the chimneys of the roof with the upper and lower part of the courtyard. For the walls, arches and vaults, reinforcement technologies were used for the recovery and conservative restoration in which the authenticity and integrity of the work stands out, in the adoption among the materials, FRCM fibre-reinforced composites, in glass fibre. A.R. alkali resistant, with premixed, two-component cement mortar, based on pozzolanic and fibre-reinforced binders. The house recognized as a UNESCO heritage site was partially intended by the owners for a museum, cultural events, etc. in which visitors also have the opportunity to have an interactive and immersive experience with smartphones and digital applications, for an augmented reality of the environments by Gaudi. From the modernist period there are multiple recent restoration and conservation interventions in the context of urban regeneration, such as the example of 20th century industrial architecture in Malaga known as "the ship" included in the Spanish DOCOMOMO register for the conservation of works urban. But in the context of urban regeneration with valorisation and redevelopment of buildings, in Rome, among other recent historical and modern restorations, there is that of the Mausoleum of Emperor Augustus, with a diameter of 90 m and a height of approximately 45 m, from the 28th century BC, with the arrangement of the external spaces, designed by the architect Francesco Cellini and others, and carried out by Roma Capitale with the aim above all of redeveloping the area to be used as a public space in a qualitative relationship of use of the asset represented by the monument. Therefore restoration and redevelopment interventions with innovation of materials and construction technologies with the use of FRCM which indicate the combination of a fibrous phase resistant to traction which integrates into a matrix with excellent adhesion properties to the support to be consolidated and is also called TRC (Textile Reinforced Concrete), TRM (Textile Reinforced Mortars), IMG (Inorganic Matrix-Grid composites). They are resilient materials with high mechanical performance for structural reinforcement capable of absorbing the efforts of overloads and anti-seismic insulation in concrete structures, masonry for modern and historic monumental buildings, for the conservation of cultural heritage according to the different types of intervention . In particular for the protection and safeguard against seismic action in new reinforced concrete structures and in pre-existing historic masonry ones, among the innovative anti-seismic dissipative systems the hydraulic devices stand out, appropriately chosen for each individual case, with prevented instability (Shock Transmitter Unit or STU), and in shape memory alloys (Shape Memory Alloy Device or SMAD with seismic isolation strategies. For the survey in particular typologies for a functional reorganization of the building, the use of reinforced plasters is required, the recovery of floors, and the diagnosis, starting from analysis with a collection of data, a geometric survey is defined which is followed by the processing of another into a 3D structural geometric model and subsequently into a FEM (Finite Element Method) mathematical model, in which the structures are detected and intervened on with necessary simulations.

They are obtained with the aid of automatic meshers with non-linear analysis and in the presence of the dynamics of the building in which the various expulsions of structural components have not occurred. For the application of materials in the restoration and recovery project of FRCMs, the CNR, there are CEI-UNI-CNR national regulations. Furthermore, the 2019 Italian guidelines of the Superior Council of Public Works are in force: Guideline for the identification, qualification and acceptance control of inorganic matrix fibre-reinforced composites (FRCM) to be used for the structural consolidation of existing buildings, [38] with a valid contribution to the round robin drafting of the RILEM TC 250-CSM between European laboratories and universities.

The European standards and technical specifications to which the fibres, threads and mortars as matrices of FRCM composites must comply are UNI EN 13002-2 and ISO 13002 for carbon fibres, UNI EN 13003-1-2-3 for aramid and PBO fibres, ISO 16120-1/4; EN 10244-2 for steel wires etc.

Therefore at a procedural level in a restoration project both for structural consolidation and for seismic and gravitational loads through the use of FRCM (Fig.25), of good chemical-physical compatibility with masonry and concrete substrates, there are also these European and national regulations in force with indicative intervention strategies for the design, execution and control of historical artefacts on the technological system of the building. In particular on the classes of foundation, elevation and containment load-bearing structures through reinforcements to masonry structures (arches and vaults, confinement of masonry pillars, floor and top curbs, wall panels), instead reinforcements on concrete structures (nodes between /pillar, confinement of pillars, floors, flexural reinforcement of beams, pillars and floor joists, shear reinforcement of beams and pillars, anti-overturning of infills, reinforced concrete partitions, of bridges and more).

For example, for anti-seismic reinforcement on the intrados and also on the extrados of vaults or masonry arches to increase tensile strength and contract the opening of the hinges, plating with bands of nets or unidirectional fabric, in carbon fibre, is used galvanized with very high resistance, and with hydraulic lime-based mortar, while for floor kerbing an external kerbing is applied with fabric bands that surround the building or with reinforced masonry top curbs through the application of horizontal mortar joints of FRCM reinforcement networks. For concrete structures and small sections such as floor joists, PBO mesh with a high-performance inorganic matrix is applied, (Fig.25. a,b).

Mediterranean Architectural Heritage - RIPAM10
Materials Research Proceedings 40 (2024) 319-322

Materials Research Forum LLC
https://doi.org/10.21741/9781644903117-34

a   b

Fig. 25 (a,b) . Application of FRCM on vaults: a. [41], b. PBO and carbon long fiber reinforcement FRCM on reduced section concrete structures (such as floor joists) with 105 g/m² unidirectional PBO mesh and MX-PBO Concrete inorganic matrix. [42]

Conclusions

Technological innovation reveals, in all construction sectors, the application of products and systems that satisfy the performance needs of the different results in the new design process, above all with a view to valorisation and resilience in which sustainable digitalisation strategies are highlighted . In particular, the restoration project with the recovery and redevelopment of the cultural heritage, and for the reduction of seismic vulnerability and prevention against various environmental, anthropogenic and natural disasters, through an efficient design process model with innovative technologies of intervention in efficient Master Plans and networks with scheduled maintenance and timely checks and which highlights a multi/interdisciplinary participation of skills, Bodies, Universities, in scheduled maintenance and continuous checks, aim precisely at achieving these objectives in compliance with laws and regulations national and international regulations, and in line with ICCROM, UNESCO, UNFCCC, Paris Agreement, ICOMOS, New Technology, COP27 Cleantech, IT and ITC etc. They are mainly based on transmitting to future generations the historical, architectural and landscape heritage in its authenticity and integrity by interacting with users for the purpose of potential enjoyment with possible adaptive reuse of the architectural artefact, in the different distribution forms and types of settlement, in compatible urban redevelopments and regenerations, which include the revitalization of external spaces with recontextualisation and valorisation of monuments.

The restorative design intervention criteria are based on a practice of reversibility, distinguishability, minimal intervention and chemical-physical compatibility, with innovative intervention methodologies which highlight non-invasive diagnostics and monitoring with digitalized ITC technologies, for a follow-up of the state of degradation, for the protection and conservation of monuments with use of detection technologies, from the application of HBIM to the use of 3D/BIM laser scanners, photogrammetry, FEM modeling and more. Use of scanning microscopy with SEM-EDS technology, and many other detection methodologies, diversified tests for the state of absorption by capillarity for condensation, etc. structural monitoring with intelligent wireless sensor technologies, sensors such as SMOOTH, MEMS for intelligent data return.

Therefore resilience and sustainability with nanotechnologies and structural and conservative reinforcements for efficient eco-compatible interventions, with low environmental impact, to counteract the phenomena of degradation, biodeterioration from various causes including environmental ones, natural seismic disruptions, hydrogeological disruptions, anthropic activities and more with application of durable and reversible natural materials, such as for raw earth constructions. For this type of construction, research, for the purposes of conservation and restoration and also for reconstruction, is mainly aimed at the engineering of natural materials, green building, reinforced with chemical stabilizers including metakaolin, lime, natural additives, geopolymeric materials, all innovation of construction systems and use of biotechnology. Furthermore, the application of intelligent and resilient materials and technologies for the restoration and consolidation of masonry, stone and reinforced concrete monuments is widespread with the use of composite anti-seismic materials FRCM with PBO and for seismic isolation with hydraulic devices, appropriately chosen to each individual case, with prevented instability (Shock Transmeitter Unit or STU), and in shape memory alloys (Shape Memory Alloy Device or SMAD). Finally, digitalisation, digital reality (DR) and virtual reality (VR) strategies with augmented reality as an interactive experience with immersive visualizations of the user and the places visited with 360° video technology and apps. The challenge is to pass on to future generations a sustainable architectural heritage as a document of historical, architectural, artistic, archaeological and cultural heritage value, through the adoption of a procedural operational model, in which to highlight the technological innovations in the restoration intervention with the use of innovative technologies that are resilient and sustainable from both an environmental, economic and social point of view.

Summary

If you follow the "checklist" your paper will conform to the requirements of the publisher and facilitate a problem-free publication process. Thank you for your help in this matter.

Credit authorship contribution statement

Mocerino Consiglia : Conceptualization, methodology, Writing – original draft, Investigation, Data curation, Validation
Abderrahim Lahmar: Visualization, Writing –review and editing
Mohamed Azrour: Visualization, Writing –review and editing,
Abdelilah Lahmar: Writing – review and editing, visualization, Validation

Declaration of Competing Interest

The authors declare that they have no known competing financial interests or personal relationships that could have appeared to influence the work reported in this paper.

References

[1] ICCROM, QUARTERLY REPORT 1, Rome, Italy 2023. https://www.iccrom.org/iccrom-strategic-directions-and-objectives

[2] UNESCO, Saving our World Heritage Brussels, Belgium 2022. https://en.dl-servi.com/product/saving-our-world-heritage

[3] ICOMOS-Post-Trauma_Reconstruction_Proceedings-VOL1-ENGok.pdf. 2016 https://openarchive.icomos.org/id/eprint/1707/1/ICOMOS-Post Trauma_ Reconstruction_Proceedings-VOL1-ENGok.pdf ;

[4] ICOMOS-Post-Trauma_Reconstruction_Proceedings-VOL_2_Final.pdf. 2016https://openarchive.icomos.org/id/eprint/1707/3/ICOMOS-Post-Trauma_Reconstruction_Proceedings-VOL_2_Final.pdf

[5] Marrakech - https://www.marocco.org/cosa-vedere-marocco/marrakech/cosa-vedere-marrakech/

[6]C.Brandi Teorie del Restauro, Einaudi Editore ,Torino, Ed. 2, 2000.

[7] V. Bokan Bosiljkov, A. Padovnik, T.Turk, Conservation and Restoration of Historic Mortars and Masonry Structures, Springer HMC 2022. https://link.springer.com/book/10.1007/978-3-031-31472-8?source

[8] Ministero della Cultura, Codice dei beni culturali e del paesaggio, ai sensi dell'articolo 10 della legge 6 luglio 2002, n. 137, Decreto Legislativo 22 gennaio 2004, n. 42 https://www.beniculturali.it/mibac/multimedia/MiBAC/documents/1226395624032_Codice2004.pdf

[9] UNESCO, Guidance and Toolkit for Impact Assessment in a World Heritage Context UNESCO, ICCROM, IUCN, 2023. https://whc.unesco.org/en/guidance-toolkit-impact-assessments/

[10] UNESCO, The Operational Guidelines for the Implementation of the World Heritage Convention, 2023 Operational Guidelines, WHC.23/01, 24 September 2023 https://whc.unesco.org/en/guidelines/

[11] Quando una scossa da 6.3 sfregiò L'Aquila e la provincia https://www.agi.it/cronaca/news/2021-04-06/terremoto-l-aquila-abruzzo-12065249/#

[12] La Francia e l'Italia fianco a fianco all'Aquila per festeggiare la rinascita della chiesa di Santa Maria del Suffragio, https://it.ambafrance.org/La-Francia-e-l-Italia-fianco-a-fianco-all-Aquila-per-festeggiare-la-rinascita

[13] ICR Normative http://www.icr.beniculturali.it/pagina.cfm?usz=5&uid=128&umn=71

[14] G.Carbonara, Restauro architettonico: principi e metodo, Ed. Mancosu Editore, 2013

[15] CNR, ISSMC, Soluzioni innovative per la conservazione del Patrimonio Culturale-Malte, 2023 https://www.issmc.cnr.it/ricerca/linee-di-ricerca/ambiente-edilizia-ed-patrimonio-culturale/beni-culturali/malte-idrauliche-a-calce-cl-e-nhl-per-il-restauro/

[16] V.Balboni, M.Montuori, M.Zuppiroli, Premio Internazionale di Restauro Architettonico "Domus restauro e conservazione-prima edizione ", in Paesaggio Urbano, Maggioli Editore 2-2011 https://www.paesaggiourbano.org/wp-content/uploads/2019/09/PU-2011_2-x-iPad.pdf Libro PU 2_11.indb

[17] Due anni fa l'incendio della cattedrale di Notre Dame, Vatican news, 15 aprile 2021 https://www.vaticannews.va/it/mondo/news/2021-04/due-anni-fa-l-incendio-della-cattedrale-di-notre-dame.html

[18] M.Héritier , A. Azéma , D. Syvilay , E. Delqué-Kolic , L. Beck , I. Guillot I, et al. (2023) Notre-Dame de Paris : The first iron lady ? Archaeometallurgical study and dating of the Parisian cathedral iron reinforcements. PLOS ONE 18(3): e0280945. https://doi.org/10.1371/journal.pone.0280945; https://journals.plos.org/plosone/article?id=10.1371/journal.pone.0280945#pone-0280945-g001

[19] R. Roussel, L. De Luca, AN APPROACH TO BUILD A COMPLETE DIGITAL REPORT OF THE NOTRE DAME CATHEDRAL AFTER THE FIRE, USING THE AIOLI PLATFORM, ResearchGate GIUGNO 2023 https://www.researchgate.net/publication/371899955_AN_APPROACH_TO_BUILD_A_COMP

LETE_DIGITAL_REPORT_OF_THE_NOTRE_DAME_CATHEDRAL_AFTER_THE_FIRE_
USING_THE_AIOLI_PLATFORM

[20] C.Mocerino, TOP OFFICE Tecnologie intelligenti di riqualificazione. 2017 Gangemi
Editore International, november 2017, Roma- ISBN 978 884923540-L.

[21] Tomasz Jele´nski Practices of Built Heritage Post-Disaster Reconstruction for Resilient
Cities. MPDI 2018, https://www.mdpi.com/2075-5309/8/4/53/htm

[22] Design by the author C.Mocerino. S.MARIA DELLA LUCE IN TRASTEVERE DI
GRABRIELE VALVASSORI, in rivista PALLADIO, pagg.105/118, Ed.Istituto Poligrafico e
Zecca dello Stato S.p.A. 2007, Roma

[23] Photos by the author C.Mocerino

[24] Veiga, Joao Pedro and others ; Environmental influences on historical monuments: a multi-
analytical characterization of degradation materials, NASA/ADS, maggio 2020
https://ui.adsabs.harvard.edu/abs/2020EGUGA..2221804V/abstract

[25] S. Pescari a, Laurenţiu Budău a, Clara – B. Vîlceanu
Riabilitazione e restauro della facciata principale di edificio storico in muratura – Opera
Nazionale Rumena Timisoara, Elsevier, 2023

[26] Ministero della Cultura .La carta del restauro. 1972 Italia
https://soprintendenzapisalivorno.beniculturali.it/wp-content/uploads/2019/08/circ-117-del-1972-
Carta-del-restauro.pdf

[27] MotivexLab, Analisi SEM-EDS di contaminazione di superfici, Torino, Italia.
https://motivexlab.com/analisi-sem-eds/

[28] CORDIS UE, Final Report Summary - SMOOHS (Smart Monitoring of Historic
Structures), 2015 file:///C:/Users/lmoce/Downloads/CORDIS_project_212939_en%20(1).pdf

[29] G. Mastrobuoni, F. R. Nicosia, Piattaforma per il monitoraggio del Parco Archeologico del
Colosseo, ASITA 2019, http://atti.asita.it/ASITA2019/Pdf/175.pdf

[30] Photos by the author C.Mocerino

[31] Natan Sinigaglia, Distorsion in space time, 2020,
https://natansinigaglia.com/works/distortions-in-spacetime/

[32] Il team e il progetto Casa Batlló, https://www.casabatllo.es/it/si-di-noi/

[33] E.Ferraris, L'innovazione nel settore culturale museale, in Energia, ambiente e innovazione,
rivista EAI Roma Italia, n.2/ 2023

[34]A. R.Sprocati, F.Tasso, C.Alisi, P. Marconi e G. Migliore, Biotecnologie in gioco verso
processi e prodotti sostenibili per i beni culturali, ENEA in Energia, ambiente e innovazione,
rivista EAI Roma Italia, n.4/ 2016,
https://www.eai.enea.it/component/jdownloads/?task=download.send&id=376&catid=14&Itemi
d=101

[35] M. Bispo, Sostenibilità edilizia: case in terra cruda nel mondo , 5 luglio 2022
https://laterracruda.com/blogs/ecosostenibilita/sostenibilita-edilizia

[36] M. Trovato, Marocco: la Valle del Draa e le dune di Zagora, in Africa Rivista, 6 ottobre
2017, https://www.africarivista.it/marocco-la-valle-del-draa-e-le-dune-di-zagora/116659/

[37] CNR, PROGETTO COMUNE DI RICERCA Costruzioni in terra cruda: studio, recupero e
innovazione nei materiali e nelle tecniche costruttive storiche, Italia, 2019,

https://www.cnr.it/it/accordi-bilaterali/progetto/2846/costruzioni-in-terra-cruda-studio-recupero-e-innovazione-nei-materiali-e-nelle-tecniche-costruttive-storiche

[38] ACI American Concrete Institut, RILEM TC 250-CSM &ACI 549 - Guide to Design and Construction of Externally Bonded Fabric-Reinforced Cementitious Matrix (FRCM) and Steel Reinforced Grout (SRG) Systems for Repair and Strenghtening Masonry Structures, November 2020, ISBN: 978-1-64195-120-3https://www.concrete.org/Portals/0/Files/PDF/Previews/549.6R-20_preview.pdf

[39] Casa Batlló: un restauro da record, REDAZIONE GREENME, 15 luglio 2019 https://www.greenme.it/viaggi/europa/casa-batllo-restauro/

[40] Mapei per Casa Batlló.Foto e video Gallery, 2022, https://www.mapei.com/it/it/realta-mapei/dettaglio/mapei-per-casa-battlo

[41] Sistemi di rinforzo delle strutture in calcestruzzo e muratura basati sugli FRCM, rivista INGENIO, Italia 11.11.2021. https://www.ingenio-web.it/articoli/rinforzo-strutturale-i-sistemi-frcm-ruregold-per-il-consolidamento-di-una-chiesa-a-troina/

[42] RUREGOLD, Sistema di rinforzo FRCM per calcestruzzo composto da rete unidirezionale in PBO da 105 g/m² e matrice inorganica MX-PBO Calcestruzzo, Solignano (PR), Italia https://ruregold.com/it/prodotto/pbo-mesh-105/

Mediterranean Architectural Heritage - RIPAM10
Materials Research Proceedings 40 (2024) 350-357

Materials Research Forum LLC
https://doi.org/10.21741/9781644903117-36

Traditional Architectures Along the Cultural Route of James I of Aragon in the Province of Valencia: Leveraging Laser Scanning and BIM for Heritage Management

Sandro PARRINELLO[1,a*], Alberto PETTINEO[2,b]

[1] Department of Architecture, University of Florence, Italy

[2] DAda-LAB, Department of Civil Engineering and Architecture, University of Pavia, Italy

[a] sandro.parrinello@unifi.it, [b] alberto.pettineo01@universitadipavia.it

Keywords: Traditional Architecture, Cultural Routes, Laser Scanning, BIM (Building Information Modeling), Heritage Management

Abstract. This research explores the traditional architectures along the cultural route of James I of Aragon in the province of Valencia, with a particular focus on the implementation of Laser Scanning and Building Information Modeling (BIM) technologies for heritage management. The study centers on a prominent case study, the Almonecir Castle located in the province of Castellon, near Valencia, where a comprehensive 3D model has been constructed using BIM to showcase the castle's evolutionary stages from its Arab origins to its current state. The research highlights the benefits of adopting laser scanning and BIM technologies in heritage management, specifically in the context of the Almonecir Castle. By digitizing and preserving the architectural evolution of this significant historical site, these technologies play a crucial role in safeguarding cultural heritage and enhancing visitor experiences. Moreover, the study sheds light on the potential for widespread application of laser scanning and BIM in managing and conserving other traditional architectures along the cultural route of James I of Aragon in the province of Valencia and beyond. In conclusion, this research exemplifies the transformative impact of laser scanning and BIM technologies in preserving and interpreting historical architectural treasures. The study advocates for the integration of these advanced tools in heritage management practices, facilitating a more profound understanding and appreciation of the cultural heritage that has shaped the province of Valencia throughout its illustrious history.

Introduction

The present work starts and develops from cultural heritage enhancement actions carried out within the project PROMETHEUS - PROtocols for information Models librariEs Tested on HEritage of Upper Kama Sites", funded by the EU Horizon 2020 - R&I - RISE - Research & Innovation Staff Exchange program (Marie Skłodowska-Curie grant agreement No 821870). Prometheus project activities are being conducted by the University of Florence (Italy), the University of Pavia (Italy), the Polytechnic University of Valencia (Spain), and Gdansk University of Technology (Poland), in collaboration with other international corporate partners, with the aim of promoting interdisciplinary action for documenting and structuring information on architectural heritage [1]. The investigation, by delving into the historical significance of the cultural route of James I of Aragon, unravels its architectural heritage as a testament to the region's rich cultural past. Through extensive field surveys and archival research, key architectural elements that have withstood the test of time are identified, creating a foundation for subsequent digital documentation. The study introduces the implementation of Laser Scanning technology to capture precise and detailed point cloud data of the Almonecir Castle's physical structure. By employing this advanced technique, the researchers have been able to create a highly accurate 3D representation of the castle, capturing its intricate details and subtle transformations over the centuries. Subsequently, Building

Information Modeling (BIM) is used to consolidate the gathered data and construct a comprehensive digital model. This BIM model serves as a powerful tool for heritage management, enabling a dynamic representation of the castle's various evolutionary phases up to contemporary times. The integration of BIM facilitates a holistic approach to heritage preservation, promoting informed decision-making and efficient planning for restoration and conservation efforts.

The historical footprint of James I's Route

The development of cultural and landscape routes allows for the integrated management of various elements of environmental and cultural heritage. These thematic routes play a central role in the preservation of cultural heritage and tourism enhancement of the involved municipalities, promoting cooperation for the joint enhancement of territorial resources. Currently, the European Union is focusing special attention on the establishment of actions for the preservation and enhancement of cultural and natural heritage on the territory [2]. The definition of the James I Route is part of this process, aiming to develop a protocol for the documentation and representation of Cultural Heritage Routes. The Route, as it has been defined, does not follow a linear path, but is adapted to the morphology of the Valencian territory, retracing the emblematic places in the history of the province, witnesses to the conquest of the territory by James I. The definition of this itinerary not only contributes to the enhancement of the region's cultural heritage, but also offers the possibility of appreciating the historical and identitary elements that are the result of the monarch's legacy: the enclaves of Alto Palancia, where the advance toward the new territories started; the massive castles of Cullera and Morella; the monastery of Santa María and the ruins of the ancient castle in the city of Puig.

Unlike popular perception, there is no real continuity between the earlier Muslim kingdom and the later Christian kingdom after the conquest of James I [3]. The Arabs ruled Valencia from an uncertain date, generally estimated to be around 714, until 1238 [4]. During these centuries, Valencia and its region generally depended on emirs, caliphs, and kings who ruled in Cordoba, with some exceptions during specific periods [5]. The geographical, linguistic, ethnic and political differences in the territory also influenced the architectures, as they are the result of continuous transformations and historical stratifications. The Arabs first and the Christians later used their construction knowledge to transform, expand and rebuild architecture. James I's permanence in the Valencian territory had an impact on the local architecture, leaving a cultural heritage of considerable importance that can still be read today within the different historical and cultural stratifications. In the context of the research project, were identified and analyzed only a few of the most significant and interesting sites among more than 200 [6] related to the monarch's feats. These sites were categorized according to their original function, including religious buildings, monasteries, and churches, as well as defensive structures such as castles and city walls. Included the Almonecir Castle, which has been the subject of an indepth study.

Almonecir Castle through its history

Arabian influence in Spanish architecture, is reflected not only in the richly decorated palaces of Andalusia, but also in the fortified systems spread throughout the territory. Arabian construction techniques influenced the Christian way of building, particularly in the use of tapial. Located in the western area of the Sierra de Espadan, Almonecir Castle is one of the most remarkable examples in the region of work built with tapial technique [Fig.1]. Its remote location, far from main communication routes, had a major impact on the historical development of the surrounding villages, forming a natural obstacle to urban growth and at the same time giving it strategic value in different historical phases [7]. The castle's significant historical relevance in the region emphasizes its prestige as an emblematic landmark, part of a sophisticated defense system covering the entire extension of the Sierra de Espadan. This well-structured defense system has

Mediterranean Architectural Heritage - RIPAM10
Materials Research Proceedings 40 (2024) 350-357

Materials Research Forum LLC
https://doi.org/10.21741/9781644903117-36

proven to be crucial in preserving the security and integrity of local communities throughout the centuries.

Fig. 1 Views of the current state of the castle after partial conservation and restoration work.

The origins of Almonecir Castle are closely linked to the Arab conquest of the Iberian Peninsula and the origin of a new state known as Al-Andalus [8]. Initially part of the Umayyad Empire of Damascus, Al-Andalus separated in 756 CE and reached the height of its power with the Caliphate of Cordoba (929-1035 CE) [9].

Almonecir Castle in its early phase was a fortified monastery or ribat, inhabited by a community of Muslim warrior monks known as Murabitin. In the second half of the 11th century and the early years of the 12th century, poor administration and inefficient government caused great insecurity in the rural areas of Al-Andalus, especially in the regions of Murcia and Valencia. As a result, farmers were forced to leave their houses and farms and seek refuge in castles erected on hills and in strategic locations, including the Castle of Almonecir [10]. After the Conquest, King James I conceded to the royal chancellor Berenguer de Palou, bishop of Barcelona, the Castle of Almonecir together with all the lands, villages and inhabitants associated with it, as a form of reward for his contribution to the conquest of the kingdom of Valencia[11]. The Castle played a central role in the Espadan War, which broke out following the decree issued in 1525 by Charles V requiring all Muslims in the Crown of Aragon to convert to the Catholic religion or leave the Iberian Peninsula. Following the rebellion and expulsion of the Arabs from the Espadán area, Almonecir Castle gradually lost importance and its military functions became obsolete. Today the Castle constitutes an important historical testimony of the events that shaped the region over the centuries. Its ruins represent a tangible example of the military art and fortified architecture of the period, whose transformations reveal the complexity of the social, economic and political relations that have involved the Castle over the centuries.

Documentation activities and elaboration of a digital database
The scientific analysis of the castle and its architectural elements offers the opportunity to deepen understanding of the construction techniques used, the materials employed, and the various transformations. This study provides a more complete and accurate view of the castle's history, including the social, cultural and political aspects that characterised its life and role within the society of the time. Preliminary activities for the spatial knowledge of the site and the planning of survey operations started in July 2022. The field documentation actions were carried out using two different data acquisition methodologies: Range-Based, through the use of mobile laser scanner instrumentation (MLS) with LIDAR sensors to obtain highly reliable metric measurements; Image-Based (Reflex type cameras and UAVs) to acquire information related to color and surface, using SfM (Structure from Motion) photogrammetry. The documentation using the Leica BLK2GO laser scanner mobile required a series of planning actions by the operator to organize the entire data acquisition campaign. The entire fortified complex (in its accessible parts) was subdivided into different acquisition macro-zones, where for each of them scans were planned

Materials Research Forum LLC
https://doi.org/10.21741/9781644903117-36

according to the size of the area to be covered, internal environments and/or vertical connections [Fig.2]. The database consists of 8 scans (average time per scan: 5 minutes, total number of points: 242,656,283, average number of points per scan: 30,332,035).

Due to the conformation of the territory and the position of the castle, the portions of the walls on the external front were difficult to access and it was not possible to carry out acquisition paths. To integrate the external architectural data and to acquire the surrounding area, was used ground and aerial photogrammetric survey. The digital restitution of a survey carried out by aerial photogrammetry systems - oriented both to the correctness of the data and to the realistic reproduction of the morphology of the surfaces - produces a good basis for integrating the point cloud from laser scanner instrumentation [12]. The necessity to survey the Castle using UAVs arises from the need to acquire the upper portion of the Homenaje tower and the external areas of the walls. The acquisition phase was carried out through the planning of several manual flights structured on two different scales: a punctual one around the tower; and a territorial one involving the castle and part of the surrounding space. For each flight, a radial path was carried out with variations in height [Fig.3].

MLS Scans and macro-zones

Photogrammetry and macro-zones

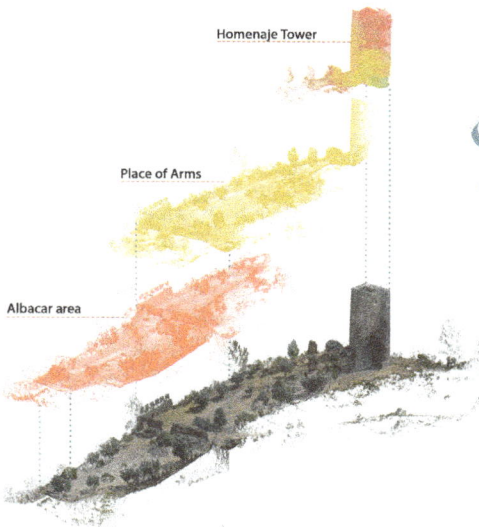

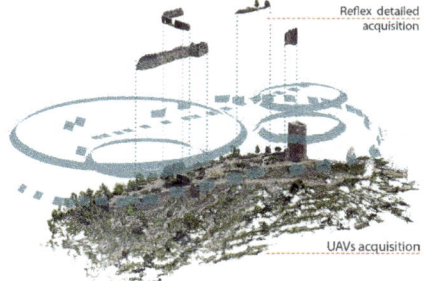

UAVs acquisition path

Fig.2-3, Scan overview using MLS instrumentation (BLK2GO), and Point cloud from photogrammetry. It is possible to see the frame sequences from drone survey.

Then, the photogrammetric data from Drone acquisition was integrated through a photogrammetric ground campaign using a reflex camera. The photogrammetric processing phase - aimed at obtaining a reliable point cloud such that it can be integrated with the point cloud from MLS - was managed using two different software: Agisoft Metashape and RealityCapture. The use of two software, which are based on different operating algorithms, made it possible to compare the generated point clouds and to choose the best data from a morphometric point of view[Fig.4]. The point clouds from the three different sources (Laser Scanner, Photogrammetry processed by Metashape and Reality Capture) were combined to obtain a single database from

353

Mediterranean Architectural Heritage - RIPAM10   Materials Research Forum LLC
Materials Research Proceedings 40 (2024) 350-357   https://doi.org/10.21741/9781644903117-36

integrated datasets. This made it possible to perform a morphometric comparison operation on the basis of the MLS point cloud [Fig.5]. The three point clouds were then integrated into a single global database from which the most suitable combination of integrations could be selected for subsequent import into the parametric modeling software. The final database consists of the data from the mobile laser scanner and the photogrammetric cloud processed in RealityCapture. In general, the discriminating factor for the choice was the best geometric quality against a nearly comparable accuracy and average deviation from the laser scanner cloud.

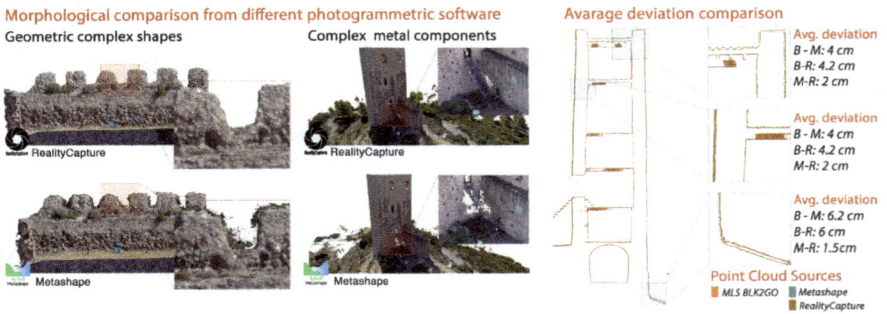

Fig.4-5, Processing results and morphometric data comparison from the different sources

Structuring the HBIM project from data survey

The methodological application of Scan-to-BIM to the Cultural Heritage field opens up a potential range of applications, allowing the integration and management data of different natures, from architectural to territorial scale. These potential applications allow the realization of tools ranging from architectural and historical analysis to urban planning and design [13]. In the architectural information models, each represented element becomes "smart" because it can contain different levels of relevant information. This research considers the application of the scan-to-BIM process in order to obtain an informative model of the castle, useful for its management, maintenance and enhancement over time. To develop this methodology, the global point cloud was treated through a cleaning process and then exported in a format compatible with the BIM platform Revit. The modeling of the castle was structured according to a dual principle: the chronological division of the construction phases, and the typological subdivision of the individual elements on an architectural and territorial scale. To this purpose were used preset parametric families already present in the software while for some specific components, it was necessary to implement the default abacus with new elements. Before proceeding to the typological modeling of the elements, were set up 'work phases' where historical evolution models were prepared based on historical considerations. The first modeling step was the topographical surface on which the architectural structures were subsequently placed. For the realization of the terrain surface it has been used the UAVs photogrammetric data: the DTM (Digital Terrain Model) was exported from the Agisoft Metashape software. Before the exportation, the model was cleaned from the anthropic elements.

The modeling phase of architectural components of the castle was carried out by isolating the individual elements following a semantic typological decomposition. The following information components were associated with each element: creation phase, demolition phase, cross section, vertical angle, and material. Each model element was renamed according to a code describing the element type and its dimensions.

Mediterranean Architectural Heritage - RIPAM10   Materials Research Forum LLC
Materials Research Proceedings 40 (2024) 350-357   https://doi.org/10.21741/9781644903117-36

The model for the management and knowledge of its historical stratification
The development of the model through a BIM platform enabled and facilitated the structuring of an abacus of the castle's typological elements. Various information (metric, constructive, historical and military function) is linked to each element. This allowed the codification of the individual elements and their possible re-use for the creation of informative models of typologically related fortifications[Fig.6].

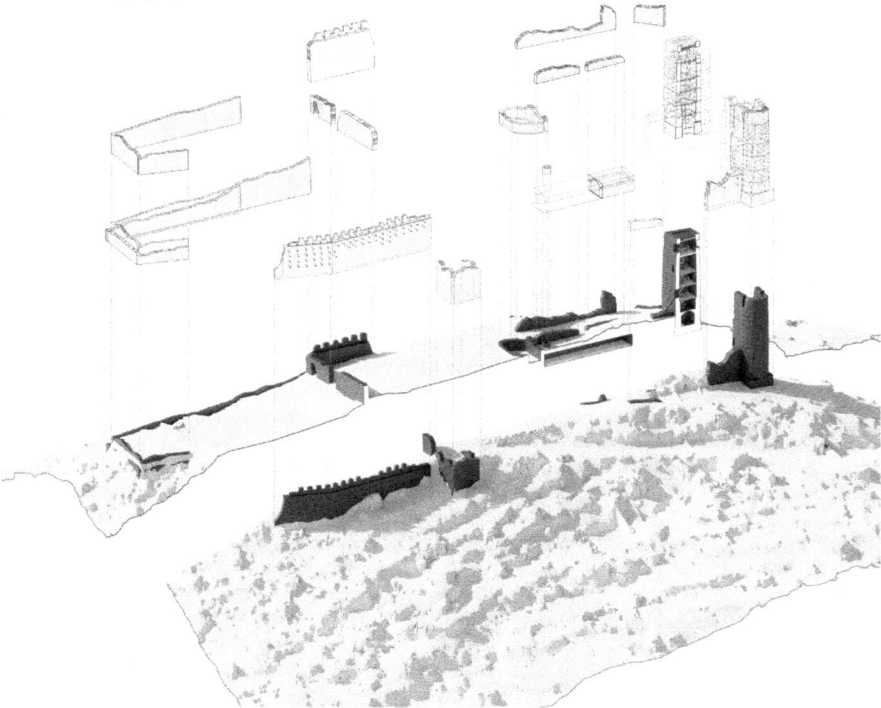

Fig. 6, Axonometric cross-section of the elements modeled in Revit. In the background the parametric model of the current state of the castle (2023), in the foreground the castle before the partial restoration work (2008).

The abacus of elements inside the Revit project collects in an organized manner all the elements used for the representation of the historical phases, allowing them to be searched and consulted according to an organised excel file. In order to realise coherent reconstructive hypotheses [Fig.7], were examined numerous coeval case studies inside the Valencian Community and in neighboring regions. The typological analysis was supported by the archaeological evidence, and hypotheses were made by mediating the information from the two sources. Subsequently, to support the modelling of the phases, through the setting and management of the phase filters in the Revit software, was structured a 'reliability map' in order to allow the real reliability of the hypotheses. The possibility of including, in addition to the geometric basis representative of the object, other levels of information, represents a useful tool to follow all the operational processes necessary for the conservation and management of the architectural asset. The BIM system makes it possible to

Mediterranean Architectural Heritage - RIPAM10

Materials Research Proceedings 40 (2024) 350-357

Materials Research Forum LLC

https://doi.org/10.21741/9781644903117-36

associate the various objects with a series of parameters that characterise them specifically for what they are and not only as surfaces.

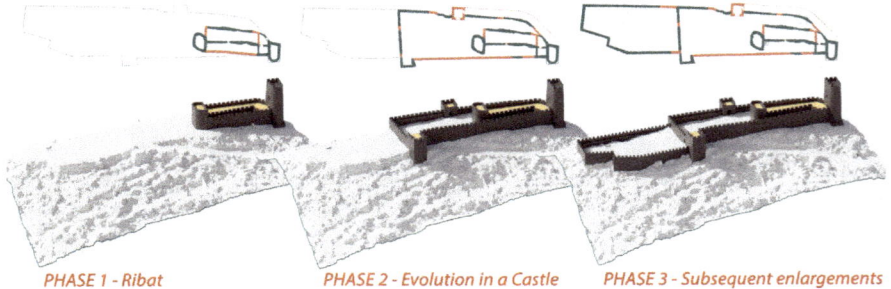

PHASE 1 - Ribat   PHASE 2 - Evolution in a Castle   PHASE 3 - Subsequent enlargements

Fig. 7, Three historical developmental hypotheses for the castle, from around X to XV century.

Conclusion

In the field of Cultural Heritage, research in the implementation of BIM procedures is still open and in an experimental phase, in relation to the different possible fields and application perspectives. The adoption and consolidated use of parametric models of the built environment represents a great achievement in the management of all aspects and practices related to the existing. In this sense, it is essential to consider the application possibilities of BIM to the entire cultural heritage, in accordance with the most advanced technologies currently available, covering all stages from acquisition and survey to management for preservation. Heritage-BIM is configured as an innovative element compared to traditional methods of management, cataloguing and valorisation of architecture. Through an information modelling methodology capable of storing and visualising geometric and documentary data in a single database, it is possible to obtain querying models in which all historical and technical information, often dispersed among different sources, is brought together. This methodology allows for the preservation of information, appropriately selected by facilitating data storage management processes, minimising information redundancy and data dispersion. The application of BIM methodologies also provides the conditions for continuous updates and necessary modifications in response to the evolution of digital systems for the preservation of historical memory. These processes, aimed at promoting the territory, play a key role in the process of enhancing and preserving the historical-architectural heritage, contributing to cultural promotion of James I's territory.

References

[1] S. Parrinello, F. Picchio, R. De Marco, A. Dell'Amico, Documenting the cultural heritage routes. The creation of informative models of historical russian churces on Upper Kama region, Int. Arch. Photogramm. Remote Sens. Spatial Inf. Sci., XLII-2/W15, (2019) pp. 887–894. https://doi.org/10.5194/isprs-archives-XLII-2-W15-887-2019; S. Parrinello, A. Dell'Amico, Experience of Documentation for the Accessibility of Widespread Cultural Heritage. Heritage (2019), 2, 1032-1044. https://doi.org/10.3390/heritage2010067

[2] B. M. Feilden, J. Jokilehto, Management Guidelines for world Cultural Heritage management, ICCROM, Rome 1998.

[3] A. U. Arteta, Orignes del Reino de Valencia, cuestiones cronologicas sobre su reconquista, VOL.I e II, Saragozza, 1981.

Mediterranean Architectural Heritage - RIPAM10   Materials Research Forum LLC
Materials Research Proceedings 40 (2024) 350-357   https://doi.org/10.21741/9781644903117-36

[4] H. Miranda, Historia musulmana de Valencia y su región: novedades y rectificaciones, 3 VOL., Ayuntamiento de Valencia, Valencia, 1970.

[5] A. U. Arteta, Orignes del Reino de Valencia, cuestiones cronologicas sobre su reconquista, VOL.I e II, Saragozza, 1981.

[6] A. M. Criado, V. G. Domenech, Comunitat Valenciana 2030, Síntesis de la Estrategia Territiorial, Generalitat Valenciana, Valencia, 2012.

[7] V. F. Martì, Torre y Castillos de la Sierra d'Espadan, Sociedad Castellonense de cultura obras de investigation historica LXXVII, Castellòn, 2008.

[8] J. P. Rodriguez, Historia del Castillo, in J. P. Rodriguez, (Eds.), El Castillo de Almonecir, Castellò, 2005, pp.19-35.

[9] P. Guichard, Al-Andalus: 711-1492: une histoire de l'Espagne Musulmane, Paris, 2000.

[10] J. P. Rodriguez, Historia del Castillo, in J. P. Rodriguez, (Eds.), El Castillo de Almonecir, Castellò, 2005, pp.19-35.

[11] 'Libre de Repartiment', ancient register in which King James I's accounting officer registered property donation at the conclusion of the conquest of Valencia.

[12] F. Picchio, Acquisition protocols for UAV photogrammetric data – Comparison in methodological SfM procedures from architectural till urban scale, in S. Barba, A. Dell'Amico, M. Limongello, S. Parrinello, S. (Eds.), D-SITE – Drones-Systems of Information on culTural hEritage – For a spatial and social investigation, Pavia University Press, Pavia, 2019, pp. 70-79.

[13] S. Parrinello, A. Dell'Amico, From Survey to Parametric Models: HBIM Systems for Enrichment of Cultural Heritage Management. In: c. Bolognesi, D. Villa, (Eds.), From Building Information Modelling to Mixed Reality. Springer Tracts in Civil Engineering. Springer, Cham, 2021, https://doi.org/10.1007/978-3-030-49278-6_6, A. Dell'Amico, HBIM and digital documentation for development reliable parametric models on complex systems. AIP Conference Proceedings 2021. https://doi.org/10.1063/5.0071096

Mediterranean Architectural Heritage - RIPAM10
Materials Research Proceedings 40 (2024) 358-362

Materials Research Forum LLC
https://doi.org/10.21741/9781644903117-37

Object Detection Method for Automated Classification of Distress in Rabat's Built Heritage

Oumaima KHLIFATI[1,a] *, Khadija BABA[2,b] and Sana SIMOU[1,c]

[1]Civil and Environmental Engineering Laboratory (LGCE), Mohammadia Engineering School, Mohammed V University in Rabat, Morocco

[2]Engineering and Environment Laboratory (LGCE), Higher School of Technology-Salé, Mohammed V University in Rabat, Morocco.

[a]oumaimakhlifati@research.emi.ac.ma, [b]khadija.baba@est.um5.ac.ma, [c]sanae.simou@gmail.com

Keywords: Heritage Sites, Object Detection Model, Damage Classification, Automatic Detection

Abstract. Rabat, the capital city of Morocco, proudly boasts a rich and complex architecture-al legacy that beautifully blends historical influences ranging from Islamic to con-temporary designs. Conserving this unique heritage holds paramount importance in safeguarding the city's distinctiveness and cultural significance. Conventional approaches to cataloging and categorization have been time-consuming and susceptible to human errors. Hence, this study aims to overcome these obstacles by creating a sophisticated object detection model to streamline the classification process. In this study, we propose an innovative deep learning-driven approach to detect and classify various degradations of built heritage. The dataset used in this study comprises numerous captured images that display diverse types of degradation, including cracks, collapse, rising damp, spalling, delamination, and lichens. Manual annotation was conducted to label the various damages present in the dataset. These labeled images were then used to train and validate the model. Multiple performance metrics were employed to assess and evaluate the model's performance, including precision and recall. Based on the results, the developed model has demonstrated excellent performance in both detecting and classifying different types of damage. This model's effective use aids local authorities in urban planning, heritage preservation, education, and tourism promotion, yielding broad implications.

Introduction

Rabat, the majestic urban center and political hub of Morocco, confidently harbors a plethora of age-old monuments. These notable locations serve as lasting testimonials to the civilizations that played a pivotal role in influencing the course of North African history, especially during the Roman and Islamic epochs. Within this collection of historical treasures are numerous locations with origins tracing back to the 8th century, including the distinguished Chellah.

The charm of Chellah captivates a growing multitude of visitors coming from varied backgrounds and age brackets, establishing it as a top-tier tourist hotspot. Yet, this extraordinary location bears the force of nature's relentless influence, undergoing both physical and chemical deterioration. The primary reasons behind the physico-chemical decline of historical structures in Chellah can be pinpointed as follows: environmental elements like rainfall, surface water runoff, and moist air, abundant in salts from the Atlantic Ocean, penetrate the materials either through infiltration or capillary ascent.

To safeguard this historical landmark from decay, it is crucial to conduct regular evaluations of these structures to ensure their protection and conservation. Promptly recognizing and categorizing any damage will streamline timely interventions and upkeep initiatives. The prevailing methods

Mediterranean Architectural Heritage - RIPAM10
Materials Research Proceedings 40 (2024) 358-362

Materials Research Forum LLC
https://doi.org/10.21741/9781644903117-37

for identifying surface damage in historical heritage sites largely rely on on-site visual inspection techniques, complemented by dedicated equipment [1]–[5]. Nevertheless, due to the rapid progress of deep learning methodologies, the previously mentioned challenges can be tackled. Several Convolutional Neural Networks (CNNs) have emerged, demonstrating outstanding precision in classification and recognition. As an example, Makantasis et al. utilized a three-layer CNN to examine cracks in tunnels [6]. Cha et al. have proposed a method for visual evaluation based on the Faster Region-based Convolutional Neural Network, with the goal of achieving nearly instantaneous concurrent identification of various types of impairments. The detection technology, formerly constrained to contemporary civil engineering, poses challenges in terms of compatibility. In this research, we employed the YOLO (You Only Look Once) method for identifying and categorizing diverse damages found in Chellah.

Dataset and Method
The dataset employed in this study includes images manually taken at Chellah using a Canon 700D camera. A total of 120 images is at your disposal, each showcasing unique dimensions influenced by the diverse scales of the captured scenes. These images specifically concentrate on various forms of impairments observed at Chellah, encompassing delamination, cracks, lichens, rising damp, spalling and partial collapse. Fig. 1 presents some samples of this dataset.

In this research, YOLOv5 is employed as the chosen model to achieve the task of object detection. Table 1 presents the structure of YOLOv5, revealing its complex design and composition. The algorithm framework is primarily divided into three essential components: The foundational network, the bottleneck layer network (Neck), and the detection layer (Head).

The foundational network includes a regular convolution module (Conv), a C3 module, and a spatial pyramid pooling module-fast (SPPF). The neck module incorporates a blend of crucial elements, encompassing a standard convolution module (Conv), a C3 module, concatenation, and an up-sampling process. The detection layer (Head) consists of three layers of detection.

Fig. 1: sample images of Chellah monument

Table 1. architecture of YOLOV5

Layers	Stride	Arguments
Backbone	Conv	[3, 64, 6, 2, 2]
	Conv	[64, 128, 3, 2]
	C3	[128, 128, 3]
	Conv	[128, 256, 3, 2]
	C3	[256, 256, 6]
	Conv	[256, 512, 3, 2]
	C3	[512, 512, 9]
	Conv	[512, 1024, 3, 2]

Materials Research Proceedings 40 (2024) 358-362

https://doi.org/10.21741/9781644903117-37

	C3	[1024, 1024, 3, 2]
	SPPF	[1024, 1024, 5]
Neck	Conv	[1024, 512, 1, 1]
	Upsample	-
	Concat	[1]
	C3	[1024, 512, 3]
	Conv	[512, 256, 1, 1]
	Upsample	-
	Concat	[1]
	C3	[512, 256, 3]
	Conv	[512, 512, 3, 2]
	Concat	[1]
	C3	[512, 512, 3]
	Conv	[512, 512, 3, 2]
	Concat	[1]
	C3	[1024, 1024, 3]
Head	Detect	-

$$(1)$$

Results and discussion

To evaluate the improved efficiency of the YOLOv5 algorithm in detecting damages within the Chellah context, precision-confidence and recall-confidence curves are employed. These curves function as vital instruments for evaluating the model's performance by examining the relationship between precision and confidence, and recall and confidence. In Fig. 2(a), the precision-confidence curve for the proposed model is depicted. This curve holds a vital role in understanding the reliability of the model's predictions. It allows us to evaluate the equilibrium between precision and confidence. As we move along the curve, there is a simultaneous increase in both precision and confidence. This signifies the model's growing confidence in its predictions and its enhanced ability to accurately identify true positive instances of damage. Fig. 2(b) depicts the display of the recall-confidence curve for the proposed model. Examining this curve enables the assessment of the equilibrium between recall and confidence. Advancing along the curve, we notice a simultaneous increase in both recall and confidence. This signifies a fortification of the model's prediction certainty, resulting in the successful detection of a higher number of authentic damage instances. As a result, the recall rate increases, emphasizing the model's improved proficiency in identifying actual damages. Furthermore, the elevation of confidence levels underscores the growing reliability of the model's predictions.

To verify the outcomes produced by the model, Fig. 3 displays the results generated by the model in identifying bounding boxes for each damage type. Based on these findings, one can infer that the model demonstrates an impressive capacity to distinguish among the various deteriorations in Chellah. This affirms its effectiveness in accurately identifying and precisely locating different anomalies across the surveyed surfaces.

Mediterranean Architectural Heritage - RIPAM10
Materials Research Proceedings 40 (2024) 358-362

Materials Research Forum LLC
https://doi.org/10.21741/9781644903117-37

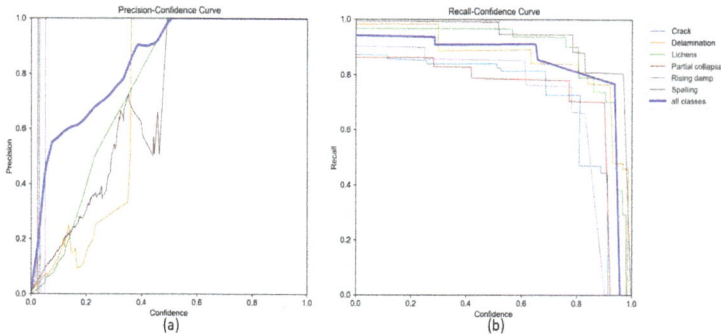

Fig. 2. (a) *Precision-confidence curve of the proposed model for the six damages, (b) Recall-confidence curve of the proposed model for the six damages.*

Fig. 3. *Illustrative instance showcasing predicted bounding boxes utilizing the YOLOv5 model.*

Conclusions

The captivating allure of Chellah draws an increasing number of tourists from diverse backgrounds and age groups, cementing its position as a premier tourist destination. However, this exceptional site carries the impact of nature's unyielding forces, facing both physical and chemical degradation. Assessing damages frequently demands the proficiency of experienced specialists, who dedicate substantial time and effort to the evaluation procedure. To overcome these constraints, this research presents a novel automated damage detection method employing a deep learning network. Utilizing the YOLOv5 object detection technique, our method strives to accurately detect and classify the various types of damage within the Chellah monument. Derived from the obtained results, the model has effectively identified and precisely pinpointed various damages existing in Chellah, overcoming challenges such as imbalanced datasets and a restricted database.

References

[1] V Gattulli, L Chiaramonte. Condition Assessment by Visual Inspection for a Bridge Management System. Computer-Aided Civil and Infrastructure Engineering. 2005;20(2):95-107. https://doi.org/10.1111/j.1467-8667.2005.00379.x

[2] LA Santos, I Flores-Colen, MG Gomes. In-situ Techniques for Mechanical Performance and Degradation Analysis of Rendering Walls. Restoration of Buildings and Monuments. 2013;19(4):255-266. https://doi.org/10.1515/rbm-2013-6606

[3] T Reyno, C Marsden, D Wowk. Surface damage evaluation of honeycomb sandwich aircraft panels using 3D scanning technology. NDT & E International. 2018;97:11-19. https://doi.org/10.1016/j.ndteint.2018.03.007

[4] M O'Byrne, F Schoefs, B Ghosh, V Pakrashi. Texture Analysis Based Damage Detection of Ageing Infrastructural Elements. Computer-Aided Civil and Infrastructure Engineering. 2013;28(3):162-177. https://doi.org/10.1111/j.1467-8667.2012.00790.x

[5] I Flores-Colen, J Brito, V Freitas. Expected render performance assessment based on impact resistance in situ determination. Construction and Building Materials. 2009;23(9):2997-3004. https://doi.org/10.1016/j.conbuildmat.2009.05.003

[6] K Makantasis, E Protopapadakis, A Doulamis, N Doulamis, C Loupos. Deep Convolutional Neural Networks for efficient vision based tunnel inspection. In: 2015 IEEE International Conference on Intelligent Computer Communication and Processing (ICCP). ; 2015:335-342. https://doi.org/10.1109/ICCP.2015.7312681

[7] YJ Cha, W Choi, G Suh, S Mahmoudkhani, O Büyüköztürk. Autonomous Structural Visual Inspection Using Region-Based Deep Learning for Detecting Multiple Damage Types. Computer-Aided Civil and Infrastructure Engineering. 2018;33(9):731-747. https://doi.org/10.1111/mice.12334

Mediterranean Architectural Heritage - RIPAM10
Materials Research Proceedings 40 (2024) 363-374

Materials Research Forum LLC
https://doi.org/10.21741/9781644903117-38

Integrated Digital Survey Methodologies for Late Medieval Fortifications in Tuscany: The Fortress of Malmantile

Giovanni PANCANI

Department of Architecture, University of Florence, Italy

giovanni.pancani@unifi.it

Keywords: Laser Scanner Survey, Malmantile, Walls, Arno Valley, Leonardo Drawing 8P

Abstract. The "castle of Malmantile", on the hills of Lastra a Signa, was built with the intention of strengthening the Florentine defenses against the rival city of Pisa. It represents one of the most interesting examples of late medieval fortification that has best preserved the entirety of the walls. It was originally built as a military outpost along the ancient road that connected Florence to Pisa. Following the downgrading of the military defense role, it became an inhabited center, a reference point for the surrounding fertile countryside. The data acquisition was carried out in three different survey campaigns, the first two carried out between 2009 and 2010 and the last, completed and updated, carried out in July 2018. The restitution work was carried out to give a complete representation of the place, its constructive characteristics and its state of conservation / decay. Exhaustive thematic maps have been extracted from the obtained documents, which have provided an adequate account of the conditions of the places.

Figure 1. Fortress of Malmantile, bird's eye view

Introduction

The fortress of Malmantile (figure1) rises on the hills that divide the Arno valley from the Pesa valley, in the municipality of Lastra a Signa. The fort represents one of the most interesting examples of late medieval fortification that has preserved the entirety of the original layout. It was originally built as a military outpost along the ancient road that connected Florence to Pisa, the

Mediterranean Architectural Heritage - RIPAM10   Materials Research Forum LLC
Materials Research Proceedings 40 (2024) 363-374   https://doi.org/10.21741/9781644903117-38

Via Vecchia Pisana, and only later, when it lost its military functions, did it become a town. The survey work was carried out with three acquisition campaigns starting from 2009 and continued the following year and ended with a campaign to complete and integrate the data carried out in 2018. The first drawings addressed by individual portions followed the order of the acquisition campaigns and have been reordered, completed and normalized. The study of the territory and the morphology of the fortress has allowed the formulation of some fascinating hypotheses about the presence of the fortification in some representations of the fifteenth century.

Historical Analysis
The fortified nucleus of Malmantile is located on the medieval route of the Via Pisana (now Vecchia Pisana) in the stretch that goes from Lastra a Signa to Montelupo Fiorentino, on the ridge that divides the valleys of the Arno river and its tributary Pesa. The settlement stands on a hill, near the Romanesque church of San Pietro in Selva and documented as early as the thirteenth century [1], whose top has a slightly rounded morphology with a large area that can be defined as flat. The morphology of the site most likely facilitated the construction of a fortress where, as Gabriele Corsani writes in "Historical Atlas of Italian cities, Tuscany 1 Lastra a Signa" there was already a military stronghold since the 12th century [1]. With an almost regular rectangular plan (fig. 3), the Fortress was built to safeguard the Florentine area with respect to nearby Pisa; it is located along a ridge path already in existence in the Etruscan era, the ancient road to Pisa, which from the hills follows the route of the Arno along the Gonfolina gorges, up to the confluence with the Pesa stream. The route, in the stretch from Montelupo to Signa, is protected by numerous fortifications and culminates with the Castle of Signa, which was placed to defend the port and the only bridge over the Arno between Pisa and Florence [2]. The end of the 14th century was marked by numerous clashes and consequent looting, as in 1380 when the troops led by Alberico da Barbiano, who were carrying out looting in the Val d'Elsa area, clashed with the Florentines [3]; or as following the assault on the village of Lastra a Signa in 1397 by the troops of Galeazzo Visconti. These events made clear the inadequacy of the defenses set up to protect the "Lastrigiana Community". In this regard, the Florentine Republic believed that the time had come to ensure a safe refuge for the inhabitants of the territory and in particular of the districts of the village of Lastra a Signa and of the people of San Pietro in Selva [4]. On 14 April 1400 the construction of the fortress of Malmantile was approved at the same time as the construction of the walls of Lastra a Signa [1]. La Roccaforte, is located in a dominant position with respect to the Arno valley, and served, as well as for the refuge of the local populations against the raids of the opposing armies, as the main sighting site and stronghold able to organize the first defense with respect to any attacks that came from the coast. There is documentation on the progress of the construction works of the fortress up to 1413, when, with the provision for their financing for a further three years, news about it was interrupted. However, in 1424 the Florentine Republic, following the surveys of the ten supervisors who had found that the work had not completed the work as there were no towers, corbels and battlements, therefore arranged for the construction of the work to be finished [4].

It is said that Brunelleschi took part in the construction of the fortresses of Malmantile and nearby Lastra a Signa, however the analysis of historical data provides us with information that is not exactly corresponding. From the documents we know that on 12 September 1426, Filippo Brunelleschi and Biagio D'Antonio were commissioned by the Cathedral Workers to evaluate the work done "murorum portarum et anpiportarum castri slabs" to weld the workers [4]. So it seems that the intervention of the architect of the Dome of the Cathedral was limited to an "administrative test". However, between 1424 and 1426 we have no news about it, therefore we cannot exclude Brunelleschi's involvement in the construction of the walls of the two Isatrigian fortifications, even if these do not present innovative ideas, but are limited to implementing the knowledge hitherto adopted for military architecture [4].

Mediterranean Architectural Heritage - RIPAM10   Materials Research Forum LLC
Materials Research Proceedings 40 (2024) 363-374   https://doi.org/10.21741/9781644903117-38

The cause of the loss of importance of the fortified axis, from Lastra a Signa to Malmantile, were: the slowness with which the works on the fortification of Malmantile were carried out; the type of construction adopted, however very similar to that used for the contemporary walls of Lastra a Signa, typical of late medieval buildings; a late completion of the work, which came when the rival city was also subdued, after the fall of the Florentine Republic in 1530 [1].

In the second half of the eighteenth century, the definitive reopening of the ancient roadway that ran along the Arno along the Gonfolina gorges caused the downgrading of the ridge route towards Pisa. Downgrading which, added to the loss of strategic-military importance of the fortified garrison of Malmantile, accentuated the isolation position of the fortification which from this moment on remains used as an agricultural village, within which there will not be a significant increase for a long time building and demographic [1].

Towards the end of the nineteenth century some houses began to be built both inside and outside the walls, although in 1896 Guido Carocci Regio Inspector of the Superintendence of Antiquities and Fine Arts of Florence, classified the walls of Malmantile, together with those of Lastra a Signa, as "buildings of local importance". [1]. However, the greatest manufacturing development will take place after World War II in the immediate vicinity of the Castle along the routes of the old Pisana road.

In 1969, in an attempt to stem a growing erosion of the territory around the walls with the D.M. of 10 January, the fortress and the surrounding area are subjected to landscape restrictions.

Figure 2. point of the laser Scanner survey, raster image of the projection in true size of the plamimetry of the Fortress of Malmantile

Typological Analysis
The late-Gothic fortress appears to be typologically in line with the fortifications of the fourteenth-century "Terre Nuove": it responds in fact to the military needs related to the defense of the markets that the Florentine Republic had felt the need to fortify. The defensive functionality was in line with the medieval tradition as regards the thickness of the walls, such as to be designed only to structurally support the elevation of the walls and not to resist the artillery shells that will make their entry into the military equipment only from the 16th century [4]. Malmantile Castle has a

track of about 123x70 meters, describes an almost perfect rectangle and its perimeter is 373 meters (fig. 3). The fortress is arranged longitudinally along the axis of the road to Pisa in a direction from east to west. The six towers, without internal façade, all protrude on average about 3 meters from the perimeter of the walls, one for each of the four corners and one in the middle of each of the long sides. The two doors in the center of the short sides, also protruding from the walls, are connected by the only road in the fortification. On the sides of the road far from the walls are the buildings of the oldest construction, while the subsequent superfetations often leaned against the walls, both on the external and internal walls.

The perimeter of the walls of the castle of Malmantie is made of loose stones, however its state of conservation is not the best. Of the protruding defensive apparatus, built with corbels with stone brackets and pointed arches in bricks and machicolations with alternate arches for the plumbing defense, there is only a short section on the west front, on the north side of the Pisan gate (fig. 3).

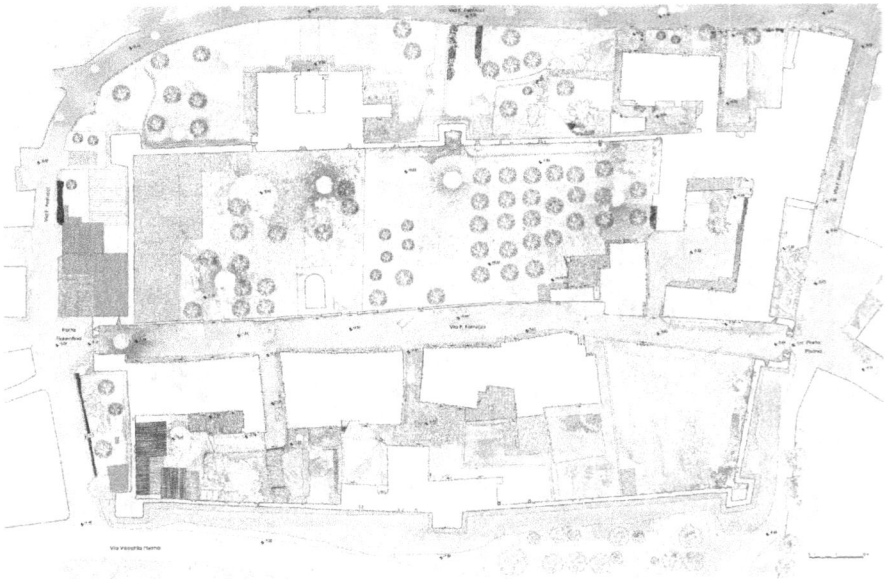

Figure 3. Plan of the architectural survey of the Fortress of Malmantile

Survey Project

Born from the experiences of surveying and enhancing the territory of Lastra a Signa, realized thanks to the collaboration of the local municipal administration, the survey project of the fortress of Malmantile is consequent to the survey of the contemporary walls of Lastra a Signa [5].

The survey lasted for some years using multiple technologies, which harmonized in a product that could be defined as integrated digital survey [6].

Started with a first campaign in 2009 during which the external perimeter of the walls was analyzed, it continued in 2010 with the entire internal perimeter and the survey of the buildings. In 2018, a further campaign was carried out to detect all the points that for any reason had not been detected in previous campaigns. Finally, in 2021 a drone flight was carried out to create an accurate photogrammetry and acquire some bird's-eye images necessary for comparisons with Leonardo's Landscape drawing.

Mediterranean Architectural Heritage - RIPAM10

Materials Research Proceedings 40 (2024) 363-374

Materials Research Forum LLC

https://doi.org/10.21741/9781644903117-38

Figure 4. Front of the Nord Front (C-C' section) of the Fortress of Malmantile, from top to bottom, Photoplane, Wirfreme, Material Analysis and the state of conservation.

Many instruments were used in the laser scanner campaigns;

In 2009, during the survey of the external perimeter of the walls, an old instrument was used, the Leica HDS ScanStation1, a device that was also quite dated at the time, very slow but very precise, and able to create a reflectance data [7] very reliable. This type of instrumentation, due to its slowness and therefore to the modest working speed, made it possible to carry out no more than 15 scans a day which, to optimize times, had to be carried out with the least possible overlap. This condition, to allow the registration of the point cloud with sufficient reliability margins, made it necessary to create a topographic polygon.

The survey of the interior of the walls of 2010 was made with a more modern and faster instrument the Leica HDS6000 which allowed greater amounts of work and therefore a higher number of scans in the days we had available, in fact, it was possible do without the topographical survey, also considering the points of support obtained in the previous year's campaign (Fig. 2).

For the photogrammetric survey of the many fronts of the internal buildings and the walls of the fortress, 2D photo-straightening techniques were used [8], since the photomodeling and 3D photogrammetry software were still not very mature, therefore they were often incomplete and ineffective to use for such an extensive survey [9].

The representations, all made at a scale of 1:50, were performed with great accuracy by calibrating, first of all, the photographic images on the orthoimages (scaled rasterizations of the true-size projections of a certain section of the point cloud), thus creating a series of photoplanes perfectly coinciding with the data coming from the point cloud [10]. On the calibrated ortho-photoplanes, associated with the ortho-images, the sections and elevations were drawn with an accurate description of all the wall faces and all the details present on the fronts of the buildings and the fortress. The floor plans and the roof plan were drawn with the same technique[11].

Mediterranean Architectural Heritage - RIPAM10   Materials Research Forum LLC
Materials Research Proceedings 40 (2024) 363-374   https://doi.org/10.21741/9781644903117-38

Figure 5. Front of the East Front (G-G' section) of the Fortress of Malmantile, from top to bottom, Photoplane, Wirfreme, Material Analysis and the state of conservation.

The survey shows the sections / elevation of the two fronts of the road inside the fortress, sections A-A 'and B-B'; all the external fronts of the fortress were also designed, sections C-C ', D-D', E-E ', and G-G '(Fig. 4, 5, 6,); finally an internal cross section, the H-H 'section (Fig. 7). Four thematic maps were made for each section: the ortho-photoplan which allowed to appreciate both the state of conservation of the materials but also the texture with which the wall faces were made; the wireframe drawing [12], in which the warping of the ashlars and the composition of the fronts with all the elements that contributed to its realization have been explained; in addition, the most evident materials and degradations were identified.

Mediterranean Architectural Heritage - RIPAM10
Materials Research Proceedings 40 (2024) 363-374

Materials Research Forum LLC
https://doi.org/10.21741/9781644903117-38

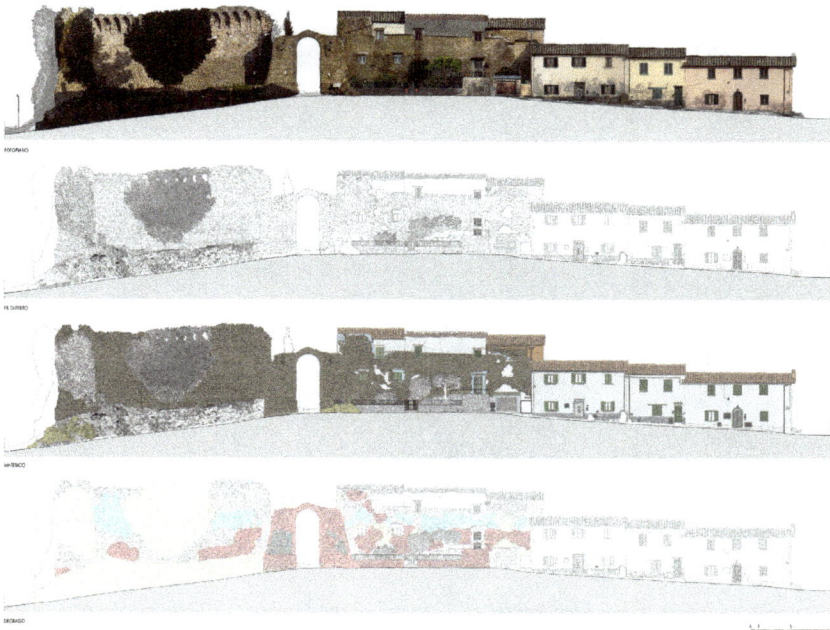

Figure 6. Front of the West Front (D-D'' section) of the Fortress of Malmantile, from top to bottom, Photoplane, Wirfreme, Material Analysis and the state of conservation.

With the 2018 survey, made with a Z + F 5006 laser scanner, twin instrument of the Leica HDS6000, the survey was completed by making all those scans that had not been performed in the previous surveys. Above all, the scans along the external circuit of the fortress were carried out since the small number of scans made during the first campaign did not allow to acquire many gorges and shadow areas and details which, once detected, made it possible to obtain a complete geometric model of the Malmantile castle [13].

The scans from the three survey campaigns that lasted from 2009 to 2018 were registered with a new rigid roto-translation operation following the last survey. The resulting point cloud was subjected to certification and testing in relation to compliance with minimum error and data reliability requirements. This operation consists in carrying out several certain sections for each individual partial registration and numerous horizontal and vertical sections on the general registration. In these sections, the so-called section wires (which belong to different scans) are checked by checking that they have a coincident trend, and if this does not happen, the maximum distance is not greater than mm. 1.5 [7]. The certification of the point cloud allows to obtain elaborations made on certain data and therefore sufficiently reliable to perform subsequent elaborations and considerations.

Figure 7. An internal cross section (H-H' section), of the Fortress of Malmantile, from top to bottom, Photoplane, Wirfreme, Material Analysis and the state of conservation.

The last phase of this survey was the flight made with a Jpi Mavic 2Pro drone in December 2021, it was necessary to take bird's eye images of the fortress of Malmantile (Fig. 8), useful images to verify some observations regarding the similarity between the fortification on the cobblestone hills and the walled city featured in Leonardo da Vinci's Landscape drawing. During this flight, the photogrammetric survey of the entire fortress and the surrounding context was also carried out using SfM acquisition techniques [14]. The images were assembled using 3D Zephir software with which a 3D model was made, from which a beautiful high definition aerial image was extracted (Fig. 9).

Figure 8. (above) Point cloud model of drone aerial photographs processed using SfM

Figure 9.(above right) Aerial view of Malmantile photo by drone and SfM techniques

Mediterranean Architectural Heritage - RIPAM10   Materials Research Forum LLC
Materials Research Proceedings 40 (2024) 363-374   https://doi.org/10.21741/9781644903117-38

Final Remarks and Conclusions

In the first place:

The perimeter of the fortification of Malmantile describes an almost regular rectangle, which together with the arrangement of the towers, induces some observations and reflections. In the bird's eye view (c.1503-4 - Windsor, Royal Library, RCIN 912685) for the project of the large diversion canal of the Arno (Fig. 10), Leonardo makes a representation of the fortress of Malmantile, which, although concise, is perfectly recognizable in its salient elements: the rectangular layout with towers at the four corners and intermediate towers on the two long sides. By observing the Landscape drawing preserved in the Cabinet of Drawings and Prints of the Uffizi Gallery inventory 8P (Fig. 11), It is possible to see how the fortress represented on the left side of the landscape presents itself as a fortified village, with rectangular walls, towers at the corners and sectional towers on the long sides. So also in this drawing we can see some similarities with the fortress of Malmantile, as Roberta Barsanti also finds in her essay for the Leonardian celebrations for the 500th anniversary of the death of the genius Vinciano in 2019 [15]. These similarities do not allow us to think that the city represented in drawing 8P is Malmantile, since other characteristics, such as the keep on the right corner of the fortress or the density of houses inside, would make it impossible. However, we cannot fail to consider that Leonardo, in order to go from Florence to Vinci, had to pass through the Pisan road, therefore from Malmantile. It should be taken into consideration that drawing 8P was made in 1473 when Leonardo was 21 years old, and the castle on the Lastrigiane hills, precisely because it was on the route he took to go from Vinci to Florence, was certainly one of the places he had visited. That the small fortress had attracted Vinciano's interest can be deduced from the Royal Library map, as he had represented it despite the strategic context of the regional context, both for the changed geopolitical conditions and for the introduction of gunpowder , appeared to be less interesting in the panorama of the defenses of the Florentine Republic. Therefore, the layout of the Landscape drawing which refers to the regular layouts of the walls of the Terre Nuove, built in fourteenth-century Tuscany, which inspired the fortification of Malmantile (Fig. 14), built in the first half of the fifteenth century, may have attracted the young man's attention. Leonardo, so as to grasp some salient elements that he may have represented in the famous Landscape drawing.

Figure 10.. (above) Leonardo da Vinci, Detail of the bird's eye view drawing, 1503-4 , Windsor, Royal Library, RCIN 912685, for the project of the Arno's great diversion canal, the arrow indicates the fortress of Malmantile.

Figure 11. (above right) Leonardo da Vinci, Landscape drawing, Cabinet of Drawings and Prints of the Uffizi Gallery inventory 8P.

Mediterranean Architectural Heritage - RIPAM10   Materials Research Forum LLC
Materials Research Proceedings 40 (2024) 363-374   https://doi.org/10.21741/9781644903117-38

Secondly:

It is common opinion that the fortress of Malmantile, over the centuries from its construction to today, has suffered some collapses and has lost a significant part of the defensive structures such as corbels, cantilevered walkways and battlements. This deduction is probably due to the presence of these garrisons in the contemporary walls of Lastra a Signa. Despite the poor state of conservation of the stone walls, where there are no adjacent buildings, the aforementioned cantilevered defenses are almost always present. The two fortifications were built in parallel to meet the same military needs, therefore it is quite likely to have interpreted the documents in which the conclusion of the work was requested, in particular the resolution of 1424 not only as evidence of the desire to prune the two fortifications but also how to have succeeded, which is certainly true for Lasta a Signa but probably not entirely true for Malmantile. From a careful reading made by Gioia Romagnoli of the resolution of 1424, it is clear that the ten supervisors appointed by the Florentine Republic to survey the state of the works of the castles of Lastra a Signa and Malmantile, report that the two fortifications lacked corbels and battlements, and there they arranged for their construction. However, in subsequent resolutions we no longer find reference to the Malmantile Castle and we do not find any trace of payments even in the final balance of Paolo di Matteo Fastelli Petriboni [4].

The careful study of the elevated wall faces allows us to observe how, the corbels present only on the west side of the walls, in the north portion with respect to the western door (Fig. 5), have their shutter around the altitude of about m. 8.60 with respect to the reference altitude 0.00 located at the southern corner of the Pisana gate, and reach the maximum altitude of m. 1120 (Fig. 12). The rest of the walls, except for the towers, does not exceed the altitude of m. 8.60 (Fig. 13). These observations, together with the payment documents not found, lead us to think that the corbels and battlements were not demolished by historical events and neglect, but that they were never built. As we have seen, the fortress is built at a time when military techniques and geopolitical assets change, the long continuation of the construction of the fortress of Malmantile, and the high costs incurred, at a certain point, probably, did not justify further expenses, reason why, it was probably decided to permanently interrupt the construction work.

Figure 12. (above) Malmantile, Porta Pisana with the stretch of walls where the corbels are still present and part of the walkway.

Figure 13. (above right) Detail of the defensive apparatus protruding, made with corbels with stone shelves and pointed arches in brick and embrasures with alternate arches for the defense leaping west front, near the Pisana door.

Materials Research Forum LLC
https://doi.org/10.21741/9781644903117-38

Conclusions

This study has allowed us to advance two hypotheses, one certainly fascinating, which however is based exclusively on observations that may be plausible, but which do not have documents to confirm what has been stated, but above all remain exclusively in the field of possible but not demonstrable events.

In the second case, the study and observation of the elevated walls, which was cross-referenced with the research done by Gioia Romagnoli for "studies on the heritage of Lastra a Signa for his The Middle Ages in the hills south of Florence, provides us with a plausible explanation with respect to the story of the construction of the Malmantile Fortress.

Figure 14. The Fortress of Malmantile as seen from the hills to the south

References

[1] G Corsani. (1993), Lastra a Signa. Atlante storico delle città italiane, Toscana, 1, Firenze, Buoninsegni Editore.

[2] A. Baldinotti, R. Barsanti (2003), Capitolo III. Il Ponte, in (a cura di Ciuffoletti Z.), Storia della comunità di Signa, Volume II, L'identità Culturale, Firenze Edifir, pp31-34.

[3] S. Ammirato (1853), Historie Fiorentine. Torino.

[4] G. Romagnoli (2000), Il patrimonio artistico di Lastra a Signa – Architettura Civile, in (a cura di Moretti I.), Il Medioevo nelle colline a sud di Firenze, Firenze, Polistampa, pp. 168-178.

[5] S. Bertocci, G Pancani., A. Cottini (2020), La cinta muraria di Lastra a Signa: metodologie di rilievo digitale integrato, In, Navarro Palazón J: García-Pulido L.J., Defensive Architecture of the Mediterranean, Granada:Universidad de Granada, Universitat Politècnica de València, Patronato de la Alhambra y Generalife, pp. 255-262.
https://doi.org/10.4995/FORTMED2020.2020.11498

[6] G Verdiani. (2012), Le metodologie e le strumentazioni per il rilievo laser scanner, in Bini M., Bertocci S. (a cura di), in Manuale di rilievo architettonico e urbano, Capitolo 5.1, Novara, De Agostini Scuola S.p.A., 2012, pp. 169-174.

Mediterranean Architectural Heritage - RIPAM10
Materials Research Proceedings 40 (2024) 363-374

Materials Research Forum LLC
https://doi.org/10.21741/9781644903117-38

[7] G. Pancani (2017), La città dei Guidi: Poppi. Il costruito del centro storico, rilievi e indagini diagnostiche, Firenze: Edifir, Firenze, p. 147.

[8]M Docci., , D. Maestri (1994). Manuale di rilevamento architettonico e urbano, Roma-Bari: Laterza Editore.

[9] R De Marco.;, S Parrinello. (2021). Digital surveying and 3D modelling structural shape pipelines for instability monitoring, in, historical buildings: a strategy of versatile mesh models for ruined and endangered heritage, ACTA IMEKO, vol. 10, pp. 84-97, ISSN:2221-870X. https://doi.org/10.21014/acta_imeko.v10i1.895

[10] G. Pancani (2011), Lo svolgimento in vera grandezza delle volte affrescate delle sale dei quartieri al piano terreno di Palazzo Pitti a Firenze, in, Il Disegno delle trasformazioni, atti delle Giornate di Studio , Napoli 1-2 dicembre 2011, Clean Edizioni, Napoli 2011.

[11] G Pancani., S Galassi., L Rovero., L Dipasquale., E., Fazzi ?G Tempesta. (2020), Seismic vulnerability assessment of the triumphal arch of Caracalla in Volubilis (Morocco): past events and provisions for the future, "International Archives of the Photogrammetry, Remote Sensing and Spatial Information Sciences, XLIV-M-1-2020, 435–442. https://doi.org/10.5194/isprs-archives-XLIV-M-1-2020-435-2020

[12] M Ceconello. (2003), Strumenti e tecniche di visualizzazione, in (a cura di Gaiani M.), Metodi di prototipazione digitale e visualizzazione per il disegno industriale, l'architettura degli interni e i beni culturali, Milano, Edizioni Polidesign, pp. 511-580.

[13] G. Pancani; M. Bigongiari (2020), Digital survey for the structural analysis of the Verruca fortress, in PROCEDIA STRUCTURAL INTEGRITY, vol. 29, pp. 149-156. https://doi.org/10.1016/j.prostr.2020.11.151

[14] F. Fratini (2023), The Rocca Vecchia fortress in the Gorgona island (Tuscany, Italy): building materials and conservation issues, in Bevilacqua M. G. , Ulivieri (a cura di), Defensive Architecture of the Mediterranean, Vol. XIII / (Eds.), Pisa University Press (CIDIC) / edUPV. https://doi.org/10.12871/9788833397948121

[15] R Barsanti. (2019), Il Paesaggio di Leonardo del 1473 - Studi e interpretazioni, in Barsanti R. (a cura di), Leonardo a Vinci, Alle origini del genio, Firenze, Giunti Editore, p.135-149.

Keyword Index

www.ingramcontent.com/pod-product-compliance
Lightning Source LLC
Chambersburg PA
CBHW071318210326
41597CB00015B/1262